Renewable Fuels

I0131961

Rajni Kant

Former Chief Engineer
Uttar Pradesh State Electricity Board, Lucknow
and
Former Consultant and Adviser to
Various Design and Engineering Organisations in India

and

Keshav Kant

Former Professor of Mechanical Engineering
Indian Institute of Technology, Kanpur
Uttar Pradesh, India
and
Former Professor Emeritus, Department of Mechanical Engineering
Pranveer Singh Institute of Technology, Kanpur, UP, India

CRC Press
Taylor & Francis Group
Boca Raton London New York

CRC Press is an imprint of the
Taylor & Francis Group, an **informa** business

A SCIENCE PUBLISHERS BOOK

Cover credit: The cover has been designed using image/assets from Freepik.com.

First edition published 2024
by CRC Press
6000 Broken Sound Parkway NW, Suite 300, Boca Raton, FL 33487-2742

and by CRC Press
4 Park Square, Milton Park, Abingdon, Oxon, OX14 4RN

© 2024 Taylor & Francis Group, LLC

CRC Press is an imprint of Taylor & Francis Group, LLC

ISBN: 978-1-032-05983-9 (hbk)
ISBN: 978-1-032-05984-6 (pbk)
ISBN: 978-1-003-20012-3 (ebk)

DOI: 10.1201/9781003200123

Typeset in Times New Roman
by Radiant Productions

Dedicated to the memory of
Our Beloved Mother (Late)
Mrs. Ram Mohini
An Ideal and Exemplary Mother

Our Beloved Father (Late)
Mr. Mathura Prasad
A Versatile Scholar, Teacher, Educationist and Guide

Our Beloved Brother (Late)
Mr. Chandra Kant
A simple person with Noble Ideas and Sacrifice for the Family

and Beloved Sister (Late)
Ms. Indira Vati Saxena
A Dedicated Teacher and a Talented Artist

Foreword

Current interest in global warming and potential disasters arguably caused by the resulting climate change has fueled accelerating research and development activity in renewable energy resources. Over the past five decades, it is estimated that the global energy consumption has more than quadrupled; much of it has been derived from use of non-renewable fossil fuels leading to emission of greenhouse gases. The energy sector causes up to two-thirds of the greenhouse gas emissions. Reduction of carbon-dioxide emission to net zero by 2050 will still result in 1.5°C rise in the average global temperature, which is already highly undesirable. A truly massive multi-disciplinary effort requiring huge capital investment by both public and private sectors with international cooperation is required to meet this goal.

It is heartening to note significant progress is being made around the world to mitigate greenhouse gas emissions. For example, in 2030, the economy will be some 40% larger than it is in 2022, but the energy consumption will be 7% lower according to several estimates. This requires increased utilisation of renewable energy resources such as solar and wind energy. Hydro power and nuclear energy are two largest resources of low-carbon electricity, but fossil fuels still contribute the most to the overall energy consumption. Many countries have effectively decoupled energy emissions from economic growth. Many countries in the EU have over the period 2000–2014 reduced carbon-dioxide emission by 10 to 30% while their GDP increased 10 to 40%. UK and USA have reduced their greenhouse gas emissions in the energy sector by 20% and 6%, while their GDP rose by 27% and 28% respectively. This has been achieved by improved efficiency of the processes involved, but a part of it is via use of renewable energy resources.

This book covers in a very comprehensive, multi-disciplinary and concise manner an in-depth coverage of diverse renewable fuels. This is a theme of critical importance for reduction of greenhouse gases in a sustainable way. The technological challenges faced are not only scientific in nature, but they often face formidable economic and government policy issues. The key is to make the technologies cost-effective. The goal is to achieve sustainable economic growth without a drastic rise in the carbon footprint of energy-based processes in industry and transportation as well as in other sectors, which consume significant amounts of energy. It is noteworthy that some have argued in favor of a "degrowth" model, which is uncoupled from economy; this option is clearly not widely acceptable.

The authors of this book, which serves a dual role as textbook as well as a handy reference book, have done an admirable effort which lasted over five years. It is a highly readable book with necessary references for in-depth study by interested readers. It covers a wide range of topics covering the whole range of renewable fuels that are developed as well as being developed around the world. As the readers will quickly find out the range of sustainable fuels is very large as is the range of feedstocks that are used. Without repeating the detailed outline of the book content by the authors in the Preface, suffice it to state that they deserve our whole-hearted compliments for undertaking this massive effort to produce this valuable compendium. I do fervently hope that postgraduate curricula in engineering as well as sciences will include this subject as a core or an elective subject depending on the discipline. This book is highly recommended for acquisition by academic as well as industrial libraries supporting education, production and R&D in the energy sector in particular.

Arun S. Mujumdar
Former Professor of Chemical Engineering, McGill University, Montreal, Canada &
Professor of Mechanical Engineering (Retired), NUS Singapore
www.arunmujumdar.com

Preface

Electricity generation using fossil fuels is the number one source of greenhouse gases and the leading cause of industrial air pollution, which takes a heavy toll on our environment by polluting air, land, and water. Sulfur dioxide (SO_2) and nitrogen oxides (NOx) released into the air by fossil-fuel power plants, vehicles, and oil refineries are the biggest cause of acid rain today, according to the EPA. In fact, two-thirds of SO_2 and one fourth of NOx found in the atmosphere come from electric power generators. Twenty-nine percent of nitrogen oxides (NOx) react with sunlight to create ground-level ozone and smog, which can cause lung inflammation, asthma attacks, etc., because the nitric acid, sulfurous acid, and sulfuric acid react with ammonia in air to form solid crystals smaller than 2.5 microns, which become the nucleation sites for particle growth. Particulate matter is especially harmful to people with lung and heart disease. Mercury is a highly toxic metal that is released from coal-fired power plants and accumulates in biological organisms, where it is constantly recycled in the environment as it moves up the food chain.

The importance of renewable fuels, which are plant-based fuels, has increased because of the need to curb the emission of greenhouse gases (GHGs). These gases accumulate in the earth's atmosphere and have already started causing damage to the environment and changes in global climate. The primary objective for deploying renewable fuels is to advance economic development, improve energy security, improve access to energy, and mitigate climate change. They reduce the dependence on imported fuels and conserve nations' natural resources. While renewable fuels when burnt also produce carbon dioxide (CO_2), a greenhouse gas that contributes to global warming like nonrenewable fuels, the regrowth of plants to the original extent absorbs an equivalent amount of CO_2. Therefore, the utilization of renewable fuels can ideally be a **carbon zero operation**.

Replacing the fossil-fuel infrastructure will take time. Strong, consistent support is required globally to demand clean energy for all. Renewable energy can reduce the negative impact of fossil fuels on the environment.

The authors realized the need for writing a book on 'Renewable Fuels' highlighting the urgent requirement for them and the methods and technologies already developed as well as those in process of being developed to utilize them successfully. The authors have taken almost five years to study the various technologies, referring to an enormously large number of references on the subject, and organizing the material on renewable fuels in the form of this book. Studying the literature on the subject was a very rewarding experience for the authors. It inspired them to delve more into the depths of the subject to inspire and motivate the readers to accomplish further work in the area.

This book is aimed at professionals, consultants, and people in the academia working in this field so that they can acquire an adequate understanding of the subject in a systematic manner and can refer to the various references given at the end of each chapter for additional information. At the same time, the book is structured as a textbook so that it can be used for undergraduate and graduate courses on the subject. The material related to the questions at the end of each chapter is covered in the text. The authors very much hope that the professionals specializing in this area, research workers working in this field, course instructors as well as students will find the book useful.

Biomass like crop residues, forest wastes, animal wastes, human wastes, municipal solid wastes, etc., may degenerate into methane, CO_2, and N_2O. It may also catch fire and generate CO_2 along with other pollutants including some N_2O. Methane and N_2O are more powerful greenhouse gases than CO_2. Planned combustion does not create methane, but creates CO_2. Alcohols, biodiesel, and many liquid fuels can be made from lignocellulosic biomass and therefore biomass may be used in most useful and planned manner to keep in check the untoward emissions and also avoid the loss of an energy resource.

In order to fulfil the ever-increasing need for wood, availability of land is a problem, because of the land required for agriculture and developmental activities. All waste lands or non-agricultural lands must be used for plantations. A positive policy is required for conservation of forests and replantation of degraded forest area. Measures are required for giving thrust to plantations of suitable trees for various soils and climatic conditions. Agroforestry should be taken up on a large scale to fulfil the local needs of timber, fuel, fruits, etc. For fuel needs, the fast-growing trees suitable for particular area/soil/climate may be planted.

The restoration of sodic lands and mined out land should be done for planting of trees. Bioremediation and phytoremediation of soils would be very useful for land restoration and plantation. Oilseed bearing trees suitable for the specific area may be grown on non-agricultural and waste lands to provide vegetable oil for production of biodiesel.

Alcohol can be made from sugar or sugar containing materials, starch or starch containing materials, and even lignocellulosic biomass. This would not interfere with food production. The process for producing ethanol from biomass through biological fermentation involves pre-treatment, hydrolysis, fermentation, and product purification. The hydrolysis converts 'Carbohydrate polymers' $(C_6H_{10}O_5)_n$ like 'cellulose and starch', into 'monomeric sugars (glucose, fructose)'. After the hydrolysis, many of the microorganisms may ferment carbohydrates under anaerobic conditions to ethanol. The microorganisms may be bacteria, yeast, or fungi.

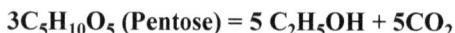

$C_6 H_{12} O_6$ (Glucose/Hexose) = $2C_2H_5OH$ (ethanol) + $2CO_2$ (Carbon dioxide)

$3C_5H_{10}O_5$ (Pentose) = $5 C_2H_5OH$ + $5CO_2$

Initial purification of ethanol is done by distillation by which ethanol with a purity of 96.6% (maximum) can be obtained. Further drying or purification may be performed by using hygroscopic materials or membrane separation/pervaporation/vapor permeation.

There is a great potential for producing ethanol from widely available cheap raw material containing cellulose. There are a number of technologies available for such production. Efforts are being made world over to commercially produce ethanol from cellulosic materials.

Butanol can also be manufactured from plant resources and therefore it is considered as a potential renewable biofuel and can be used in internal combustion engines. It can be blended with gasoline in much higher percentage than ethanol without any modifications to the engine. It may be stored on board on hydrogen fueled vehicles having fuel-cells using hydrogen. The on-board reforming butanol can provide hydrogen for fuel cells of the vehicle. Butanol now is being produced by bacterial fermentation of a variety of feed stocks. Bioethanol production plants can be modified for production of bio-butanol.

Biodiesel can be made from various oils and fats, which are simple lipids. Animal fats and vegetable oils are typically esters of free fatty acids . There are many Oil seeds producing trees for nonedible oil, which along with waste vegetable oils should be used for making biodiesel so that it does not interfere with food. The process of converting oil to biodiesel consists of the following steps:

- Vegetable oils are converted into 'methyl esters' by 'transesterification'.
- 'Transesterification' removes 'glycerol' from triglycerides and thus reduces the viscosity of the resulting product called 'Biodiesel'.

Various processes of making biodiesel are:

i. Base transesterification catalyzed by alkalis

ii. Acid transesterification catalyzed by acid. It is advantageous for waste/used oils.

iii. Enzymatic transesterification. It gives very good yields from used oils and fats

iv. Reactive extraction of biodiesel. It is done by treating the ground oilseeds with methanol.

v. Catalyst-free transesterification. It uses supercritical methanol at high temperatures and pressures in a continuous process.

Some algae are very rich in lipids which are a good source of biofuel. Oil yield per acre from microalgae may be thousands of times of oil yield per acre from soybeans. Algae can be harvested several times in a short time frame and can grow 20–30 times faster than food crops. Algal oil does not contain sulphur. Algae can be harvested by several methods. Growth parameters, growth phases and problems in growing algae are mentioned. It can be produced efficiently in photobioreactors (PBR) under controlled conditions. Various methods of oil extraction from algal biomass are discussed in the text.

Fungi also is a source of oil. Occurrence of oil bearing fungi, conditions for increasing their lipid content, and oil extraction methods from fungi are also given.

Lignocellulosic material can be converted to liquid fuels by several methods including **dimethyl ether.** Methods include the Fischer-Tropsch process for converting syngas to liquid fuels; biochemical route; fast pyrolysis; direct thermochemical liquefaction of algal biomass, and using algal hydrogenation.

Dimethyl ether (DME) a synthetic fuel, may be a prospective future fuel for vehicles. Methods for its production from lignocellulosic biomass including direct and indirect synthesis, use of diesel blended with DME in vehicles and present limitations in its exclusive use are highlighted.

Methane is a powerful GHG, but is a very good fuel. It can be obtained both from renewable and non-renewable sources. Renewable sources of methane include: (i) anaerobic decay/biological or chemical processes from renewable materials, (ii) thermal gasification of organic materials, and (iii) landfills (generally from municipal solid wastes (MSW)).

Methanization of organic wastes, fermentation kinetics, methods of biogas production, generation of landfill gas from MSW, etc., are discussed. Various aspects of production of gaseous fuels through thermal gasification of lignocellulosic waste materials/biomass are also discussed.

Nonrenewable sources are: methane recovery from gas hydrates from oceans and that from coal mining. The recovery of methane from these sources should be accomplished with a sense of urgency because they are the potential dangers to the environment. If the methane from these sources comes out in atmosphere inadvertently, it would cause greater greenhouse effect and accelerated global warming than its planned utilization and conversion to CO_2. From convenience point of view, it is advantageous to convert methane obtained from any source to liquid fuel (GTL Technology). Various reforming processes are used.

Hydrogen on combustion yields only water or steam and no GHG is evolved so it is absolutely clean and pollution free fuel. Hydrogen is not available in nature as a gas and it has to be produced by releasing it from water or other compounds of hydrogen. Characteristics and properties of hydrogen are quite favourable for its use as a green fuel, but in practice there are many challenges in production and use of H_2. Reasons for promotion include: abundant availability as element and as a by-product from some industries; use as fuel in fuel cells/I.C. engines; and ensured availability even after the fossil fuels get exhausted. Hydrogen can be used in fuel cells, possibly in aircraft and jet engines as backup or auxiliary power units onboard and rocket engines and in vehicles as a hybrid fuel *(Hypthane)* by blending it with CNG up to 30% or by the onboard generation of hydrogen from metal hydrides or onboard reforming of butanol.

Several technologies for separation of hydrogen from gasification products are available. Various methods of storage of hydrogen including absorption/adsorption, hydride formation, underground storage, material failures by hydrogen embrittlement in pipelines and vessels, hydrogen leak detection, pipeline integrity monitoring, and environmental concerns are also discussed.

Er. Rajni Kant and *Professor Keshav Kant*

Acknowledgements

Writing a book is much harder than I thought, but much more rewarding than I could have ever imagined. None of this would have been possible without the joint effort of the two authors, who had to spend very long hours for several years in studying the subject material, sorting it out and choosing the subject matter. Our main focus was on the reader- as a professional, a research worker in the field and also a fresh student, who would like to get into this area and acquire relevant information on the subject of his choice. The book is written in a text book mode so that it can be used as a text for UG/Graduate courses at the University level and also as a reference book for the Professionals and research workers. The whole effort has been to present the matter lucidly for the reader. A University Professor and a research worker will find the book very useful in picking the required subject matter and relevant relatively recent references. Although this period of our lives was filled with many ups and downs, but we decided to bring this work to a successful conclusion.

Both of us would like to thank Lord Krishna for His Blessings in inspiring us to do our best and share this information with the Academia, Professionals and the Students.

We would like to express our sincere gratitude to Professor Arun S. Mujumdar for many useful suggestions from time to time during the course of writing this book.

We would like to thank specially (Late) Dr. Iver Simonsen- Former Energy Specialist at Pulp and Paper Research Institute of Canada, Pointe Claire, Quebec for many useful suggestions. Very special thanks to my former research Associate and colleague from McGill University—Dr. Hesham Seoud for his suggestions.

We would like to sincerely thank everyone in the Publishing team of Science Publishers specially Mr. Raju Primlani Acquisitions Editor for guiding me throughout at different stages of the book.

We would like to express our sincere thanks to former Research Associates Mr. Himanshu Gupta, Mr. Anand Shekhar and Mr. Ravi Kumar in the Department of Mechanical Engineering of Indian Institute of Technology Kanpur, India for their enormous help in organizing the references and figures of the book and readily giving me miscellaneous help as and when needed.

Lastly, we would like to thank our brother—Dr. Krishna Kant, Professor in the Computer and Information Systems (CIS) Department at Temple University, Philadelphia, and Director of the Temple site of the NSF for his enormous moral support and tremendous help throughout. Thanks are also due to Er. Neeraj Kant-son of the second author for his help in many ways from time to time.

Er. Rajni Kant and Professor Keshav Kant

Contents

$$\boxed{\textbf{C}\text{HAPTER } \textbf{1}}$$

Introduction

1.1 Background [19,20,21,22]

There has been continuous rise in fuel consumption in the world due to factors such as a rise in the global population, the adoption of a modern way of living with high carbon lifestyle and high carbon economy, the setting up of large number of industries and fossil fuel-based electric power stations, an increased need of energy for water pumping for irrigation/agriculture, domestic and industrial consumption, etc. The fuel combustion which is required to provide energy for domestic, industrial, transport, agricultural and power sectors, also increases the carbon dioxide level in the atmosphere, which is the main component of greenhouse gases, although other greenhouse gases are also significantly damaging. The pollution caused by the burning of fossil fuels is also responsible for many other problems.

Fossil fuels are not renewable and thus, their reserves on earth are finite. We have to tap inexhaustible sources of energy like solar energy or wind energy or renewable fuels for sustaining growth and fulfilling the needs of human beings. The growth or development can only be sustained if global resources are managed successfully to perpetually satisfy changing human needs without impairing the environment and while conserving natural energy resources.

Deforestation has a taken place on a large scale to vacate lands for various industrial, power, irrigation, housing, mining and road projects, as well as for growing food. The need for fuel wood and forest products is also responsible for deforestation. Natural causes like forest fires, earthquakes, floods, etc., cause large-scale destruction of forests as well. The world's Forest cover has been degraded to some degree in the past few decades, because of many causes like various anthropogenic activities, animal grazing, unauthorized wood cutting, fire, growing demand for housing, industries, power projects, irrigation projects, roads, and railways, etc. As per estimates by various authorities, the demand for timber (logs) and fuel wood is projected to increase in the future while supply is to remain flat, which would lead to a significant supply deficit.

Collecting dead/fallen branches and litter from forests is widely practiced in rural areas for their subsistence. This is not of big concern, but the unauthorized cutting of trees for fuel wood or timber from natural and plantation forests has a very bad long-standing impact on forest ecosystems. This is because plants absorb carbon dioxide* from the atmosphere, sequester the carbon in their structure and release the oxygen. This process has largely suffered a setback due to destruction of forests.

*[**Note**: Plants capture CO_2 from the atmosphere by way of a **photosynthetic reaction**, in which **plants and algae** use energy from sunlight to combine carbon dioxide (CO_2) from the atmosphere with water (H_2O) in the

*soil to **form carbohydrates** in the presence of **chlorophyll**. Thus, plants efficiently use solar energy to provide us fuel. The photochemical reaction takes place as follows:*

$6\ CO_2 + 6H_2O$ (in the presence of sunlight & chlorophyll) $= C_6H_{12}O_6 + 6\ O_2$

That is, $=$ Glucose + Oxygen

$6\ n\ CO_2 + 5\ nH_2O$ (in the presence of sunlight & chlorophyll) $= (C_6H_{10}O_5)_n + 6\ nO_2$

That is, $=$ Carbohydrates/starch/cellulose + Oxygen

*It may be mentioned that the light energy absorbed by **chlorophyll** is grabbed by very efficient subcellular components of plants called **chloroplasts** through a series of steps.]*

Forests/plantations/trees have three functions: ***regulative, productive, and protective.***

(1) The ***regulative functions*** include: carbon dioxide sequestration and oxygen release, control of floods, reducing of extremes in atmospheric temperature, help in maintaining groundwater level and absorption of pollutants from air, water, and land.

(2) The ***productive functions*** include: production of fuel, timber, latex, resins, herbal medicines, and other useful products.

(3) The ***protective functions*** include: checking of soil erosion and sedimentation of water reservoirs/canals/river beds, conservation of soil moisture, rainwater harvesting to ensure a consistent groundwater level, preservation of ecology, protection against wind, dust, desertification, noise, radiation, etc.

By creating large-scale tree plantations for fulfilling the demand of wood and its controlled exploitation, the ecological deterioration happening presently can be arrested. It may be borne in mind that if the rate of consumption of any resource material exceeds the rate of its renewal, it may become extinct. It is, therefore, imperative that the rate of consumption of wood should always be well below the rate of its growth, otherwise the imbalance would worsen the situation. It would be a positive step in the direction of greenhouse gas control.

Deforestation has contributed a great deal to the rise in the level of greenhouse gases in the atmosphere, resulting in global warming. Presently, there is an urgent need to reduce the emission of greenhouse gases from all sources and processes, exploitation of sources of energy which do not emit carbon dioxide, using of fuels which are carbon neutral like biomass, carrying out large-scale plantation of trees, using energy economically everywhere, using only the most efficient processes in all sectors, expediting carbon dioxide sequestration/carbon dioxide capture and the non-atmospheric disposal of CO_2. The planting of trees, aquaculture, and various green practices are very useful to check global warming.

The wood and biomass generated by plants are renewable. A part of biomass and plant litter can be converted to compost manure for use in growing food crops, trees, and other plants while the balance can be directly used as fuel or can be converted to other convenient biofuels. The whole cycle of using plant-based fuels and growing plants/crop/trees in (in equivalent quantities) may become an almost *'zero carbon operation'*.

The large industries and thermal power stations are using up the fossil fuels like coal, lignite, petroleum oils and natural gas. There is a finite reserve of these fossil fuels in the world and are going to exhaust one day. The renewable energy sources (like solar energy, wind energy, biomass energy, etc.) are therefore to be tapped before it is too late. The use of renewable fuels is an effort in this direction.

According to the Global Resource Assessment 2015, the world had 4,128 million ha of forest in 1990 (which was 31.6% of the global land area); by 2015 this area had decreased to 3,999 million ha (which works out to 30.6% the of global land area). This is a large change indicating a net loss of some 129 million ha of forest between 1990 and 2015, about the size of South Africa. The top ten countries reporting the greatest annual net loss of forest area in 2010–2015 according to the above report were: Brazil, Indonesia, Myanmar, Nigeria, United Republic of Tanzania, Paraguay, Zimbabwe, Democratic Republic of the Congo, Argentina, and Bolivia. Some of the countries have since taken positive steps to check the trend of the reduction of forest cover, such as encouraging tree plantation on a large scale, and the results are encouraging [13]. The top ten countries ranked according to the largest forest cover according to this report are indicated in Table 1.1.

The International Energy Agency (IEA) is predicting a major surge in CO_2 emissions from energy this year, as the world rebounds from the pandemic. According to IEA, although the total energy emissions for 2021 will still be slightly lower than those in 2019, the CO_2 will rise by the second-largest annual amount on record. The use of coal in Asia is expected to be a key contributor: the IEA says it will push global demand up by 4.5%, taking it close to the global peak seen in 2014. However, renewable energy is also booming, with green sources set to supply 30% of electricity this year.

A constellation of satellites will be flown into space this decade to try to pinpoint significant releases of climate-changing gases—in particular, carbon dioxide and methane. Prototypes will launch in 2023 with more satellites to follow from 2025. The initiative is being led by an American non-profit organisation called Carbon Mapper. It will use technology developed by the US space

Table 1.1. Top Ten Countries Ranked According to Forest Area 2015 [13].

Se. No.	Country	Forest in thousand ha	Percent of land area of the country	Percent of global forest area
1.	Russian Federation	814,931	48	20
2.	Brazil	493,538	58	12
3.	Canada	347,069	35	9
4.	USA	310,095	32	8
5.	China	208,321	22	5
6.	Congo	152,578	65	4
7.	Australia	124,751	16	3
8.	Indonesia	91,010	50	2
9.	Peru	73,973	58	2
10.	India	70,682	22	2
	Total	2,686,948		67

Table 1.2. Countries reporting the greatest annual forest area gain (2010–2015).

Se. No.	Country	Annual Forest Area Gain	
		Area (000 ha)	% of 2010 Forest Area
1.	China	1542	0.8 2
2.	Australia	308	0.2
3.	Chile	301	1.9
4.	United States of America	275	0.15
5.	Philippines	240	3.5
6.	Gabon	200	9.7
7.	Lao People's Democratic Republic	189	1.1
8.	India	178	0.3
9.	Viet Nam	129	0.9
10.	France	113	0.7

Table 1.3. India's Forest Cover (2011) [2].

Se. No.	Land Use/Forest Cover Category	Area km²	Percent of Geographical Area
1.	Total Forest Cover*	692,027	21.05
a.	Very Dense Forest	83,471	2.54
b.	Moderately Dense Forest	320,736	9.76
c.	Open Forest	287,820	8.75
2.	Scrub	42,176	1.28
3.	Non-forest Area	2,553,060	77.67
4.	Total Geographical Area	3,287,263*	

* includes 4,662 sq. km area under mangroves
Source: India State of the Forest Report 2011, and Forest Survey of India, Ministry of Environment
Notes:

1. *The total forest and tree cover of India is 80.9 million hectare which is 24.62 percent of the geographical area of the country. As compared to the assessment of 2019, there is an increase of 226100 ha in the total forest and tree cover of the country. 17 states/Union Territories have above 33 percent of the geographical area under forest cover.*
2. *Total carbon stock in country's forest is estimated to be 7,204 million tonnes and there an increase of 79.4 million tonnes in the carbon stock of country as compared to the last assessment of 2019. The annual increase in the carbon stock is 39.7 million tonnes.*

Source: https://pib.gov.in/PressReleasePage.aspx?PRID=1789635

agency over the past decade. The satellites—20 or so—will be built and flown by a San-Francisco-based company, Planet.

Lockdowns all over the world because of the Covid-19 pandemic resulted in the biggest drop in demand for energy since World War Two. This drop reduced carbon emissions by around 6% in 2020 due to all kinds of restrictions. But present IEA predictions indicate that the energy demand is booming in the developing world and coal will be playing a key role. The demand for coal is likely to be close to the global peak seen in 2014—and that has implications for efforts to monitor climate change. Global carbon emissions are set to jump by 1.5 billion tons in 2021—driven by the revival of coal use in the power sector. The emissions predictions for 2021 would be even worse, if demand for oil increased to pre-Covid-19 levels.

Overall green energy sources will provide 30% of electricity generation, the highest level since the beginning of the Industrial Revolution. China is likely to account for almost half the global increase in renewable electricity this year. Use of renewable energy on a large scale is a welcome effort to mitigate climate change but continuing massive investments alongside in coal and gas will contribute to enormous CO_2 emissions.

1.2 Renewable Fuels

1.2.1 General Introduction to Renewable Fuels

Renewable fuels are derived from plants, various other organisms, or even water (hydrogen from water). These do not include fuels created by fossilization of biomass over long periods such as petroleum, natural gas, coal, lignite, etc. There are two types of renewable fuels namely, 'primary' and 'secondary'. Biofuels such as hydrogen produced by electrolysis of water can play a very important role in reducing the use of fossil fuels and also in reducing net greenhouse gas emissions. Biomass as an energy resource has only been partially utilized so far and there is enough scope for deploying this resource efficiently.

The importance of sustainable renewable fuels including plant-derived secondary fuels has very much increased due to environmental problems. *If we preserve the wealth of forests even now, before they are further destroyed, the* forest waste *can prove to be a very significant resource for making second-generation renewable fuels. Similarly, crop residues and other agricultural biomasses can be good resource for bioenergy.*

Various organisms like bacteria, fungi, algae, protozoa, rotifers, worms, etc., play an important role in soil fertility. Bacteria, fungi, and algae play an important role in the production of secondary biofuels from plant material. Annexure A at the end of this chapter gives an introduction about these organisms.

1.2.2 Examples of Primary Renewable Fuels

- Wood
- Plant foliage and residues, and forestry wastes
- Coppice from willow (grown in North America and Europe for woody biomass) and high-density woody coppice produced from poplars and eucalyptus, etc.
- Grasses like hay, switchgrass (*Pancium virgatum*), reed canary grass, miscanthus (a fast-growing perennial grass which is native to Asia), etc.
- Crop residues like straw, corn stover, wheat straw, paddy straw, stalks of millets or other crops, maize cob, maize husk, rice husk, rye husk, etc.
- Straw, shells, husk, stalks from oilseed crops, pulses, etc.
- Stalks and other remains of cotton, jute, sisal, manila, etc.
- Bagasse from sugar cane
- Coconut shell, coconut husk, remains of coconut tree, etc.
- Peels of fruits and vegetables, and remains of their crops/trees
- Tobacco, coffee husk, etc.
- Other agricultural wastes, etc.
- Nonrecyclable wastepaper and cardboard, unusable packing wood, etc.
- Carbon containing refuse, solid/semisolid/liquid wastes
- Animal dung (of cows, buffaloes, camels, horses, donkeys, sheep, and other animals)
- Halophytes like *Salicornia brachiata, Alkali sacaton*, crowfoot grass, iodine bush, Palmer saltgrass, etc., which are salt-tolerant and drought-resistant plants [See Para 2.2.6(e)]
- Dried water hyacinth (*Eichhornia crassipes*),* which is an abundant water weed (see note below).

Seaweeds like *Macrocystis pyrifera* (giant kelp, which have massive leaves that can be periodically harvested), *Laminaria japonica* (Oriental kelp which is exploited for food in Japan and China), *Sargassum* sp. (brown type algae; some of these species harbor nitrogen-fixing bacteria, thus requiring little fertilizer application), *Gracilaria tikvahiae* (red algae capable of growing in closed or semi-closed cultures), green algae (some of which have very high biomass production rates), etc. Also, please see Para 3.6.5. [12].

- It may be noted that many microalgae apart from seaweeds grow in saltwater as well as freshwater. Many types of microalgae are a source of lipids and are now considered as an abundant source of oil for conversion to biodiesel (see Chapter 5 and also Annexure A of this chapter).

***Note on water hyacinth as a biomass resource for fuel:* [15,39]**

Because of eutrophication, the ponds and lakes are generally found clogged by the weed Eichhornia crassipes (commonly known as water hyacinth), which is considered to be a highly problematic invasive species. The plants do not have any natural enemies and thrive unrestrictedly in the water environment and spread far and wide. This water weed has become an ecological plague, suffocating lakes, diminishing dissolved oxygen and the survival rate of fish, and hurting the local economies in almost all the countries in temperate and hot regions of the world.

The best form of management of water hyacinth is its prevention. If prevention is no longer possible, it is best to treat the weed infestations when they are small to prevent them from being establishing. Early detection and a rapid response are required. Control is generally best applied over the least-infested areas before dense infestations are tackled with consistent follow-up. If possible, the flow of nutrients from the surrounding catchments should be minimized. Small- scale infestations can be controlled manually. Larger infestations should be tackled with harvesters, but the running costs are high.

Chemical control by herbicides is only effective in the short term and needs to be repeated over a long period. The use of herbicides creates objectionable water pollution. Biological control by insects (like moths, other beetles, grasshoppers), mites and fungi (rusts) is possible. Additionally, biological control with agents like weevils (Neochetina bruchi and Neochetina eichhorniae) is successful.

In short, this weed is very difficult to control and needs to be completely eradicated from surface water sources. It is therefore necessary to utilize its existing population in some way. Like other aquatic plants, this plant also absorbs heavy metal from water. The organic fertilizer made from water hyacinth is highly alkaline and can only be used on acidic soils.

Some communities have been making use of weeds on a very limited scale for various purposes. For example, because of its extremely high rate of development, Eichhornia crassipes is an excellent source of biomass which can produce biogas. One kilogram of this dried weed can give 370 liters of biogas with a caloric value of 22 MJ/m³. The weed has to be first chopped and then dried. The dried matter can be gasified like any other organic matter and the resulting gas can directly be used as fuel or can be converted to liquid fuel/hydrogen/other biofuels, or chemicals.

The labor involved in harvesting water hyacinths can be greatly reduced by taking advantage of the prevailing wind direction.

1.2.3 Availability of Forest Wastes for Bioenergy Production

Depending mainly on climatic factors and soil structure and composition, different types of crops are cultivated in the world. These may be wheat; paddy; millets like maize, sorghum/jowar, pearl millet/bajra; rye; various pulses; oilseeds like sunflower, safflower, mustard, sesame, soybeans, ground nut, oil palm, flaxseed; coconut; tobacco; coffee; vegetables; fruits; fiber crops like cotton/jute, spices, banana, sugar cane, jute, sisal, manila, etc.; or other crops.

The availability of agricultural crop residues for bioenergy production differs from one agricultural region to another in the world. It may be mentioned that the estimates of available forest wastes (foliage, leaves, etc.) for bioenergy production from residues should account for ecological constraints and alternative uses of the total waste production. A certain amount of forest wastes falling on the ground in the forest or plantation is required for ecological uses, the most important of which are for maintaining soil organic carbon (SOC) and soil productivity, checking wind and water erosion, improving soil structure, etc. This amount should not be less than 1000 t/km² of forestry residues from final harvest of the forest [13]. Regarding non-foliage residues produced during intermediate operations, 50% of theoretical potential may be assumed useable.

Part of forest waste may be used by communities living near a forest as fuel for cooking, and some of it may be required as cattle feed. The actual availability of forestry waste for generating bioenergy would differ from biome to biome, and place to place and there may be a great amount of variation. It would depend on natural climatic conditions, and various factors such as cattle rearing, use of cattle for agriculture, cheap availability of cooking gas/coal as domestic fuel, presence of industries nearby producing straw boards or other products from residues, economic conditions in the area, etc. As per a 2011 WHO report, about 50% of the domestic fuel needs of agricultural communities are fulfilled by crop residues [14]. Also, there are technical limits to the collection of residues from crops for use as a bioenergy resource.

In most countries, there is enough surplus agricultural biomass available even after use by the rural people for various purposes including fodder for animals, compost formation, soil ecological needs, etc. However, as the population's needs increase, crop production will need to keep pace and therefore, crop residues and processing residues are expected to continue to increase.

1.2.4 Controlled Biomass Utilization—A Fire-preventing Measure

An advantage of gathering dry leaves, dry grass, and foliage biomass from under the trees in forest or other plantations and using this biomass for purposes of fuel or making compost is preventing forest fires. Most incidents of fires in forests have taken place due to dry leaves/dry grass/dry foliage lying under the trees catching fire. The fire can take place due to extraneous reasons like humans smoking in the forest, fires being lit for cooking or heating by human beings, or even by spontaneous combustion on slow oxidation by atmospheric oxygen. The incidents of destruction of forests and plantations by fire occur all over the world every year.

1.2.5 Examples of Biomass Production by Some Trees

- **Coconut:** Coconut trees generate biomass residues like wood, fronds (leaves), husks, and shells. The tree density may be of the order of 120 trees per hectare. Each tree may yield about 12–14 dry leaves/fronds weighing about 1.5 kg. Coconut trees have a long, productive life varying between 50 to 100 years and therefore not much can be said regarding the amount of woody biomass available by replacing the trees. [On a wet basis, coconuts may consist of: copra (28–30%); water (22–25%); husks (33–35%); hard shell (12–15%).] The coconut tree density may be about 140 to 160 trees per hectare and each tree may yield 3 to 7 kg copra per year. Per hectare per year, about 2,220 kg of dry husks and 1,040 kg of dry shells may become available.

- **Oil Palm:** The useful life of a palm tree is about 25 to 30 years. The tree density is about 120 to 142 trees per hectare and tree trunks and fronds (leaves of palm) become available during the replanting of oil palm trees. The coconut tree would yield about 500–600 kg stem wood and 120 kg fronds per tree when replaced after 25–30 years. Besides these, fiber, shells, and empty bunches are generated during processing, resulting in about 1,853 kg of fiber, 2,780 kg of shells, and 1,483 kg of empty bunches per hectare as dry matter.

- **Rubber:** The useful life of a rubber tree may be about 25–35 years. Yield per hectare of wood on replacement of tree after 25–35 years would be about 180 cubic meter (with density of 0.72 tons per cubic meter with a moisture content of 60 %).

Note: As an example, the increasing trend of the production of biomass (from 1994 level to 2010 in an agricultural country (India) is shown in Table 1.4). The approximate biomass production in India may be 700 million tons per year, while the potential for production of biomass secondary biofuel may be about 100 million tons per year.

1.2.6 Alternative Uses of Lignocellulosic Biomass [2]

It is to be specifically considered when deciding about the available potential of biomass for conversion to secondary biofuels that there are some alternative and competing uses of biomass as mentioned in Table 1.5.

1.2.7 Secondary Renewable fuels/Second-generation Fuels [27]

Secondary renewable fuels are fuels derived from plant sources like oilseeds, vegetable oil and fats, crops/grains, agricultural crop residues, forest residues, organic wastes, sugary residues, any other type of biomass, algae, fungi, etc. Even water can be resource for a renewable fuel like hydrogen. The main examples of secondary renewable fuels derived from the abovementioned sources are:

(i) *Ethanol/ethyl alcohol (C_2H_5OH)*

(ii) *Butanol (C_4H_9OH)*

Table 1.4. Example of Field-Based Crop Residue and Processing Residues in India [30].

Crop residues	Production in million tons	
	1994	**2010 (projected)**
Cotton stalk	19.39	30.79
Rice straw	214.35	284.99
Wheat straw	103.48	159
Sugarcane tops	68.12	117.97
Maize stalk	18.98	29.07
Soybeans	12.87	34.87
Jute stalk	4.58	1.21
Groundnut straw	19	23.16
Processing-based residue		
Rice husk	32.57	43.31
Rice bran	10.13	13.46
Maize cob	2.59	3.97
Maize husk	1.90	2.91
Coconut shell	0.94	1.50
Coconut husks	3.27	5.22
Ground nut husk	3.94	4.80
Sugarcane bagasse	65	114.04
Coffee husk	0.36	0.28

Table 1.5. Alternative and Competitive Uses of Biomass.

Source	Material	Competitive Use
Agriculture and product processing	Wheat and paddy straw, stalks and cobs of millets including maize, etc.	Cattle feed, composting
	Straw, husk, shells of oilseeds, and pulses, etc.	Cattle feed, composting
	Sugarcane biomass including bagasse	Burning by sugarcane industry to raise steam and generate power
	Oil cakes after oil extraction, rejected fruits, peels	Animal feed, fish feed, composting
	Edible oil from oilseeds	Human consumption
	Non-edible oil from oilseeds crops or tree-borne oilseeds	Medicinal uses, soap making
	Rice husk	Burning, composting
	Rice bran	Source of edible oil, vitamin B_1, animal feed
	Residue after milling and screening of flours of grains	Animal feed
	Stalk of pulses, fiber plants like cotton, jute, sisal, manila	Domestic fuel
Forestry	Wood	Timber for buildings, furniture, plywood, pulp and paper, packing, etc.
	Wood residue, bark, leaves	Soil conditioning, mulching, burning as fuel, composting
	Wood shavings, sawdust, chips	Fiber boards, pulp and paper
Wastes	Wastes of paper, cardboard, furniture parts, packings	Recycling, pulp and paper, burning
Seaweeds	Seaweeds	Composting, extraction of food supplements. Some seaweeds are edible and are used for human and animal consumption

(iii) *Biodiesel mainly methyl esters* *(RCOOCH₃); derived from vegetable oils*
[**'R' in the formula for methyl ester is alkyl group such as CH₃, C₂H₅, C₄H₉, etc.]*

(iv) *Biodiesel derived from algae oil, fungi oil*

(v) *Renewable liquid fuels (like methanol, butanol, dimethyl ether abbreviated as DME, renewable diesel) derived from lignocellulosic biomass through gasification route*

(vi) *Methane/biogas/landfill gas, gaseous fuel through biomass gasification*

(vii) *Hydrogen*

1.2.8 Sorghum Stalks as a Source of Secondary Fuel

Sorghum is a genus of flowering plants in the grass family Poaceae. It is cultivated in warm climates worldwide and naturalized in many places, notably in Australia, USA, African countries, and many Asian countries, including India, and islands in Indian and Pacific Ocean. One sorghum species is grown for grain, while many others are used as fodder plants. It is either intentionally cultivated or allowed to grow naturally in pasture lands. Sorghum requires less water compared to other grain crops.

While sorghum is mostly cultivated for grain and fodder, *it is interesting to know that the stalks of sorghum yield 15 to 20% fermentable sugar, which can be converted to alcohol.* Sorghum is a potential alternative to meet the growing demand of bioethanol. The crop can be grown in semiarid areas as well, because it needs much less water as compared to sugarcane. Though commercial and large-scale production of sorghum has not been realized yet to its full potential, a few sugar-producing mills and corporates have developed open-pollinated varieties and photoperiod-sensitive hybrids, enabling year-round production. For example, Tata Chemicals, India, has started production. Research in this direction is going on in many countries by organizations such as ICRISAT and the National Research Centre for Sorghum.

Apart from the sugar content of sorghum stalks, its lignocellulosic biomass can also be used for producing biofuels like other biomasses.

1.2.9 Savings in Greenhouse Gas Emission due to the use of Biofuels [6,27,34]

As quoted in The EC Renewable Energy Directive 2009/28/EC, there is a net saving in greenhouse emissions for biofuels like biodiesel, bioethanol, and biogas. The extent of savings in greenhouse emissions would depend mainly on:

- Direct or indirect land use change.

- Details of biomass feed stock and its cultivation especially with respect to fertilizer application rates, type of fertilizer (say chemical fertilizer, biofertilizer, compost manure, etc.), manufacture of the type of fertilizer used, and the emission of nitrous oxide from soil.

- Transportation involved and transportation needs of fuel/energy.

- Processes and technology used in the conversion of biomass to fuels.

- Source of processing energy (heat and electricity) for conversion of biomass to biofuels.

- Amount of energy used in all the operations involved.

- Methods of calculation of greenhouse gas emissions and choice of co-product allocation procedures.

1.2.10 *Various Secondary Biofuel Production Routes from Plant Sources* [16,29]

There are various methods for making useful secondary renewable fuels/biofuels from plant materials, as shown in Figure 1.1. It is clear from the figure that various types of fuels and also chemicals can be obtained from resources based on seeds, fruits, corn, vegetable oil, foliage, wood, forest wastes, agricultural residues, and other types of biomass waste products.

In addition to the secondary fuels already mentioned, the following secondary fuels can also be obtained from plant resources:

Dimethyl ether (DME), mixed alcohols (containing methanol, ethanol, propanol, butanol), synthetic diesel, syn gas (consisting of carbon monoxide and hydrogen), etc.

There are various ways for obtaining these fuels, such as chemical processes, biochemical processes, gasification process, etc. Some fuels are synthesized after gasification. It may be mentioned that any sugar component can be converted to alcohol by fermentation technology, while any vegetable oil can be converted to biodiesel by transesterification technology. The biomass gasification route gives greater freedom for synthesizing the various types of fuels mentioned above.

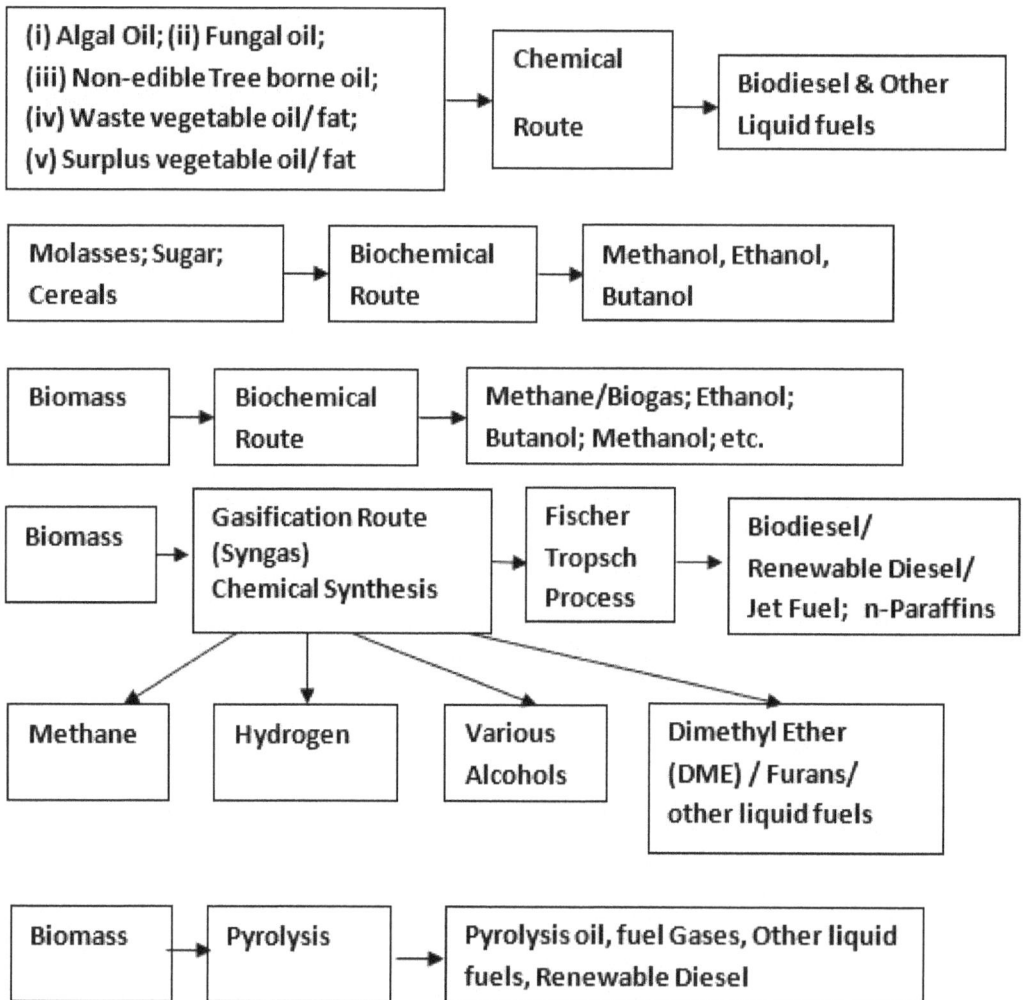

Fig. 1.1. Various Routes for Production of Secondary Biofuels.

However, only a section of biomasses is available for secondary biofuel production because they are used for many other things, difficulty in collection, and destruction by fire. The collection of biomass itself requires a lot of human labor and involves volume reduction and transportation costs.

The sugarcane juice or molasses conversion into ethyl alcohol by fermentation is one of the oldest technologies known to mankind, but now the technology for conversion of even the lignocellulosic materials like tree litter, various types of straws/ stalks, crop residues can be a source for alcohols (and even biodiesel).

Notes:

i. *Apart from secondary biofuels, lots of bioproducts or bio-based products and chemicals can be produced from renewable biological sources such as plants, seaweed, and algae. High-value co-products can be produced from woody biomass during the early steps in the conversion process to fuel. These chemicals can be used to make paints, plastics, solvents, packaging, pharmaceuticals, cosmetics, and even nutritional supplements and textiles.*

The chemicals which can be made from plant sources include: acetic acid (found in road de-icing salts, paint, and solvents); ethyl acetate (found in nail polish remover and cosmetics, and used to decaffeinate coffee); ethylene (a gas used to make plastic objects such as bags and water bottles, and used to ripen fruit, etc.).

ii. *It may be mentioned that pyrolysis can occur at both high temperatures or moderate temperatures. Torrefaction is a form of pyrolysis at temperatures typically ranging between 200–320°C.*

1.2.11 *R&D Efforts for Producing Second-generation Fuels* [5,11]

Considerable research activities are taking place mainly in North America, Europe, and a few developing countries (e.g., Brazil, China, India, and Thailand). In China and Brazil, the pilot plants for producing second-generation biofuels are already operating.

India and European Union are cooperating in research areas and innovations regarding next generation biofuels as per DBT-ICT Centre of Energy Biosciences, Institute of Chemical Technology, Mumbai, India. A number of companies like Indian Oil Corporation, Tata Chemicals, India Glycols, and Fermenta Biotech, and some institutes and universities are taking part.

The key to economic lignocellulosic ethanol production through a biological route is the effective combination of pretreatment, hydrolysis, and fermentation of lignocellulosic material. Presently, pretreatment determines the efficiency of successive processing steps to convert cellulose and hemicellulose to hexoses and pentoses. Pretreatment overcomes the recalcitrant nature of lignocellulose by opening its structure and enabling enzymes to access the cellulosic fraction. Further research in this field is needed to improve the efficiency of pretreatment, reduce the cost of enzymes used for hydrolysis by modifying the enzymes and combining hydrolysis with fermentation, and developing microorganisms that are able to effectively utilize both hexoses and pentoses to produce ethanol. Although, the consolidation of steps in bioprocessing of pretreated biomass is being used even now, further research will enhance the conversion efficiencies. This research is continuing on these issues as well as on the individual biocatalytic steps as well as the chemical engineering involved. It is expected that the improvements will go on and gaps in research will continue to be addressed because of the thrust of research on producing second generation fuels economically.

In the thermochemical conversion of biomass into secondary fuels, Fischer-Tropsch (FT) synthesis has shown high potential to improveme the economics of the process. Further development is required in the area of achieving '*longevity and robustness of the catalysts*' as well as more '*cost-effective cleaning-up of the syngas.*' Better and cheaper *catalysts for FT synthesis of the gases* derived from renewable resources are needed for increasing the desired liquid fuel yields.

Good gas clean-up is required so that the catalyst is not deactivated by inhibitors and catalysts from biomass should be comparable in cleanliness to that derived from natural gas. This becomes extremely important in case of synthesis of certain fuels.

If the bio oil is processed in existing refineries after suitable modifications, it would reduce the capital costs and infrastructure costs. Further developments in this direction would be necessary.

Development of stabilizers for the oils so that they are not too viscous during transport, and improvements in catalysts so that more stable oils are formed, should be endeavored towards.

In the case of DME, suitable infrastructure will have to be developed for delivering DME as a transportation fuel, which is gaseous under atmospheric pressure and temperature ranges. Vehicle modifications will also be required. DME has lower energy density and therefore it requires a fuel storage tank twice the size of a conventional onboard diesel fuel tank for the same driving distance [5].

The world over, new developments and research are being carried out. Developing countries like India are also making constant efforts in this direction, where the emphasis on research and development (R & D) is in the following fields:

- Production of green diesel through the hydro treatment route
- Lignocellulosic ethanol
- Finding high-yielding microalgal strains with a higher content of algal oil
- Collection/harnessing microalgae for oil extraction
- Algal growth in open ponds
- Development of enzymes and fermentation strains of microorganisms
- Major R&D projects focused on biomass pyrolysis
- Biomass cracking
- Improvement of TBO (Tree-borne Oilseeds) yields.

1.3 Life Cycle Analysis of Greenhouse Gas Emissions [23,24,25,26,28,34,38,40]

The real purpose for which renewable fuels are recommended is that the total greenhouse gas (GHG) emissions over a long period of time should reduce and the global warming threat is combated. Therefore, the assessment of the life cycle of greenhouse gas emissions must be done when the replacement of an existing fuel is considered. This assessment calculates the amount of greenhouse gases that are released per unit of fuel. This includes emissions and/or carbon sequestration as well as any land use changes as a result of growing of biofuel crops. The EPA was required to use a life cycle assessment in its evaluation of the Renewable Fuel Standard.

The life cycle emission accounting evaluates and reports on the full life cycle GHG emissions associated with raw materials extraction, manufacturing, or processing, transportation, use, and end-of-life management of goods or service. The life cycle perspective accounts for all emissions connected to goods or services, regardless of which industrial or economic activities or sectors produce these emissions (energy, mining, manufacturing, or waste sectors) and when these benefits occur over time. A most comprehensive, up-to-date life cycle assessment model has been developed for greenhouse gases, regulated emissions, and energy use in transportation (GREET) at Argonne National Lab for various fuels (including biofuels and petroleum-based fuels).

1.3.1 Comparison of Life Cycle GHG Emissions of Renewable Fuels with Petroleum Fuels

It is interesting to note that the life cycle greenhouse gas emissions for biofuels/renewable fuels were evaluated in the US [*under the EPA renewable fuel standard program (RFS2) regulatory impact analysis*] in percentage in comparison to the life cycle GHG emissions of matching petroleum products, as shown in Table 1.6.

The table compares the life cycle greenhouse gas (GHG) emissions of various biofuels with those of the petroleum fuels they replace, as assessed by the US Environmental Protection Agency. The biofuels compared included corn ethanol, sugarcane ethanol, switchgrass ethanol, corn butanol, soybean biodiesel, and waste grease biodiesel. The GHG emissions benefit of ethanol varies according to the feedstock used to produce it, with the greatest benefit achieved by ethanol produced from switchgrass. ***The negative GHG emissions*** from energy crops, including a number of grass resources (herbaceous crops) such as switchgrass (*Panicum virgatum*), and miscanthus, ***indicate a net sequestration of carbon into the soil and biomass***. Biofuel from energy crops can reduce GHG emissions by 101 to 115 percent. Corn stover, a residue from corn, can reduce GHG emissions by 90 to 103 percent. Life cycle emission levels of biodiesel vary according to the feedstock, with biodiesel produced from waste grease resulting in a greater GHG emissions benefit than emissions from biodiesel produced from soybean biodiesel. Energy-intensive practices are increasingly being used to extract oil and produce gasoline. Therefore, as biofuels replace more carbon-intensive gasoline, they provide even greater emission reductions. Thus, when comparing biofuels and gasoline, it is possible to see that the emissions gap between the two is wide.

It may be mentioned here that the blending of ethanol with petrol or biodiesel with petroleum diesel improves the combustion in engines because of the extra oxygen content in them and helps to lower carbon monoxide, hydrocarbon, and smoke emissions. Table 1.6 gives a lot of insight into the justification of using secondary renewable fuels in place of petroleum fuels as regards the purpose of lowering greenhouse emissions (which create global warming).

An interesting study on life cycle GHG emissions for different light-duty vehicles and fuel pathways was carried out in the US. The comparison of life cycle GHG emissions from different vehicles by Nick Nigro and Shelley Jiang for the Center for Climate and Energy Solutions (US) is shown in Table 1.7. The study led to the following conclusions which are very informative [31]:

(1) Battery electric vehicles have the lowest GHG emissions (but vary greatly depending on the* 'fuel mix of electricity'), followed by hybrid electric vehicles run on a combination of electricity and ethanol, followed by hydrogen fuel cell vehicles. This includes the consideration of 'emissions from production, transportation and consumption of fuels'.

['Fuel mix of Electricity': Composition of electricity generated by different fuels (say, % generated by coal + % generated by oil + % generated by natural gas + generated by nuclear energy + % generated by hydro energy + % generated by any other low carbon source.)]*

Table 1.6. Comparison of Biofuel Life Cycle Greenhouse Gas Emissions with Those of Petroleum/Gasoline [9].

Fuel →	P. Gasoline	Corn Alcohol	SC Alcohol	Swg. Alcohol	Corn Butanol	P. Diesel	Soybean Bio-diesel	WG Bio-diesel
Percent of Petroleum GHG Emissions	100	79	39	−10	69	100	43	14

Abbreviations used: P.—Petroleum, SC—Sugarcane, Swg. —Switchgrass, WG—Waste grease
Notes: Emissions data is based on the analysis by the US Environmental Protection Agency for the Renewable Fuel Standard Program (RFS2).

Table 1.7. Life Cycle GHG Emissions from Different Light-Duty Vehicles & Fuel Pathways [4,17,32].

Type of vehicles/Fuels	Conventional	Hybrid	PHEV	Fuel Cell Vehicle	Battery Electric Vehicle
Gasoline	442.58	343.81	302.71	--	--
E85	328.85	212.27	224.5	--	--
Natural Gas	389.42	--	--	--	--
Hydrogen	--	--	--	218.44	--
Electricity	--	--	--	--	196.30

(2) The lowest-emitting plug-in hybrid electric vehicle (PHEV) amongst all the vehicles and fuels considered is the gasoline-powered PHEV (when recharged from a very low carbon grid). It would have far lower emissions than the lowest-emitting gasoline hybrid vehicle. PHEVs fueled by biomass and cellulosic ethanol are also very low emitting, even if they are partly powered by electricity from a more carbon intensive grid (e.g., an off-peak US average grid, with about 50% coal, 29% natural gas, and 20 % nuclear energy).

1.3.2 Impact of Substituting Biodiesel for Petroleum Diesel [39]

1. *A report from NERL (US) regarding the effect of a 100% biodiesel substitution for petroleum diesel in an urban bus would:*

- Reduce CO_2 emission by 74.5%
- Almost eliminate SOx from tailpipe emissions
- Reduce life cycle CO emission by 32%
- Reduce life cycle SOx emission by 35%
- Reduce life cycle total particulate matter by 8%
- Increase life cycle NOx emission by 13%
- Increase non-methane hydrocarbon emission by 35%
- Decrease methane emission slightly

The report also says that blending petroleum diesel with 20% biodiesel would bring down CO emissions by 15.7%. [33]

2. *Biodiesel sourced from neem oil when substituted for petroleum diesel led to a decrease in CO_2 emissions and smoke, while increasing NOx.* [31]

3. *Biodiesel sourced from palm oil, when substituted for petroleum diesel, led to a decrease in CO_2 emissions, polyaromatics, particulates, and hydrocarbons.* [40]

Biodiesel is a much cleaner fuel compared to petroleum diesel and conforms to EPA (US) certification. The use of biodiesel in a conventional diesel engine not equipped with new diesel after treatment results in a substantial reduction of unburned hydrocarbons, carbon monoxide, and particulate matter compared to emissions from diesel fuel. In addition, the exhaust emissions of sulphur oxides and sulphates (major components of acid rain) from biodiesel are essentially eliminated compared to those emitted from higher sulphur diesel.

Of the major exhaust pollutants, both unburned hydrocarbons and nitrogen oxides are ozone- or smog-forming precursors. Emissions of nitrogen oxides may be slightly lower, depending on the duty cycle of the engine and the methods of controlling the NOx generation used.

Based on engine testing, using the most stringent emissions testing protocols required by EPA for certification of fuels or fuel additives in the US, the overall ozone-forming potential of the spectated hydrocarbon emissions from biodiesel was nearly 50 percent less than that measured for diesel fuel.

New technology diesel engines (i.e., those with PM traps and SCR technology required for on-road applications in the US after 2010) reduce the emissions of both PM and NOx with B20 over 90% than those emitted in the 2004 model and make NTDEs as clean or cleaner than either gasoline- or natural gas-fueled engines. The use of low carbon B20 in NTDEs make them the clean, green technology of the future.

1.3.3 *Biodiesel is Better for Human Health than Petroleum Diesel*

Diesel exhaust has highly harmful impact on human health because of the SOx, particulate matter, polycyclic aromatic hydrocarbons (PAH), and nitrated PAH compounds, etc., it emits. Polycyclic aromatic hydrocarbons (PAH) and nitrated PAH compounds are identified as potential cancer-causing compounds. The particulate matter, SOx, in emissions is linked to asthma, chronic obstructive pulmonary disease (COPD), and other lung and throat diseases, while carbon monoxide is a poisonous gas. All these emissions are greatly reduced when using biodiesel rather than petroleum diesel fuel in the engines. Even the blended petroleum fuels with biodiesel have less emissions than pure petroleum diesel.

1.3.4 *Impact of Biodiesel on Mitigating Global Warming*

Using biodiesel is the best greenhouse gas mitigation strategy for today's medium- and heavy-duty vehicles. A 1998 biodiesel lifecycle study, jointly sponsored by the US. Department of Energy and the US. Department of Agriculture, concluded that biodiesel reduced net carbon dioxide emissions by 78% compared to petroleum diesel. This is due to biodiesel's closed carbon cycle. *The CO_2 released into the atmosphere when biodiesel is burned is recycled by growing plants, which are later processed into fuel.*

1.3.5 *Biofuel Policy and Renewable Fuel Standard*

Regarding the replacement of petroleum fuels by biofuels, policy and standards have been framed in order to get savings on greenhouse gas emissions. The US has issued a Renewable Fuel Standard, while the European Union has issued two policy directives explicitly targeted to reduce GHG emissions. Brief notes about the US. Renewable Fuel Standard and European Union policy directives targeting reduction of GHG emission are mentioned below:

Note on US Renewable Fuel Standards [18,27,34,36]

The following points regarding the US Renewable Fuel Standard are worth mentioning here:

(a) **The Renewable Fuel Standard (RFS)** *is a US federal program that requires transportation fuel sold in the United States to contain a minimum volume of renewable fuels. The RFS originated with the Energy Policy Act of 2005 of the US and was expanded and extended by the US. Energy Independence and Security Act of 2007 (EISA). The RFS requires renewable fuel to be blended into transportation fuel in increasing amounts each year, escalating to 36 billion gallons by 2022. Each renewable fuel category in the RFS program must emit lower levels of greenhouse gases (GHGs) relative to the petroleum fuel it replaces.*

(b) **Requirements of the US Renewable Fuel Standard Regarding Emissions:** *The US Environmental Protection Agency (EPA) administers the program of Renewable Fuel Standard (RFS) and establishes the volume requirements for each category based on the EISA-legislated volumes and fuel availability. The US EPA tracks compliance through the Renewable Identification Number (RIN) system, which assigns a RIN to each gallon of renewable fuel.*

Entities regulated by RFS include oil refiners and gasoline and diesel importers. The volumes required of each obligated party are based on a percentage of its petroleum product sales. Obligated parties can meet their renewable volume obligations (RVOs) by either selling required biofuels volumes or purchasing RINs from parties that exceed their requirements. Failure to meet requirements results in a significant fine.

(c) **Mandated Greenhouse Gas Reduction by Renewable Fuels of Various Categories:** *Each year, the RFS program requires the sale of specified volumes of renewable fuels according to the categories below. EISA established life cycle GHG emissions thresholds for each category, requiring a percentage improvement* *relative to the emissions baseline of the gasoline and diesel they replace. The reduction in greenhouse gas emission by the amount of renewable fuel replacing the petroleum diesel/gasoline must be according to the following guidelines.*

i. Conventional Biofuel: *Any fuel derived from starch feedstock (e.g., corn and grain sorghum). Conventional biofuels produced from plants built after 2007 must demonstrate a 20% reduction in life cycle GHG emissions.*

ii. Advanced Biofuel: *Any fuel derived from cellulosic or advanced feedstock. This may include sugarcane- or sugar beet-based fuels; biodiesel made from vegetable oil or waste grease; renewable diesel co-processed with petroleum; and other biofuels that may exist in the future. Nested within advanced biofuels are two subcategories: cellulosic biofuel and biomass-based diesel. Both biomass-based diesel and cellulosic biofuel that exceed volumes in their respective categories may be used to meet this category. Fuels in this category must demonstrate a life cycle GHG emissions reduction of 50%.*

iii. Biomass-based Diesel: *A diesel fuel substitute made from renewable feedstocks, including biodiesel and non-ester renewable diesel. Fuels in this category must demonstrate a life cycle GHG emissions reduction of 50% (see Para 5.12.3, 5.12.4, and 5.12.5).*

iv. Cellulosic Biofuel: *Any fuel derived from cellulose, hemicellulose, or lignin—nonfood-based renewable feedstocks. Fuels in this category must demonstrate a life cycle GHG emissions reduction of at least 60%.*

Note on European Union (EC) Biofuel Policy Directives Targeting Reduction of GHG Emissions [3,6,10,34]

The Renewable Energy Directive and Fuel Quality Directive impose requirements that biofuels should meet certain sustainability criteria. These criteria also apply to bioliquids for heat and power, but not to other forms of renewable energy such as solid biomass. These cover the greenhouse gas emissions savings from using the fuels, and the types of land that may be converted to produce biomass for biofuels production. There are also conditions on European feedstock production on cross-compliance with agricultural sustainability rules.

Both Directives set a minimum threshold of 35% greenhouse gas (GHG) savings compared to fossil fuels that must be achieved by a biofuel to be eligible for support under the Member State renewable energy policies. This 35% threshold did not apply to facilities that were already in operation by 23rd January 2008 until after 1st April 2013. In effect from 1st January 2017 onward, this GHG savings threshold rose to 50%. Beginning 1st January 2018, installations starting production on or after 1st January 2017 must meet 60% or higher GHG savings compared to fossil fuels.

The Directives define a lifecycle methodology (detailed by Article 7d (1) and Annex IV in the Renewable Energy Directive) to calculate greenhouse gas emissions from biofuel production. Based on the methodology, the European Commission has calculated default emissions for different biofuel production pathways. Regulated entities reporting biofuel under the Directives may in general report that it has the default carbon intensity without providing any additional information to Member States. The Directives also include values for typical emissions, which are in general lower than default emissions. Regulated entities that are able to provide additional information about their production processes will be permitted to report based on typical emission values. The Directive also allows regulated entities to provide process specific information to generate a different lifecycle emissions intensity value.

At present, the Directives do not account for indirect land use change (iLUC) in their life cycle calculations of GHG savings, which include only direct emissions. The Directives do, however, recognize iLUC as an issue; they required the European Commission to submit a report reviewing the impact of iLUC on GHG emissions by the end of 2010 along with, if appropriate, a proposal to account for it. The European Parliament and the Council were required to vote on any such iLUC proposal by the end of 2012. Both of these deadlines have been missed. See the following section on the European Commission's iLUC proposal.

EC Legislation Relating to Biofuels [3,27]

Amendment of the Fuel Quality Directive and Renewable Energy Directive—Directive (EU) 2015/1513, the "iliac Directive", was published on 9th September 2015 in the official Journal of the European Community. This directive limits the way Member States can meet the target of 10% for renewables in transport fuels by 2020, bringing an end to many months of debate. There will be a cap of 7% on the contribution of biofuels produced from 'food' crops, and a greater emphasis on the production of advanced biofuels from waste feedstocks. Member States were then supposed to include the law in national legislation by 2017, and show how they were going to meet sub-targets for advanced biofuels.

Renewable Energy Directive 2009/28/EC [6]

- *Approval by European Parliament on 17th December 2008*
- *By 2020, 20% share of RES in final energy consumption, 20% increase in energy efficiency*
- *10% target for RES in transport in each Member State*
- *National Renewable Energy Action Plans were required by June 2010*
- *Burden sharing for RES targets except transport*
- *Harmonized approach with the Fuel Quality Directive*
- *No biofuel feedstock from carbon rich or biodiverse land*
- *EC has to report on compliance with environmental and social sustainability criteria of major biofuel exporting countries*
- *Minimum GHG reduction for biofuels by 35% and 50% from 2017 on; 60% for new installations from 2017 onward; for plants operating in January 2008, GHG requirements will start in April 2013 (amended in Directive (EU) 2015/1513)*
- *Bonus of 29 g CO_2/MJ for biofuels from degraded/contaminated land*
- *EC proposal for incorporating indirect land use changes by the end of 2010; special clauses for plants built before 2013 (see Directive (EU) 2015/1513)*
- *Biofuels from waste, residues, non-food cellulosic material, and lignocellulosic material will count twice for RES transport target*
- *Mass balance approach for certification of sustainability*
- *EC will negotiate bilateral and multilateral agreements*
- *Establishment of a committee for sustainability of biofuels.*

Amendment to the Fuel Quality Directive (2009/30/EC)

Directive 2009/30/EC *amending* **Directive 98/70/EC** *on environmental standards for fuel (Fuel Quality Directive) aims at:*

- *Further tightening environmental quality standards for a number of fuel parameters.*
- *Enabling more widespread use of ethanol in petrol.*
- *Introducing a mechanism for reporting and reduction of the life cycle greenhouse gas emissions from fuel.*
- *Reduction in life cycle GHG emissions from the energy supplied. A binding target of 6% as the first step while leaving open the possibility for increasing the future level of ambition to 10%.*
- *To that effect, in a 2012 review, the Commission will need to assess a further increase of the ambition level of 2% from other technological advances, such as the supply of electricity for use in transport. A further 2% is envisaged to be achieved by the use of CDM credits for flaring reductions not linked to EU oil consumption.*
- *Incorporation of sustainability criteria for biofuels used to meet GHG reduction requirement. Creation of a specific committee working jointly with the RED to coordinate the energy and environment aspects in future development of biofuel sustainability criteria.*
- *Reduction of sulphur content in inland waterway fuel in one step to 10 ppm by 1st January 2011.*
- *Phasing in of 10% ethanol (E10) petrol: To avoid potential damage to old cars, continued marketing of petrol containing maximum 5% ethanol guaranteed until 2013, with the possibility of an extension to that date if needed.*
- *Derogations for 'petrol vapor pressure' for cold summer conditions and blending in of ethanol are subject to Commission approval, following an assessment of the socio-economic and environmental impacts, in particular, on air quality.*

- *Increase of allowed biodiesel content in diesel to 7% (B7) by volume, with an option for more than 7% with consumer information.*

Key elements of the amendment

The amount of biofuels produced from 'food' crops as part of the contributors to the target of 10% renewables in transport will have a cap at 7%. The other 3% will come from different counted alternatives:

- *Biofuels from Used Cooking Oil and Animal Fats (double counted)*
- *Renewable electricity in rail (counted 2.5 times)*
- *Renewable electricity in electric vehicles (counted 5 times)*
- *Advanced biofuels (double counted)*
- *Benchmark for the share of advanced biofuels in the transport sector of 0.5%*

The agreement also includes the reporting and publishing of data on ILUC-related emissions on both the national and European level.

Member States have to transpose the directive into national legislation by mid-2017, and establish the level of their national indicative sub-targets for advanced biofuels.

Other EC and European Parliament policy activity on biofuels

On 16th October 2013, the European Parliament Intergroup on 'Climate Change, Biodiversity and Sustainable Development' held a workshop entitled, "The future of Biofuels as alternative fuel for the transport sector." A summary report and presentations are now available.

On 24th January 2013, the EC published COM (2013) 17 Clean Power for Transport: A European Alternative Fuels Strategy, which encompasses biofuels as well as LNG, SNG, electricity, and hydrogen. See also the press release Europe Launched on Clean Fuel Strategy. The strategy document advocates support for sustainable advanced biofuels produced from lignocellulosic feedstocks and wastes, as well as algae and microorganisms. It recommends no further public support for first generation biofuels produced from food crops after 2020.

1.4 Heat Content of Renewable Fuels

The heat content of a fuel is an important parameter and should be known before deciding to use it. The approximate net calorific values of some renewable fuels or their components are mentioned in the Table 1.8 [1 kcal = 4.1868 kJ].

Wood Properties

Freshly cut wood contains 50–60% moisture. Moist wood is elastic; its brittleness increases when its moisture evaporates. Dry wood normally consists of 40–50% cellulose, 25–35% hemicellulose, 20–35% lignin, 1–2% pectin, and traces of starch. Wood when burnt emits less carbon dioxide than coal.

The proximate analysis of air-dried wood would be: The volatile matter here includes moisture 75–80%, fixed carbon 20–25%, and mineral matter 0.6–1%. The mineral matter in ash comprises calcium, sodium, potassium, magnesium, phosphorus, and silicon. The average ultimate analysis of wood (moisture-free and ash-free basis) would be approximately: Carbon 50–53%, Hydrogen 5.7–6.3%, Oxygen 39–43.5%, Nitrogen 0.1–0.2%, and Sulphur, less than 0.05%. The ultimate analysis for an estimate may be taken as: C (50%), H (6%), O (44%).

Pyrolysis of dry wood may approximately yield: Water 25%, Carbon monoxide 18.3%, Carbon dioxide 11.5%, Hydrogen 0.5%, light hydrocarbons 4.7%, tar 20%, and char 20%.

Table 1.8. Approximate Net Calorific Values of Some Renewable Fuels or Their Components.

Se. No.	Renewable Fuels	Calorific Value
1.	Wood (air-dried)	12 to 15.5. MJ/kg (Net)
2.	Wood (green)	10.5 MJ/kg (Net)
3.	Refuse	9 GJ/tonne
4.	Wheat straw/Barley straw	12.38 MJ/kg (Net)
5.	Millet/Rye/Oats Straw	12.39 MJ/kg (Net)
6.	Sorghum straw	12.38 MJ/kg (Net)
7.	Rice husk	19.33 MJ/kg net (2637–2829 kcal/kg)/(GCV 3460 kcal/kg)
8.	Rice Straw	16.02 MJ/Kg (Net)
9.	Maize Stalk	5.25 MJ/kg (Net)
10.	Maize cobs	16.28 MJ/kg (Net)
11.	Bagasse	8.94 MJ/kg (Net)#
12.	Cassava stalk	17.5 MJ/kg (Net)
13.	Ground nut husk and shells	15.66 MJ/kg (Net)
14.	Paper	3890 to 4444 kcal/kg 7000 to 8000 Btu/lb
15.	Sawdust	13.8 MJ/ kg (Net)
16.	Coffee pods	13.8 MJ/kg (Net)
17.	Cow dung cake	6 to 8 MJ/kg
18.	Cocoa pods	16 MJ/kg (Net)
19.	DMF (2.5-Dimethyl furan)	31.5 MJ/litre
20.	Ethanol (Ethyl alcohol C_2H_5OH)	26.79 MJ/kg(Net)
21.	Methanol (Methyl alcohol CH_3OH)	19.76 MJ/kg(Net)
22.	Butanol (Butyl alcohol C_4H_9OH)	33.1 MJ/kg (Net)
23.	Biogas ($CH_4 + CO_2$)	21.6 MJ/M³
24.	Methane (CH_4)	50.01 MJ/kg (Net)
25.	Ethane (C_2H_6)	47.5 MJ/kg (Net)
26.	Propane (C_3H_8)	46.37 MJ/kg (Net)
27.	Butane (C_4H_{10})	45.7 MJ/kg (Net)
28.	Hydrogen (H_2)	120 MJ/kg (Net)
29	Carbon monoxide (CO)	10.11 MJ/kg (Net)
30.	Hydrogen sulphide (H_2S)	15.24 MJ/kg (Net)
31.	Ammonia (NH_3)	18.67 MJ/kg (Net)
32.	Hydrogen cyanide (HCN)	24.3 MJ/kg (Net)
33.	Lignin	25.1 MJ/kg
34.	Fat & Vegetable oil (Approximate)	39.8 MJ/kg
35.	Dimethyl ether (LHV)	28.430 MJ/kg
36.	Renewable Diesel (LHV)	43.7–44.5 MJ/kg

* **Composition of Wheat Straw (approximate):** Carbon 49.6%, Hydrogen 6.2%, Oxygen 43.6%, Nitrogen 0.6%, Sulphur Nil, Ash 4.75 %.

The bulk density of sawdust would be 219–227 kg/m^3 and that of wood shaving would be 150–160 kg/m^3. As a rule of thumb for biomass gasification, it may be assumed that 1kg of biomass gives 3 m^3 of wood gas of density 1 kg/m^3.

1.5 Hybrid Fuels [7,42]

Apart from those mentioned above, there are some hybrid fuels which are mixtures of some of the above fuels. Although ethanol has been mixed with petrol in small percentages for a long time, petrol/gasoline is not called a hybrid fuel. An example of a good hybrid fuel is 'Hypthane', which is a mixture of natural gas and a small percentage of hydrogen. This is described in Para 7.22, Chapter 7.

1.6 Gas from Oceans and Coal Mines

Although methane recovered from coal mines and from gas hydrates in oceans is not a renewable fuel, it is environmentally advantageous to recover it for use as a fuel and would be considered a green practice. If global warming continues, the methane hydrates may break up and methane will leak into the atmosphere. Methane is a far more powerful greenhouse gas than carbon dioxide. The coal mine methane needs tapping primarily because it is a safety hazard during the mining of coal, and secondly, because its leakage into the atmosphere would create a more powerful greenhouse effect than when it is recovered and burnt as fuel. This subject has been included in Chapter 6 (see Paras 6.16 and 6.17).

1.7 Biomass as Fuel [8]

Waste packaging wood, nonrecyclable paper waste, waste straw/grasses, other agricultural wastes, forestry waste, and refuse containing sizable amounts of combustible matter should not be destroyed by free burning. These wastes should either be converted to useful products, secondary fuels, compost manure, etc., or should be at least utilized as fuel to generate heat or power in an efficient manner. Biomass can be used alone or mixed with coal in boilers specially designed for such fuels. Normally, grate type boilers were used for such fuels, but now specially designed fluidized bed combustion (FBC) boilers are also being used for burning wastes/refuse.

It may be pointed out that liquid and gaseous fuels can be obtained from biomass; a topic that is analyzed in subsequent chapters.

1.7.1 Problems in the Use of Biomass as Fuel

(i) It may be mentioned that biomass should be air-dried and stored under a cover to avoid *the growth of fungi and spores and decomposition*. Uncontrolled and unattended decomposition of biomass and growth of *mold and bacteria* are potential health hazards.

(ii) Biomass has a high volatile content and significant quantities of combustible volatiles are released around temperatures of 180 ºC. The volatile content is in the range of 50 to 80% in various biomasses. Self-heating due to slow oxidation by atmospheric oxygen and subsequent *spontaneous combustion* should be guarded against.

(iii) The fire risks in use of biomass as fuel is much more than with any other solid fuel.

(iv) The burning of biomass does leads to environmental problems such as the generation of *oxides of sulphur, oxides of nitrogen, carbon monoxide, volatile organic compounds (VOCs), hydrogen chloride gas, dioxins, and suspended particulate matter. The solutions regarding reducing the emissions of these pollutants are similar to those for other fuels*.

(v) The emission of oxides of nitrogen such as nitric oxide (NOx) and nitrogen dioxide is mostly dependent on the temperature of the furnace. The higher the temperature of burning, the higher is the NOx content. It may be mentioned that nitrous oxide, which is a greenhouse gas, is formed during low temperature burning. Due to the higher moisture content of biomass, the burning temperatures for different types of biomass may be lower than that of other solid fuels under similar conditions. This may result in conditions favorable for the formation of nitrous oxide.

(vi)   The sulphur and chlorine contents of biomass are responsible for the emission of Sulphur dioxide, Hydrochloric acid gas, and other compounds of chlorine. Biomass, in general, has more chlorine, but less sulphur than coal.

The use of *fluidized bed combustion (FBC) boilers* has the advantage of drastically reducing the emission of oxides of nitrogen because of lower furnace temperatures, and emission of oxides of sulphur and other acidic gases because of the use of lime/limestone as the bed material in the furnace, which absorbs the acidic gases.

(vii) By maintaining proper combustion control over the fuel in the boiler, the emission of carbon monoxide and volatile organic compounds/hydrocarbons can be drastically suppressed.

(viii) The ash content of biomasses is low (generally less than 5%), while the moisture content widely varies. The ash deformation and fusion temperatures for certain types of biomass may be quite low (for example, the deformation temperature of barley is in the range of 750–800°C).

The low ash fusion and deformation temperatures (which is due to higher amounts of potassium, sodium, and phosphorus) cause problems of ash agglomeration and slagging of boiler heating surfaces. The ash agglomeration and clinker formation may cause serious problems for any boiler, particularly for chain grate stoker boilers. The fluidization of the bed may be adversely affected by ash agglomeration both in the chain grate as well as in the fluidized bed combustion type boilers.

(ix)   On release of volatile contents of biomass at low temperatures, the balance material is very reactive and may cause corrosion and fouling of heating surfaces. The slag deposits may be very dense and the molten ash may also attach itself to refractory materials and also on the heating surfaces in the boiler. Under certain conditions, the volatiles also may condense on the cold heating surfaces, causing fouling. It is therefore necessary to choose the proper type of biomass for burning in a boiler and also provide anti-slagging equipment/measures in the boilers. *[The slag removal from the bed material must be done on a regular basis in FBC boilers.]*

(x)   Biomass has a low ash content. The bulk of particulate matter can be removed by use of cyclonic dust separators/multiclones. The balance of the fly ash has a very fine particle size and the particulate emission can be easily controlled within the prescribed limits by use of bag filters. Electrostatic precipitators are not generally found suitable for biomass boilers. For small installations, wet scrubbers can be used. It may be mentioned here that the heavy metal content in biomass fuels is lower as compared to the similar content found in fossil fuels like coals and oils.

(xi)   Raising Bulk Density of Biomass for Transportation
The densities of most types of biomass are quite low, which makes their handling and transportation difficult and costly. Such wastes should be air-dried and hydraulically pressed to substantially increase their bulk density for easy transportation. The compressed fuel can then be transported to the industrial units, which may utilize them as such or convert it into liquid fuels or gaseous fuels.

(xii) Briquetting of Biomass (Agricultural and Forestry wastes)
Every year, millions of tons of agricultural waste, forestry waste, residues waste, and industrial waste is disposed of by burning or is destroyed by wildfires. The uncontrolled burning causes more air pollution than controlled burning. Technology is available for converting agro-waste

and forestry waste into higher bulk density fuel briquettes even without the addition of binders. *(For example: Advance Hydrau-Tech Pvt. Ltd., Delhi, is equipped with this technology.)*

(xiii)   Apart from the ease of transportation, the briquettes of air-dried biomass have other advantages such as low ash content, higher fixed carbon, consistent burning, uniform combustion, eco-friendly use, and higher calorific value (approaching 4000 kcal/kg). Briquettes can also be used for heating applications for domestic and industrial use. This fuel can be used in the gasifier for gasification of biomass to produce gaseous fuel or other products.

(xiv)   The industries which may use this fuel for thermal applications include: (a) textile units and spinning mills; (b) leather industry; (c) brick making units, (d) dyeing units; (e) food processing units; (f) milk plants; (g) ceramic industry, (h) refractory industry; (i) solvent extraction plants; (j) rubber plant, (k) chemical plants, etc.

(xv)   Most of the carbon-containing solid wastes (biomass fuels) can be converted to briquettes. These include the following:

(a) Sugarcane bagasse, leaves, and the upper portion of cane, (b) rice husk, (c) paddy straw, (d) wheat straw and other straws, (e) coir pith, (f) sawdust/bamboo dust, (g) wood chips and other types of wood waste, (h) husks of palm, coconut, soybean, coffee, etc., (i) seedcase and seed shells, (j) stalks of cotton/sunflower, etc., (k) waste of tobacco and tea, (l) forestry waste, (m) tree droppings/leaves/foliage, (n) grasses and weeds.

(xvi)   Fuel Preparation for Pulverized Fuel/FBC boilers
To use biomass as fuel, the biomass should be shredded and air-dried. Further milling can be done to the required size by hammer mills for use in pulverized fuel (PF) boilers or fluidized bed combustion (FBC) boilers.

1.7.2 *Co-firing of Biomass and Coal* [8]

The co-firing of coal with biomass is quite a popular way of using biomass as fuel. There can be four options for milling and co-firing of coal and biomass as mentioned below:

1.   The biomass and coal can be pulverized together and fired together through the existing coal firing equipment. *(This option may be only possible for kernels or some hard biomass substances, but not for herbaceous biomass/stalks/grasses, etc. The performance of coal mills would certainly deteriorate in a co-milling process.)*

2.   Direct injection of pre-milled biomass through the existing coal firing system and burners. *(The biomass would have to be milled at a separate location in this option. This pre-milled biomass is injected through the walls of the furnace. No additional combustion air (secondary air) is provided for the biomass. This technique is applicable in case of corner-fired or wall-fired boilers only.)*

3.   Separate milling of biomass and coal and separate firing equipment/burners for biomass and coal. *(This is a costly option and may require major modification, but it may be difficult to find space in existing installations for this type of modification. This option can be used in new installations.)*

4.   Separate milling of biomass and coal, and utilizing biomass as a reburn fuel in a reburn type of boiler. *(This option is good for new installations.)*

Co-firing Biomass and Coal Involving Gasification of Biomass

Gasifying the biomass and using the resulting gases for co-firing with coal. (This option is good for future installations specially designed for this purpose or for an existing boiler already having a facility for firing natural gas and coal, in which the natural gas can easily be replaced by gasifier gases. [For details of biomass gasification, see Para 6.14].)

1.7.3 *Typical Example of Utilization of Rice Husk in the Rice Industry* [35]

Rice husk is quite abundantly available as agricultural waste in rice-growing countries and can be used for the generation of gaseous fuel or liquid fuel. Rice husk can also be used directly as fuel for FBC boilers for the generation of steam. Sand with a particle size in the range of 0.25–0.55 mm can be used as the bed material in the FBC boiler.

The approximate typical composition of rice husk comprises' 30% cellulose, 25% hemicellulose, 12% lignin, 34% others and ash.

The typical proximate analysis of rice husk is: moisture 9.6%, volatile matter 66.1%, fixed carbon 9.2%, ash15.1%, and a gross calorific value (GCV) of about 3460 kcal/kg.

The typical bulk density of rice husk is 117.5 kg/m^3 and the particle size is 1.95–2.28 mm.

Rice husk can be used as fuel to generate steam for process needs of industries and/or in a steam turbine generator unit for electric power generation. A typical scheme for utilization of rice husk as fuel in a rice mill for producing steam for processes and electricity generation is illustrated as a block diagram in Figure 1.2.

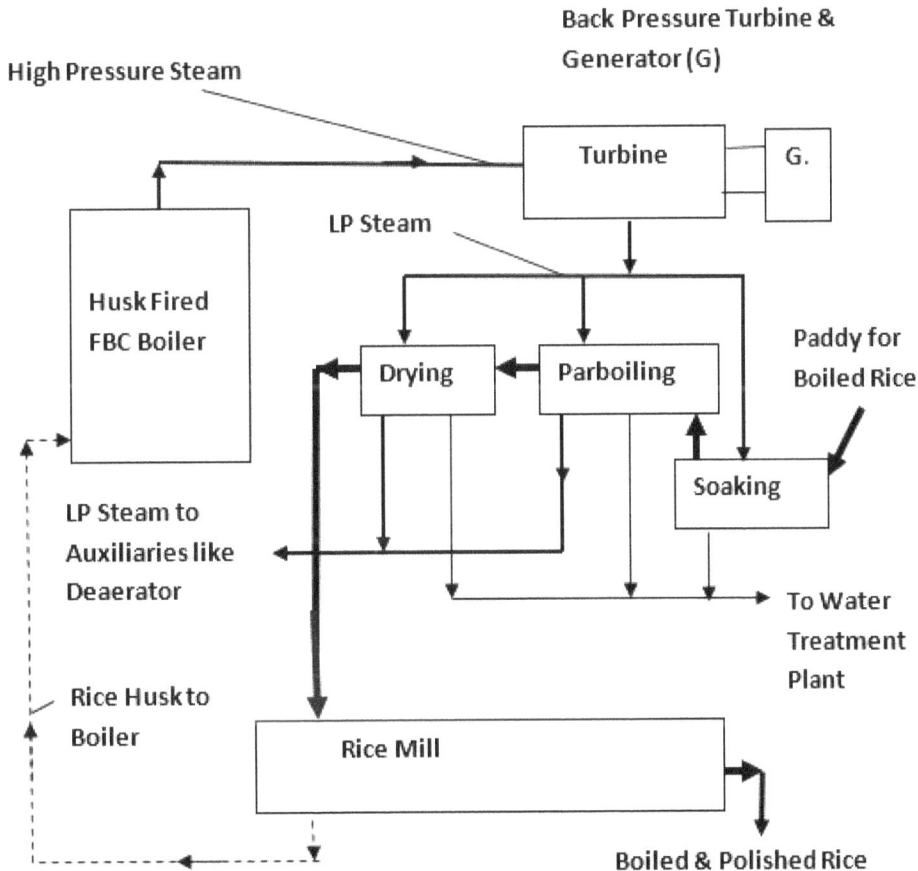

Fig. 1.2. LP steam—Low Pressure Steam; SCAHF—Steam Coiled Air Pre-Heater.
Block Diagram for Rice Husk Fired FBC Boiler and Back Pressure Turbine—Generator.

1.7.4 *Municipal Solid Waste as a Renewable Fuel Resource* [43]

Municipal waste can also be treated as a renewable resource for fuel as it is generated every day and contains lots of organic materials (biomass).

Thousands of tons of municipal solid waste (MSW) is generated every hour; the average rate of MSW generated per capita per day being anywhere between 0.45 and 0.7 kg. Sewage as well as MSW can generate biogas (see Chapter 6). The incineration of solid waste having very significant content of combustible matters may also simultaneously generate some electricity by utilizing the calorific value of the waste. However, while incinerating waste, it is important to keep in mind that the emission of pollutants is minimized and reduced below the permitted values and also that the leachate from ash should not cause water pollution.

The garbage/solid waste should not be burnt in the open before collection. The best way would be to segregate MSW into various components before collection. The portion intended for use as fuel should be shredded and dried and then fired in a boiler to raise steam which may be used for power generation or for other needs.

The organic portion of municipal solid wastes can also be converted to gaseous or liquid fuels like any other biomass. When it is used as landfill material, it will generate landfill gas (a kind of biogas).

Waste combustion is particularly popular in countries such as Japan where land is a scarce resource. Denmark and Sweden have been leaders in using the energy generated from incineration of waste for more than a century; the energy is used for heating or producing electricity. A number of other European countries like Luxembourg, the Netherlands, Germany, and France also use the energy of waste by using waste-fired boilers or incinerators. *Elaborate emission control (along with effluent control) equipment and systems must accompany the waste-firing boilers and ancillary systems in order to prevent environmental pollution.*

MHI Japan has used a unique system (as described below). The system seems attractive for reducing air pollution along with other types of pollution from municipal solid waste and also to generate energy.

MHI System of Utilizing Municipal Solid Waste [36]

In this system, the waste is first shredded and dried on a drying feeder by high temperature sand in a fluidized bed combustion (FBC) boiler, that is, in the char combustion chamber. The temperature maintained on the feeder is 200–250°C. The water vapor and chlorine content of the waste are ejected and homogenized. The waste along with sand is fed into a thermal cracking gasifying furnace and the hot sand from the fluidized bed combustion boiler is also further added for supplying heat for thermal cracking and a temperature of 400–450°C is maintained in the gasifying furnace. Two products are formed here namely, cracked gas and char. The char so produced has very little chlorine content, and along with sand, is fed to the fluidized bed combustion boiler where the char burns and generates heat. The cracked gas (as a heat source) is fed to a melting furnace of the vertical swirling type, where ash separated by bag filters from the flue gases of FBC boiler is melted to form slag. The non-combustibles like iron, aluminum, and other metals are discharged from thermal cracking furnace in a non-oxidized state. The sand and small-size particles discharged from the thermal cracking furnace and as well as in some quantity from FBC boiler, is fed back into the FBC boiler's combustion chamber. The flue gas from the melting furnace is sent into a FBC boiler for its heat utilization. The steam raised in the FBC boiler is commonly used for driving the 'steam turbine generator set' for electricity generation. The standard steam water cycle is used. A simplified flow diagram of the system is shown in Figure 1.3.

For controlling air Pollution, the flue gases from the FBC boiler are finally passed through gas cleaning equipment comprising of:

(i) Dust removing bag filter—for removal of suspended particulate matter.

(ii) Reacting bag filter—for removal of acid gases; (slaked lime is dosed in the flue gas before the inlet to the reactive bag filter).

(iii) Selective catalytic reactor (SCR)—for removal of NOx before going to the stack.

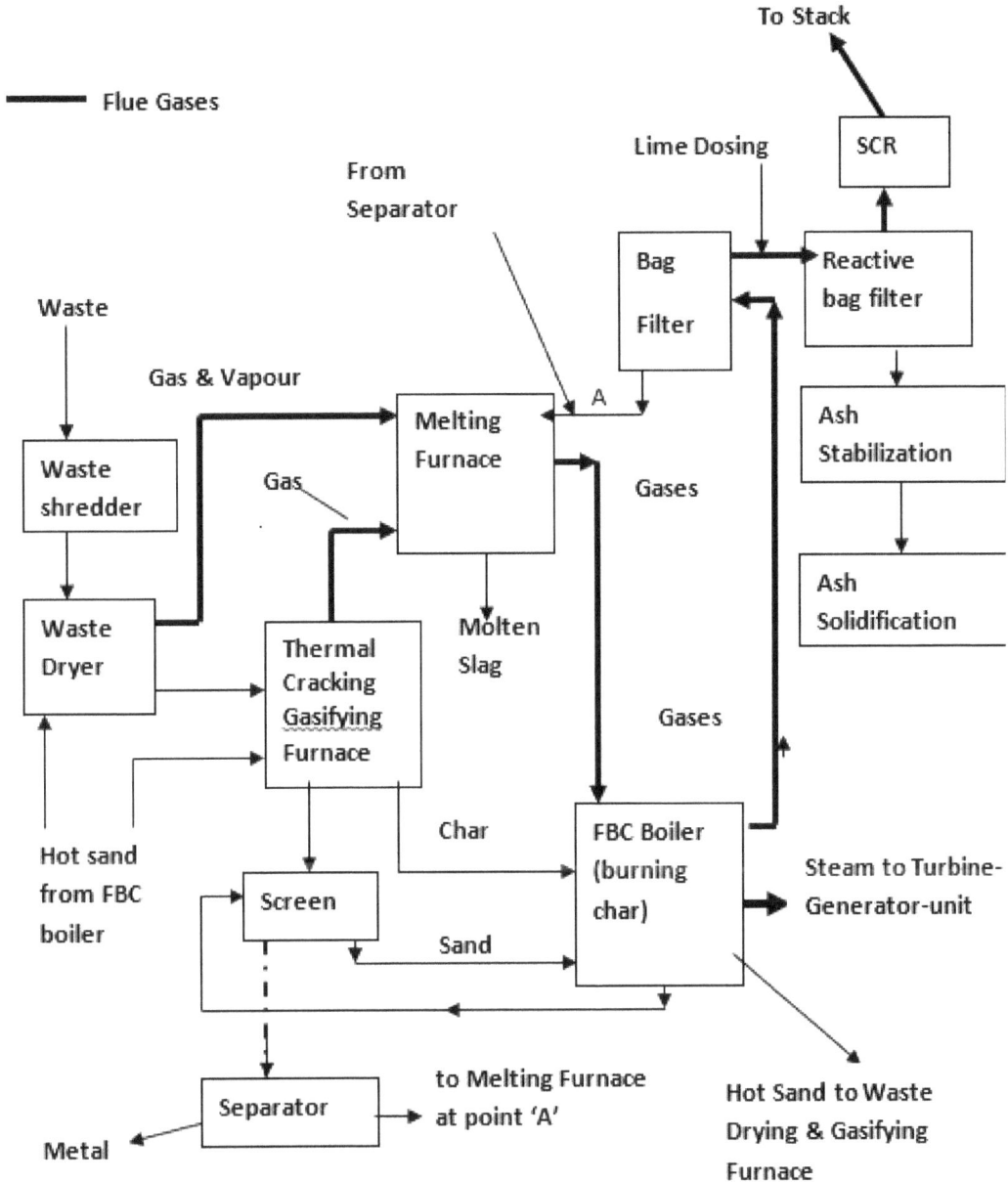

Fig. 1.3. Municipal Solid Waste Incineration with Thermal Cracking, Ash Melting and Power Generation.

No gases or vapors leave the system without cleaning. The emissions of oxides of sulphur and nitrogen (SOx and NOx), hydrochloric acid gas, and dioxins are all very well controlled in the above system and are much below the permitted levels. The ash from the 'dust-removing bag filter' is led to a melting furnace to form slag while the ash from 'reactive bag filter' is stabilized and solidified. The process prevents water pollution by leachate and reduces air pollution due to fine dust. The gaseous pollutants like NOx, SOx, HCl, and VOC are reduced by the bag filter and reactive bag filter.

1.8 Plastic Waste as a Source of Renewable Fuel [41]

A lot of plastic waste is generated every day and can be regarded as a renewable source of fuel or energy. Some of it is recyclable. Only recycling some of the plastic waste or even leaving it as it is can create pollution problems.

A catalyst system has been invented* for depolymerization of plastic wastes for its eventual conversion to *coke, liquid fuel, and gaseous fuel.* Such a process would consist of:

a. Carrying out '*random depolymerization*' of plastic waste in a specially designed reactor in the absence of oxygen and in the presence of coal and certain catalytic additives, the maximum reaction temperature being 350°C.

b. *Fractional distillation* for separating various liquid fuels by way of difference in their boiling points.

As oxygen is absent and the temperature is moderate, the formation of dioxins is ruled out. The process has been designed as a continuous process. It is claimed that all types of plastic wastes including CDs, floppies, laminated plastics can be used as the source material in the process.

The process can be utilized as a continuous process and works out to be cheaper than the batch process used by the Japanese company M/s Ozmotec to convert plastic waste to liquid fuel by pyrolysis.

[The invention was made in GH Raisoni College of Engineering Nagpur, India, by Prof. Alka Umesh Zadgaonkar, Dept. of Applied Chemistry, and a live laboratory demonstration of the process was also conducted.]

1.9 Sewage and Industrial Liquid Waste as a Source of Renewable Fuel [37]

The sewage or municipal liquid waste is a renewable resource for the 'generation of biogas', while the sludge (from the process of biogas generation) is a resource for making 'compost/manure'. Some industrial wastes with high organic content can also be treated as a renewable resource. The industrial liquid wastes from dairy industries, distilleries, and fruit and food processing plants, etc., are good resources for generation of biogas. The liquid wastes from industries like paper and pulp mills, tanneries, etc., should be first processed for recovery of chemicals before using them for biogas generation or for any other purposes. The liquid effluent left post the biogas generation can be used for irrigation after a suitable treatment, depending on the contents of the effluent. (Also see *Para 2.1.5 and 2.1.7 and Para 6.3.*) Sewage water can even be used for growing algae for the production of algal oil and its conversion to biodiesel. (See Note in Para 5.2.)

1.10 Summary

The importance of renewable fuels, which are plant-based fuels, has increased because of the need to curb the emission of greenhouse gases, which are accumulating in the atmosphere and have already started causing damage to the environment and changes in climate. Renewable fuels when burnt also produce carbon dioxide, similar to nonrenewable fuels such as coal or petroleum oil. However, the regrowth of the plants to the original extent also leads to an absorption of an equivalent amount of carbon dioxide. Ideally therefore, the utilization of renewable fuels can be a carbon zero operation.

Biomass like crop residue, forest wastes, animal wastes, human wastes, municipal solid wastes, etc., if not put to proper use, may degenerate in a natural way to generate methane, carbon dioxide, and other pollutants, including nitrous oxide (N_2O). It may also catch fire and cause emission of carbon dioxide along with other pollutants, including N_2O. It may be emphasized here that methane and nitrous oxide are much more powerful greenhouse gases than carbon dioxide. Planned combustion does not create methane, but creates carbon dioxide which is a much less powerful greenhouse gas than methane. Even alcohols, biodiesel, and other types of liquid fuels or chemicals can be made from lignocellulosic biomass. It is therefore recommended that the biomass may be used in a most useful manner and in a planned manner to keep in check the untoward emissions which would be more dangerous and also amount to loss of an energy resource.

The renewable fuels may be categorized in two broad categories, namely primary and secondary. While the primary renewable fuels are various types of biomasses, the secondary renewable fuels are alcohols, biodiesel, biogas, hydrogen, etc., derived from plant-based materials, animal wastes, algae, or fungi. Even the municipal solid wastes and sewage can fit into the category of renewable resource for secondary biofuels.

The first chapter provides a general introduction to renewable fuels, with examples, various methods of production of secondary renewable fuels. It also discusses the life cycle analysis of greenhouse gas emissions of such fuels. Utilization of primary renewable fuels like biomass, rice husk, municipal solid waste, plastic waste, sewage, etc., have been discussed.

ANNEXURE-A

1. Bacteria [Source: https://microbiologysociety.org/our-work/75th-anniversary-a-sustainable-future/circular-economy/circular-economy-case-studies/bacteria-to-develop-sustainable-infrastructure.html]

Bacteria are tiny, single-cell organisms (most of them 0.5 μm to 20 μm in size). The cell wall gives a characteristic shape to the cell; for example: round, rod-shaped, spiral. There can be variation in the shape of some bacteria depending on the culture conditions, which is termed as pleo-morphism. Certain species are further characterized by the arrangement of cells in cluster chains, or discrete pockets. They can form pigments, which impart a characteristic color. The cytoplasm of bacteria may also contain many granules of stored carbohydrates, lipids, etc. Mobility of some bacteria is facilitated by hair-like appendages called "flagella".

Bacteria reproduce by a division process called binary fission. Under adverse conditions, certain microorganisms produce spores which may survive dryness, temperature extremes, and can germinate upon return to a favorable environment for them. Other microorganisms form spores at a particular stage in their normal life cycle. Under some conditions, many bacterial species become surrounded by gelatinous material, which provides means of attachment and protection from other organisms. Sometimes a common gelatinous covering called "slime" covers many cells.

Generally, bacteria grow best between 20°C and 40°C, but there are some species which thrive at extremes. According to temperature ranges, they have been classified as:

- Psychrophilic (cold-loving) bacteria grow best between 4 to 10°C.
- Mesophilic bacteria act in the temperature range of 20 to 40°C.
- Thermophilic bacteria act in the temperature range of 50 to 55°C.

Based on a nutrient's requirements, bacteria are classified as heterotrophic or autotrophic, although several species function both as heterotrophic and autotrophic microorganisms. Autotrophic bacteria use carbon dioxide as a source of carbon and oxidize inorganic compounds for energy such as:

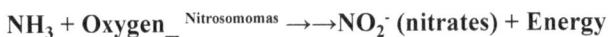

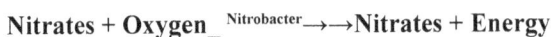

NH_3 + Oxygen_ $\xrightarrow{Nitrosomomas} \rightarrow NO_2^-$ (nitrates) + Energy

Nitrates + Oxygen_ $\xrightarrow{Nitrobacter} \rightarrow$ Nitrates + Energy

H_2S + Oxygen_ $\xrightarrow{Thiobacillus} \rightarrow H_2SO_4$ + Energy

Heterotrophic bacteria use organic compounds as an energy and carbon source for synthesis. Instead of heterotrophic, a commonly used term is "saprophyte", which refers to an organism that lives on dead or decaying organic matter. Heterotrophic bacteria are grouped into three classes:

- Aerobes – require free dissolved oxygen
- Anaerobes – do not require oxygen
- Facultative – bacteria that use dissolved oxygen, but can also respire and multiply in its absence. (For example, *Escherichia coli*, a common coliform is a facultative bacterium.)

Generally, Iron bacteria are autotrophic which oxidize inorganic ferrous (iron) as a source of energy. These filamentous bacteria occur in iron-bearing water and deposit the oxidized iron as $Fe(OH)_3$ in their sheath. All species of Iron bacteria like Leptothrix and Crenothrics may not be strictly autotrophic, however they are truly iron-accumulating bacteria and thrive in water pipes conveying water containing dissolved iron and form yellow or reddish slimes. When such bacteria die, they may decompose, imparting foul tastes and odors to the water.

Ferrous Salt + Oxygen $\xrightarrow{\text{Iron bacteria}}$ **Ferric Salt + energy.**

There are bacteria which degrade oily substances and phenolic compounds and consume heavy and toxic metals (like nickel, mercury, selenium). Nitrogen and phosphorus are generally nutrients for all bacteria.

Paratrophic bacteria are parasitic bacteria and require living material such as proteins. These are pathogenic, cause diseases because they depend on the living host for their nutrition.

There are however many transient forms of bacteria. There are bacteria in soil, while others may be present in water, wastewater, air, or any material. They may also develop in food. Many kinds of bacteria are present in our digestive system which promote digestion. In nature, various bacteria perform various useful functions like taking nitrogen from air and providing it to plants, aiding their growth. Many bacteria are useful to human beings and utilized for biotechnological processes in manufacturing various substances like chemicals, enzymes, etc.

The microorganisms take part in the formation of deposits on surfaces of water tanks and pipes. In settling tanks for water, the number of bacteria decreases during the first two days of settling, and then algae propagates in the tank. When algae die, it is destroyed by putrefactive bacteria. The dissolved oxygen content is depleted; the oxidation potential drops, and the taste and smell of the water deteriorate. Microbiological processes occur in settling tanks and receptacles of water. It is connected to the decomposition of organic substances in anaerobic conditions (lack of oxygen).

Phytobenthos (plants collectively living at the bottom of water) often develop on the sludge surface in settling tanks. When this vegetable matter dies, it gives an unpleasant odor to the remaining water. When natural water is passed through slow filters, a biological film is formed on the surface of the filtering media/sand. This film consists of microorganisms (usually nitrifying bacteria, and sulphur bacteria), phytobenthos (under water plant communities), and tiny animal communities. The depth of life in this film normally does not exceed a few centimeters. The microflora develops on fast filters as well, in warm season. The amount of 'ammonical nitrogen' decreases and 'nitrites/nitrates' increase in summer.

The process of settlement and deposits formed on water takes place in water supply systems and wastewater systems depending on their quality.

Many causative agents of infectious diseases are transmitted to man and animal through the medium of water, air, food, or other materials. When the parasitic bacteria of a particular type get into a living being's body, they adapt themselves to new conditions and cause the corresponding disease. Waterborne causative agents (organisms) may be responsible for diseases like typhoid, paratyphoid, dysentery, cholera, infectious jaundice, hepatitis of all kinds, plague, infestation of worms (helminth), etc.

2. Algae [Source: https://www.sciencedirect.com/science/article/pii S0306261911000778]

Algae are a large diverse group of organisms that are difficult to be defined. Historically, they have often been grouped together with plants; a group of lower plants containing chlorophyll, comprising a primitive structure with stalks, leaves, or roots. In a broad sense, algae can be defined as simple photosynthetic organisms that use sunlight to convert water and carbon dioxide into sugars and oxygen. Algae can occur both in freshwater, and seawater. Algae can be unicellular, multicellular formations, and colonies. Additional pigments (over and above the commonly found green chlorophyll) may be present in algae. Algae perform various useful functions in nature; the most important being the absorption of carbon dioxide and the release of oxygen. They are utilized in the manufacture of various useful substances like oil, vitamins, proteins, etc. A few types of algae are mentioned below.

Algae are incredibly important to our global ecology and bio-geochemistry. More than two billion years ago, the production of oxygen by algae photosynthesis dramatically changed the Earth's atmosphere and enabled the evolution and diversification of complex organisms that breathe oxygen (like us). Although they are not as conspicuous as plants in our everyday lives, phytoplanktons, which are microalgae, in the ocean currently do about as much photosynthesis each year as all the plants on land. Some of the phytoplanktons end up sinking to the bottom of the ocean, taking carbon with them. So, without algae, the carbon dioxide concentration in the atmosphere would be even higher than it already is. Microalgae are of many varieties and types.

Some microalgae have high oil content. The importance of algae has increased greatly because algae oil can be produced from them. The algae oil, being free from sulphur, is a good raw material for the production of biodiesel.

a. Blue-Green Algae (Cyanophyceae)

This type of algae may be unicellular or multicellular with a characteristic cell structure, not having a typical nucleus or chromatophore. The protoplast of blue-green algae is differentiated into peripherally colored layers and a central part. The assimilating pigments are chlorophyll, phycoerythrin (a red pigment in algae), phycocyanin (dark blue pigment in algae), and carotene (pink or reddish yellow). Depressions contain special bodies called 'endoplasm' where 'chromatin' substance is found. The cells of this algae have gas vacuoles (small cavities). 'Blue-green algae are widespread in nature—it can be found in freshwater, saltwater, soils, rocks, arctic area, etc., due to their extraordinary stability in unfavorable conditions and unexacting requirements for nutrition.

b. Green Algae (Chlorophyceae)

Green algae is very common type of algae. This class of algae has a number of variations. They can be in a vegetative state. The cells of green algae are different than those of the blue-green algae (cyanophyceae). Green algae have a cellulose case, a vacuole with a cell juice, differentiated nucleus, and chromatophore in a cup, band, plate, or granular form. Green algae contains chlorophyll (the green pigment) and carotene. They reproduce both by asexual and sexual (with the formation of motile zoospores) methods. The latter are the parts of the cell after its division, provided with a flagellum. The shapes of Chlorophyceae cells are varied (spherical, crescent-shaped, triangular, and irregular). Most Chlorophyceae are mononuclear.

The class of Chlorophyceae comprises 'Protococcous (*Protococcineae*)' algae. These are green, unicellular organisms living independently or in colonies, and are non-motile when in the vegetative state. Some genus have a fixed number of cells. The product of assimilation is starch. The simplest 'Protococcous' is '*Chlorella vulgaris*'.

Green algae also comprise Volvox organisms, the greater part of which is covered with a case of pectin and cellulose. Chromatophore is pure green, parietal, cup-shaped, and with a thickened base.

It covers the lower part of the cell. Sometimes chlorophyll is masked with haematochrome and the algae are red. For example, *Haematococcus nivalis*, a red algae is found in polar regions (colour: snow red). It gives red color to Arctic and Alpine regions.

Cells are of various shapes, such as oval, pear-shaped, egg-shaped, spindle-shaped, etc. Volvox algae are widespread. They are found in pure as well as in dirty water bodies, in various pits, pools, and other depressions filled with rainwater. They occur in soil. Part of them are saprophytic, that is, they live on dead or decaying organic matter.

Diatoms (*Diatomeae*) algae consist of two overlapping shells. The shells are not closed and can be set apart. The protoplasm is arranged in thin layers by the walls to form a protoplasmic bridge inside many species. The rest of the cell space is filled with the juice. There is only a single nucleus. Chromatophore, in addition to chlorophyll, contain brown pigments and the algae are therefore yellowish or dark brown and are of various shapes (such as granules, plates, etc.).

The products of assimilation are oil, volutin (a substance found in granular form in the cytoplasm of various cells and contributes to the formation of chromatin), etc. The diatoms replicate by vegetative division or by auxo-spores. In the vegetative division, each part receives a mother shell, while the missing shell grows into a new one during the growth of the cell.

Diatomaceous algae are widespread in nature and propagate into large masses in water (both freshwater and saltwater). The diatoms of the Pennales group can be found on ground. Diatoms can be used as fodder.

3. Fungi [Source: https://www.fungimag.com/spring-08-articles/sustainability.pdf]

Fungi are characterized by simple vegetative bodies from which reproductive structures are elaborated. All fungal cells possess distinct nuclei and at some stage in their life cycles, produce spores in specialized fruiting bodies. Fungi contain no chlorophyll and therefore, require sources of complex organic molecules for growth. Many species of fungi grow on dead organic material, while others are parasites. Many varieties consume carbohydrates, inorganic nitrogen, and salts. There are thousands of species of fungi. They can be broadly divided into four classes:

- Phycomycetes.
- Basidiomycetes.
- Fungi Imperfecte.
- Ascomycetes.

Mushrooms and toadstools are large, highly organized basidiomycetes, which have gills usually arranged radially to a central stalk. Amongst the ascomycetes are commonly occurring 'aspergillus and penicillium' molds; the latter are frequently seen blue-green growth on decaying citrus fruits. Yeasts also belong to Ascomycetes class. The Phycomycetes group have a general structure similar to that of green algae but possess no chlorophyll. Fungi imperfecte is a little- understood group. It is encountered in industrial microbiology.

Both fungi and bacteria are usually quite abundant in the area around the roots of the plant. They benefit from nutrients like simple sugars exuded by the plant roots into soil. The activities of fungi and bacteria create chemical and biological changes in the area around the roots (rhizosphere), so that the nutrients are available for plant uptake. The organic matter is decomposed by the fungi and they make inorganic nutrients available. Fungi are responsible for a number of biochemical reactions/changes in the natural environment. Fungi are also responsible for many diseases in plants and animals. Some fungi have a high content of oil, which is a good raw material for the production of diesel oil.

4. Yeast [Source: https://www.sciencedirect.com/science/article/pii/ S0944501317308674]

Fungi which develop predominately in the form of unicellular elements are called 'yeast'. These are unicellular organisms surrounded by a cell wall possessing a distinct nucleus. Yeast belongs to the class of Ascomycetes fungi. Generally, yeast reproduces by a budding process (a small new cell is pinched off the parent cell). Under certain conditions, however, an individual yeast cell may become a fruiting body producing four spores. Yeast is used in various biochemical processes.

Fungi and yeast can develop in water/moist food materials/moist woody material/decaying material. Fungi and yeast can grow in slightly alkaline environments but in acidic media, their growth rate is high.

Yeast produces enzymes and therefore, has many uses in the promotion of biochemical reactions. Some fungi have lipid content and may become a source of oil.

5. Protozoa [Source: https://www.ncbi.nlm.nih.gov/books/NBK8325/]

Protozoa are unicellular animal organisms. Most of these are hundreds of times bigger than bacteria. They have a soft, flexible, and relatively brittle outer cell membrane, but they have no solid case. The cell membrane usually consists of chitin-like compounds and does not contain cellulose. Inside the case, there is a nucleus and cytoplasm. The latter contains vacuoles which perform various functions. For example, the role of a stomach is performed by the food vacuole. Dissolved nutrients pass from it into the cytoplasm and are consumed by the organism through its vital processes. Other vacuoles accumulate metabolites, which are then excreted. The cytoplasm also contains granules holding nutrients which are used whenever the organism lacks them.

Protozoa of the Mastigophora type (having flagella) reproduce by binary division through longitudinal axis, while protozoa of the Infusoria type (having cilia) undergo transverse division.

Protozoa breathe in dissolved oxygen in water, discharging CO_2 and consume only soluble substances. Protozoa can turn into cysts which can live in the absence of moisture for long periods (sometimes even several years). Some protozoa are pathogenic.

Sludge in wastewater and the biological film created on a wastewater system may contain sarcodina, flagellate, ciliata, and suctoria types of protozoa.

Sucking infusoria (that is, suctorial) attach themselves to plants, mollusks, and sludge flocs by a non-contractile stalk. These tentacles are very strong. A small suctoria can simultaneously hold several ciliated infusoria, each of which sometimes exceed the size of the infusoria of its prey. Using tentacles, uctorial suck in the liquid contents of its prey's body. Suctoria feed mainly upon ciliated infusoria, but sometimes onalgae, animals, and vegetable remnants also.

There is a species of infusoria called 'sedimentators', which they ingest food by sedimentation of suspension. Protozoa occur everywhere, in sewage, sludge, excrements, soils, dust, river waters, lakes, and oceans, aerobic sewage treatment plants.

Protozoa are active participants in the mineralization of organic substances in natural and artificial conditions of sewage treatment. Some protozoa are causative agents of various diseases in many animals and humans.

6. Rotifers [Source: https://bbrc.in/bbrc/papers/pdf%20files/ Volume%204%20-%20No%201%20-%20Jun%202011/12.pdf]

Rotifers (Rotatoria) are representative of the animal kingdom and have a more complicated structure than protozoa. Rotifers are coated with a translucent but strong testa (shells). The wider anterior opening in the testa comprises of its head crowned with a rotary apparatus, whose cilia are set in constant motion. On its bottom, the testa ends with a narrow opening from which emerges a pedicle, a muscular growth with a sedimented chitin shell.

Its digestion system consists of a mouth which widens to a muscular throat, a masticatory apparatus intended to grind food, an oesophagus, and a stomach. The stomach leads to a narrow intestine connected with the excretory organ. In rotifers, the blood supply and respiratory systems are absent. The rotifers cause sediments by their method of consuming food. Using their cilia, the attached organisms give a spindle-like movement to water, and the funnel is directed into the mouth of the animal from its narrow end. Protozoa, bacteria, and organic substances get into the rotifer through the funnel. Most rotifers have eyes (red spots).

Rotifers are aerobes and are sensitive to oxygen deficiency. The highest temperature that Rotatoria can withstand is 50°C. When the conditions are unfavorable, the rotifers form a cyst, and the head and the leg get pulled into the testa. Rotifers are sensitive to the changes in the active reaction of the medium, and are indicator-organisms, characterizing the efficiency of the aerobic sewage treatment plant in operation.

Worms are a higher form of organisms than rotifers; both rotifers and worms are also helpful in degrading waste.

QUESTIONS

1. *Enumerate different services performed by forests/trees.*

2. *What is a biofuel?*

3. *What is lignocellulosic biomass?*

4. *What is the connection between "burning" petroleum and climate change?*

5. *Why does blending of ethanol with gasoline improve the combustion of gasoline?*

6. *How do biodiesel emissions compare to petroleum diesel?*

7. *What is meant by life cycle greenhouse gas emissions? Compare the life cycle greenhouse emissions of a few biofuels with that of petroleum fuels.*

8. *What is the Renewable Fuels Standard (RFS-2) and why is it important for biofuels?*

9. *Why does burning biofuel have less of an effect on atmospheric carbon dioxide than burning petroleum, even though both fuels produce carbon dioxide when they burn?*

10. *How do biodiesel emissions compare to petroleum diesel? Is biodiesel better for human health than petroleum diesel?*

11. *Is the use of biodiesel helpful in mitigating "global warming"?*

12. *Differentiate between renewable and nonrenewable fuels and give examples of primary and secondary renewable fuels. Name the primary fuels available.*

13. *Write short notes on (i) water hyacinth, (ii) sorghum as a source of biofuel.*

14. *Are there some renewable standards? What are the essentials of such a standard?*

15. *What are the sustainable feedstocks for production of biofuels? Comment on 'The seaweeds can be a source of biomass for production of biofuels'. What is the difference between microalgae and seaweeds?*

16. *What are the different routes by which biofuels can be produced from biomass?*

17. *Can you utilize all the biomass produced for production of biofuel? What are the other essential uses of biomass such as forest waste and crop residues for which the provision must be made before you can use them for biofuel?*

18. *What are the alternative competitive uses of biomass?*

19. *How can biomass be utilized as a fuel? Give examples. How would the collection of dry biomass from under the tree for utilization help in preventing forest fires?*

20. *What are the potential problems which may be faced in use of biomass as fuel in boilers? Indicate the solution to these problems.*

21. *What are the options available for the co-firing of coal and biomass and what are the merits and demerits of each option?*

22. *Why is it important for the environment to utilize the methane from gas hydrates of the sea and coal-mining methane?*

23. *Describe the ways to utilize municipal solid wastes as fuel.*

24. *Comment on 'the sewage, and industrial liquid waste can be utilized as a source of renewable fuel and also as irrigation water'.*

25. *Distinguish among different organisms such as bacteria, fungi, yeast, algae, protozoa, and rotifers. How are they related to generation of biofuels?*

References

1. Annual report of ministry of non-conventional energy. Government of India 2001–2002.
2. Bioenergy Resource Status in India - working paper May 2011. Prepared for DFID by the PISCES RPC Consortium. M.S. Swaminathan Research Foundation, Chennai, India.
3. Biofuel legislation in EU. http://www.biofuelstp.eu/biofuels-legislation.html.
4. Calster, Geert Van, Wim Vandenberghe and Leonie Reins. Research Handbook on Climate Change Mitigation Law, Edward Elgar Publishing, USA.
5. Dimethyl Ether (DME) fact sheet.
6. EC Renewable Energy Directive 2009/28/EC of European Parliament and Council of 23rd April 2009 on the promotion of the use of energy from renewable sources and amending and subsequently repeating directive 2001/77/EC and 2003/30/EC. Brussels Belgium: European Commission.
7. Eriksson, E.L.V. and E.MacA. Gray. September 2017. Optimization and integration of hybrid renewable energy hydrogen fuel cell energy systems – A critical review. Applied Energy. 202(15): 348–364. https://www.sciencedirect.com/science/article/abs/pii/S0306261917306256.
8. etipbioenergy.eu/value-chains/conversion-technologies/conventional-technologies/biomass-co-firing.
9. Environmental Protection Agency (EPA) Renewable Fuel Standard Program (RFS2). Regulatory Impact Analysis [http://www.afdc.energy.gov/data/10328].
10. European Union Biofuel Policy. http://transportpolicy.net/index.php?title=EU:_Fuels:_Biofuel_Policy.
11. Gaps in the Research of 2nd Generation Transportation Biofuels--Analysis and Identification of Gaps in Research for The Production of Second-Generation Liquid Transportation Biofuels. http://www.biofuelstp.eu/downloads/Gaps_in_research_of_2nd_generation_transportation_biofuels.pdf.
12. Gerard, V.A. 1989. Seaweeds. *In*: Kitani, O. and C.W. Hall (eds.). Section 1.3.2. Biomass Handbook, Gordon and Breach, NewYork.
13. Global resource assessment report (second edition). 2015. Food and Agriculture Organization of The United Nations Rome. 2016. (http://www.fao.org/3/a-i4793e.pdf) and (http://www.uncclearn.org/sites/default/files/inventory/a-i4793e.pdf).
14. Graham, R., A.E. Harvey, M.F. Jurgensen, T.B. Jain, J.R. Tonn and D.S. Page-Dumroese. 1994. Managing Coarse Woody Debris in Forests of the Rocky Mountains. United States Department of Agriculture - Forest Service, Washington DC, USA (http://onlinelibrary.wiley.com/doi/10.1111/gcbb.12285/pdf).
15. http://keys.lucidcentral.orgfact sheet-Echhorniacrassipes.
16. http://www.biofuelstp.eu/factsheets/dme-fact-sheet.html.
17. (a) http://www.c2es.org/publications/lifecycle-greenhouse-gas-emissions-different-light-duty-vehicle-fuel-pathways-synthesis. (b) https://www.epa.gov/ghgemissions/sources-greenhouse-gas-emissions
18. http://www.afdc.energy.gov/laws/RFS.
19. https://www.bbc.com/news/science-environment-56805255
20. https://www.bbc.com/news/science-environment56762972#:~:text=A%20constellation%20of%20satellites%20will,profit%20organisation%20called%20Carbon%20Mapper.
21. https://www.iea.org/reports/global-energy-review-2021.
22. https://www.energy.gov/sites/prod/files/edg/media/BiofuelsMythVFact.pdf.
23. https://www.eesi.org/articles/view/biofuels-versus-gasoline-the-emissions-gap-is-widening.
24. https://www.mjbradley.com/sites/default/files/MJBA_Role-of-Renewable-Biofuels-in-a-Low-Carbon-Economy.pdf.

25. https://archive.ipcc.ch/pdf/special-reports/srren/SRREN_FD_SPM_final.pdf.
26. https://www.mofa.go.jp/files/000498436.pdf.
27. https://www.regi.com/products/transportation-fuels/renewable-diesel.
28. http://sustainabilitycommunity.springernature.com/posts/engineering-carbon-neutral-singapore.
29. International Journal of Emerging Technology and Advanced Engineering (International DME Association Washington, D.C). https://www.aboutdme.org/index.asp?bid=234#Q6.
30. Lali, Arvind. November 2010. Next Generation Biofuels India Scenario and Areas for Cooperation. DBT-ICT Centre for Energy Biosciences, Institute of Chemical Technology, Mumbai, India-EU and Member States Partnership for a Strategic Roadmap in Research and Innovation New Delhi, India.
31. Nabi, M.N., M.S. Akhter and M.M. Zaglul Shahadat. Feb 2006. Improvement of engine emissions with conventional diesel fuel and diesel – biodiesel blends. Bio-resource. Technology. 97(3): 372–378.
32. Nigro, Nick and Shelley Jiang. July 2013. Lifecycle Greenhouse Gas Emissions from Different Light Duty Vehicle and Fuel Pathways. A Synthesis of Recent Research, Center for Climate and Energy Solutions, USA.
33. Sheehan, John, Vince Camobreco, James Duffield, Michael Graboski and Housein Shapouri. May 1998. Life cycle inventory of biodiesel and petroleum diesel for use in an urban bus. Final report, NREL/SR-580-24089 UC Category 1503. National Renewable Energy Laboratory, Golden, CO.
34. Singh, Devendra, K.A. Subramanian and M.O. Garg. January 2018. Comprehensive review of combustion, performance and emissions characteristics of a compression ignition engine fueled with hydroprocessed renewable diesel. Renewable and Sustainable Energy Reviews. 81(2): 2947–2954. https://doi.org/10.1016/j.rser.2017.06.104.
35. Srinivasa Rao, K.V.N. and G. Venkata Reddy. April–June 2005. Combustion studies of rice husk in fluidized combustion boiler. Water & Energy International Journal. CBIP, New Delhi, India.
36. Terasawa, Yoshinori, Shizuo Yasuda, Hirotoshi Horizoe, Jun Sato and Yoshinori Gotou. Jun. 2001. Commercialization of MSW Incineration system with direct melting, by thermal cracking for high efficient generation of electricity. Mitsubishi Heavy Industries, Ltd. Technical Review. 38(2): 82–86.
37. Tsai, Wen-Tien. August 2012. An analysis of the use of biosludge as an energy source and its environmental benefits in Taiwan. Energies. 5: 3064–3073; doi:10.3390/en5083064.
38. US EPA – EP A420- F- 07 035. April 2007.
39. Wikipedia, the free encyclopaedia. *Eichhornia crassipes*.
40. WHO. 2011. Global Database of Household Air Pollution Measurements [Online]. Available at: http://www.who.int/indoorair/health_impacts/databases_iap/en/(accessed 10 December 2014).
41. Wong, S.L., N. Ngadi, T.A.T. Abdullah and I.M. Inuwa. October 2015. Current state and future prospects of plastic waste as source of fuel: A review. Renewable and Sustainable Energy Reviews. 50: 1167–1180. https://doi.org/10.1016/j.rser.2015.04.063.
42. Yan, C.S. et al. 2007. A new alternative fuel for reduction of polycyclic aromatic hydrocarbon and particulate emission from diesel engines. J. Air Waste Manag. Association. 57: 465.
43. Zhang, Dong Qing, Soon Keat Tan and Richard M. Gersberg. Aug. 2010. Municipal solid waste management in China: status, problems and challenges. J. Environ. Manage. 91(8): 1623–33. PMID: 20413209. DOI: 10.1016/j.jenvman.2010.03.012.

CHAPTER **2**

Wood
A Primary Energy Source

2.1 Energy Plantations—Features and Land Availability

2.1.1 General

Wood has been used as a fuel and for other purposes from times immemorial and the demand for it continues to increase. Increasing plantation of trees is essential to fulfil the demand for wood and to make up for the reduction in forest cover. There is a shortage of land because of the need to grow more food, build houses, roads, industries, power plants, etc., to cater to the ever-growing population. However, any land that is spare or degraded and is not used for agriculture or other purposes can be used for growing trees/plants for fuel wood, seeds for extracting oil, or other useful products. Thus, the agroforestry practice is very useful for tackling wood shortage in rural areas.

Trees grown, especially, for fuel wood or biomass, come under the definition of energy plantations. The plants for such plantations should therefore be selected and nurtured for maximum energy yields per unit area at the lowest cost. These plants should have a short rotation period and should be planted densely. Such plants are cropped at regular intervals depending on the specific species; the roots in certain species may be left in the ground so that the plant may sprout again.

An energy plantation should preferably have the following features:

a. Faster growth and early maturity (say 4–7 years in the case of trees)
b. High yield
c. Naturally regenerate, and should sprout easily and spread fast
d. Thrive in the natural weather conditions of the region
e. Readily reseed

Additionally, accelerated growth systems are available through cloning, tissue culture, species manipulation and genetic engineering.

It may be mentioned here that even certain grasses can be called energy plantations as they can be a source of secondary renewable fuels. (In case of grasses, the period of maturity may be as small as 4 to 6 months.)

2.1.2 Land Availability for Plantations

Land availability for plantation in addition to already designated forest land is difficult, but for this purpose, we can always find suitable land and spaces in villages, towns, and cities, along roads, canals, reservoirs, railway lines, within industrial units and colonies, etc. A lot of land may be lying waste where mining of minerals or coal/lignite is taking place or near petroleum drilling sites, stone quarries, etc. This land can be restored for tree plantation or agriculture. Wasteland utilization for plantations is mostly possible through suitable restoration practices. It is important to note that each type of wasteland may require a specific restoration procedure. Different types of wasteland include:

a. **Water-eroded land** such as ravines lands, waterlogged areas, marsh lands.

b. **Degraded land** formed by industrial activity like coal mining, mining for various ores, stones, etc., brick manufacturing, beneficiation systems for coal and various ores, disposal of ash of coal/lignite, and implementation of various constructing activities.

c. **Wind-eroded lands** such as shifting sand dunes, wind-eroded areas with extreme moisture stress, coastal sand dunes, etc.

d. **Hot arid and semiarid land** as found in western and central India (Gujarat, Rajasthan, Madhya Pradesh, etc.).

e. **Land within degraded forests**, that is, there are many degraded forests where trees should be planted, and tree-density should be restored/increased as required.

f. **Landfills containing solid waste**, that is, an area where the solid waste from cities and industries is dumped on a lot of land. This land needs restoration. In case this land cannot be used for building construction, it can be used for tree plantation.

g. **Saline lands/sodic land**, that is, lands rendered uncultivable due to very high alkalinity.

Note: Marginal Land

A land which is presently not being used due to its poor soil and other unconducive features for both agricultural and industrial uses, is called marginal land. For agriculture, it cannot provide minimum return to justify production without additional help in terms of subsidies, but it could provide acceptable returns using new technologies and crops. All marginal lands though can be used for tree plantations.

2.1.3 Measures Required for Giving Thrust to Plantations

The following measures can give a thrust to the plantations:

Large-scale planting of trees: Large-scale plantations on all available land (which is not being used for agriculture or for other purposes) should be done and proper care and maintenance of plantations should be undertaken so that the plants do not die. It is well known that every year, millions of new saplings are planted, but these most of these saplings soon die due to lack of adequate care. It is necessary to encourage the planting of trees alongside *roads, railway lines, rivers, streams, and canals, and on other unutilized lands* under state, corporate, institutional, or private ownership. Green belts should be created in urban/industrial areas as well as in arid tracts. Such a program will help to check erosion and desertification as well as improve the microclimate. For this, appropriate policy initiatives, financial incentives, organizational setups, citizen partnerships, etc., would be required. For speedy results, a mix of fast-growing trees and other varieties of trees should be planted.

- **Practicing agroforestry:** Agroforestry (i.e., forestry along with agriculture) should be practiced to fulfil the need of rural communities for fuel wood and timber as well as to provide secondary benefits of having income through fruits, seeds, and other products.

- **Practicing social forestry:** The tree development in villages should be taken up as indicated in Para 2.5.3 of this chapter.

- **Appropriate amendments in laws:** Provisions are required in land laws to facilitate and motivate individuals and institutions to undertake tree-farming and the growing of legumes on their own land. Degraded lands should be made available for this purpose either on lease or on a tree-ownership basis (that is, the tree, if grown, would be owned by grower).

- *Planting leguminous varieties of plants*: Growing trees and other plants of leguminosae varieties, which fix atmospheric nitrogen and make the soil richer should be preferred and planted in between.

- *Utilization of acclimatized species for particular areas*: Tree planting and greenery development should be preceded by development of data on indigenous and acclimatized species of the local area, with focus on the particular climatic conditions and presence of environmental pollutants. In such data, plants having attribute of pollution abatement and ecosystem protection/ development should be identified. Suitable varieties and types of plants are readily available, but people handling the actual plantation work, rarely, have accurate knowledge about it. Botanists should therefore be consulted for choosing the most suitable varieties of trees or plants for various situations. The selection of trees for planting from the data mentioned above should also take into consideration the local soil condition, availability of irrigation facilities, rate of growth of the tree species (whether the tree species is fast growing or slow growing), and needs of the community.

- *Utilization of biomass for making manure*: Biomass/agricultural waste should be converted to 'compost manure', providing much-needed manure for use in the plantations. The surplus compost can be used for agriculture where it can be used as a substitute for chemical fertilizers or at least decrease their use, which have adverse effects on soil in the long run.

- *Controlled felling of trees*: Unauthorized felling of trees and avoidable tree felling in the name of development must be eliminated. In developmental projects, like industrial, power and irrigation projects, and housing projects only the absolutely necessary number of such trees should be cut. Compensatory forestation for the number of trees cut should be implemented. It may be suggested that this plantation be maintained through an agency appointed and supervised by project authorities under the overall supervision of the government's forestry department.

- *Better prevention of damage by fire*: Fires continue to destroy plantations and even agricultural crops; therefore, more precautions should be taken against the occurrence of fire hazards and more efforts are needed for the isolation of any accidental fire when it occurs. Providing efficient firefighting services to douse and isolate fires is highly important in the interest of boosting plantations.

- *Sustainable management of soil (See Para 2.1.4)*
- *Utilization of landfill area for plantation (See Para 2.1.8)*
- *Utilization of alkaline/sodic lands for boosting plantation (See Para 2.2.3)*
- *Utilization of acidic soils/podsol areas (See Para 2.2.5)*
- *Utilization of mined-out areas for plantations (See Para 2.4)*
- *Utilizing 'tissue culture technique' for rapid production of nursery plants (See Para 2.6)*
- *Rainwater harvesting and watershed management for benefit of plantations (See Para 2.7)*

It is worth mentioning that all types of lands including polluted lands should be utilized for planting vegetation by sustainable management of soils. It is necessary to implement a sustained waste land development and reclamation strategy as a government policy. Except for barren rocky areas and snow-covered hills and glaciers, most of the wasteland can be reclaimed and used for plantations. The techniques of bioremediation and phytoremediation for polluted soils are very successful and should be used extensively.

2.1.4 *Sustainable Management of Soil* [15,60,63,87]

Soil should be managed in a sustainable manner as it forms the base of vegetation/agriculture/ plantations to fulfill the need for food, feed, fiber, fuel, medicines, carbon sequestration, etc., of an increasing population. Soil comprises a complex and dynamic ecosystem and forms the habitat of bacteria, fungi, protozoa, microarthropods, and small animals. The physical and chemical properties of soil are attributed to nonliving components of soil, but living organisms also affect changes in its properties to some extent. The nature of the soil varies across different geological and geographical landscapes and is affected by climate, land use, and various other physical and chemical factors. It is alarming that one-third of soils all over the world have been degraded as estimated by experts and the arable land and productive land per person is fast decreasing. (On a side note: The **year 2015** was declared as the '**International Year of Soils**').

Main reasons for the degradation of soil

Soils get degraded because of the following reasons:

- Over cropping
- Excessive use of chemical fertilizers
- Lack of organic manure
- Flooding, waterlogging, or flood irrigation (that is over irrigation)
- Use of pesticides
- Putting non biodegradable materials in the soil
- Soil erosion and landslides
- Mining activities
- Making of mud bricks
- Removal of topsoil (the soil layer that is most fertile) for any human activity
- Pollution of land by industrial wastes, petroleum products, ash, e-wastes, chemicals, wastes containing heavy metals/toxins, etc.
- Irrigation of land by heavily polluted water

Methods for sustainable management of soil

Some methods to manage soil in a sustainable manner are mentioned below:

- Planting a mix of leguminous and non-leguminous plant species.
- Crop diversification/polyculture/multiple cropping and changing crop rotation.
- Using organic soil amendments like manure and compost from animal wastes (*animal dung, urine, waste feed, etc.*), biodegradable organic wastes, instead of 'chemical fertilizers'.
- Use of 'bio pesticides*' instead of 'chemical pesticides.'
- Leaving crop residues and foliage of trees in the soil, ploughing back green manure (legumes plants).
- Wherever required, using soil additives (*see Para 2.2.3 and 2.2.5 about plantation in alkaline/ sodic lands and in acidic soils/pod sol areas*).
- Use of PGPR (*Plant Growth Promoting Rhizobacteria* see Para 2.1.7).
- Avoiding waterlogging and making proper slopes on the land. *Proper surface drainage and subsurface drainage play an important role in the prevention of over flooding (which leads to land getting degraded).*
- Rainwater harvesting, making ponds at many places and watershed management *(The drainage water from the land should be led to a pond from where it can be used when needed).*
- Using proper irrigation techniques like drip irrigation instead of flood irrigation to avoid water wastage and spoiling of land due to overirrigation.
- Bioremediation of soil pollutants (*when microorganisms are exclusively employed to use or consume these pollutants as a source of energy* (see Para 2.8)) and the use of biofertilizers.

- Phytoremediation of soil pollutants (*when certain plants are grown on polluted land and absorb/ uptake these pollutants* (see Para 2.8)).

*Note on biopesticides [55,81,105,120]

(1) Azadirachta indica *Out of the many plants with bioactive properties,* Azadirachta indica *(Neem) is a good pesticidal option. This plant contains a compound named 'azadirachtin', which is a tetra-nortri-terpenoid compound. It adversely affects the harmful insects by making changes in their hormonal system, feeding activity, reproduction, and flying ability. Azadirachtin has low toxicity for mammals and degenerates rapidly in the environment. It has few side effects on non-target species and beneficial insects. Although Azadirachtin is present in all the parts of the* Azadirachta indica *tree, the seeds of the plant have the highest concentration. There are several commercial products containing this compound like Azatin XL, Biodeme, Amazing, etc.*

When the biomass of Azadirachta indica is applied to the soil, some plant species absorb the compounds present in it through their roots and translocate it through the plant tissue resulting in protection from insects.

Raw neem oil emulsified with water is used in many ways to protect crops from pests. It is composed of a complex mixture of biologically active compounds. More than 25 other active compounds are found in it, for example, Myelination, Vipul, Salinan, etc. Its extremely bitter flavor and taste makes many insects stop feeding on host species if the oil is applied to it. Mites, whiteflies, aphids, and other types of soft-bodied insects suffocate when they come into contact with neem oil.

The derivatives from the tree can influence nearly 400 species of insects. Azadirachta indica based bio-pesticides are very effective against many plant pests and insects like leaf-miners, caterpillars, cut-worms, fruit flies, parasites, rice pest, brown plant hoppers, tungo virus, leafhopper, potato moths, weevils, flour beetles, bean seed beetles.

(2) Trichoderma Fungi: *The Trichoderma sp. of Fungi acts as a biocontrol for parasitic fungi.*

**(3) Lantana Camara and Clerodendrum infortunatum *(Bhant/Bhat/Bhantaka):* ** *Herbal pesticides are also obtained from* Lantana Camara *and* Clerodendrum infortunatum *(Bhant/Bhat/Bhantaka).*

(4) Jatropha: *Crude extracts of Jatropha leaves via ether, butanol, and hexane can control 'termites' and thus this plant acts like a biopesticide. (See note in Para 4.10, Chapter 4).*

**(5) Leaf extracts of (i) Calatropis procera (Akdo), and (ii) Withania somnifara *(Ashwagandha)*, ** *have been found effective in inhibiting leaf spot fungi in mango. The former is more effective than the latter.*

2.1.5 Biofertilizers and Compost: Advantages of their Use [18,19,49,50,73]

Biofertilizers

Biofertilizers are natural fertilizers that generally act as microbial inoculants of bacteria, algae, and fungi, combined with nitrogenous matter of plant and animal/human waste. They contain living cells of different microorganisms which have the ability to convert nutritionally important elements from unavailable (un-assimilable) form to a form available (assimilable) to the plants through biological process. Hence, the bio-fertilizers augment the availability of nutrients to the plants. They may also have nitrogen fixers like (Rhizobium, Azotobacter* for legume crop, Azolla for lowland paddy, Acetobacter for sugar cane, and Azotobacter and Agospirellium for non-legume crop), phosphate-dissolving bacteria (PSB), and 'Mycorrhizae fungi. 'The strength of biofertilizers is determined by two basic parameters, namely—the number of cells and the efficiency of microorganisms to fix nitrogen or solubilize phosphates.

Note on Azotobacter biofertilizers: *Azotobacter is a free-living, gram-negative, aerobic, nitrogen-fixing bacterium. The Azotobacter biofertilizer contains a very large number of live Azotobacter bacteria and because of their <u>nitrogen-fixing characteristics</u>, this fertilizer can be used on any non-legume crop over a short, medium, or long duration to provide nitrogen to the plant. These bacteria also secrete certain <u>growth-promoting hormones such as indole acetic acid, gibberellic acid, and cytokines,</u> which promote vegetative growth and root development.*

Liquid Biofertilizers

Occasionally, liquid biofertilizers are used by farmers. These are liquid formulations containing the dormant form of desired microorganisms and their nutrients along with substances that encourage the formation of resting spores or cysts to ensure a longer shelf life of the plant and a greater tolerance to adverse conditions. The dormant form, on reaching the soil, produces a fresh batch of active cells, which grow and multiply by utilizing the available carbon source in the soil or from root exudates.

Conversion of Organic Waste into Compost [26]

All the biomass/waste generated in agriculture, agroforestry, animal husbandry or otherwise should preferably be converted into compost. The vermicomposting technique—in which worms, mainly earthworms are used to form biomanure—can also be used. Even aquatic weeds like water hyacinth (*Eichhornia crassipes*), water lettuce (*Pistia stratiotes*), ditch bank weed, and cattail (*Typha angustata*) can be converted into good quality vermincompost within two–three months. These composts contain nitrogen, phosphorus, potassium, etc.

Composting is a process which converts waste into biomanure/fertilizer which may have the useful microorganisms contained in bio-fertilizers.

Note: Use of dried plant material of **Lantana xamara**—*It may be of interest to mention here that the addition of even 3% of the dried plant material of Lantana camara (Ghaneri/Hesike) has a good effect on the growth and yield of wheat crops. This plant has phytohormones such as salicylic acid and gibberellic acid. The use of dried material of Lantana camara in the field can replace the chemical fertilizers for better ecofriendly growth and yield of wheat crops.*

Composting Process [3,81,102]

Municipal solid wastes and sludge from municipal sewage treatment can be composted into organic fertilizer, which is a good soil conditioner and nutrient source for plants/agricultural crops. The animal wastes and agricultural wastes are also good source materials for making compost manure.

In compost manure formation, the biological decomposition of organic solid waste takes place in presence of moisture. The excess of moisture inhibits the transfer of oxygen and forces the process to become anaerobic, while excessive ventilation lowers the moisture content and inhibits the biological decomposition. However, in aerobic decomposition, both oxygen and moisture are necessary components for microbiological activity, where carbon dioxide, water, and humus substances are produced. Even in this situation, excessive ventilation can dry out the compost and lowers the temperature.

Composting is the biological decomposition of organic solid waste to a relatively stable end product. The decomposition is brought about by bacteria and fungi. At the start of process, the temperature is low and mesophilic bacteria are active, but due to bacterial action, the temperature rises (say 45–60°C) and thermophilic bacteria become active. Normally, the fungi which are active may be 'Geothrichium', 'Penicillium', 'Rhizopus, 'Aspergillus', and 'Mucor' but the dominance and the presence of various fungi would depend on the nature and composition of organic matter.

Due to bacterial action, organic acids are produced in the initial stage, causing a fall in the pH value, but due to mixing and aeration, which is done with the aid of machinery (during composting process) the pH value increases. When composting is done without the aid of machinery (manual systems) and where mixing and aeration is not possible, the pH value is corrected by the addition of lime or other buffers.

Mechanical composting systems have mixing mechanisms and aerating mechanisms for accelerating the composting process. Stable fertilizer can be produced within few weeks, which takes months in manual systems.

It is well known that carbohydrates and trace elements are also necessary for microbial action. The required ratio of carbon, nitrogen, and phosphorus (CNP) is approximately 100:5:1. An addition of resource of nitrogen and phosphorous to the 'process of formation of compost' accelerates it and also adds to the value of fertilizer. For compost to be correctly and efficiently formed, it is necessary to have the right amount of carbon, nitrogen and phosphorous. A higher amount of carbon slows down the process, while a higher amount of nitrogen causes unpleasant odors due to the release of excessive nitrogen compounds like ammonia and nitrous oxide.

Spent wash discharged from distilleries and organic press-mud cake-like wastes from industries can be composted by treatment with *Trichoderma viridie*. The Trichoderma*—inoculated compost is found to be very good for plant growth.

*[*Trichoderma-inoculated compost*—Trichoderma species is a potent biocontrol agent against parasitic fungi and therefore, acts like a biopesticide. The *Trichoderma* sp. produce a wide variety of enzymes. For example: '*Trichoderma reesei*' produces cellulases and hemicellulases and *Trichoderma longibratum* produces xylanases, which are useful in the enzymatic hydrolysis of cellulosic materials into monosaccharides/glucose. '*Trichoderma koningii*' is useful in biosorption of hexavalent chromium.]

Advantage of Using Compost over Chemical Fertilizers

It has been well established that the use of chemical fertilizers deteriorates the soil in the long run. Each succeeding year, the soil would need increasingly greater quantities of chemical fertilizer to maintain the same productivity level. Moreover, the manufacturing of chemical fertilizers is a very energy-intensive process. Increased energy production requires the burning of more fuel. The production of energy as well as the chemical fertilizer process adds to air pollution, water pollution, greenhouse gas emissions, and global warming. But with the exclusive use of compost/biofertilizers, we reduce these adverse effects on the environment. Figures 2.1(a) and (b) show the environmental advantage in using compost over chemical fertilizers.

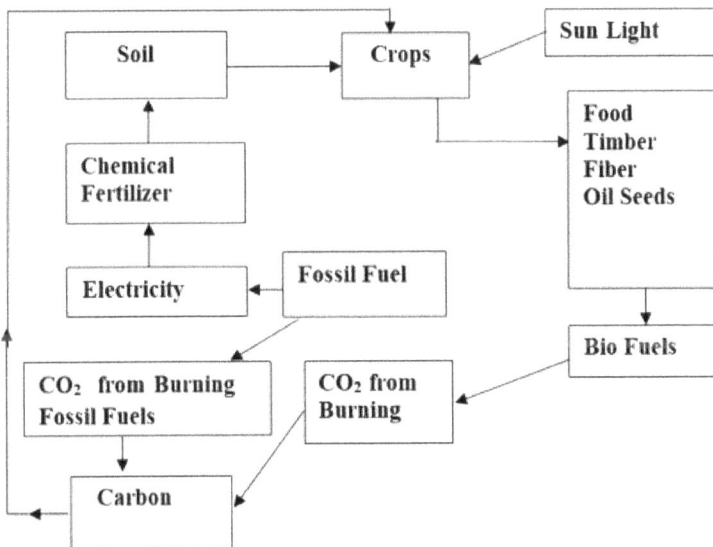

Fig. 2.1a. Use of Chemical Fertilizer (Higher Emissions/Environmental Effects).

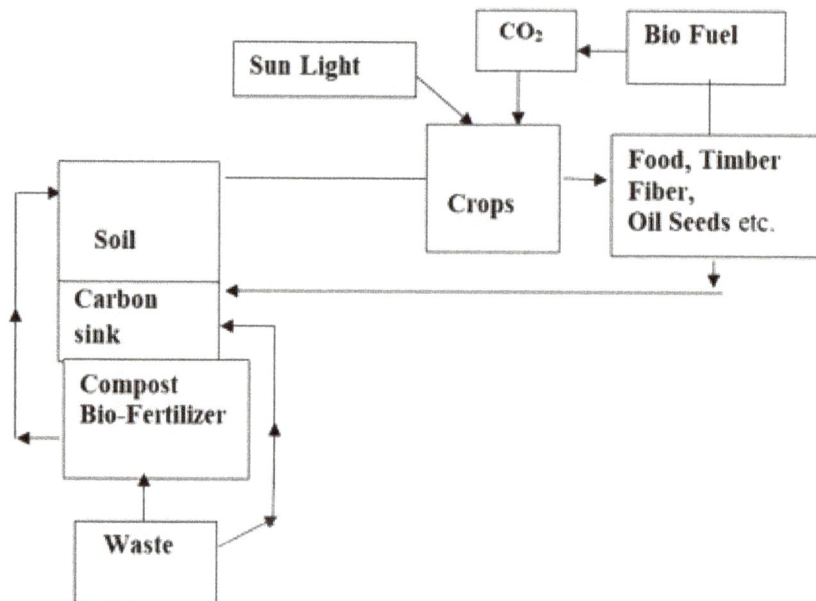

Fig. 2.1b. Use of Compost/Biofertilizer from Biomass (Lower Emissions/Environmental Effects).

Safety from Pathogenic Pollution in Composting

From the point of view of pollution and spreading of diseases, it may be mentioned that most of the pathogens (disease-causing microorganisms), except a few may not be able to survive at the temperature of 60°C reached during the composting process for longer than half an hour. There are antagonistic processes caused by antibiotic-inhibiting organisms and fungi which may kill most of the pathogens. Composting is therefore thought to be safe for producing useful manure.

Methane Emission During Composting

It may be noted that during composting, apart from aerobic decomposition, some anaerobic decomposition also takes place, forming methane, which is released into the atmosphere (methane is a more powerful greenhouse gas than carbon dioxide). But a major portion of carbon remains sequestered in the soil and the overall conversion of waste to compost and its use as a fertilizer to replace chemical fertilizers is environmentally friendly despite some emission of methane.

Phosphorousan Important Constituent of Soil

Apart from nitrogen, phosphorous is a very important element for the growth of organisms and plants. This element is, generally, in short supply. It is found in rocks in certain areas. It may be present in very small quantities of dust and its gaseous phase in the atmosphere temporarily before it mixes with soil or water. Phosphorous is continuously washed into oceans and therefore, soil becomes deficient in phosphorous. In order to promote the growth of plants, therefore, it is necessary to add additional phosphate fertilizer to the soil. A phosphorous cycle on earth is depicted in Figure 2.2.

2.1.6 PGPR as a Natural Alternative to Chemical Fertilizer [41,69,80,93,100]

The 'Plant growth–promoting rhizobacteria' referred to as 'PGPR' offers an ecofriendly approach to stimulate the growth of plants. Some species of soil bacteria present around the roots of plants stimulate plant growth by various biochemical mechanisms. The PGPR perform various functions such as: (i) Increasing availability of plant nutrients, (ii) stimulating growth through phytohormones, (iii) degrading organic pollutants, and (iv) production of antibiotics and antifungal metabolites.

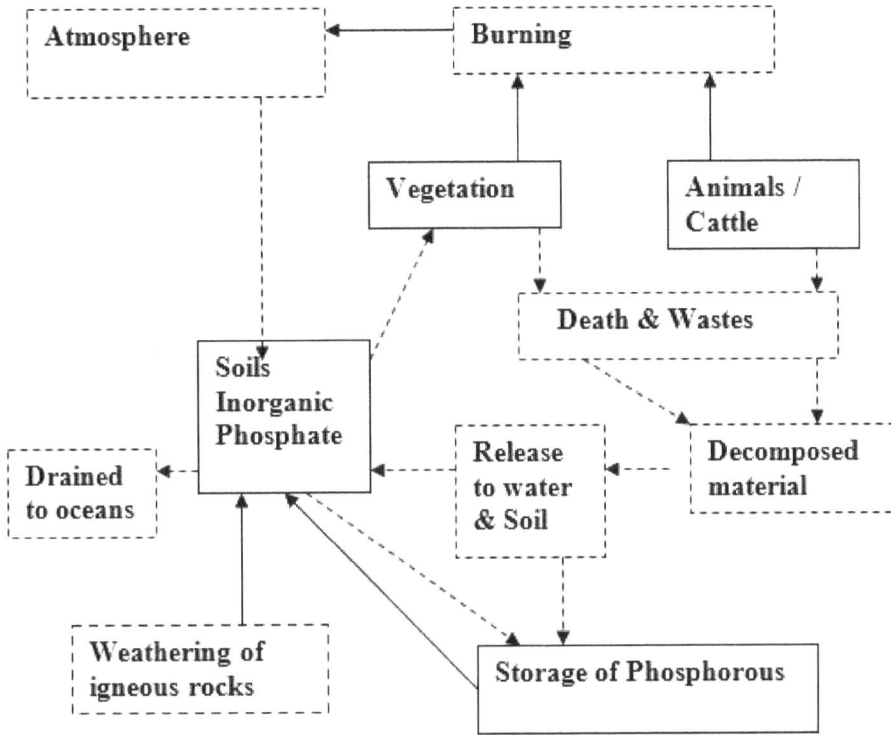

Fig. 2.2. Phosphorous Cycle on Earth.

Many PGPR have been identified which include: *Pseudomonos fluorescens, P. putida, P. aeruginosa,* and *Bacillus subtilis* and other Bacillus species. The PGPR can be a substitute for chemical fertilizers.

2.1.7 Sewage Water Utilization for Plantation [2,4,8,17,64]

It may be mentioned that sewage water after primary treatment (*i.e., gravity separation, aeration, storage in mechanically aerated lagoons for minimizing objectionable contents*), if available, can be utilized for irrigating plantations. The sewage water contains lots of plant nutrients and therefore, the use of sewage water would be beneficial in plantations.

The use of sewage water *without primary treatment is undesirable,* because the high organic load in untreated sewage water may cause a problem of excessive accumulation of organic matter, salts, heavy/toxic metals, and microorganisms (including plant pathogens) in the top layer of the soil, resulting in the deterioration of its texture, decrease in water percolation in clay loam soils (*having more than 33% clay and less than 36% sand*), and improper aeration of soil. It may also lead to some plant diseases and cause toxicity of crops due to presence of excessive toxic elements in them, rendering them unsuitable for use.

After the recovery of methane/biogas from industrial wastewater such as distillery spent wash after anaerobic treatment, the effluent can be advantageously used for the irrigation of plants after partial treatment and dilution.

2.1.8 Restoration of Landfills for Plantations [52,61,97]

Landfills where municipal solid waste (MSW) has been disposed of are a major eyesore and this land needs to be restored and utilized. Plantation of trees in the area is the best option if the same is not going to be utilized otherwise. However, there are some problems in utilizing the landfills such as:

- Risk of fire hazard in the area because of the emission of landfill gas due to anaerobic decomposition of embedded organic wastes inside landfills. The landfill gas can catch fire as it has about 50% methane (and balance of the gas is carbon dioxide with high contamination by carbon monoxide, compounds of sulphur, nitrous oxide, halogens, ammonia, etc.).
- Rainwater percolation through the solid waste stored on the land will generate leachate by extracting soluble components, which will pollute the land and have adverse effects on the plants. The leachate will pollute groundwater unless some initial precautions are taken before ground is used for dumping municipal solid wastes.
- Ammoniacal nitrogen and chemically reduced metal ions like iron, manganese, aluminum, etc., may exert harmful effects on plants.

Restoration

It is advisable to extract the landfill gas by erecting vertical wells and gas piping, a gas abstraction plant with a gas blower and a gas flare, a passive gas barrier, and an improved final soil cover. The landfill gas/biogas should be sent for utilization. It may also be attractive to extract the compost/manure formed from the garbage for use in agriculture or plantations (Also see Chapter 6, Para 6.9).

Suitable varieties of trees/plants according to the climate and soil in the area should be planted. Proper drainage and a leachate collection system should be provided. The collected leachate can be used for irrigation after the rainy season, as the leachate has a high content of plant nutrients. The leachate may require some treatment before use in some cases.

2.2 Tree Plantation

2.2.1 *Fast-growing Trees* [38,46,113]

Fast-growing trees are needed to mitigate the consequences of the destruction of forest and plantations. These trees will also fulfil the demand for fuel and firewood. Fast-growing trees may vary in different climates and regions according to the terrain, soil, and other conditions. However, some popular types of fast-growing trees include:

Acacia: It is a fast-growing tree found in tropical and monsoon climates. The growth is very good throughout the year in regions with rainfall and warm temperatures throughout the year, such as tropical regions. However, it can grow well even in monsoon regions where the rains are only seasonal, but winters (with minimum temperatures above freezing point) and hot summers occur.

Acacias species suitable for use as <u>firewood</u> include: *A. melanoxylon, A. acurninata, A. dealbata, A. decurrens, A. elats,* and *A. mearnsii.* [1,47]

It is a *leguminous tree and makes the soil richer*. Acacia (genus Acacia) comprises about 160 species of trees and shrubs in the pea family (Fabaceae). Acacias are native to tropical and subtropical regions of the world, particularly Australia (where they are called 'wattles') and Africa, where they are well-known landmarks on the veld and savannas. They are present in all terrestrial habitats, including rainforests, woodlands, grasslands, coastal dunes and in arid areas. [1]

[One *Acacia falcate* tree in Sabah, Malaysia, managed to gain an astounding 35 feet (10.7 m) height in just 13 months, which comes out to just over an inch (2.7 cm) per day. [113]]

Poplar: Populus or poplar species are *fast growing* and include the cottonwood and aspen trees. Populus is a genus of 25–35 species of deciduous flowering plants in the family Salicaceae, native to most of the Northern Hemisphere. Some varieties are suited to temperate climates and some are more suited to subtropical and tropical regions with a hot climate.

In the United Kingdom, the poplar (as with fellow energy crop, willow) is typically grown in a short rotation coppice system for two to five years (with single or multiple stems), then harvested and burned—the yield of some varieties can be as high as 12 oven-dry tons per hectare every year. It is a good *firewood* plant.

Some poplar varieties to grow for *firewood* recommended by the Michigan State University Extension, (US) are 'Carolina', 'Robusta', and 'Raverdeaux'. Some poplars can grow as high as 30 m (100 feet).

The various varieties of poplar are broadly as follows:

- Populus section Populus—aspens and white poplar (circumpolar subarctic and cool temperate regions, and mountains farther south, white poplar: warm temperate regions).
- Populus section Aigeiros—black poplars, some of the cottonwoods (North America, Europe, western Asia; temperate regions).
- Populus section Tacamahaca—balsam poplars (North America, Asia; cool temperate region).
- Populus section Leucoides—necklace poplars or bigleaf poplars (Eastern North America, eastern Asia; warm temperate region).
- Populus section Turanga—subtropical poplars (Southwest Asia, east Africa; subtropical to tropical regions).
- Populus section Abaso—Mexican poplars (Mexico; subtropical to tropical regions).

Cottonwood poplars are often wetlands or riparian trees. The aspens are among the most important boreal broadleaf trees. *They consume a vast amount of water from the ground and are not recommended for areas experiencing low groundwater levels.* [46,83]

Eastern Cottonwood: A Populus deltoids variety (a tree), which grows best on moist well-drained sands or silts near streams, often in pure stands. It is a very fast-growing tree, commonly found in North America, and grows along streams and on bottomlands from southern Quebec westward into North Dakota and southwestern Manitoba, south to central Texas, and east to northwestern Florida and Georgia, but is absent in higher elevation areas (Appalachian areas). It has been used primarily for core stock in manufacturing furniture and for pulpwood, but now there is a considerable interest in cottonwood for producing energy biomass (fuel), *because of its high-yield potential and coppicing ability.* [29]

Eucalyptus: It is grown in many countries in tropical and subtropical climates, especially in Australia, Southeast Asia, Brazil, some African countries, etc. Some species of eucalyptus also grow in temperate climates like Spain, Portugal, Italy, Turkey, Greece, South Africa, etc. It has also been introduced in many temperate zones and tropical/subtropical/monsoon zone countries.

All the Eucalyptus species, in general, *are much faster growing than most other trees in cultivation except some species* (such as *E. pauciflora* and its subspecies, *E. gregsoniana, E. vernicosa, E. stricta*, etc. whose growth is slow). Some of the noteworthy species of eucalyptus which are fast-growing trees, with high yield and needing a reasonable amount of rainfall are *Eucalyptus globulus, E. fastigata, E. carnaldulensis, E. saligna, E. obliqua, E. nitens,* and perhaps *E. muelleriana.*

Eucalyptus can, depending on the particular species, as well as certain climatic conditions, reach six or seven meters in only six years, and contribute 20 to 25 tons of firewood per hectare each year. Eucalypts will coppice many times before they have to be removed and replaced. The first crop can be harvested (depending on the species, climate, and other factors affecting tree growth) within about seven or eight years; subsequent cuttings can be taken from coppiced growth, on five to six-year rotations. [47]

Transgenic eucalyptus is an artificially developed plant, created by *splicing Brassica genes with eucalyptus genes.* These trees grow 30% faster than other types of natural eucalyptus, and can gain 4.9 m (16 feet) a year. Within just five years, some plants may manage to top 30 m (100 feet). It may be suitable as firewood. [32,113]

Giant Sequoia [35]: Sequoia trees are best known for their mammoth size, but they are also very *fast growing*. They can gain 1.2–1.8 m (4–6 feet) per year for around 10 years, and then can keep adding on around 0.6 m (2 feet) per year for the next 30 years. **Giant** sequoias grow to an average height of 76–84 m (250–275 feet) and 5–7 m (15–20 feet) d.b.h. (diameter at breast height). A record tree height has been reported to be 95 m (310 feet) in height and 11 m (35 feet) d.b.h. The oldest-known giant sequoia based on ring count is 3,200 years old. Sequoiadendron giganteum (*giant sequoiais is the sole living species in the genus Sequoiadendron, and one of three species of coniferous trees known as redwoods, classified in the family Cupressaceae in the subfamily Sequoioideae*).

Giant sequoia groves mostly are on granite based residual and alluvial soils. They are generally found in a humid climate characterized by dry summers and frosty winters. The elevation of the giant sequoia groves generally ranges from 1,400–2,000 m (4,600 to 6,600 feet) in the north, up to 1,700–2,150 m (5,600 to 7,050 feet) to the south. Giant sequoias generally occur on the south-facing sides of northern mountains, and on the northern faces of more southerly slopes. Trees can withstand temperatures of –31°C (–25°F) or colder for short periods of time, provided the ground around the roots is insulated with either heavy snow or mulch.

The natural distribution of giant sequoias is restricted to a limited area of the western Sierra Nevada, California. Giant sequoia cultivation is very successful in the Pacific Northwest from western Oregon north to southwest British Columbia, with a fast growth rate. It is successfully grown in most of western and southern Europe, the Pacific Northwest of North America north to southwest British Columbia.

Black Locust:: *Robinia pseudoacacia*/black locust is a fast-growing deciduous tree that <u>withstands drought and tolerates poor soil</u> conditions and can resist low temperatures even up to about –2°C (–35°F). Although it is said to increase soil fertility, but the fixing of nitrogen by this tree is not fully established.

The decay resistance of black locust is legendary. In wet conditions, it may last 500 years, but in dry conditions it may last even 1500 years. It was promoted for tree plantation in Europe, especially for use of <u>its wood for shipbuilding</u>. Its calorific value is also high. The black locust (*Robinia pseudoacacia*) grows to a height of 15 m (50 feet) throughout the United States. It has a straight trunk and undergoes <u>rapid growth</u>, making it a <u>suitable candidate for firewood production</u>. The black locust tree is tolerant of a wide range of soil and growing conditions. It sprouts readily from the stump as well as from the roots, and has the ability to form dense thickets if not managed properly. [46,115]

Green Ash: *Fraxinus pennsylvanica*/'*Fraxinus excelsior*'/green ash can be grown in a variety of soils throughout the *United States*. It is tolerant of extreme cold and heat and grows fast, up to 18 m (60 feet) tall. It spreads readily from the large number of seeds produced in the fall. Green ash produces *a good fire when still wet*. Ash, however, is susceptible to the insect emerald ash borer beetle. [46,116]

Catalpa: The Catalpa speciose/catalpa tree grows best in moist, well-drained soil but is tolerant of a wide range of growing conditions throughout the *United States*. It's considered a fast-growing, weedy tree in landscapes, and is suitable for <u>firewood</u> production. The catalpa tree grows rapidly to be 21 m (70 feet) tall. [46]

Thornless Honeylocust: *Gleditsia triacanthos*/thornless honeylocust, a *fast-growing* tree, with an eventual height of 23 m (75 feet) and is native to *North America* and is good for *firewood.* It is adaptable to a wide range of growing conditions and soils throughout the United States, including extreme cold and heat. [46]

Paulownia trees: These trees are native to the *temperate zone of Asia* and are <u>*fast growing and can be used for firewood*</u>. Paulownias, particularly Paulownia tomentosa, also called "empress tree", are now also grown in the *United States*. The trees provide good timber for furniture.

Bamboo: Bamboo grows incredibly quickly, which is one reason it is often used to make sustainable, ecofriendly products. Bamboo species are being cultivated in temperate zones as well as in tropical/subtropical zones. Replanting bamboo is fairly easy thanks to the swift growth rate of the plant. The spreading root structure allows one rootstock to produce several shoots, permitting horizontal growth. Some species can literally grow 10 cm per day. [113]

Fast Growing Trees Monsoon Region in South Asia including the Indian Subcontinent

Fast growing trees of this region include: Populus sp. (Poplar), Eucalyptus sp., Acacia sp. (Kiker, Babool), Lemon, *Terminalia belerica* (Bahara), *Terminalia chibula/T. cirina* (Harra), *Pisidium guajava* (Guava), *Butea monosperm* (Palas/Dhak), *Peltophorum pterocarpum* (Gulmohar/Rusty shield bearer), *Moringa oleifera* (Sahjan), *Mangifera indica* (Mango), Sawani, Subabul, Hafarwari, *Emblica officinalis* (Amla/Aonla), *Pithecellobium dulce* (Jangaljalebi), *Sarca indica* (Ashok), Casuarina (Vilayati Saru, Jungli Jhao, Jangli Saru, Savukku); Khejri or Sami, Fig, *Nyctanthes arbortristis* (Harshingar), *Ficus glomerata* (Goolar), *Pongamia pinnata* (Karanji /Karanja), *Albizia lebbeck* (Sirsi), *Syzygium cuminii/S. jambos* (Jaman), *Gauzuma ulmifolia*, *Grevillea robusta*, *Acacia tortilis* (Babool, Kiker), *Athocephalus indica* (Kadamba), *Cordia dichotama* (Lasoda), *Erythrina indica* (Coral Tree/Mandara/Panjira/Pangri/Pharad/Dadap/Dholdhak), etc.

The fast-growing shrubs and other plants include Bougainvillea, Madar, Kasunda, Bhant, Behaya, Ghaneri Kamni, Red Kaner, Chandni, Yellow Kaner, Rangoon Creeper, Sadabahar, Prosopis juliflora/Prosopis cineraria (Shami), Arythrina, Leucaena, etc.

2.2.2 *Good Firewood Plants* [46,47,116]

Firewood Trees of Temperate Zone

The following plant species are good for use as firewood. The wood of some of these trees can also be used as timber and for furniture.

Poplar species including eastern cottonwood; thornless honey locust; catalpa; black locust; some Eucalayptus species; Osage Orange (*Maclura pomifera*) [*Osage Orange: The tree is for temperate climate and is drought resistant and can survive in poor soils*]; Ironwood (*Ostrya virginiana*); Honeylocust (*Gleditsia triacanthos*; Bitternut Hickory (*Carya cordiformis*); Mulberry (Trees from the Moraceae family); Ash (*Fraxinus* sp.) botanical name '*Fraxinus excelsio*r' Madrone (*Arbutus menziesii*); Hickory (*Carya* sp.); Oak (*Quercus* sp.); Paulownia trees [*it can adjust to almost any soil conditions as long as the soil is well-drained*]; *Beech (*Fagus* sp.) [*grows in cool to warm climate, requires moist and rich soil, mature height about 25–30 m*]; Sugar Maple (*Acer saccharum*)). [47,115,116]

*** Note on Beech tree**: Beech (Fagus) is a genus of deciduous trees in the family Fagaceae, native to temperate Europe, Asia, and North America. The southern beeches (Nothofagus genus) previously thought to be closely related to beeches, are now treated as members of a separate family, Nothofagaceae. They are found in Australia, New Zealand, New Guinea, New Caledonia, Argentina, and Chile (principally Patagonia and Tierra del Fuego).*

The European beech (Fagus sylvatica) is the most commonly cultivated, although there are few important differences between species aside from detail elements such as leaf shape.

The common European beech (Fagus sylvatica) grows naturally in Denmark and southern Norway and Sweden up to about the 57th–59th northern latitude.

East Asia is home to five species of Fagus, only one of which (F. crenata) is occasionally planted in Western countries. Smaller than F. sylvatica and F. grandifolia, Japanese beech is one of the most common hardwoods in its native range.

Firewood Trees for Subtropical, Tropical, or Monsoon Zones in Asia
In general, the wood, which is not good for timber or cannot be used as timber or furniture wood, is used as firewood in these regions because of shortage of timber.

There are some species of Acacia such as—*A. melanoxylon, A. acurninata, A. dealbata, A. decurrens, A. elats* and *A. mearnsii*; Eucallyptus species such as—*E. Tereticomis, E. Camaldulensis, Acacia 'Cupressiformis'; Acacia leucophloea, Acacia Planifrons E. Globulus, E. Robusta, E. Hybrid,* and *E. Grandis;* Casuarinas species especially *Casuarinas equisetifolia; Mangifera indica* (Mango); *Ficus rumphii* (Paakar); *Wendanlandia* (Bankat); *Tamarindus indica* (Tamarind); *Prosopis juliflora* (Mesquite/Honey mesquite); *Prosopis cineraria; Cordia dichotama* (Lasoda), *Pithecellobium dulce* (Jangaljalebi), Acacia named 'Cupressiformis' is like 'Cypress'; *Acacia leucophloea, Acacia planifrons.*

2.2.3 *Plantation on Alkaline/Sodic Lands—Soil Alkalinity/Sodicity* [14, 74, 96]

The salinity of water may primarily be caused by salts from rocks, minerals, and soil. The absence of surface drainage, opposing slopes, and clayey ground causes inflow of surface run-off into depressions, oversaturation of soil, and elevation of the water table.

Climatic factors like a semi-arid climate with large temperature variations, and low or erratic rainfall may cause concentration of salts. During summer, there is a capillary rise of water from the shallow water table through the fine pores of silt clay. The evaporation of water leads to the precipitation of salts in the nearby surface soil. The continuation of this process makes the soil hard and impervious by depositing salts in the sediment pores, thus rendering soil unsuitable for agriculture. The insufficient rains are unable to leach out the salts from the soil and thus, salt concentration continues to increase in the soil.

There are two most important causative factors for soil alkalization such as:

- Over-irrigation
- Lack of drainage

Deposits of salts in the soil cause soil alkalization and make the land barren. Large tracts of such barren lands are seen in plains, especially in areas where there is canal irrigation.

Large tracts of land the world over are affected by salt, and are termed as sodic lands.* These *lands* have poor soil structured-chemical properties. The soil has a high pH value, *exchangeable sodium percentage*, and *sodium absorption ratio*, and very low fertility.

It affects the plant growth directly through ion toxicity and nutrient deficiencies and indirectly through increased soil dispersion and decrease in infiltration. These soils, generally, have a very low organic carbon content (SOC—soil organic content).

Rehabilitation of sodic soils through agroforestry increases the soil organic carbon and nutrients pool, thus increasing the land's fertility.

Note: (In India, about 6.73 million hectares of land are affected by salt, out of which about 3.77 million hectares have been designated as sodic land).

Soil Classification

Table 2.1a shows the classification of soils based on salinity and sodicity, while Table 2.1b shows the characterization of soil salinity and sodicity for alluvial and vertisol soils. *(Vertisols have low hydraulic conductivity, low porosity/low drain ability, and low permeability, and a tendency to crack and swell. Black cotton soil is a type of vertisol.)*

The Table 2.1c shows the terms used to describe the alkalinity and sodicity characteristics of soil and irrigation waters.

Table 2.1a. Classification of Soil—sodicity/salinity—according to the Indian System. [14]

Soil Class	Electrical Conductivity (EC) dS/m	Exchangeable Sodium Percentage (ESP)	Sodium Adsorption Ratio (SAR)	pH Value
Normal	< 4	< 15	< 13	< 8.2
Saline	> 4	< 15	< 13	< 8.2
Sodic	< 4	> 15	> 13	> 8.2
Saline-sodic	> 4	> 15	> 13	> 8.2

Table 2.1b. Characterization of Soil Salinity and Sodicity for Alluvial and Vertisol Soils.* [14]

Se. No.	Class of soil Salinity and Sodicity	Salinity		Sodicity of soil
		Electrical Conductivity dS/m	pH value	Exchangeable Sodium Percentage (ESP)
1	Normal	< 4 (< 2)	< 8.2	< 15 (< 5)
2	Slight	4–8 (2–4)	8.2–9.0	15–40 (5–10)
3	Moderate	8–10 (4–8)	9–9.8	40–60 (10–20)
4	Strong	16–32 (8–16)	9.8–10.5	60–70 (20–30)
5	Very strong	> 32 (> 16)	> 10.5	> 70 (> 30)

*Note: (a) The above characteristics are derived from a soil saturation extract in water.
(b) The figures in brackets are for vertisol soils.

Table 2.1c. Various Terms used with Respect to the Salinity & Sodicity of Irrigation Waters.

Sr. No.	
1.	Electrical conductivity (EC) measured in µS/cm or in dS m^{-1} (Micro Siemens/cm) or in Deci-Siemens/meter) [This is also referred to as salinity]
2.	Sodium Absorption Ratio (SAR) = (Na) / [{(Ca + Mg)/2 }$^{1/2}$] [This is also referred to as Sodicity]
3.	Sodium percentage (SP) = 100. [(Na + K)/(Na + K + Ca + Mg)]
4.	Residual Sodium Carbonate (RSC) = (CO$_3$ + HCO$_3$) – (Ca + Mg)
5.	Sodium to Calcium Activity Ratio (SCAR) = (Na)/{(Ca)$^{1/2}$}
6.	Magnesium /to Calcium Ratio: [Mg/Ca] Ratio
7.	Magnesium Hazard (Mg – Hazard) = 100.[(Mg)/(Ca+ Mg)]
8.	Exchangeable Sodium Ratio (ESR) = [Exchangeable Na]/[Exchangeable (Ca +Mg)] [For salt affected soils, ESR and SAR may have an approximately linear relationship; for example, ESR = 0.15 (SAR)]

Note:
I. In the above formulae, all concentrations of ions are expressed as m.eq. per litre except for Item No.8.
II. Leaching Factor is written as L.F.
III. For Item No. 8, all units are in meq/100 g.
IV. Exchangeable Sodium Percentage is written as ESP percentage = 100.[(ESR)/(1+ESR)]

2.2.4 *Methods for Rehabilitating Sodic Lands and Plantations* [23,48,75,85,109]

Rehabilitation of such soils can be done in the following manner:

- Soil testing should first be carried out.
- Soil structure and the moisture-holding capacity should be improved through the application of biochar and soil amendments.

- Improve drainage.

- Leaching the soil with thorough irrigation after salt exposure. Flush out salt through the soil by adding 2 inches of water over a 2–3 hour period, and stopping if runoff occurs. Repeat this treatment three days later if salt levels are still high.

- Necessary nutrients may be added to the soil through organic manure, biofertilizers, and compost.

- 'Press-mud' waste from sugar mills can be used for the treatment of sodic lands.

- Some cyanobacteria species like Nostoc calcicola along with gypsum have been used for the remediation of such soils. *[The cyanobacteria* Nostoc calcicola *is tolerant to halides and can fix nitrogen. It grows well in sodic soils and decreases the pH value of soil significantly when applied in combination with gypsum (i.e., gypsum inoculated with this bacterial culture is applied) as per experiments.]*

- Mulching to prevent evaporation and the subsequent build-up of salt in the soil.

- Suitable varieties of green manure plants/grasses, etc., should be grown and ploughed back into the soil. If the climate is temperate (temperature not below 0°C), *Sesbania aculeate* (Dhaincha), sun hemp, cluster bean (*Lasiurus scindicus* sp.), and cluster bean (*Cymbopogon jwarancusa*) are good green manure plants.

- Plant suitable trees for soils in tropical/ subtropical/monsoon climate regions are such as: *Casuarina glauca* (Khejri/Sami/Jangli Saru/Vilayati Saru), *Leucena leucocephala* (lasobaval/ vilayati baval), *Prosopis juliflora* (Mesquite/Honey Mesquite), *Acacia nilotica* (Kiker/Babool), *Psidium* (Guava), *Kochia scoparia* (bui/kauraro), *Azadirachta indica* (Neem), *Madhuca latifolia* (Mahua), Coconut, *Syzygium cuminii/S.jambos* (jamun), Fig, *Carrissa carandas* (Karonda), *Prosopic dulc*e, Pongamia, *Acacia auriculiformis* (kiker/babool), *Albizia procera*, *Terminalia arjuna* (Arjun), eucalyptus, *Dalbergia sissoo* (Shisham), *Pongamia pinnata* (karanji/karanja), *Tamarindus* (Tamarind/Imli), *Phyllanthus* sp. (Kanocha/Kanochha/Nalausereki/Jar-Amla/ Bhui-Amla/Kilkkanelli, etc. They may be planted in between crops or at least on the boundary.

It has been found that *Prosopis juliflora/Prosopis dulce* (Mesquite/Honey Mesquite/Shami, Jhand), *Acacia auriculiformis/Acacia nilotica* (Babool/Kiker), *Albizia lebbeck/Albizia Amra/Albizia julibrissu/Albizia procera* (sirsi/lallei/lal sirsi/Krishna shirsha), *Bauhinia* sp. (Amli/Karmia/Kattra/ Kairwal/Kaliar/Raktakanchan/Koiral/ Kachnal/Kachnar/Asundro/Phalgu/ Banraj/ Sevela Kanchan), Azadirachta indica (Neem), are good for sodic soil remediation. *Prosopis juliflora* (mesquite/honey mesquite) speedily aids the decomposition and mineralization of plant litter. *Dalbergia* sp. (Sisso/ Shisham) and *Terminalia* sp. (Har and Bahayra) can also grow on this land in-between other plants. These plants can also be used for the complete afforestation of the land in these climatic regions.

- Subsequent fertilizing only when a soil test or plant symptom indicates that fertilizer is needed, and then only at rates recommended by soil analyses and fertilizer labels.

- The salt-affected lands can thus be utilized for carbon sequestration both in soil and plants. It is possible to develop green cover on such lands with diverse plant communities consisting of herbs, shrubs, and trees. *Leguminous plant species contribute more significantly to land renewal and increasing fertility as compared to non-leguminous species. However, phosphorus availability increases in the soil planted with non-leguminous varieties.* The quality of land improves after some years of raising these trees. The soil organic carbon, exchangeable calcium, and cation exchange capacity (CEC) increase with time, corresponding to plant growth, whereas, the pH value and exchangeable sodium percentage (ESP)/exchangeable sodium decrease markedly. Thus, these plants can be used for the reclamation of sodic land.

- After partially reclaiming sodic soils, certain non-traditional crops like amaranth, and some medicinal plants like chamomile, vetiver, asparagus, mucuna, desmodium, damask rose, and floricultural crops like marigold and tuberose can also be grown in between the planted trees for boosting the income of the farmer, if the climate of the region is favorable to such plants.

- The sodic lands of the *subtropical regions* can also be utilized for growing lemongrass which yields essential oil of medicinal value. Lemongrass can grow on *acidic as well as alkaline lands (with pH value range of 4.3–8.9). Lemongrass grown on alkaline/sodic land has more oil content than the grass grown on non-sodic land.* The chemical composition of oil from lemongrass from sodic land and that from non-sodic land would slightly differ. There are also some other salt-tolerant crops like sugar beet, barley, date palm, etc., which can be grown in-between planted trees.

2.2.5 *Plantation in Acidic Soils/Podsol Areas* [59,66,122]

In some areas, the nature of soil is acidic and the fertility is low as *Podsol, which is a base-deficient ash-grey acidic soil,* is abundantly present. The pH value of the soil is quite low (even less than 5.5). If the climate permits, large green plants like conifers, pines, spruces, cedars (deodar), and cypresses can thrive on such soils. The formation of acidic soils may be due to the following mechanism:

Organic acids may form during humidification under cold, humid conditions and they react with minerals in the soil/rocks to dissolve the basic cations potassium (K), sodium (Na), calcium (Ca), magnesium (Mg), which get leached out of the soil, leaving it acidic and deficient in iron and aluminum, but richer in silica. Podsol or acidic soils tend to have low availability of essential elements for agriculture like nitrogen, phosphorus, potassium, sulphur, manganese, boron, etc.

The acidic soils/podsols can be treated for restoration by the mineral dolomite, a carbonate of calcium and magnesium that can neutralize the soil acidity, after which the utilization of land can be done for plantation and agriculture in the normal way in accordance with climatic conditions.* *[For example: Podsol is found in Sikkim, India.]*

2.2.6 *Trees Suitable for Saline Lands with Adequate Moisture Content in Hot Climates (Inhospitable Soils and Climates)* [9,10,23,43,117]

Some trees and shrubs have special qualities like surviving on saline land. Different types of trees are recommended for different degrees of salinity as mentioned below:

(a) *For High Saline Land* – Date palm, Zizyphus Jujuba (Ber), Tamarindus indica (Tamarind), *Eucalyptus camalclulensis, Eucalyptus halophylla, E. occidentalis, Casuarina cunninghamiana, Cordia dichotama* (Lasoda), *Butea monosperm* (Palas/Dhak), *Diospyros melanoxylon/D. tomentosa* (Tendu), *Tamarax aphylla* (Farash), *Kochia[4] scoparia* etc. Some herbs like Latjira, Bhat-kataiya Morang, Shail kanta, etc. can grow on soil that is highly saline. *Salicornia brachiata* can also be grown.

Chinese medicinal materials like *Glycyrrhiza uralemis* and *Epheara intermedis* are adaptable to saline soil and can boost the farmer's income. Saline areas can also grow *Medicago saliva*, which makes the area suitable for growing fresh grasses.

(b) *For Moderately Saline Land*: In addition to plants mentioned above in (a), the other varieties of suitable trees are: *Azadirachta indica* (Neem), *Madhuca latifolia* (Mahua), *Cocos nucifera* (Coconut), *Syzygium cuminii/S. jambos* (Jamun), Fig, *Carrissa carandas* (Karonda), *Casuarina cunninghamiana*, Eucalyptus sp. like E. leucoxylon, *E. microcarpa, E. melliodora, E. occidentalis, Eucalyptus polyanthemos, Prosopis juliflora*, Pongamia, etc. (Some fruits like cherry, phalsa, pomegranate, etc. can also be grown.) *Carob (Ceratonia siliqua)* [Fabaceae], *native to the Mediterranean, is a leguminous evergreen tree with moderate salinity and good drought-tolerance.*

(c) *For Low Saline Condition*: In addition to all the plants mentioned above in (a and b), other varieties of suitable trees are: *Emblica officinalis* (Amla/Aonla), *Pisidium* (Guava), *Mangifera indica* (Mango), *Eucalyptus occidentalis*, *Aegle marmelos* (Bael), etc. (Grapes can also be grown.) **[117]**

(d) ***Suitable Plants for Growing on Soil with Ash*** *(Redundant ash ponds or already filled ponds with ash, overlaid by soil, and pits filled with good soil on the ash pond):* Acacia (Babul), Subabul, *Terminalia arjuna* (Arjuna), *Zizyphus nummularia* (Jharberi), *Zizyphus jujuba* (Ber), *Cordia dichotama* (Lasoda), Behya, etc.

(e) ***Plants for Estuarine Areas Affected by Salinity:*** *Halophytes Salicornia brachiata:* halophytes are a group of plant, which are salt tolerant and drought resistant and can be used as food, fodder, fertilizer, in papermaking, and for many other purposes. This group of plants grow in transitional zones between coastal mangroves and terrestrial plants. *Salicornia brachiata* is an important plant, which is also called *"Sea asparagus"* and grows along the coastal estuarine habitats of tropical, subtropical, and temperate climate zones in many countries of the world including India, where it is found abundantly in the Godavari estuary in Seema Andhra. It can be cooked and consumed as food or even pickled and used as fodder for sheep, goats, and cattle. Its seeds contain oil of good edible quality as they are high in PUFA (polyunsaturated fatty acids) and is similar to safflower oil. This plant's foliage and oil can be a *potential source of biofuels, including biodiesel.*

2.2.7 *Trees Grasses and Shrubs Suited to Saline Soils in Temperate Zones* [5,117]

1. Following trees are good for saline soils in temperate climates (like that in east USA):
   * Red buckeye (*Aesculus pavia*) deciduous tree
   * Black walnut (*Juglans nigra*) deciduous tree, good for salt spray locations
   * Black locust (*Robinia pseudoacacia)* deciduous tree, good for salt spray locations
   * Japanese tree lilac (*Syringa reticulate*) Deciduous tree, good for salt spray locations
   * Honeylocust (*Gleditsia triacanthos*) deciduous tree, good for salt spray locations
   * White fringe tree (*Chionanthus virginicus*) deciduous tree
   * White ash (*Fraxinus Americana*) deciduous tree, good for salt spray locations
   * Common persimmon (*Diospyros virginiana*) deciduous tree, good for salt spray locations
   * Golden rain tree (*Koelreuteria paniculate*) deciduous tree, good for salt spray locations
   * Sweetbay magnolia (*Magnolia virginiana*) evergreen tree
   * Japanese black pine (*Pinus thunbergiana*) evergreen tree, good for salt spray locations
   * White poplar (*Populus alba*) seciduous tree, good for salt spray locations
   * Carolina cherry laurel (*Prunus caroliniana*) deciduous tree
   * Argan tree (*Argania spinosa*) [Sapotaceae] is an extremely drought- and salt-tolerant evergreen tree that grows in the bushlands of southwestern Morocco.

Grasses and Shrubs Suitable for Saline Lands in Temperate Zones

1. Alkali Sacaton (*Sporobolus airoides*) [Poaceace] is a grass, commonly found on sandy alkaline soils in Mexico and the US.
2. Crowfoot grass (*Dactyloctenium aegyptium*) [Poaceace], commonly found in Africa and the Middle East.
3. Iodine bush (*Allenrolfea occidentalis*) [Chenopodiaceae] is a succulent evergreen bush, somewhat similar to Salicornia, growing in the inland salt marshes and arid salt lands of the western US.

4. Palmer saltgrass (*Distichlis palmeri*) [Poaceace], native to the southwest US and Mexico, is commonly found growing in muddy tidal flats, salt marshes, and estuarine habitats with elevated salt concentrations.

5. Sea oats (*Uniola paniculata*) [Poaceace] is an attractive dune grass native to the coastal regions of the southeastern US, Mexico, and the Caribbean.

6. Wild rice (*Zizania aquatica*) [Poaceace], native to North America, is a tall aquatic grass that grows in the shallow brackish waters of rivers, ponds, and lakes. (Wild rice was an important food staple for the Native Americans, high in protein (17%) and low in fat (1.2%)).

7. Agave (*Agave* spp.) [Agavaceae] is a salt- and drought-tolerant succulent shrub (lily family) of the dry open deserts of the Americas. Evidence suggests that A. deserti and A. murpheyi have been an important source of high energy food for over 9,000 years. [9,10,23]

2.2.8 *Some Common Plants that Grow on Salt-affected Soils in Temperate Zones*

1. *For very highly affected soil* [23]
Black greasewood (*Sarcobatus vermiculatus*); Inland saltgrass (*Distichlis stricta*); Nuttall's alkaligrass (*Puccinellia nuttalliana*); Beardless wildrye (*Leymus triticoides*); Shore arrowgrass (*Triglochin maritima*); Red glasswort (*Salicornia ruba*); Seepweed (*Suaeda depressa*); Pickleweed (*Salicornia* spp.).

2. *Highly affected soil* [23]
Alkali cordgrass (*Spartina gracilis*); Slender wheatgrass (*Elymus trachycaulus*); Saltbush species (Atriplex spp.); Winterfat (*Krascheninnikovia lanata*); Alkali bluegrass (*Poa juncifolia*); Alkali sacaton (*Sporobolus airoides*); Foxtail barley (*Hordeum jubatum*); Cinquefoil species (*Potentila* spp.). [23]

3. *Salinity Tolerance of Some Tree/Shrubs* [23]

Tree/shrubs	Salinity EC tolerance micromhos/cm	Tree/shrubs	Salinity EC tolerance micromhos/cm
Ash, Green	12	Buckthorn, Sea	15
Hawthorn	13	Honeylocust	8
Honeysuckle	10	Lilac common	12
Peashrub siberian	10	Pine Austrian	11
Pin scotch	9	Saltbush	18
Winterfat	12		

4. *Crops Which Tolerate Salinity Levels* [23]

Crops	EC range micro moh/cm	Tolerance
Barley	8–16	High tolerance
Sugar beet	7–13	Moderate tolerance
Safflower	6–10	Moderate tolerance
Wheat	7–8	Low tolerance
Oats	4–8	Low tolerance
Corn	3–6	Low tolerance
Beans	1–2	Low tolerance

5. *Phytoremediation of Saline Soils by Use of Halophytes* [67,68,84]

Apart from accumulation of salt from the saline habitat, many of the halophytes are capable of remediating toxic metals and can grow and yield some 'halophytes, which are edible and may be used for forage, and as oilseed crops. They not only can remediate the salt-contaminated soils, but also provide food, fodder, fuel, and industrial raw material and increase the income of the farmers. The halophytes biomass can also be utilized for making gaseous/liquid fuels. The examples of such halophytes are given below:

- Suaeda fruticosa accumulate sodium and other salts. *S. maritime* and *Eluvium portulacastrum* accumulate even greater amounts of salts in their tissue and thus gives higher remediation rates for the saline and sodic land.

- Juncus rigidus and *J. acutus* used in Egypt provided a favourable response in the reduction of the EC of soil. *(The EC of soil had a 50% saturation decreased from 33 to 22 dS m⁻¹ in a single growth period.)* Also, amshot grass was found to reduce the exchangeable sodium percent of the surface layer of the soil.

- *Portulaca oleracea* removes salts very efficiently.

- *Arthrocnemum indicum, Suaeda fruticosa,* and *Sesuvium portulacastrum* seedlings grown on a saline soil and significantly reduced the soil salinity, especially sodium salts.

- Salt accumulator halophyte species such as *Tamarix aphylla, Atriplex nummularia,* and *A. halimus* have also been found to decrease salinity.

- The salt-tolerant species *Leptochloa fusca* (L.) Kunth (kallar grass) was found to decrease the salinity/sodicity and pH value up to the depth of 100 mm of soil. [43]

6. Effect of Salt on Plants

Exposure to salt spray can cause stem and foliage disfigurement, reduced plant growth, and often, plant death. The plants are exposed to salt by aerial salt drift from seas or lakes. The direction of wind and frequency of storms are important factors in aerial salt drift. In de-icing operations, salt is added to melt away the ice or snow and this is harmful to the plants and vegetation, especially near the roads, sidewalks, parking lots, etc.

Note: *The seaweeds which grow in sal water in sea and estuaries can be an important source of biomass from which biofuels can be derived.*

2.2.9 *Plants for Hot Arid and Semiarid Regions* [13,34,42]

The trees which have deep, widespread root spread system can survive in arid/semiarid regions. In hot arid regions, trees like *Prosopis cineraria* (Shami), *Tecomella undulata* (Rugtrora), *Prosopis juliflora* (Mesquite/Honey mesquite) *Zizyphus nummularia* (Jharberry), date palm, *Acacia tortilis,* Acacia Senegal (kumta/khor), Acacia Seyal, etc., can be grown.

In hot semi-arid regions, trees like *Acacia nilotica, Acacia tortilis* (a promising species for desert greening) thrive. The jojoba is another promising species of economic value which has been found suitable for planting in these areas and has also proved to be a promising species for desert greening. *Dalbergia sissoo* (Sisu/Shinshupa/Shisham), *Proscopis cineraria* (Shami), *Azadirachta indica* (Neem), *Zizyphus nummularia* (Jharberry/Bordi), *Prosopis juliflora* (Mesquite/Honey Mesquite), *Zizyphus jujuba* (Ber), Mango, *Terminalia belerica* (Behera), Kaththa, *Pithecellobium dulce* (Jangaljalebi), small *Emblica officinalis* (Amla), Chirunji, *Jatropha curcas, Albizia lebbeck* (Siris), *Kochiaᵃ scoparia, Albizzia amara* (Lallei/Krishna Sirisha), *Butea Monosperm* (Dhak), *Acacia* Sp. Such As *A. mimosaceae, A. caesalpiniaceae, And A. fabaceae* (drought-tolerant Sp.), *Argania spinose* (Argan tree, a drought- and salt-tolerant evergreen tree of the bushlands of southwestern Morocco).

As per the Central Institute of Medicinal and Aromatic Plants (Lucknow, India) lemongrass (Suvarna), which contains about 1% essential oil, can grow in semi-arid regions, if some soil moisture is maintained. All the plants mentioned for growing in hot, arid regions can grow well in semi-arid regions.

Note: Kochia[4] is a hardy, salinity-resistant plant that can be widely used as emergency forage for livestock. It can grow in the desert ecosystem by using saline water. It has good tolerance to an elevated level of salinity up to 23 dS/m. It controls oxidative stress and continues to grow.

2.2.10 *Temperate Zone Desert Vegetation* [25,46,112]

Temperate deserts occur in the western United States, Argentina, and Asia (like the Great Gobi Desert) and other places in the world; North America's temperate desert is the Great Basin Desert. The desert's indigenous vegetation includes cacti like Prickly Pear Cactus, Christmas Cactus, Desert Spoon Cactus, Ball Cactus, and *Mammillaria parkinsoii*. In some places, where some moisture is available, some drought tolerant bushes grow.

In the *Sonoran Desert* (which is a North American desert covering large parts of the southwestern United States in Arizona and California and northwestern Mexico in Sonora), a cactus exclusive to the area (the Giant Saguaro Cacti) grow, forming almost a forest, and providing shade for other plants and nesting places for desert birds. Saguaro grow slowly, but may live up to about 200 years. The surface of the trunk is folded like a concertina, allowing it to expand, and may hold tons of water after a good downpour.

In North America's temperate desert, the Great Basin Desert, big Sagebrush (*Artemisia tridenta*) is common, growing more than 0.5–3 m tall with a deep taproot 1–4 m in length, coupled with laterally spreading roots near the surface [12], while the red osier dogwood (*Cornus sericea*) grows along streams. *Ephedra virdis* (Mormon tea) is a drought-tolerant garden plant, but another species, *Ephedra nevadensis* (Nevada Mormon tea), is more drought-tolerant than green *Ephedra viridis*, which grows in locations with more moisture.

Where more moisture is available and temperature does not go below 2°C, *Robinia pseudoacacia*/black locust may be grown, as it can withstand drought and poor soil conditions.

Trees grow mainly at higher elevations or along streams where more moisture is available. Rocky Mountain juniper *(Juniperus scopulorum)* can grow up to 9–12 m tall.

2.2.11 *Plants for Dusty Areas in Hot Climates*

 (i) *Areas with High Cement Dust Pollution:* Trees like *Butea monosperm* (Palas/Dhak), Gulmohar, *Madhuca indica* (mahua), *Terminalia arjuna* (Arjun), *Tamarindus Indus* (Imli), shrubs like Bougainvillea, Behaya, and red /yellow Kaner, etc., survive in large amounts of cement dust pollution found around cement factories.

 (ii) *Suitable Plants for Roadside Plantation:* The plants suitable for roadside plantation include: Neem, Satwin, Satavar, Peepal, Sheesham, Tamarind/Imli, Kadamb, Jangal Jalebi, Pakkar, *Ficus glomerata* (Goolar), Mango, Jaman, Palas/Dhak (*Butea monosperm*), Gulmohar, *Madhuca indica* (Mahua), *Terminalia arjuna* (Arjun), *Sarca indica* (Ashok*), Morus alba* (Mulberry/Shahtut, Bamboo, Ber, Khjoor, Babool), *Athocephalus indica* (Kadamba), etc.

 (iii) *Low-Height Plants for Road Dividers:* Plants suitable as road dividers should normally have a low height. These may include: Kaner, BottleBrush, China Rose, Chandni, Bougainvillea, Harshingar, Satavar, Rubber Plant, *Morus alba* (Mulberry/Shahtut), *Sarca indica* (Ashok), *Athocephalus indica* (Kadamba), etc.

2.2.12 Trees Suited to Tropical/ Subtropical Climates and the Monsoon Climate in Asia

Trees for Tropical, Highlands (Humid, Subhumid zones) include:

JackFruit, *Mangifera Indica* (Mango), *Moringa oleifera* (Drumstick), Eucalyptus Sp., Teak (Sagwan), Sal, Ficus Regiosa (Peepal), Cashew, *Mimusops elengi* (Maulsari/Bakul), *Athocephalus indica* (Kadamba), *Cassia fistula* (Amaltas), Coconut (especially in coastal areas), Bamboo Sp., *Casuarina equisitifolia, Cordia dichotama* (Lasoda), Teak (Sagwan), Etc.

Trees for Tropical Plains include:

Azadirachta indica (Neem), Mango, *Tamarindus indica* (Tamarind/Imli), *Artocarpus heterophyllus* (JackFruit), Coconut, Sapota, *Acacia nilotica, Acacia leucophloea, Acacia plniformis, Ailanthus excelsa, Casuarinas* sp. (Jangli Saru/Khejri), *Eucalyptus* sp., *Cordia dichotama* (Lasoda), *Tectona grandis* (Teak/Sagwan), *Ficus regiosa* (Peepal), *Prosopis chilensis/Prosopis cineraria* (Shami), *Cassia fistula* (Amaltas), *Mangifera indica* (Mango), Etc.

Example of Some Trees with Special Capabilities

i.   Trees having a high capacity to absorb carbon dioxide include Teak or Sagwan, Amla, Eucalyptus, Neem, *Albizia lebbeck* (Siris/Shirisha), *Ficus regiosa* (Peepal), Gular, *Moris alba* (Mulberry/Shahtoot), *Pongamia pinnata* (Karanja), etc.

ii. Trees having good acid neutralising potential include Amaltas, *Ficus regiosa* (Peepal), Lemon, Bougainvillea, etc.

iii. Trees having noise-reducing capability include: *Ficus regiosa* (Peepal), *Eucalyptus* sp., *Tamarindus indica* (Tamarind), *Azadirachta indica* (Neem), Bamboo, etc.

iv. Rubber plant absorbs formaldehyde vapours.

v.   Medicinal Trees: There are thousands of plants with medicinal properties and it is not possible to mention them all here. A few examples of medicinal trees are: *Terminalia arjuna* (Arjun), *Dalbergia sissoo* (Sisu/Shinshupa/Shisham), *Azadirachta indica* (Neem), Terminalia belerica (Bahara), Terminalia chibula/T. cirina (Harra), Bauhinia variegata (Kachnar), *Emblica officinalis* (Aonla/Amla), *Syzygium cuminii/S. jambos* (jamun), *Cassia fistula* (Amaltas), *Aegle marmelos* (Bel), *Athocephalus indica* (Kadamba), etc. [21]

vi. Leguminosae varieties of trees (these trees take nitrogen from atmosphere and make the soil fertile), include: *Dalbergia sissoo* (Sisu/Shisham), *Cassia fistula* (Amaltas), *Sarca indica* (Ashok), *Pongamia pinnata* (Karanji/Karanja), *Erythrina indica* (Coral Tree, Mandara/Panjira/ Dadap/Dholdhak), *Proscopis spicigera* (Shami/Jhand), *Butea monosperm* (Dhak), *Acacia* sp., (Babul/Kikar), *Albizzaia* sp.(Sirsi, Lallei/ Krishna Sirisha, Lal sirsi), *Bauhinia* sp. (amli/ karmia/kattra/kairwal/kaliar/raktakanchan/koiral/kachnal/kachnar/asundro/phalgu/banraj/ sevela kanchan), casuarina, etc.

Note on Household Plants with Special Qualities

Although the household plants have no role in providing fuel, they are important from environmental reasons. NASA has published a list of 50 suitable household plants. Some of these plants can remove chemical vapors and resist insects, and are easy to maintain. Some of these important plants are:

• Areca palm which is a natural humidifier.

• Bamboo palm which is a good air purifier

• Rubber plant (potted) and dracaena which absorb formaldehyde vapours,

• Gerbera daisy whose flowers absorb benzene and formaldehyde, trichloroethylene, vapours

• Peace lily, which is good at removing benzene vapours

• Jharbera removes benzene and trichloroethylene

Note on Examples of Plants Suitable for Different Regions in India

- ***Indo-Gangetic Plains area of Punjab****:* Acacia nilotica/Acacia catechu *(Kiker, Babool)*, Dalbergia sissoo *(Shisham)*, Zizyphus sp. *(Bordi/Ber)*, Butea monosperm *(Dhak)*, Grewia asiatica *(Phalsa)*, Anogeissus latifolia *(Dhaura/ Dhaoya/Dhava)*, Populus deltoids *(Popular)*, *Eucalyptus* sp. Ficus religiosa *(Peepal)*, Ficus rumphii *(paakar)*, Terminalia arjuna *(Arjun)*, bauhinia variegata *(Kachnar)*, Athocephalus indica *(Kadamba)*, Grewia villosa *(Dhodhan/ Kharmati)*, Grewa tenax *(Gango/Achu), etc.*

- ***Indo-Gangetic Plains area of Haryana:*** Acacia nilotica/Acacia tortilis (kiker, babool) Dalbergia sissoo (sisu /shisham), Eucalyptus hybrid, Azadirachta indica (neem), Prosopis cineraria (khejri or sami), Populus deltoids, *Zizyphus* sp. (bordi/ber), Ficus religiosa (aswattha/peepal), Ficus rumphii (paakar), Mangifera indica (mango), citrus fruit trees, Bauhinia variegata (Kachnar), Terminalia arjuna (Arjuna), Grewia villosa (dhodhan/kharmati), Grewa tenax (gango/achu), etc.

- ***Indo-Gangetic Plains of Western and Central Uttar Pradesh:*** Acacia nilotica, Acacia catechu, Zizyphus jujuba *(Ber)*, Pisidium (guava), *Neem*, Populus deltoids, *Eucalyptus* sp., Morus alba *(Mulberry/Shahtut)*, Albizia lebbeck *(Siris/Shirisha)*, Syzygium cuminii *(Jamun)*, Mangifera indica *(Mango)*, Emblica officinalis *(Amla/Aonla)*, Pisidium *(Guava/Guajava)*, Aegle marmelos *(Bel)*, Athocephalus indica *(Kadamba)*, Cassia fistula *(Amaltas)*, Ficus religiosa *(Aswattha/ Peepal)*, Deris indica *(Gonj/Aru/Panlata)*, Mimusops hexandra *(Khirni)*, Mauratiana, bamboo spp., Moringa oleifera *(DrumStick)*, Ficus rumphii *(Paakar)*, *Bargad, Teak (Sagwan)*, Terminalia arjuna *(Arjuna)*, Bauhinia variegata *(Kachnar)*, etc.

- ***Indo-Gangetic Plains of Eastern Uttar Pradesh and Bihar:*** Acacia nilotica, Dalbergia sissoo *(Shisham/Sisu)*, Zizyphus jujuba *(Ber)*, Jamun, bel, *JackFruit,* Pisidium *(Guava)*, Azadirachta Indica *(Neem)*, Morus alba *(Mulberry/Shahtut)*, *Eucalyptus* spp., Bombax ceiba, Techtona grandus, Cassia fistula *(amaltas)*, Mangifera indica *(mango)*, *Letchi,* Emblica officinalis *(Aonla/ Amla)*, Wendanlandia *(Bankat)*, *Bamboo* spp., Madhuca indica *(Mahua)*, Moringa oleifera *(DrumStick)*, Tamarindus indica *(Tamarind/Imli)*, Cordia dichotama *(Lasoda)*, Teak *(Sagwan)*, Shorea robusta *(Sal/Sakhu)*, Ficus religiosa *(Aswattha/Peepal)*, Ficus rumphii *(Paakar)*, ficus bengalensis *(Barga*d*)*, Terminalia arjuna *(Arjuna)*, Bauhinia variegata *(Kachnar)*, Athocephalus indica *(Kadamba)*, etc.

- ***Central India Plateau and Bundelkhand:*** Acacia leucophloea, Acacia nilotica, Dalbergia sissoo *(Sisu/Shisham)*, proscopic cineraria *(Shami), Ber, Jaman,* Mangifera indica *(Mango)*, *Jangal Jalebi*, Small Emblica Officinalis *(Amla), Catechu (Kaththa)*, Butea monosperm *(Dhak), D. tomentosa (Tendu)*, azadirachta indica *(Neem)*, albizia lebbeck, madhuca indica *(Mahua)*, anogeissus pendula *(Dhaura/Dhavada/Dhava)*, Diospyros melanoxylon/Diospyros tomentosa *(Tendu), Cheruanji, Bamboo* sp. *(Dendrocalamus)*, Prosopis juliflora *(Mesquite/ Honey Mesquite))*, Madhuca indica *(Mahu*a*)*, Cordia dichotama *(Laso*da*)*, Ficus religiosa *(Aswattha/Peepal)*, Terminalia arjuna *(Arjuna)*, Bauhinia variegata *(Kachnar)*, Albizzia amara *(Lallei/Krishna Sirisha)*, etc.

- ***Deccan Plateau:*** *Acacia nilotica, Acacia leucophloea, Acacia ferruginous, Prosopis juliflora* (Shami/Jhand/Kalisam)*, Hardwickia binata, Azadirachta indica* (Neem)*, Albizzia lebbeck* (Siris)*, Bamboo* spp. (especially Dendrocalamus)*, Casuarina equisitifolia* (Jangli Saru/Vilayati Saru)*, Dalbergia sissoo* (Sisu/Shisham)*, Proscopis cineraria* (Khejri/Shami)*,/)*, Mango, *Mimusops hexandra* (Khirni)*, Bauhinia variegata* (Kachnar)*, Terminalia arjuna* (Arjuna)*, Cassia fistula* (Amaltas)*, Butea monosperm* (Dhak), etc.

- ***Gujarat, Rajasthan:*** Acacia nilotica, *Dalbergia sissoo* (Sisu/Shisham), Proscopis cineraria (Khejri). *Azadirachta indica* (Neem), *Zizyphus nummularia* (Jharberi), *Zizyphus jujuba* (Ber), *Mangifera indica* (Mango), *Pithecellobium dulce* (Jangaljalebi), *Albizzia lebbeck* (Sirsi), *Bamboo* sp., Date Palm.

Where irrigation facilities are available in Gujarat and Rajasthan, they can also grow poplar spp., *Morus alba* (Shahtut), *Ficus* sp./*Ficus glomerata* (Gular)/*Ficus rumphii* (Pakkar), *Grewia villosa* (Dhodhan/Kharmati), *Grewa tenax* (Gango/Achu), *Palm* sp., *Eucalyptus* sp., etc.

Some examples of other good plants for indoor use are mentioned below:

   i.   *Agave angustifolia marginatus*/Caribbean agave
  ii.   *Spathiphyllum wallis*/Peace lily
 iii.   *Aspidistra elatior* Variegata
  iv.   *Cycas revolute/*Sago palm
   v.   *Epipremnum aureum*/Golden patho
  vi.   *Ficus Benjamina 'Nuda'*/Weeping fig, Ficus lyrata/ Fiddle leaf fig
 vii.   *Neoregelia spectabills*/ Fingernail plant
viii.   *Nephrolepis tuberose*/Boston fern
  ix.   *Pepromia obtusifolia*/Pepromi
   x.   *Philodendron bipinnatifidum*/Heart leaf philodendron
  xi.   *Ployscias balfouriana*/Balfour
 xii.   *Aralias*/Marginatus
xiii.   *Portulacaria afra*/Jade plant
 xiv.   *Raphis excelsa*/Broad leaf lady plant
  xv.   *Ruscus hypoglossum*/Mouse thorn
 xvi.   *Sunsevieria trifasciata*/Snake plant
xvii.   *Schefflera actinophylla*/Queensland
xviii. Umbrella tree
 xix.   *Syngonium podophyllum*/Goosefoot plant

2.3 Agroforestry

2.3.1 Advantages of Agroforestry

Agroforestry is practiced in all countries to some extent as there are a lot of advantages gained from this practice. Some of the benefits of agroforestry (i.e., planting trees along with raising crops) are mentioned below:

- Provision of fuel, fruits, herbal medicines, shade, timber to farmers
- Soil stability
- Help to improve the microclimate
- Nitrogen fixation by leguminous trees and making the soil fertile (see Figure 2.3)
- Release of phosphate by trees' mycorrhizae
- Tree litter improves soil structure, and also increases fertility along with releasing nitrogen and phosphate, as mentioned above
- Certain trees phytoremediate the soil
- Adds to the income of farmer and improve the economy of farming itself.

2.3.2 Agroforestry in Hot Arid and Hot Semi-arid Areas [94]

Agroforestry has special advantages for arid and semiarid regions with hot climates because of the shortage of fuel, food, and other benefits of growing plants.

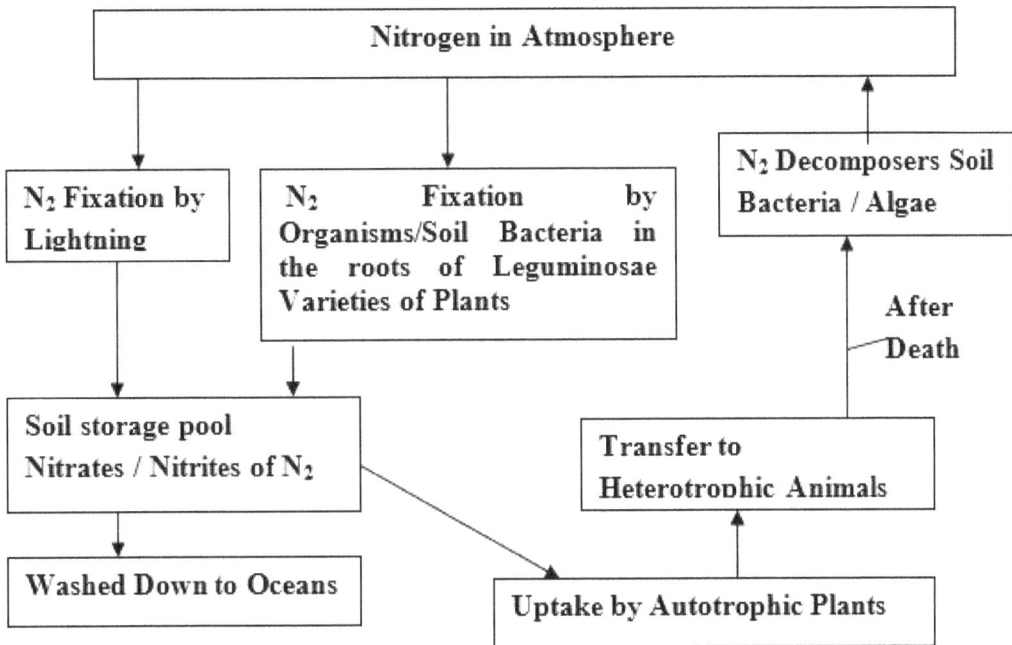

Fig. 2.3. Nitrogen Cycle.

In hot desert areas having sand dunes with annual rainfall of less than 100–50 mm, trees/shrubs like Calligonum-Haloxylon-Leptadenia in association with pearl millets and cluster bean (*Lasiurus scindicus* sp.) can be grown.

In areas with rocky, gravelly pediments having annual rainfall between 150–200 mm, trees/shrubs like the Ziziphus-Capparis tree can be grown in association with pearl millet, green gram, moth bean, cluster bean (*Cymbopogon jwarancusa*), *Aristida* sp., *and Cenchrus ciliaris* (buffel-grass or African foxtail grass). Tree species like *Prosopis cineraria* and *Zizyphus numularia* can be grown along with the grasses of the species *Cenchrus cilieria, Lasiurus sindicus*, etc.

The agri-horticulture system where jujube (Ziziphus mauritiana) is intercropped with arid legumes, for example, clusterbean/mothbean/greengram may be adopted by areas receiving rainfall of more than 250 mm.

2.3.3 Agroforestry Practice in Temperate Climate Countries—Some Examples [37]

In northern China, poplars are grown along with vegetable crops, peanuts, soybeans, sesame, indigo, cotton, while in Italy and Yugoslavia, poplars are grown with traditional cereal/ tuber crops. In the Mississippi Delta (U.S.A), poplar deltoids are grown with cotton and soybeans, while in southern Ontario (Canada) corn and soybeans are grown along with poplar in a temperate tree based agroforestry intercropping system. In Britain, the planting of Scotch pine (Pinu. Sylvestris) has been suggested.

2.3.4 Agroforestry Strategy in the Developing Countries of Asia/Africa/South America

Due to the rapid destruction of forests in these countries and the increase in population as well as their needs, agroforestry has a special charm for these countries. They should adopt the following strategy *as a step towards sustainable growth.*

- Forests should not be denuded for fuel wood and most of the fuel requirement in rural areas should be met by tree plantations on the agricultural land, along with other crops. Such plantations may also partly meet the requirements of timber, bamboo, fruits, fodder, herbal medicines, and other products. Boundaries of fields are generally chosen as areas for planting trees.

- It is better to plant the indigenous species growing in forests of the region in the degraded forests and not the other species so that the natural vegetation is restored. (Avoid agroforestry species in the forest, as we may lose some useful products which we get from forests.)

- The saline soil patches of agricultural land can be used for growing suitable trees as already suggested.

- Social forestry should be taken up on an intensive scale.

- The use of organic manure/compost/biofertilizers is always better for the several reasons already stated.

- Agroforestry should be practiced in agriculture in various regions.

Suitable Trees for Agroforestry in Tropical/ Subtropical Regions

The following trees may be a good choice for agroforestry in tropical and subtropical climates, but the choice of trees should be based on the actual needs of community. (Some of these varieties may be good for semi-arid regions as pointed out in Para 2.2.11.)

Eucalyptus sp., *casuarina equisetifolia* (Jangli Saru/Vilayati Saru), *Crataeva roxburghii* (Varna), *Terminalia arjuna* (Arjun), *Azadirachta indica* (Neem), *Artocarpus heterophyllus* (Kathail/Jackfruit), Mango, *Shorea robusta* (Sal/Sakhu), *Sagaun syzygium cumini* (Jamun), *Aegle marmelos* (Bel), *Emblica officinalis* (Amla), *Litchi chinensis* (Litchi), Guava (Pisidium), *Zizyphus jujuba* (Ber), Coconut, *Anacardium occidentale* (Cashew), *Cryptostegia grandiflora* (Rubber), Oil Palm, Date Palm, *Prosopis juliflora, Tecomella undulate* (Rugtrora), *Bauhinia* sp. (Amli/ Karmia/Kattra/Kairwal/Kaliar/Raktakanchan/Koiral/Kachnal/Kachnar/Asundro/Phalgu/Banraj/ Sevela Kanchan), *Acacia sengal, Acacia albida, Acacia nilotica, Albizia lebbeck* (Siris), *Dalbergia sissoo* (Shisham), *Butea monosperm* (Dhak), *Albizzia amara* (Lalle/Krishna Sirisha), *Cassia fistula* (Amaltas), *Casuarina* sp., *Bambusa nutans, Alnus nepalensis* (Udis/Kohi), *Boehmeria regulosa, Ficus glomerata* (Goolar), *Ficus roxburghii, Sesbania rostrata, S. cannabin, Sarca indica* (Ashok), *Pongamia pinnata* (Karanji/Karanja), *Erythrina indica* (Coral Tree, Mandara/Panjira/Dadap/ Dholdhak), *Prosopis cineraria, Proscopis spicigera* (Shami/Jhand), etc.

It is also very important in agroforestry to include *leguminosae varieties* of trees like Dalbergia sissoo, Cassia fistula (amaltas), Sarca indica (ashok), Pongamia pinnata (*Karanji/ Karanja*), Erythrina indica), Proscopis spicigera (*Shami/Jhand*), Butea monosperm (*Dhak),* Acacia sp., *Albizzaia* sp., etc.

In regions where ground levels are very high, the popular species may be included, but never in regions suffering from lack of rains or low ground levels.

The slow-growing oilseed-producing trees like *Madhuca indica* (mahua) and *Calophyllum inophyllum* (Sultana Chumpa/Punnaga/Undi) are preferred in agroforestry by farmers. [75,96,109]

Suitable Species for Providing a Major Impetus Forward in Agroforestry

From point of view of fast growth and benefits such as producing fuel wood early, the use of the following species of trees should be encouraged in agroforestry:

Eucalyptus sp., *Casuarina Equisitifolia, Acacia* sp., *Bamboo* sp. and if, groundwater level is high and rains are above normal, *Populus deltoids* sp. (*Poplar* sp.), and the *Prosopis Cineraria* sp.

The important characteristics of these trees are mentioned below:

(i) *Eucalyptus* sp.

The eucalyptus trees absorb water from a deep strata of soil and the surface soil does not get affected by moisture stress. It is an evergreen tree and improves the physical and chemical condition of soil. These trees can survive in adverse climatic conditions—in areas with a rainfall range of 400 mm–4000 mm, in coastal regions, plains, and highlands up to 2000 meters. In low temperate climate regions with winter rainfall, it can grow to a height of even 100 meters. The tree is a source of eucalyptus oil. The calorific value of the wood of this tree is 4700–4800 kcal/ kg. The wood has a specific gravity of 0.539–0.64. The wood has straight fibers and many uses apart from being fuel. The eucalyptus can be cut for use any time after 6–8 years. For dry and poor sites, a dwarf and multi-stemmed variety of eucalyptus called "Melees" has been developed. Popular eucalyptus species in India are: *E. Tereticomis, E. Camaldulensis, E. Globulus, E. Robusta, E. Hybrid,* and *E. Grandis*. (Also see eucalyptus in Para 2.2.1.)

(ii) *Prosopis cineraria* (Shame/Hand/Kalama)

Prosopis cineraria is an evergreen tree with light foliage whose maximum normal height is 12 meters and maximum normal girth is 1.2 meters. It may attain a girth of 0.8 meters in about 30 years. It can survive in dry climate with rainfall as little as 100–250 mm. The older plants are drought-resistant but the young plants require some watering. The plant can grow in Punjab, Uttar Pradesh, Rajasthan, Gujarat, Tamil Nadu, Karnataka, etc. It can grow in areas with higher rainfall in south India, but dry climate is preferable for it. It can grow in association with other plants in black cotton soil also.

Two species namely *Prosopis chinless* and *Prosopis cineraria* can be grown even in sem-iarid areas of tropical and subtropical regions.

(iii) *Prosopis Juliflora* (Mesquite/Honey mesquite)

These trees are salt-tolerant, drought-tolerant, fast-growing trees, and can provide firewood to the poor even in semi-arid areas. They can survive in low rainfall areas of 50 mm and above. These trees can attain a height of 10 meters and a diameter of 1.2 meters.

(iv) *Casuarina Equisitifolia* (Jangle sari/Vilayet Sari/Khari)

The casuarina species grow in hot semi-arid and subhumid areas of medium altitudes up to say 1400 m, with rainfall of 200–3500 mm on coarse, well-drained soils and sand dunes having subsoil moisture content. Although the tree is native to *Australia, Bangladesh, Brunei, Cambodia, Fiji, Indonesia, Malaysia, New Zealand, Papua New Guinea, Philippines, Samoa, Solomon Islands, Thailand, Tonga, Vanuatu, Vietnam*, it is grown in other parts of the world with similar climatic conditions. Casuarinas grow in impoverished soils too and are *excellent for firewood*. Casuarina has the ability to provide nutrition to nitrogen-fixing bacteria and therefore, increases the fertility of the soil, though it is not a legumin.

C. equisetifolia is found on sand dunes, in sands alongside estuaries and behind fore-dunes and gentle slopes near the sea and can tolerate salt spray and inundation with seawater in high tides. It is tolerant to both the alkaline soil and calcareous soil, but not to prolonged water- logging. These trees are widely cultivated in coastal areas. They can grow well in poor soils, improve the soil, and act as check for flying sand if two or three rows are grown and provide firewood. For these reasons, it is a good for agroforestry. [16]

(v) *Acacia* (Kiker/Babu/Ayla/Khari/Retha/Biswas/Kuma)

Acacia nilotica is a fast-growing, multipurpose tree, which is also suitable for a semi-arid climate. It can grow 2–3 cm in diameter every year. The wood is strong and has a specific gravity of 0.67–0.68.

The calorific value of its sapwood is 4800 kcal/kg and that of its hardwood is 4950 kcal/kg. The subspecies *Acacia indica* a has high production of seeds for easy multiplication. Most of the Acacia species, especially, '*Acacia senegal*' are nitrogen-fixing species and are useful for improving the poor soils and are useful as agroforestry trees.

A mixed plantation of acacia and other trees is advantageous because acacia can provide nitrogen to the other trees. A subspecies of acacia named 'Cupressiformis' is like the Acacia 'Cypress' in appearance and has a narrow, erect crown and is popular with farmers as an agroforestry tree in many areas.

Acacia tortilis, Acacia senegal, and *Acacia seyal* are widely distributed in arid and semiarid zones. (In India, these species are especially widespread in Maharashtra, Gujarat, Rajasthan, Madhya Pradesh, Uttar Pradesh, Haryana, and Punjab.)

For tropical, humid plains, acacia's other varieties like *Acacia leucophloea*, Acacia Planifrons are good, while for tropical highlands, Acacia lebbeck and Acacia ferruginous may be preferred (such as for the Deccan Plateau in Peninsular India).

(vi) Bamboo

Bamboo species grow in temperate climates as well as in tropical, semitropical regions. *Bamboo* is well-known and useful in many ways, though it does not provide fuel until it is scrapped after use. The important use of bamboo is in temporary constructions, barricades, furniture, etc., because of its straight fiber and strength. It is a good material for papermaking too.

The principle species of bamboo which grow in warm and hot climates are:

a. *Bambusa*—Its growth is widely distributed.

b. *Dendrocalamus*—Next widely distributed species, especially in semi-arid highlands.

c. *Arundinaceous, M. baccifera, D. Strictus*—Grows in cool to warm humid climates.

d. *Oxytenanthera, Dinochloa*, and *Bambusa*—Widespread in tropical regions.

e. Other genera which also grow in these regions are *Sinobambusa, Chimonobombusa, Gigantochloa, Indocalamus,* and *Oclandra.*

(vii) Poplar*

The poplar or more accurately Populus tree species is fast growing. They grow straight and have many varieties. In semitropical and tropical climates, the following species grow:

- Populus section Turanga – subtropical poplars (Southwest Asia, east Africa; subtropical to tropical regions)
- *Populus ilicifolia* – Tana River poplar (East Africa)
- Populus section Abaso – Mexican poplars (Mexico; subtropical to tropical regions)
- *Populus guzmanantlensis* (Mexico).
- *Populus mexicana* – Mexico poplar (Mexico)
- *Populus laurifolia* – Northwest Himalayas (at 2400–4000 meters)
- *Populus ciliate* – Temperate and subtemperate regions of Himalayas (at 1200–3500 meters)
- *Populus gamblei* – Eastern Himalayas (at 400–1300 meters)
- *Populus jaquemontiana* – Eastern Himalayas (at 1500–3200 meters)
- *Populus alba* – Western Himalayas (at 300–1200 meters)
- *Populus deltoids* – This is a very popular variety of Poplar species all over the Indo-Gangetic plains. It can grow up to 25 meters in height and have a girth of 100–130 cm in about 12 years, but requires irrigation facilities. It can be cropped after 6–12 years. Its wood is light, having a specific gravity of about 0.441. It is good for making match-sticks.

- It may be used as an agroforestry tree at places where the groundwater level is very high. Poplar species* of trees require large quantity of water and they cannot be grown in water-deficient areas.

Note: Popular trees can act as hydraulic pumps as their roots can reach down to the water table and establish a dense root mass that takes up large quantities of water. Poplar trees can transpire between 50 and 300 gallons of water per day out of the ground. Plants like poplar rapidly uptake large volumes of water to contain or control the migration of subsurface polluted water and soil pollutants to ground water. This is known as hydraulic control of the pollution of groundwater.

2.3.5 *Increase in Soil Fertility by Agroforestry*

Soil fertility can be regained by agroforestry in tropical and subtropical regions by planting suitable species of plants and vegetation, especially leguminous varieties. The trees and crops of leguminosae varieties fix nitrogen (N) and increase the amount of nitrogen in the soil. Through increased litter, they influence phosphorus (P) cycling to cause greater release of nitrogen and phosphorous (N & P). These plants also increase nitrogen mineralization rates.

Azadirachta indica (Neem) trees have been found to produce good quantity of biomass even in nitrogen- and phosphorus-stressed soil, such as the soil in central India (Raipur). It has also been found that in the Himalayan region, planting 'Alnus – cardamom' aids the return of the nutrients to the soil.

Integration of trees and crops are found to enhance the utilization of water, particularly during floods, as the crops use the surface moisture while the roots of tree carry the excess water deeper into the soil, apart from utilizing it themselves. It is a fact that the biological yield, moisture, and nutrient status of soil under the canopies of trees are better. For example, it has been seen that when there is mixed cultivation of barley (Hordeum vulgare) with certain trees in arid regions of Haryana (India), the yield of barley is higher. The highest increase in the yield of barley recorded was when it was grown with Prosopis cineraria trees (86%), while the increase in yield with Acacia albida trees was 57.9%, and 48.8% with Tecomella undulata trees.

In another case, the mixed plantation of certain varieties of plants on abandoned agricultural land at an altitude of 1200 m in the central Himalayas restored soil, which had been facing nutrient stress. In mixed agroforestry, the kharif crop (rainy season crop) gave higher yield even when the tree litter was the only source of manure. The trees planted in this case were *Alnus nepalese's* (Duis/Koi), *Albizia lebbeck* (Sarsi/Sharita), *Bohemia regulons, Dalbergia sissoo* (Shisha), *Ficus glomerate* (Goslar), and *F. Roxburgh* as these trees have a high rate of release of nitrogen and phosphorous (N & P) during the rainy season.

In semiarid/arid regions in subtropical regions, it is traditional to grow pearl millet, lintels, or oilseeds in fields alongside trees like *Prosopis cineraria* (khari) and *Zizyphus nummular* (bordia/jabari) as they find it better than growing solo.

Growing ginger in the interspaces of Ailanthus trophies trees has been found to cause better rhizome development than when ginger is grown solo.

Bambusa nutans (bamboo) has the potential to increase soil nutrient binding during soil restoration in shifting agriculture.

In rice-cultivating areas, Acacia nilotica trees are found to increase the soil fertility of depleted soil over a 10–15-year period.

2.4 Plantation In Mined-Out Areas [101,110,119]

Mining of minerals, coal, lignite, stones, or other resources extensively damage the land because of the destruction of vegetation, microbial communities, soil texture, the soil's chemical composition, and plant nutrients. The mining affects the soil fertility very adversely.

There are several mined-out areas the world over, which should be restored by transforming them into plantations. These areas can become a boon for land starved forestation drives by using proper methods. In coal mining, the land is devoid of soil carbon and therefore these areas provide an excellent opportunity to sequester carbon in both the soil and vegetation, and the plantation in the degraded area would also add to the efforts towards arresting global warming, besides providing for renewable fuel needs.

It may be mentioned that the topsoil (*up to 100 cm depth*) of the land which is mined, is the most fertile part of the soil. It gets significantly damaged during mineral extraction. It should be put forward to the mining authorities that the topsoil should be stripped and stockpiled separately. However, even doing this can lead to adverse consequences for the microbial communities in the stockpiled topsoil.

This topsoil, if available, should be used for top dressing during land restoration. The stones and rocks of the 'overburden' should be removed as far as feasible and stacked separately for use as material for the construction of roads and buildings, etc.

Drainage should be taken care of while erecting dumps so that the land is not spoiled by improper drainage and leachate, which may contain pollutants. These may trickle down to groundwater.

It is necessary to mention here another important problem created by mining. *If pyrites are present in the overburden and dug out soil, they would get oxidized and may form sulphuric acid with moisture, lowering the pH value of the soil. If present, the carbonate minerals tend to increase the pH value to some extent. When the soil pH drops below 6, say to 5.5, the calcium and magnesium carbonates may get washed out and the soil gets further damaged as the unoxidized pyrites may further reduce the pH value. Firstly, in acidic soils, the nitrogen-fixing bacteria and other useful microbial population is reduced and secondly, the acid formed due to moisture may dissolve metals like aluminium, manganese, etc., in the soil excessively, causing toxicity. As a consequence of this, the nutrients like nitrogen and phosphorous would not be available to plants, inhibiting metabolic process and plant growth.*

Procedure for Restoration of Mined-out Land

The procedure to be followed for restoration of mined-out areas may be outlined as follows:

a. At first, it is essential to carry out an 'ecological survey' of major vegetative association and natural succession, that is, identifying pioneering indigenous/local species of grasses, herbs, shrubs, crops, and trees.

b. The soil characterisation of mine spoils/overburdens/dumps should be done. Soil testing should be done with a view to find limiting factors for plant growth. Soil and mine spoils may contain some toxic metallic compounds*/ingredients for which plant species tolerant to those ingredients would have to be selected.

c. The essential part of any land restoration plan is consolidation, grading, drainage, and restoration of topsoil. The slopes of dumps should be stabilized against landslides and soil erosion by proper stage-wise consolidation. For larger dumps, two or three stair-type formations can be used. For stabilization, mechanical means may be used. Back filling may be done where necessary and topsoil should be spread which should have been kept separately during mining. To check soil erosion and ensure the stabilization of slopes, perennial grasses with sturdy root mat may be used as anchors. The use of metal tolerant temperate grasses like Agrostis capillaries, Festuca rubra, etc., may be used.

d. It may be necessary to apply the required soil amendments according to soil characteristics. Acidic mine soils can be effectively neutralized once they have been spread again at the reclamation site by applying either cement kiln dust containing lime (CaO) or limestone ($CaCO_3$). Lime application rates must account for both past and future pyrite oxidation in order to maintain neutral soil pH levels over time. Lime addition is a common method to decrease the heavy metal mobility in the soils and their accumulation in the plant as it increases the pH of soil.

e. Productivity of soil can be increased by adding various natural amendments such as sawdust, wood residues, sewage sludge, animal manure, biofertilizers, compost, etc., as these amendments stimulate the microbial activity, which provides the nutrients (N, P) and organic carbon to the soil. Ploughing and mulching is a necessary part of restoration.

f. Plant species should be selected according to the result of ecological survey carried out earlier, and the actual soil characteristics and trials in nurseries set up on site. The initial selection should also keep in view the capability of plant for colonising in a degraded area, fixing atmospheric nitrogen (leguminosae varieties), conserving soil, and tolerating the soil ingredients present. Preference should be given to the grasses having fodder value, plants of leguminous varieties having economic value, medicinal value, and timber and firewood value. Fruit trees, oilseed-bearing trees, trees which form canopies, plants for the benefit of wildlife, or plants required by community may be selected. All the plant selected should have ability to survive well under the prevailing conditions of the degraded land and should, preferably, have a fast growth rate.

Afforestation of these wastelands by oilseeds-bearing trees like *Jatropha curcas* Pongamia pinnata, etc., may also prove to be an efficient utilization of such lands for extracting renewable fuels.

Proper species selection for any portion of a mine site is determined by its location on the landscape, because the landscape's position influences (like direction and steepness of the slope of the dump) the availability of soil moisture and sunlight.

It is always better to use suitable grasses, shrubs, and trees according to their suitability to stabilize the slopes.

g. The use of specific planting techniques required for selected species and the appropriate application of moisture and manure should be implemented.

h. Watering should be done throughout the initial periods at regular intervals. Domestic effluents/sewage if available can be used for irrigating the plants after the primary treatment.

i. Water harvesting and using water conservation techniques should be employed so that the plants survive by using natural rainwater. The stored rainwater can be used for irrigation so that no extra water should be needed.

j. Plants should be protected against destruction by animals, pests, and others.

k. All the plant litter should be used for making compost and should be applied to the soil regularly without burning the same.

l. The soil should be tested year after year to monitor the changes in its characteristics. When the soil has become ready for more economical plantations, new varieties of crops or trees may be introduced slowly in the course of time.

It may be mentioned that there are some plants which are accumulators of some toxic metals: (a) the Indian mustard plant accumulates cadmium, lead, gold, chromium, copper, nickel, selenium, zinc, etc.; (b) the sunflower plant accumulates heavy metals like copper, manganese, zinc, cadmium, lead, nickel, including radioactive metals like uranium, strontium, cesium-137; (c) amaranthus (Sadanatiya/Ramdana/Chaulai /Marasa/Chumlisag/Katilichaurai) can remove heavy metals like cadmium, chromium, copper, nickel, lead, zinc, uranium; (d) the rapeseed plant can remove silver, selenium, chromium, mercury, lead, and zinc; (e) Highland bentgrass removes toxic metals like aluminum, manganese, lead, zinc, and arsenic; (f) *Pteris vittata* (Chinese brake fern) removes arsenic; (g) *Ajanus cajan* (Arhar Dal Plant) can take up heavy metals; (h) the guava plant can take up heavy metal like chromium. Planting of such varieties of suitable plants may be considered in the land restoration program. *[Also see Para 2.8 on Bioremediation and Phyto-remediation].*

Note: Examples of plants used in restoration of vegetation in mined-out areas of India
The following are the examples of plants which successfully restored the vegetation in the mined-out area of various regions in India:

- *Limestone mine spoils of outer Himalayas:* Salix tetrasperm *(Baishi/Panijama)*, Leucena leucocephala, Bauhinia retusa *(Kandla/Semla/Kural)*, Acacia catechu, Ipomoea carnea, Eulaliopsis binata, Chrysoogon fulvus, Arundo donax *(Baranal/Gahanal/Bansi)*, Agave Americana *(Kantala)*, Pennisetum purpureum, Erythrina subersosa *(Pangra/Thab), etc.*

- *Limestone mined-out areas of Madhya Pradesh:* Jatropha curcas, Pongamia pinnata *(Karanja)*, Ailanthus excelsa *(Maharukha), and* Withania somnifara *(Asgand/Asvagandha), etc.*

- *Bauxite mined areas of Madhya Pradesh*: Grevillea pteridifolia, Eucalyptus camaldulensis, Shorea robusta *(Sal), etc.*

- *Coal mine spoils of Madhya Pradesh*: Eucalyptus hybrid, Eucalyptus camaldulensis, Acacia auriculiformis, Acacia nilotica, Dalbergia sissoo *(Shisham)*, Pongamia pinnata *(Karanja)*, Bambusa, Peltephorum pterocarpum *(Gulmohar)*, Bambusa *(Bamboo), etc. [Bambusa shows very high accumulation of metals like Fe (iron), Cr (Chromium), Zn (zinc)].*

- *Lignite mine spoils of Tamil Nadu*: *Eucalyptus species,* Leucaena leucocephala, *Acacia,* Agave Americana *(Kantala), etc.*

- *Iron ore wastes of Orissa:* Leucena leucocephala, Albizia lebeck *(Shirisha/Siris).*

- *Haematite, Magnetite, and Manganese spoils of Karnataka:* Albizia lebeck *(Shirisha/Siris).*

- *Mica, Copper, Tungsten, Marble, Dolomite, and Limestone mine spoils of Rajasthan:* Acacia tortilis, Prosopis juliflora, Acacia senegal, Salvadora oleodes *(Baha pila/Pilu/Kankhina)*, Tamarax articulate *(Lal Jav/Magiya - main /Farash)*, Zizyphus nummularia *(Jharberi)*, Grewia tenax *(Gango/ Gwangi)*, Cenchrus setigerus, Cymbopogon *(Gajni/Rousaghas)*, Cynodon dactylon *(Dhub/Dubh)*, Sporobollis marginatus, *and* Dichanthium annulatum.

- *Rock phosphate mine spoils of Musoorie, Uttranchal*: Pennisetum purpureum, Saccharum spontaneum *(Kans/Kasha)*, Vitex negundo *(Nirgundi/Nirgandi)*, Rumex hastatus, Mimosa himalayana, Buddleia asiatica *(Newarpati/Ban)*, Dalbergia sissoo *(Shisham)*, Acacia catechu, Leucena leucocephala, *and* Salix tetrasperma *(Baishi/Varuna/Panijama), etc.*

2.5 General Essential Provisions in Forest Policies for Sustainable Growth [22,36,82,89,111]

2.5.1 *Basic Objectives of Forest Policy*

The basic policy objectives should be as follows:

- Maintenance of environmental stability through preservation and where necessary, restoration of the ecological balance that has been adversely disturbed by serious depletion of the forests of the country.

- Conserving the natural heritage of the country by preserving the remaining natural forests with the vast variety of flora and fauna, which represent the remarkable biological diversity and genetic resources of the country.

- Checking soil erosion and denudation in the catchment areas of rivers, lakes, reservoirs in the interest of soil and water conservation, for mitigating floods and droughts, and for the retardation of siltation of reservoirs.

- Checking the expansion of sand dunes in the desert areas of and along the coastal tracts.

- Substantially increasing the forest/tree cover in the country through massive afforestation and social forestry programs, especially in all denuded, degraded, and unproductive lands.

- Every country has to aim to ensure that a certain percentage of its land is under green cover, that is, covered by trees, forest, or plantations. The extensive tree cover should be maintained in hills and mountains in order to prevent erosion and land degradation and to ensure the stability of the fragile ecosystem. The cover should be extended to semiarid and arid areas to arrest the march of deserts.

- Meeting the requirements for fuel wood, fodder, minor forest produce, and timber of the rural and tribal populations.
- Increasing the productivity of forests to meet essential national needs.
- Encouraging the efficient utilization of forest produce and maximizing substitution of wood.
- Creating a massive people's movement with the involvement of women for achieving the abovementioned objectives and to minimize pressure on existing forests.
- Ensuring environmental stability and the maintenance of the ecological balance, including atmospheric equilibrium, which are vital for sustenance of all lifeforms, (human, animal, and plant). The derivation of direct economic benefit must be subordinated to this principal aim.

2.5.2 *Essentials of Forest Management* [33,118]

 (i) Existing forests and forest lands should be fully protected and their productivity should be improved. The forest and vegetal cover should be increased rapidly on hill slopes, in catchment areas of rivers, lakes, and reservoirs and ocean shores and, on semiarid and desert tracts.

(ii) Diversion of good and productive agricultural lands to forestry should be discouraged in view of the need for increased food production.

(iii) For the conservation of total biological diversity, the network of national parks, sanctuaries, biosphere reserves, and other protected areas should be strengthened and extended adequately.

(iv) Provision of sufficient fodder, fuel, and pasture, especially in areas adjoining forest, is necessary in order to prevent the depletion of forests beyond the sustainable limit. Since fuel wood continues to be the predominant source of energy in rural areas, the program of afforestation should be intensified with a special emphasis on augmenting fuel wood production to meet the requirements of the rural people.

(v) Minor forest produce provides sustenance to the tribal population and to other communities residing in and around the forests. Such produce should be protected, improved, and their production enhanced with due regard to generation of employment and income.

(vi) In order to meet the growing needs for essential goods and services which the forests provide, it is necessary to enhance forest cover and the productivity of the forests through the application of scientific and technical inputs.

Production forestry programs, while aiming at enhancing the forest cover, and meeting the needs of country, should also be oriented to narrowing, by a certain set time target, the increasing gap between the demand and supply of fuel wood. No such program, however, should entail the clearfelling of adequately stocked natural forests. Nor should exotic species be introduced, through public or private sources, unless long-term scientific trials undertaken by specialists in ecology, forestry, and agriculture have established that they are suitable and have no adverse impact on the native vegetation and environment.

Wildlife Conservation

Forest Management should take special care of the needs of wildlife conservation, and forest management plans should include prescriptions for this purpose. It is especiallycrucial to provide "corridors" linking the protected areas in order to maintain genetic continuity between artificially separated subsections of migrant wildlife.

2.5.3 *Afforestation, Social Forestry, & Farm Forestry Programs* [27,33,44,92,107]

- A massive need-based and timebound program of afforestation and tree planting should be taken up with particular emphasis on fuel wood and fodder development, on all degraded and denuded lands including forest and non-forest land. It is necessary to encourage the planting of trees alongside roads, railway lines, rivers and streams and canals, and on other unutilized lands under State/corporate institutions or private ownership. Green belts should be created in urban/industrial areas as well as in arid tracts. Such a program will help to check erosion and desertification as well as improve the microclimate.

- Village and community lands, including those on foreshores and environs of ponds/tanks, not required for other productive uses, should be taken up for the development of tree crops and fodder resources. Technical assistance and other inputs necessary for initiating such programs should be provided by the competent authorities for the programs.

  The revenues generated through such programs should belong to the village council (panchayat) where the lands are vested with them; in all other cases, such revenues should be shared with the local communities in order to provide them an incentive. The vesting, in individuals, particularly from the economically weaker sections (like landless labour, small and marginal farmers, tribals, etc.) of certain ownership rights over trees, should be considered. The beneficiaries should be entitled to use fruits and output and made responsible for their security and maintenance.

- Economic motivation may be provided to individuals and institutions to undertake tree farming and grow fodder plants, grasses, and legumes on their own land. The degraded lands should be made available for this purpose either on lease or on the basis of a tree ownership scheme to public.

2.5.4 Diversion of Forest Lands for Non-forest purposes

(i) Forest land or land with tree cover should not be treated merely as a resource readily available to be utilized for various projects and programs, but as a country's asset which requires to be properly safeguarded for providing sustained benefits to the entire community. Diversion of forest land for any non-forest purpose should be subject to the most careful examinations by specialists from the standpoint of social and environmental costs and benefits. Construction of dams and reservoirs, mining and industrial development, and expansion of agriculture should be consistent with the needs for conservation of trees and forests. Projects which involve such diversion should provide funds in their investment budget for regeneration/compensatory afforestation.

(ii) New project or schemes which interfere with forests that clothe steep slopes, catchments of rivers, lakes, and reservoirs, geologically unstable terrain, and such other ecologically sensitive areas should be severely restricted and alternatives should be worked out.

  No new project in the forest land can be taken up without prior approval of competent authorities after all the design plans—including working methods and management/monitoring plans, assessment of benefits, and the environmental impact of the project—have been submitted to them. All possible measures to protect the forests from damage caused by the construction activities of the approved project must be taken by the construction agency under the supervision of forest authorities.

(iii) Beneficiaries who are allowed mining and quarrying on forest land and in land covered by trees should be required to repair and revegetate the area in accordance with established forestry practices. No mining lease should be granted to any party, private or public, without a proper mine management plan appraised from the environmental angle and enforced by adequate machinery. For mining purposes, additional conditions like maintaining a safety zone area, fencing, and regeneration, etc., and for major and medium irrigation projects, catchment area treatment plans are to be required.

(iv) Conditions for compensatory afforestation for diversion of forest land:

In order to make up for the loss of forest by diversion of forest land for non-forest projects, compensatory afforestation should be required. This would involve identification of non-forest land or degraded forest land, work schedule, cost structure of plantation, provision of funds, mechanism to ensure the utilization of funds, and monitoring mechanism, etc. The provision of funds for compensatory afforestation should be provided by the developer or project authorities.

The comprehensive scheme should include the details of non-forest/degraded forest area identified for compensatory afforestation, map of area to be taken up for compensatory afforestation, year-wise phased forestry operations, details of species to be planted. The details of the afforestation plan along with the cost structure and management and monitoring method have to be certified by competent authorities.

Land for compensatory afforestation should preferably be identified in the proximity of reserved forest or protected forest in the same district, but if the land is not available there, it should be identified elsewhere in the state and agreed and approved by competent authorities. In certain cases, the provision of compensatory forest area required may be twice the land transferred.

2.5.5 Forest-based Industries

The main considerations governing the establishment of forest-based industries and the supply of raw material to them should be as follows:

- As far as possible, a forest-based industry should source the raw material needed for meeting its requirements, preferably by the establishment of a direct relationship between the factory and the individuals who can grow the raw material by supporting the individuals with inputs including credit, constant technical advice, and finally, harvesting and transport services.

- No forest-based enterprise, except that at the village or cottage level, should be permitted unless it has been first cleared after careful scrutiny with regard to the assured availability of raw material. In any case, the fuel, fodder, and timber requirements of the local population should not be sacrificed for this purpose.

- Forest-based industries must not only provide employment to local people on a priority basis, but also involve them fully in planting trees and generating raw material.

- Natural forests serve as a gene pool resource and help to maintain the ecological balance. Such forests should not, therefore, be made available to industries for undertaking plantation and for any other activities.

- Farmers, particularly small and marginal farmers, ought to be encouraged to grow wood species required for industries on available marginal/degraded lands. These trees may also be grown along with fuel and fodder species on community lands (not on lands required for pasture purposes), and by forest department/corporations on degraded forests not earmarked for natural regeneration.

- Industries should be encouraged and financially motivated to use alternative raw materials.

2.6 Plant Tissue Culture [58,62]

In order to rapidly produce nursery plants for supplying the plants for fulfilling the requirements of tree plantation, the plant tissue culture technique can be employed. One of the most important aspects of tissue culture technology is the rapid regeneration of forest plants. The technology is briefly described below.

Tissue culture is a technique whereby small pieces of the viable tissue of the plant (*which are called 'explants'*) are isolated from parent plants and grown in a defined nutritional and controlled environment for a prolonged period under aseptic conditions (*conditions where decay or putrefaction is prevented*). For successful plant culture, it is best to proceed with some explant rich in undermined cells, for example, those from the cortex (*bark or skin of plant between epidermis and the vascular bundles*) or stem, because such cells are capable of rapid proliferation (*growth and extension by multiplication of cells/production of shoots that may become new plants*). The explants for tissue culture technique may be taken from leaves, buds, root apex, shoot apex, nodal segments, or germinating seeds. These explants are transferred onto the suitable culture media under aseptic conditions where they grow into an undifferentiated mass of cells called 'callus'. To develop a callus from the explant, it is essential that the nutrient medium employed should contain phytohormones such as auxins*-cytokines. The hormones vary for different tissue explants from different parts of the same plant and for the same explants from different genera of plants. The typical medium used in plant tissue culture technology is the *Murashige-scoog* (*MS*) medium, although other mediums are also employed. [*auxins — indole-3-acetic acid 'IAA'*]

Various Stages in Plant Tissue Culture

There are four stages in plant tissue culture, namely:

- Initiation of culture;
- Multi-application/Subculture stage;
- Development and differentiation of subculture;
- Test tube to field.

i. **Initiation of culture:** The fungal or bacterial spores in plant material are sterilized using chlorine water, mercuric chloride, or other sterilizing agents. *(The sterilization is essential because of the fact that the culture medium contains mineral salts, vitamins, sugars, phytohormones, etc. on which the microbial contaminants can grow faster than the culture itself and destroy the culture process.)* The culture is then incubated under controlled conditions at about 25°C in light.

ii. **Multi-application or the subculture stage:** After two or three weeks, the explants show visible growth by forming callus or differentiated organs like shoots, roots, or complete plants depending upon the composition of the medium. If it is desired to obtain a large number of plantlets from the callus, it is taken out and cut into small pieces and each piece of the callus is transferred to a fresh medium in separate test-tube where the piece of callus would develop into a big mass of callus. This type of subculturing may be repeated if required.

iii. **Development and differentiation of subculture:** The subcultured callus is now processed for further development and finally, for differentiation. The concentration of phytohormones in the medium are altered for inducing the callus to differentiate. A very high cytokine to auxins ratio induces shoot formation. This is referred to as *'organogenesis'*. Alternatively, a structure called an 'embryo' may develop if the media concentration is so altered. This process is called *'somatic embryogenesis'*. If the hormonal conditions are correctly balanced, the entire plantlet can be induced to grow on the culture medium. This process is called *'regeneration'*.

iv. **From test tube to field:** In the last stage when the fully grown plants are to be obtained, the test tube–rooted plantlets are first subjected to acclimatization so that they can adjust to the field condition when planted in the field. The plantlet is taken out of the rooting medium and is washed slowly with running water for an hour or more to ensure that no piece of agar is left on the surface of the plantlet. The plantlet is put to low minimal salt medium for 24–48 hours and then transferred to a pot that contains auto-sterilized mixture of clay soil, core sand, and leaf moults in 1:1:1 proportion. The pot containing the plantlet is covered generally with transparent polythene to maintain the humidity. These conditions are maintained for 15–20 days. The plantlet can then be transferred to soil and normal agricultural practices would be followed.

2.7 Rainwater Harvesting and Watershed Management for Benefit of Plantations [11,39,40,71,86,114,121]

The survival of plants in the initial stages is dependent on proper care and watering. If there is no water nearby, it becomes very difficult to fulfil the needs of watering. If a water body, lake, or pond does not exist nearby, rainwater harvesting and watershed management by storing the rainwater in pond should be done. The water so stored would serve the purpose of plantations in the initial stages and also fulfil the needs of the nearby community, animals and birds, and would help agriculture. It would also improve overall ecology of the location.

a. *Rainwater Harvesting*

Rural areas generally use traditional methods of rainwater harvesting. Their main motive of rainwater harvesting in these areas is to facilitate irrigation for agriculture and the use of water for domestic and drinking purposes. This practice is also useful for recharging groundwater. Many of the traditional structures created by communities include ponds, tanks, stepwells, covered tanks, etc., with the primary aim of tackling drinking water problems in arid and semiarid areas. More of the similar structures and facilities may be necessary to store rainwater for the use of irrigating newly planted trees in forestlands or other lands used for plantations. These facilities can also be used for agricultural crops and pisciculture.

b. *Drainage and Diversion of Run-off into Existing Surface Water Bodies*

Waterlogging is the cause of soil deterioration/soil salinity. Subsurface drainage in general improves the environment, but its impact on the regions downstream needs to be taken care off.

It requires construction of drains up to a pond/tank or existing waterbody/lake/pond/tank. The free flow of storm runoff into these tanks and water bodies must be ensured. The storm runoff may be diverted into the nearest tanks or depression, which will create additional storage.

Biodrainage holds very promising possibilities in controlling high groundwater levels as well as soil improvement. Biodrainage generally has a good effect on salt balance in the soil. Large-scale plantation of suitable plants/trees would utilize the excess water and salt nutrients.

c. *Cascading Water Use*

Cascading water use, that is, 'upstream surplus water supplies down streams requirement' is an old concept, especially prevalent in South India. Chains of tanks have been constructed traditionally (for example in Telengana region, Karnataka, Tamil Nadu in India) at regular intervals at various heights in the slopy terrain on the principle that surplus from the upper tanks fills the lower tanks, consequently filling the whole chain of tanks and also developing an overall soil moisture regime for sustaining vegetation throughout the year. Digging of tanks along the course of small rivers is quite successful in utilizing surplus floodwater (during monsoons) for vegetation during the dry season, thus serving the purpose of rainwater harvesting in the countryside. Special measures need to be taken so that the water reaches the tail-end regions but doesn't flood them. A focused view on all flow pathways, including pumped return flows, is to be taken. The minimization of water losses from channels and reservoirs is important. The management of water quality during this cascading process is also important.

d. *Watershed Management*

The principle of collecting and using precipitation from a catchment surface is a longstanding technology followed in some countries, which is regaining popularity. *For example, as early as the third millennium BC, farming communities in Baluchistan (now in Pakistan) and Kutch (India) impounded rainwater and used it for irrigation dams.*

A watershed, also called a drainage basin or catchment area, is defined as an area in which all water flowing into it goes to a common outlet. People and livestock are deeply connected to the functioning of the watershed and their activities affect the productive status of watersheds and vice versa. Additionally, the different phases of the hydrological cycle in a watershed are also dependent on the various natural features and human activities.

Hydrologically, a watershed is an area from which the runoff flows to a common point in the drainage system. Every water stream has an associated watershed, and small watersheds aggregate together to become larger watersheds. Water travels from the higher level to the downward location and joins a stream of similar strength, where it then forms a one order higher stream.

The stream order is a measure of the degree of stream branching within a watershed. Each length of stream is indicated by its order (for example, first-order, second-order, etc.). The start of a stream, with no other streams flowing into it, is called the first-order stream. Multiple first-order streams flow together to form a second-order stream, which flow into a third-order stream and so on. The stream order describes the relative location of the reach in the watershed. Identifying the stream order is useful to understand the availability of the amount of water in the reach and its quality. It is also used as a criterion to divide larger watersheds into smaller unit.

Moreover, the criteria for selecting the watershed size also depends on the objectives of the development and terrain slope. A large watershed can be managed in plain valley areas or where forest or pasture development is the main objective. In hilly areas or where intensive agriculture development is planned, the size of watershed preferred is relatively small.

e. *Advantages of Watershed management*
  i. Improves land productivity by extending the irrigation facility and by conserving the proper moisture content of soil all the year round.
  ii. Prevention of floods and water logging.
  iii. Controlling problems of drainage, salinity, and alkalinity.
  iv. Reducing soil erosion and reservoir siltation.
  v. Collection of surplus runoff and its use for irrigation or other purposes.
  vi. Recharging of groundwater and arresting the plummeting groundwater level.

f. *Engineering and management tools for the sustainable development of water resources in semiarid rural areas*
It is important from the point of view of tree planting as well as agroforestry that water resources should be developed properly and used most efficiently.
  i. Use of appropriate available technologies in the venture.
  ii. Catchment-based water resource planning and decentralized development system and watershed management.
  iii. Rainwater harvesting, creating ponds, reviving natural lakes, and ponds.
  iv. Integration of the social and economic development together with soil and water conservation along with plantation development.
  v. Local community participation in development, operation, and maintenance with adequate financial assistance from the government.
  vi. Productive use of water by utilization of microirrigation technologies like drip irrigation. (See **Notes** below).
  vii. Creating a suitable management information system for the development, operation, and maintenance.

Notes:
  i. *If the area falls under canal command, canal automation may be insisted upon so that the water reaches the tail-end regions. Wherever feasible, the canal may be covered by solar panels to reduce evaporation losses and to generate electricity for the local needs of community.*
  ii. *It would be better to plan to transfer water to arid/semiarid regions from water-rich regions to boost plantation as well as agriculture.*

iii. *For use during initial seeding, some patented products are available for use in arid and semiarid areas.* **Pusa Hydrogel**, *a patented product of the Indian Agricultural Research Institute (IARI), New Delhi, is an indigenous semi-synthetic super absorbent polymer and is said to have a remarkable ability of reducing the irrigation requirement of crops besides improving yield and quality. It comes in a powder form and is mixed with the seeds at the time of sowing. The powder converts into gel which absorbs water 300 times its size. The gel slowly releases water to keep the soil moist for the next 15 to 20 days. About 2 kg of the product would be required for one hectare of land.* [79]

iv. **Fog Harvesting:** *It has been suggested by certain scientists that even dense fog can be harvested in some areas in winters to provide water. The technology is simple. A fine nylon-mesh about 4x4 meter sizes is erected on steel frames. The fog droplets coalesce on the open mesh screen and flow because of gravity. This can yield up to 4 to 7 liters per square meter per day of potable water.*

2.8 Bioremediation and Phytoremediation of Soil [77,90]

2.8.1 Technique and Processes [53,65,91,106,123]

The contaminated soil or polluted soil requires remediation for quality productivity and sustainability. Remediation of contaminated soils can be done *in situ* by inoculating suitable microorganisms which consume/convert/degrade the pollutants, or by planting selected plants which take up the pollutants from soil. Both the processes are bioremediation processes, but the remediation by plants has now got a separate name designated to it, that is, 'phytoremediation'.

Phytoremediation may be applied wherever the land/soil or static water is polluted. Phytoremediation has been used successfully in the restoration of abandoned metal mine workings, reducing the impact of contaminants in soils, water, or air. Contaminants such as metals, pesticides, solvents, explosives, and crude oil and its derivatives can be mitigated by phytoremediation. Many plants such as mustard, rapeseed, species of amaranth, alpine pennycress, hemp, pigweeds, sunflower, brake fern (Pteris vittata), poplar, pine trees, etc., have hyper-accumulating properties vis-à-vis different toxic substances.

A number of processes are involved in phytoremediation, such as:

(a) *Phyto-sequestration:* The phytochemical complex formation in the root zone reduces the fraction of contaminant that is bioavailable. Transport protein inhibition occurs in the root membrane, that is, transport **proteins** associated with the exterior **root membrane** can irreversibly bind and stabilize contaminants on the **root** surfaces, preventing contaminants from entering the plant. Vacuolar storage in the root cells also occurs, where contaminants can be sequestered in the vacuoles of root cells.

(b) *Phytoextraction:* It refers to the uptake and concentration of substances from the environment into the plant biomass.

(c) *Phytostabilization:* It refers to reducing the mobility of substances in the environment, for example, by limiting the leaching of substances from the soil.

(d) *Phytotransformation:* It refers to the chemical modification of environmental substances as a direct result of the plant metabolism, often resulting in their inactivation, degradation (phytodegradation), or immobilization (phytostabilization).

(e) *Phytostimulation:* It refers to the enhancement of soil microbial activity for the degradation of contaminants, typically by organisms that associate with roots. This process is also known as rhizosphere degradation.

(f) *Phytovolatilization:* It refers to the removal of substances from the soil or water with release into the air, sometimes as a result of phytotransformation to more volatile and/or less-polluting substances.

(g) *Rhizofiltration:* It refers to filtering water through a mass of roots to remove toxic substances or excess nutrients. The pollutants remain absorbed or adsorbed to the roots.

Examples of the plants which can remediate soil and remove toxic metals are mentioned in Table 2.2. Also, see Para 2.2.8 for *Phytoremediation of Saline Soils by Use of Halophytes*.

2.8.2 *Bioremediation of Petroleum Polluted Soil* [20,91,95]

Some of the land which could have been used for agriculture or planting trees may be lying spoiled due to pollution by various types of petroleum products. In this connection, it is interesting to know that bioremediation of the soils polluted by oils/petroleum products and even oil sludge can be done by some microorganisms. The technique can be successfully used for restoration of such soils for renewable fuel production.

Microorganisms which can degrade petroleum components in the natural environment have been isolated. Examples of these organisms are:

- Nocardia, Pseudomonas, Acetobacter, Flavobacterium, Micrococcus, Anthrobacter Corynebacterium.

- A number of oil-degrading fungi, bacteria, and actinomycetes are found in soils irrigated by refinery wastewater. As per investigations carried out (ref # 38) on changes in the physio-chemical and microbiological status of agricultural soils irrigated with Mathura's oil refinery's effluent, it was reported that the irrigated soils with such an effluent would possess oily ferrous odour, alkaline pH, higher amount of TOC (say 0.319%), humus (0.43%), nitrogen (0.15 5%), oil and grease (0.024%), and presence of higher alkanes (C-17 to C-27) [98]. It was also reported that the oil-containing effluents induced a number of oil degrading "Fungi, Bacteria and Actinomycetes" in the soil such as: (i) *Aspergillus flavus,* (ii) *Cunninghamella elegans,* (iii) *Hormoconis resinae,* (iv) *Fusarium oxysporum,* (v) *Micrococcus flavus,* (vi) *Pseudomonas aeruginosa,* (vii) *Sarcina lutea,* (viii) *Bacillus subtilis,* (ix) *Achromo bacter* sp., (x) *Streptomyces* sp.

- Trichoderma application promotes plant growth and controls fungal phyto-pathogens such as *A. solani, R. solani,* and *E. oxysporum*. It is also useful for the bioremediation of diesel-contaminated soil. It uses diesel as a carbon source.

a. *Scientists Turned Waste Grease into Manure*

The scientists of Motilal Nehru National Institute of Technology, Allahabad, Prayagraj, UP, India have discovered new technique for the safe, economical, and ecofriendly disposal of waste grease. A consortium of four bacteria, harvested from soil contaminated with grease obtained from various sources, were used for converting the waste grease into green manure. The technology is cheap, quick, and hassle–free. The manure so obtained can be made available for use as solid or liquid product. The manure when mixed in soil has been found to be effective for a number of crops. [45]

b. *Examples of Microorganisms Useful for Remediation of Oil-Contaminated Soils*

Some other examples of microorganisms which can be used for remediation of oil- polluted soils are mentioned in Table 2.3 along with their suppliers.

2.8.3 *Some Examples of Remediation of Polluted Soils/Land*

a. Remediation of Polluted Soils with Pesticides/Herbicides/Toxic Organics

There are certain microorganisms which degrade such pollutants in the rhizosphere ecology of plants (roots). The rhizosphere enhances the population and activity of these microorganisms. In the rhizosphere, the diverse species of heterotrophic microorganisms get together at high-population densities, and may enhance the stepwise transformation of xenobiotics (like pesticides, polychlorinated biphenyls (PCBs), and herbicides (for example, Atrazine)) by consortia or provide a right environment conducive to genetic exchange and gene rearrangements.

Table 2.2. Removal of Various Toxic substances by Plants. [102,123]

Se No.	Contaminant	Plants for Phytoremediation	Remarks
1.	Aluminum	*Agrostis castellana* (Highland Bentgrass), *Hordeum vulgare* (Barley), *Vicia faba* (Horse Beans/Bankla)	Highland bentgrass removes Al, As, Mn, Pb, Zn
2.	Aluminum & silver	*Brassica napus* (Rapeseed plant), *Brassica juncea* (Indian Mustard)	Rape seed plant removes Al, Cr, Hg, Pb, Se, Zn
3.	Arsenic	*Agrostis castellana* (Highland bentgrass), *Pteris vittata* L. (Ladder brake fern or Chinese brake fern)	
4	Chromium	*Brassica napus* (Rapeseed plant), *Brassica juncea* (Indian mustard), *Helianthus annuus* (Sunflower) *Medicago sativa* (Alfalfa), **Amaranthus* (Amaranth), *Sutera fodina, Dicoma niccoliera, Leptospermum scoparium*	Indian mustard removes Cd, Cr, Cu, Ni, Pb, Pb, U, and Zn
5.	Copper	*Brassica juncea* (Indian Mustard), *Ocimum centraliafricanum* (Copper plant), *Helianthus annuus* (sunflower), **Amaranthus* (Amaranth)	Sunflowers remove Cu, Mn, Zn, Cd, U, Pb, Ni, Cs-137, Sr
	Manganese	*Brassica juncea* (Indian Mustard*), Helianthus annuus* (Sunflower), *Agrostis castellana* (Highland Bent Grass)	Highland bentgrass removes Al, As, Pb, Zn
6.	Lead	*Agrostis castellana* (Highland Bent grass*), Brassica oleracea (*Ornemental Kale and Cabbage, Broccoli*), Brassica napus* (Rapeseed Plant), *Brassica juncea* (Indian Mustard), *Thlaspi caerulescens/Brassicaceae,* (Alpine Pennycress or Alpine Pennycrest, Alpine Pennygras*s, Triticum aestivum* (Common Wheat), **Amaranthus* (Amaranth)	
7.	Selenium	*Chara canescens Desv. & Lois* (Muskgrass). *Brassica napus* (Rapeseed plant), *Brassica juncea* (Indian Mustard)	Rapeseed also removes Ag, Cr, Hg, Pb, Zn, while Indian mustard removes Cd, Cr, U Cu, Ni, Pb, and Zn
8.	Zinc	*Agrostis castellana (*Highland bentgras*s), Brassica napus (*Rapeseed*), Brassica juncea (*Indian Mustard*), Helianthus annuus (*Sunflower*), Thlaspi caerulescens/ Brassicaceae (*Alpine Pennycress Or Alpine Pennycres*t), **Amaranthus (*Amaranth*)	
9.	Nickel	*Alyssum argenteum, Thlaspi oxyceras, Alyssum davisianum, Alyssum eriophyllum, **Amaranthus (*Amaranth*), **Sebertia acuminate (a tree)*	**The Sebertia acuminata tree is a hyper-accumulator of nickel
10.	Cadmium	*Brassica juncea (*Indian Mustard*), Crotalaria juncea (*Sunn, Hemp*), Vallisneria americana (*Tape Grass*), Thlaspi caerulescens/Brassicaceae (*Alpine Pennycress or Alpine Pennycrest*), Helianthus annuus (*Sunflowe*r), **Amaranthus (*Amaranth*)	Crotalaria juncea (sun/hemp) is high in phenolics
11.	Cobalt	*Thlaspi caerulescens /Brassicaceae, (*Alpine Pennycress Or Alpine Pennycrest*),*	
12.	Cesium-137 (Cs-137) radioactive	*Pinus* sp. *(*Monterrey Pine Tree, Ponderosa Pine Tree*)*	Can remove Sr-90, MTBE Trichloroethylene, Petroleum hydrocarbons
13.	Strontium Sr-90 (radioactive)	*Liquidambar styraciflua* (American Sweet Gum Tree*), Liriodendron tulipifera* (Tulip Tree)	
14.	Uranium (radioactive)	*Oak tree, Amaranthus* (Amaranth), *Helianthus annuus* (Sunflower), *Brassica juncea* (Indian Mustard)	Amaranth can remove Cd, Cr, Ni, Cu, Pb, Zn

Table 2.2 contd. ...

...Table 2.2 contd.

Se No.	Contaminant	Plants for Phytoremediation	Remarks
15.	Benzene	*Ficus elastica (*Rubber Plant, Indian Rubber Bush)	
16.	Hydro-carbons	*Festuca Arundinacea* (Tall Fescue), *Pinus* sp. (Pine tree), *Cynodon dactylon* (Bermuda grass), Clover grass, Alfalfa, Popular tree, Juniper fescue	Poplar tree can remove trichloroethylene

*** Note on Amaranthus (amaranth):** The genus Amaranthus (an Indian sp.) is unique in having species which are used for grain, vegetable, and ornamental purposes. Amaranthus sp. can adapt to a wide range of climatic and soil conditions and have superior nutritional quality of grain with high protein (12 to 19%). It has complementary aminoacid profiles (lysine 5 to 7%), easily digestible starch, presence of cholesterol lowering fractions in the seed oil and high carotene (vitamin A) contents in leaves. It is cultivated in Nepal, Bhutan, Mexico, Ecuador, Bolivia, Peru, India, etc. (The grain amaranth is called by various names in India such as ramdana, marscha, ganhar, and lathe.) The plant sp. has been used by humans since prehistoric times (about 4800 BC). At least 10 of this plant species give grain. All amaranth species have the capacity of phytoremediation of polluted soils especially lead, nickel, cadmium, zinc, chromium, copper, and even uranium.*

****Note on the nickel hyper-accumulator, the 'Sebertia acuminata *tree':** The Sebertia acuminata tree is astonishing accumulator of nickel, to the extent of 20% of its dry body weight and the tree is not harmed by nickel. The tree is native of rain forest of New Caledonia, an island near Australia. It is possible to plant this tree in some regions of India for removing nickle from soils. [Environews July 1998].*

Nitroaromatic compounds are used as herbicides, explosives, pharmaceutical and industrial chemicals and many lands are polluted by them. It is environmentally important to remove them from the soil.

Phytoremediation is an environmentally desirable approach for remediating contaminated soils, which may work due to both the uptake and transformation by the plant as well as by the general rhizosphere effect, which enhances the microbial community capable of biodegrading the contaminants.

Biostimulation is a successful method for remediation of soils and other metrics contaminated with a wide range of agrochemicals (pesticides and herbicides) and other toxic organics. The use of appropriate soil amendments can enhance the biodegrading potential of indigenous soil microbial populations. Plant residues have biostimulation potential. For example, rice straw and ryegrass can stimulate the soil bacterial population and several enzyme activities when used for the remediation of herbicide-contaminated soils.

b. Arsenic Pollution of Soils and Remediation

The arsenic pollution of soil and of underground water is a big problem. Arsenate is a chemical analogue of macronutrient phosphate. Plants growing in arsenate-contaminated soils readily assimilate impermissible level of arsenate, which would appear in at least some quantity in the seeds like rice. Soil amendment by phosphatic fertilizers helps to curb the uptake of arsenic.

The pollution of soil by arsenic can be addressed by bioremediation by some plants like Agrostis castellana (Highland Bentgrass), *Pteris vittata* L. (Ladder Brake Fern or Chinese Brake Fern).

It may be pointed out that there are certain bacterial isolates of Gaeobacillus stearo-thermophilous which can oxidize inorganic arsenic 3 into less toxic arsenate very fast. These strains of bacteria can be used for bioremediation of arsenic.

c. *Phytoremediation of Soil Polluted with Deicing Agents*

Sometimes the soil is contaminated by 'deicing agents such as ethylene glycol or propylene glycol with some additives', which are used to remove and prevent ice and frost from accumulating on aircraft and airfield runways. Significant quantities of deicing fluids can spill on the ground and contaminate the soil and water and therefore their removal is environmentally important. Vegetation

Table 2.3. Bioremediation of Oil-Polluted Soil by Microorganisms. [91]

Microbial Consortium Required	Bioremediation Approach [20,73]	Company/Supplier Name
BiosolverR	Enhanced accelerated in situ remediation of HC_s, based contamination, contains water-based biosurfactant/bioremediation accelerant that offers superior desorption and/or solubilization of a wide range of contaminants in the soil matrix.	Westward Chemical Corporation Massachusetts
Micro-BacR	This product comprises of microorganisms targeting PAH_s, VOC_s, high molecular weight alkanes, PCB_s, and chlorinated compounds. A special combination of bacterial strains was adapted for this product to concentrate on remediating situations involving benzene, toluene, ethylbenzene, and xylenes (BTEX). (http://micro-bac.com/mission/)	Micro-Bac International Inc, Texas
MicrocatR-SH/SL	A concentrated blend of biodegradable surfactants in easy-to-use liquid (viscous) from, wetting, and solubilizing agents specifically formulated for bioremediation programs, which is diluted and sprayed over the soil or sludge to be remediated. Microcat-SH, for heavier asphaltic oils and Microcat-SL is for lighter hydrocarbons.	Bioscience Inc, Bethlehem
Fyrezyme	Biodegradable biostimulant utilized for rapid degradation of petrochemical contamination, contains no foreign bacteria, but rather stimulates indigenous bacteria to degrade such products.	Fyre Zyme Southeast, Inc., Florida
A⁺ Microbes B Microbes C Microbes D Microbes Z Microbes F Microbes	Provides a wide range of microbial blends that are acclimatized for various types of contamination; special and custom blends are available for unusual contamination.	Albaster Corp Inc.
Nzyme BW-5 Enzyme HC-6 Enzyme OC-7	Multipurpose bioremediation blend, odour controller, and biocleaner, contains enzymes and bacteria, and a solution with biodegradable detergents. Bacteria in this blend are able to sustain growth over a wide range of temperature from 4°C–45°C for a great variety of sources such as phenol, benzene, toluene, other aromatic hydrocarbons (HCs) with hydroxylated, nitrogenated groups, octane, lubricating oils, mineral oils, etc. Degrades and digests complex and otherwise non-biodegradable compounds such as HCs, detergents, phenols, oils, grease, papers, etc.	Enviroturn Systems, Canada
Oilzapper*	Bacterial consortium developed at TERI, New Delhi by mixing five bacterial strains, which could degrade aliphatic, aromatic, asphalt bitumen, and NSO (nitrogen, sulphur, and oxygen compounds) fractions of crude oil and oily sludge.	Strain Biotech Ltd., Hyderabad, India.

Note on Oil Zapper [91]

Microbial Biotechnology Lab. of TERI (India) with Indian Oil Limited has developed a unique bacterial consortium which degrades oily sludge and crude oil's aliphatic, aromatic, asphaltene and nitrogen, sulphur and oxygen components efficiently. This bacterial consortium has been named "Oil Zapper". These micro-organisms were obtained from hydrocarbon-contaminated sites using enrichment methods. The oil zapper consortium developed can be immobilized with a suitable carrier material, 'corn cob' (woody axis of a maize ear) which is biodegradable. The immobilized culture is put into sterile polybags and sealed aseptically (not liable to decay/ putrefy/poisoning) so that the consortium survives under ambient conditions. If the oil zapper along with nutrients is used, contaminated land (with oil/sludge) can be restored. The method has been tried at both the Mathura and Barauni oil refineries for the restoration of contaminated lands.

can reduce the volume of aircraft deicing agents in the environment. A mixed culture of cold-tolerant plant species and grasses (not trees as they would interfere with the working of the runway) could be planted alongside airport deicing areas and runways to help enhance the biodegradation of glycol-based deicing agents in the soil (which have potential to cause water pollution).

d. *Phytoremediation of Soil Polluted with Trichloroethylene*

Trichloroethylene has been widely used for a variety of purposes like anesthesia, dry cleaning agents, degreasing agents etc. This chemical is carcinogen. It has polluted water and thereby, soil, in many locations. Poplar species trees (especially the clones of Populus trichocarpa and Populus deltoids) can clean up soils as well as the underground water up to the reach of their roots. When the soil is cleaned up, the land can be used for other trees and crops.

e. *Phytoremediation of Soil Polluted with Benzo(a)pyrene & Hexa-chloro-biphenyl*

These chemicals are carcinogenic. The polluted land/soil by these chemicals can be remedied by planting tall fescue grass (Festuca arundinacea Screb.), which results in the increased mineralization of benzo(a)pyrene and hexa-chloro-biphenyl. The soil so remedied would become useful for planting trees as well as for agriculture.

f. *Phytoremediation of Soil Polluted with Hydrocarbons/Petroleum /Diesel oil*

Hydrocarbons, petroleum, and diesel oil, etc. can be removed from the soils by Festuca arundinacea (tall fescue), *Pinus* sp. (pine tree), Cynodon dactylon (Bermuda grass), clover grass, alfalfa, poplar tree sp., and Junniper fescue.

2.9 Summary

This chapter has been devoted to wood (a primary renewable energy source). In order to fulfil the ever-increasing need for wood, the availability of land is becoming a problem, because of the requirement of land for agriculture (for the production of food) and for development activities. All the waste lands or non-agricultural lands, vacant land on the sides of roads, canals, etc. should be made use of for plantations. The sustainable management of soil is discussed in detail. Measures required for encouraging the creation of plantations and examples of trees suited to various soils and climatic conditions have been mentioned. Agroforestry should be taken up on a large scale to fulfil the local needs of timber, fuel, fruits, etc. For fuel needs, the fast-growing trees suitable for a particular area/soil/ climate may be planted.

The restoration of sodic lands and mined out land should be done so that trees can be planted. Sodic land restoration should be done by suitable soil amendments (like biochar, cyanobacteria species such as Nostoc calcicola along with gypsum, organic manure, biofertilizers, compost, press mud from sugar factories, green manure plants). Salt-tolerant leguminous plants/ trees and useful grasses may be grown on the land. Soil organic carbon, exchangeable calcium, and the cation exchange capacity (CEC) increases with the time corresponding to plant growth, whereas, pH value, and exchangeable sodium percentage (ESP)/ exchangeable sodium decrease markedly. The land becomes fertile in a few years; any trees or crops can then be grown as per the requirement of the community. The land restoration of mined-out areas involves back filling, consolidation, grading, drainage, restoration of topsoil, slope stabilization, tier-type formation for higher dumps, top dressing with fertile soil, adding soil nutrients through organic manure, compost, biofertilizers, and then planting of suitable grasses of fodder value and leguminous varieties of plants/trees, which are tolerant to the metallic ingredients in the soil and can survive in the climate of the region.

For rapidly producing nursery plants for fulfilling the requirements of tree planting, the plant tissue culture technique can be employed.

Rainwater harvesting and watershed management may be very useful for the proper restoration and plantation in any area. Bioremediation and phytoremediation of soils is very useful for land restoration and plantation.

Forest conservation is an important issue from several points of view, including sustainably fulfilling the need for fuel wood, timber, other forest products, besides arresting the rise of CO_2 content in atmosphere. A positive policy is required for the conservation of forests and the replantation of degraded forest area in view of shrinking forest areas due to pressure on land requirements for developmental activities and growing more food.

In order to boost the supply of vegetable oil for the production of biodiesel, oilseed-bearing trees suitable for the specific area may be planted. Such trees may be grown on non-agricultural land and waste lands. Some examples of oilseed-bearing trees are: Oil Palm, Coconut, Pongamia pinnata, Azadirachta indica, Madhuca indica, Calophyllum inophyllum, Jatropha curcas, Soapnut Tree, Olive, etc.

ANNEXURE-B

Some Botanical Names of Trees

Bougainvillea, *Erythrina stricta* (**Madar**), *Cassia sophera* (**Kasunda**)
Clerodendrum infortunatum (**Bhant**), Behaya, *Lantana. linn/Lantana indica* (**Ghaneri**)
Murraya paniculate (**Kamni/Machula**), *Nerium indicum/Nerium oleander* (**Kaner**), Chandni
Quisqualis indica (**Rangoon Creeper**), Sadabahar
Prosopis juliflora/Prosopis cineraria (**Shami**)
Arythrina, Leucaena, etc. Azadirachta Indica (**Neem**)
Dalbergia sissoo (**Shisham**), *Pongamia pinnata* (**Karanja**), Bambusa
Peltephorum pterocarpum (**Gulmohar**)
Moringa oleifera (**Drumstick**), *Tamarindus indica* (**Tamarind/Imli**)
Cordia dichotama (**Lasoda**), Teak (**Sagwan**), *Shorea robusta* (**Sal/Sakhu**)
Ficus religiosa (**Aswattha/Peepal**), *Ficus rumphii* (**Paakar**), *ficus bengalensis* (**Bargad**)
Terminalia arjuna (**Arjuna**), *Bauhinia Variegata* (**Kachnar**), *Athocephalus indica* (**Kadamba**)
Pithecellobium dulce (**Jangaljalebi**)
Nyctanthes arbotristis (**Harshingar**), *Zizyphus nummularia* (**Jharberi**), *Zizyphus jujuba* (**Ber**)
Acacia nilotica/Acacia catechu, Acacia nilotica (**Kiker, Babool**), *Ficus rumphii* (**Pakkar**)
Ficus glomerata (**Gular**), *Bauhinia Variegata* (**Kachnar**), *Terminalia arjuna* (**Arjuna**)
Cassia fistula (**Amaltas**), *Diospyros melanoxylon/Diospyros tomentosa* (**Tendu**)
Emblica officinalis (**Aonla/Amla**), *Achras Sapotaceae* (**Sapota**)

QUESTIONS

1. *What is meant by fast-growing trees? Give examples of fast-growing trees.*

2. *Give examples of trees which can provide firewood in semiarid/arid zones.*

3. *Give examples of (a) plants suitable for hot arid and semi-arid zones, (b) vegetation in desert regions of temperate zones.*

4. *What are halophytes and what is their speciality? Give examples.*

5. *What is meant by sustainable management of soil? What are the causes of degradation of soils? How would you manage soil sustainably?*

6. *Enumerate the importance of the sustainable management of soils and briefly describe measures to be taken for the same.*

7. *How does the phosphorous cycle work on earth? What is the importance of phosphorous for plants?*

8. *Write short notes on the following, detailing their role in boosting plantations:*

   (a) biofertilizers, (b) compost, (c) plant-growth-promoting rhizobacteria (PGPR).

9. *Comment on 'Sewage water irrigation of plants in arid areas would be a boon.' What precautions are necessary for using sewage water for irrigation?*

10. *What are biopesticides and what are the advantages of using them?*

11. *Enumerate the advantages of using biofertilizers and compost instead of chemical fertilizers.*

12. *What is agroforestry? How is it different from forestry?*

13. *Comment on 'Soil fertilization by agro forestry'. Which type of plants should be selected for agroforestry?*

14. *What are the causes of the excessive alkalinity of soil? How would you rehabilitate sodic land? What is the role of agroforestry in this?*

15. *What are the plant species which can thrive on saline and salt-affected land and also reduce soil salinity?*

16. *What are podsol areas/(acidic soil areas)? How would you treat acidic soils?*

17. *What measures are necessary to boost plantations? Enumerate the areas which may be utilized for plantation drives.*

18. *What are the problems with mined-out areas for use of plantations? Describe the procedure for restoration of mined-out areas for plantations.*

19. *What is meant by the phytoremediation and bioremediation of soils? How does the phytoremediation and bioremediation of soil work? Give examples.*

20. *What is the importance of phytoremediation and bioremediation for the restoration of mined-out areas and/ or other areas with polluted soils?*

21. *What should be the essentials of a sustainable forest policy particularly in a developing country? Write a short note on compensatory plantation in the context of the utilization of forest land for non-forest purposes.*

22. *What is plant tissue culture? How would it be useful in boosting plantations?*

23. *Write a short note on the following measures detailing their importance for boosting plantations: (i) rainwater harvesting (ii) watershed management*

References

1. Acacia (https://en.wikipedia.org/wiki/Acacia).
2. Adriel, Ferreira da Fonseca, Uwe Herpin, Alessandra Monteiro de Paula, Reynaldo Luiz Victória and Adolpho José Melfi. March/April 2007. Review: Agricultural Use of Treated Sewage Effluents: Agronomic and Environmental Implications and Perspectives for Brazil. Sci. Agric. (Piracicaba, Braz.). 64(2): 194–209. https://doi.org/10.1590/S0103-90162007000200014.
3. Ali Khan, M.A. and K. Kashyap. 2007. Utilization of agroindustrial waste inoculated by Trichoderma viride and quality of mature compost monitored by C:N and bioassay test. Journal Agropadology.
4. Ali Khan, M.A. and Priti Kaushik. Dec. 2005. Indigenous technology for ecofriendly utilization of distillery spent effluent as fertilizer-irrigation in crops: sustainable development. Third Int. Conf. on Plants and Environmental Pollution. ISEB & NBRI, Lucknow, India.
5. Appleton, Bonnie, Vickie Greene, Aileen Smith, Susan French, Brian Kane, Laurie Fox et al. 2015. Trees and Shrubs that Tolerate Saline Soils and Salt Spray Drift. Publication 430–031. Virginia Cooperative Extension, Virginia State University.
6. Bagchi, Dr. (Mrs.) K.S. and Dr. S.S. Bagchi. July 1991. History of Irrigation in India, I – Irrigation in Ancient India (2295 BC – 11 Century AD). Irrigation and Power Journal. 48(3).
7. Behl, H.M. October 2000. Organic Cultivation. Environews.

8. Berglund, S. and P.L. Hermite (Eds.). June 1983. Utilisation of Sewage Sludge on Land: Rates of Application and Long Term Effects on Metals. Commission of European Communities. Proc. of a Seminar held at Uppsala. D. Reidel Publishing Company, Dordrecht.

9. Biosalinity Awareness Project. Listing of Halophytes & Salt-Tolerant Plants. (http://www.biosalinity.org/salt-tolerant_plants.htm).

10. Biosalinity Awareness Project. Understanding the impact of salinization and implications for future agriculture. Current Research & Development. http://www.biosalinity.org/current%20research.htm.

11. Borthakur, Saponti. October 2009. Traditional rain water harvesting techniques and its applicability. Indian J. of Traditional Knowledge 8(4): 525–530.

12. Brayshaw, T.C. 1996. Artemisia tridenta Nutall (Big Sagebrush). Trees and Shrubs of British Columbia. Page 337 (ISBN 9780774805643). University of British Columbia Press, Vancouver.

13. Breman, Hank and Jan-Joost Kessler. 1995. Woody Plants in Agro-Ecosytems of Semi-Arid Regions. Springer Verlag.

14. Bundela, D.S., Sethi, Madhurama, R.L. Meena, S.K. Gupta and Sharma, D.K. Sept. 2015. Remote Sensing and GIS for Performance Assessment of the Western Yamuna Canal System for Sustainability. Water and Energy International 72(9): 54–65.

15. Burton, M. 2010. Irrigation Management, Principles and Practices. CABI. UK.

16. Casuarina trees. [http://www.worldagroforestry.org/treedb/AFTPDFS/Casuarina_equisetifolia.PDF].

17. Chapman-King, R., T.M. Hinckley and C.C. Grier. 1986. Growth responses of forest trees to wastewater and sludge applications. *In*: Cole, D.W., Henry, C.L. and Nutter, W.L. (Eds.). The Forest Alternative for Treatment and Utilization of Municipal and Industrial Wastes. University of Washington Press, Seattle, WA.

18. Chandra, K. Organic Manures. National Centre of Organic Farming. https://ncof.dacnet.nic.in/Training_manuals/Training...in.../Organicmanures.pdf.

19. Chand, S., M. Anwar and D.D. Patra. 2006. Influence of long-term application of organic and inorganic fertilizer to build up soil fertility and nutrient uptake in mint mustard cropping sequence. Communications in Soil Science and Plant Analysis. 37: 63–76. Supply system on soil quality restoration in a red and laterite soil. Archives of Agronomy and Soil Science 49: 631–637.

20. Chandra, S., R. Sharma, K. Singh et al. 2013. Application of bioremediation technology in the environment contaminated with petroleum hydrocarbon. Ann. Microbiol. 63: 417–431. https://doi.org/10.1007/s13213-012-0543-3.

21. Chopra, R.N., S.L. Nayar and I.C. Chopra. 1956. Glossary of Indian medicinal plants. CSIR, India.

22. Costanza, Robert and Herman E. Daly. March 1992. Natural capital and sustainable development. Conservation Biology. 6(1): 1–155. https://doi.org/10.1046/j.1523-1739.1992.610037.x.

23. Dan Ogle. February 2010. Plants for saline to sodic soil conditions. Technical Note. USDA-Natural Resources Conservation Service Boise, Idaho – Salt Lake City, Utah. (https://www.nrcs.usda.gov/Internet/FSE_PLANTMATERIALS/publications/idpmstn9328.pdf).

24. Department of Land Resources. 2003. Guidelines for Hariyali. http://dolr.nic.in/ Hariyali: Guidelines.htm. DOLR, Ministry of Rural Development, Government of India, New Delhi, India.

25. Deserts. [https://en.wikipedia.org/wiki/Desert] and [https://en.wikipedia.org/wiki/Sonoran_Desert].

26. Dhawan, Gitanjali. Dec. 2005. Vermi-composting as an Eco Tool in management of some aquatic weeds. 3rd Int. Conf. on plants and environmental pollution. ISEB & NBRI, Lucknow, India.

27. Dongre, Prakash and Nagindas Khandwala. 2011. Role of Social Forestry in Sustainable Development - A Micro Level Study. Int. J. of Social Sciences and Humanity Studies 3(1). ISSN: 1309-8063.

28. Dutta, Venkatesh. April 2002. (Tata Energy Research Institute New Delhi). Bio-remedy for oil pollution. Science Reporter. CSIR, New Delhi.

29. Eastern Cottonwood: a Top 100 Common Tree in North America. [https://www.thoughtco.com/eastern-cottonwood-tree-overview-1343200].

30. Environews. October 2009, July 2011, Jan 2014, and Jan 2015. NBRI, International, Society of Environmental. Botanists.

31. Environews. 4/1998, 7/1998, 7/1999, Oct 1999, 7/2000, Oct 2002, July 2003, Jan 2005. ISEB, NBRI, Lucknow, India.

32. Eucalyptus Growth Rate. [http://www.angelfire.com/bc/eucalyptus/eucgrowth.html].

33. Evans, Julian, 1988. Plantation Forestry in the Tropics- Tree Planting for the Industrial, Social Environmental and Agroforestry Purposes. Clarendon Press Oxford.

34. Fischer, R.A. and Neil C. Turner. 1978. Plant productivity in the arid and semiarid zones. Ann. Review, Plant Physiology. 29: 277–317.

35. Giant Sequoia or Sequoiadendron giganteum. [https://en.wikipedia.org/wiki/Sequoiadendron_giganteum].

36. Gillis, M., D.H. Perkins, M. Roemer and D.R. Snodgrass. 1992. Economics of development, Ed. 3, pp.xvi + 635pp., Publisher : W.W. Norton & Company, Inc. New York.
37. Gold, Michael A. and James W. Hanover. Agroforestry systems for temperate zones. file:///C:/Users/rkant/Desktop/Agroforestry%20Systems%20for%20the%20Temperate%20Zone.pdf.
38. Golberg, Jay. The Best Fast-Growing Trees to Use for Firewood. [https://www.ehow.com/list_7302666_fast_growing-trees-use-firewood.html].
39. Government of India. 1994. Guidelines for Watershed Development. New Delhi, India. Department of Land Resources, Ministry of Rural Development, Government of India.
40. Government of India. 2008. Common Guidelines for Watershed Development Projects. National Rain-fed Area Authority. Ministry of Land Resources, Government of Andhra Pradesh, India.
41. Gupta, Sapna and Ruchi Seth. Feb. 2015. Plant Growth Promoting Rhizobacteria as a tool for sustainable. 5th Int. Conf. on plants and environmental pollution. ISEB & NBRI, Lucknow, India.
42. Harper, Kimball T. and James R. Marble. A role for nonvascular plants in management of arid and semiarid rangelands. Vegetation science applications for rangeland analysis and management pp. 135–169. Chapter Part of the Handbook of vegetation science book series (HAVS, volume 14). [https://link.springer.com/chapter/10.1007/978-94-009-3085-8_7].
43. Hasanuzzaman, Mirza, Kamrun Nahar, Md. Mahabub Alam, Prasanta C. Bhowmik, Md. Amzad Hossain, Motior M. Rahman et al. 2014. Potential Use of halophytes to remediate saline soils. review article. BioMed Research International, Article ID 589341, 12 pages. [http://dx.doi.org/10.1155/2014/589341; (https://www.hindawi.com/journals/bmri/2014/589341/].
44. Highway mission, National green. Plantation across highways. NGHM.
45. Hindustan Times Dated 6th Nov 2015. [kasandeep@hintustantimes.com].
46. [http://www.ehow.com/list_7302666_fast_growing-trees-use-firewood.html].
47. http://www.grahamandrews.com/growing_trees_for_firewood.html.
48. Jain R.K., Singh Bajarang, K.P. Tripathi and Srivastava Neeta. 2002. Reclamation of a Sodic Soil through Afforestation with *Azadirachta indica* and *Pongamia pinnata*. Journal of the Indian Society of Soil Science. 50(1): 147–148.
49. Jat, R.S. and I.P.S. Ahlawat. 2006. Direct and residual effect of vermicompost, biofertilizers and phosphorus on soil nutrient dynamics and productivity of chickpea-fodder maize sequence. Research Reviews, Practices, Policy and Technology. J. of Sustainable Agriculture 28(1): 41–54. https://doi.org/10.1300/J064v28n01_05.
50. Jeyabal, A. and G. Kuppuswamy. November 2001. Recycling of organic wastes for the production of vermicompost and its response in rice–legume cropping system and soil fertility. European J. of Agronomy, 15(3): 153–170. https://doi.org/10.1016/S1161-0301(00)00100-3.
51. Kant, Rajni. 1997. Waste management in thermal power station. Water and Energy R&D conf., Vadodara. CBIP, India.
52. Kim, K.D. and E.J. Lee. July 2005. Potential tree species for use in the restoration of unsanitary landfills. Environ Manage 36(1): 1–14.
53. Kruger, Ellen L., Todd A. Anderson and Joel R. Coats. Phytoremediation of Soil and Water Contaminants. ACS Symposium series 664. American Chemical Society. Washington, DC.
54. Kumar, Akhilesh and S.K. Tewari. Feb. 2015. Utilization of Industrial Wastes for Cultivation of East Indian Lemongrass in Sodic Soil. (SV/P-10). Vth Int. conf. on plants and environmental pollution. ICPEP-5, NBRI Lucknow, India.
55. Kumar, Akhilesh and Nandita Singh. 2014. First report of *Maconellicoccus hirsutus* Green infestation on *Jatropha curcas* saplings. Phytoparasitica. 42(1): 71–73. DOI 10.1007/s12600-013-0339-4.
56. Kumar, Anil, Abhishek Niranjan, Namrata Pandey, Jai Chand, S.K. Sharma, Alok Lehri and S.K. Tiwari. Feb 2015. Effect of soil condition on oil content, chemical composition and carbon isotope ratio of lemon grass grown on sodic and non-sodic soil. 5th Int. Conf. on plants and environmental pollution. ISEB & NBRI, Lucknow, India.
57. Kumar, V., S. AlMomin, A. Al-Shatti, H. Al-Aqeel, F. Al-Salameen, A.B. Shajan and S.M. Nair. 2019. Enhancement of heavy metal tolerance and accumulation efficiency by expressing Arabidopsis ATP sulfurylase gene in alfalfa. Int. J. Phytoremediation. 1: 1112–1121. doi: 10.1080/15226514.2019.1606784.
58. Kumar, Sunil and M.P. Singh. 2009. Plant Tissue Culture. APH Publishing Corporation, New Delhi.
59. Kumar, Sushil. April 1996. Neutralization of soil Acidity in parts of Sikkim. Symposium Earth Sciences in Environmental. Assessment and Management. GSI, Lucknow, India.
60. Lal, R. 1995. Sustainable Management of Soil Resources in the Humid Tropics. United Nations University.
61. Landfill Restoration and Aftercare. 1999. Landfill Manuals, Environmental Protection Agency, Wexford, Ireland.

62. Low cost options for tissue culture technology in developing countries. August 2002. Collection of Research Papers), Proc. of a Technical Meeting organized by the Joint FAO/IAEA. Division of Nuclear Techniques in Food and Agriculture and held in Vienna. IAEA-TECDOC-1384.

63. Magdoff, F.R. and R.R. Weil. 2004. Soil Organic Matter in Sustainable Agriculture. CRC Press.

64. Mahmoud Nasr. 2016. Utilization of Treated Wastewater and Sewage Sludge in Forest Ecosystems. Forest Research: Open Access ISSN: 2168-9776. Nasr, Forest Res. 5(3). DOI: 10.4172/2168-9776.1000e123.

65. Malti, Subodh Kumar. Feb. 2007. Bioreclamation of coalmine overburden dumps--with special emphasis on micronutrients and heavy metals accumulation in tree species. Environ Monit Assess. 125(1-3): 111–22. doi: 10.1007/s10661-006-9244-3. Epub 2006 Dec 16.

66. Management and Conservation of Tropical Acid Soils for Sustainable Crop Production. March 1999. Proc. Of a joint meeting of FAO/IAEA, Division of Nuclear Techniques in Food and Agriculture, held in Vienna.

67. Manousaki, Eleni and Nicolas Kalogerakis. 2011. Halophytes Present New Opportunities in Phytoremediation of Heavy Metals and Saline Soils. Ind. Eng. Chem. Res. 50 (2): 656–660. DOI: 10.1021/ie100270x.

68. Manousaki, Eleni and Nicolas Kalogerakis. 2011. Halophytes—An emerging trend in phytoremediation. Int. J. of Phytoremediation. 13(10): 959–969. | https://doi.org/10.1080/15226514.2010.532241.

69. Meena, Kusum Narwal, Nayan Tara and Baljeet Singh Saharan. 2018. Review on PGPR: An alternative for chemical fertilizers to promote growth in aloevera plants. Int. J. Curr. Microbiol. App. Sci. 7(3): 3546–3551. https://doi.org/10.20546/ijcmas.2018.703.407.

70. Minglin, L., Z. Yuxiu and C. Tuanyao. 2005. Identification of genes up-regulated in response to Cd exposure in *Brassica juncea* L. Gene. 363: 151–158. doi: 10.1016/j.gene.2005.07.037.

71. Mishra, Prasanta K., M. Osman, Satendra and B. Venkateswarlu (Eds.). July 2011. (Collection of papers). Techniques for Water Conservation & Rainwater Harvesting for Drought Management. SAARC Training Program, Central Research Institute for Dryland Agriculture, Hyderabad, SAARC Disaster Management Centre, New Delhi and Indian Council of Agricultural Research, New Delhi.

72. Prasad, M.N.V. (University of Hyderabad) July 2011. Dept. of plant sciences. Biological re-cultivation of industrial deserts or lunarscapes' Enviro News.

73. Mondal, Tanushree, Jayanta Kumar Datta, Naba Kumar Mondal. April 2017. Chemical fertilizer in conjunction with biofertilizer and vermicompost induced changes in morpho-physiological and bio-chemical traits of mustard crop. J. of the Saudi Society of Agricultural Sciences 16(2): 135–144. https://doi.org/10.1016/j.jssas.2015.05.001.

74. Nabati, J., A. Masoumi; M.Z. Mehrjerd, M. Kafi, A. Nezami and P.R. Moghaddam. 2011. Effect of salinity on biomass production and activities of some key enzymatic antioxidants in kochia (kochia scoparia). Pakistan Journal of Botany; ISSN 0556-3321; 43(1): 539–548.

75. Nainwal, R.C., D. Singh, R.S. Katiyar, S.S. Tripathi, S.K. Sharma, S. Singh et al. 2016. Rehabilitation of sodic waste land through agroforestry system. Int. J. of Plant and Environment. 2(1& 2): 29–35.

76. Nair, P.K.K., P.K. Shaji and T. Alexander. April 2011. Eco-development in the context of environmental concerns. Environews. Int. Society of environmental botanists, NBRI, Lucknow, India.

77. Naofumi Shioi (Ed.). April 2018. Advances in Bioremediation and Phytoremediation. Open access peer-reviewed Edited Volume. DOI: 10.5772/67970.

78. National Forest Policy (India). 1988. No. 3-1/86-FP Ministry of Environment and Forests (Department of Environment, Forests & Wildlife).

79. News in Hindustan times. dated 22.06.2016. about Pusa Hydogel. Counter Droughts. Indian Agricultural Research Institute (Indian Council of Agricultural Research), New Delhi.

80. Özlem Bariş, Fikrettin Sahin, Metin Turan, Furkan Orhan. September 2014.Use of Plant-Growth-Promoting Rhizobacteria (PGPR) Seed Inoculation as Alternative Fertilizer Inputs in Wheat and Barley Production. Communications in Soil Science and Plant Analysis.45(18):2457–2467. DOI:10.1080/00103624.2014.912296.

81. Pandey, Adarsh. Jan. 2010. Antagonism of Two *Trichoderma* Species against *Alternaria alternata* on *Capsicum frutescens*. Journal of Experimental Sciences. 1(5): 18–19.

82. Pearce, David Edward Barbier and Anil Markandya. March 2000. Sustainable Development-Economics and Environment in the Third World. London, eBook Published November 2013 (eBook ISBN9781134159062), 230 pages.

83. Populus variety of trees. (https://en.wikipedia.org/wiki/Populus).

84. Qadir, M., D. Steffens, F. Yan and S. Schubert. May/June 2003. Sodium removal from a calcareous saline–sodic soil through leaching and plant uptake during phytoremediation. LDD-Land Degradation and Development. 14(3): 301–307. https://doi.org/10.1002/ldr.558.

85. Rahi, T.S., Kripal Singh, Bajrang Singh, and Lal Bahadur. 2012. Effect of Leguminous and Non-leguminous Species on Amelioration of Sodic Soil. Technofame-A Journal of Multidisciplinary Advance Research 2(1): 74–86. Restoration Ecology and Soil Science Group, CSIR- NBRI Lucknow. India.

86. Rajgor, Gail. May/June 2006. Revamped and rain water recycling. Refocus.

87. Rees, R.M., B.C. Ball, C.D. Campbell and C.A. Watson (Eds.). 2001. Sustainable Management of Soil Organic Matter. (Collection of papers). CABI Publishing.

88. Report from CIMAP. 7 July 2011. Lucknow. Times of India.

89. Repetto, Robert and Malcolm Gillis (Eds.). 1988. Public Policies and the Misuse of Forest Resources. A World Resources Institute Book. Cambridge University Press.

90. Rhodes, Christopher J. 2013. Applications of bioremediation and phytoremediation. Science Progress 96(4): 417–427. DOI: 10.3184/003685013X13818570960538.

91. Role of Microbes in Bioremediation: A review by Partha Bandyopadhyay, Arup Tiwary and Bidhan C. Patra. In Environmental Contaminants and Bioremediation by Arvind Kumar. APH Publishing Corp. New Delhi.   [https://books.google.co.in/books?id=FN8mifeULHcC&dq=Microbes+named+as+MicrocatR+SH/ SL+for+bioremediation+of+soils&lr=&source=gbs_navlinks_s].

92. Sanmat, sustainable development Partner to UN. "planting". equatorinitiative.org.

93. Sardrood, S.N.E., Y. Raei, A.B. Pirouz and B. Shokati. 2013. Effect of chemical fertilizers and bio-fertilizers application on some morpho-physiological characteristics of forage sorghum. Int. J. of Agronomy and Plant Production. 4(2): 223–231. http://eprints.icrisat.ac.in/id/eprint/11527.

94. Sharma, Arun K. Arid Zone agro forestry: Dimensions and Directions for Sustainable Livelihoods. Central Arid Zone Research Institute Jodhpur, India. (file:///C:/Users/rkant/Desktop/arid%20zone%20agroforestry. pdf.

95. Sharma, Neeta, Madhu Prakash Srivastava and Nupur Srivastava. Dec. 2010. Remediation of Cr-VI of Tannery effluent using Trichoderma., 4th Int. Conf. on plants and environmental pollution. ISEB & NBRI, Lucknow, India.

96. Sharma, D.K. and Singh, A. 2015. Salinity research in India-achievements, challenges and future prospects. Water and Energy International 58(6): 35–45.

97. Singh, Anand Narain, Akhilesh Singh Raghubanshi and J.S. Singh. June 2002. Plantations as a tool for mine spoil restoration. Current Science 82(12):1436–1441.

98. Saien, J. and F. Shahrezaei. 2012. Organic pollutants removal from petroleum refinery wastewater with nanotitania photocatalyst and UV light emission. Int. J. Photoenergy. 2012: 1–5.

99. Saini, H.S. and S.A.I. Mujtaba. 1996. Salinity hazards in southern part of Gurgaon, Haryana, Geol. Surv.Ind. Special. Pub. 48(2): 95–99.

100. Shampa, Dutta, Jayanta Kumar Datta and Narayan C. Mandal. December 2017. Evaluation of indigenous Rhizobacterial strains with reduced dose of chemical fertilizer towards growth and yield of mustard (*Brassica campestris*) under old alluvial soil zone of West Bengal, India. Annals of Agrarian Science. 15(4): 447–452. https://doi.org/10.1016/j.aasci.2017.02.015.

101. Sheoran, V., A.S. Sheoran, and P. Poonia. 2010. Soil reclamation of abandoned mine land by revegetation: a review. International Journal of Soil, Sediment and Water: 3(2): Article 13. https://scholarworks.umass.edu/ intljssw/vol3/iss2/13.

102. Singh, R.P. and M. Agrawal. 2007. Effects of sewage sludge amendment on heavy metal accumulation and consequent responses of Beta vulgaris plants. Chemosphere. 67(11): 2229–2240, Pergamon.

103. Singh, Satyendra Pratap, Arpita Bhattacharya, Shipra Pandey, Richa Shukla, Poonam C. Singh, Aradhana Mishra et al. Feb. 2015. Dynamic role of *Trichoderma* spp. In food security and altering soil microbial community in diesel fuel spiked soil using sole –carbon –source utilization profiles. 5th Int. Conf. on plants and environmental pollution. ISEB & NBRI, Lucknow, India.

104. Singh, Shweta, L.K. Sharma, Vijendra Chaturvedi, S.K. Sharma, Devendra Singh, R.C. Nainwal et al. Feb. 2015. Evaluation of diverse plant materials in sodic soil environment. 5th Int. Conf. on plants and environmental pollution. ISEB & NBRI, Lucknow, India.

105. Singh, Sukhvinder et al. Dec. 2005. The role of Lantana camara in the improvement of wheat crops in wheat rice cropping system. Int. Conf. On Plant and Environmental. Pollution. NBRI, Lucknow, India.

106. Singh, Veenus, K.D. Pandey, Shatrughna Singh and Durg Vijay Singh. Feb. 2015. Cyanobacterial modulated changes and its impact on bioremediation of Usar Sodic soils. 5th Int. Conf. on plants and environmental pollution; ICPEP-5, NBRI, Lucknow, India.

107. Social forestry in India from Wikipedia. https://en.wikipedia.org/wiki/Social_forestry_in_India.

108. Soni, D.K. April 2018. Heavy Metal Emission from Thermal Power Plant: Effect on Environment. 312 pages. ISBN-10: 620207311X. LAP LAMBERT Academic Publishing.

109. Srivastava, S., V.K. Mishra, Y.P. Singh and D.K. Sharma. Feb. 2015. Agro forestry: An approach towards carbon sequestration under sodic condition. 5th Int. Conf. on plants and environmental pollution. ISEB & NBRI, Lucknow, India.

110. Srivastava, Nishant K., C. Ram Lal and R. Ebhin Masto. October 2014. Reclamation of overburden and lowland in coal mining area with fly ash and selective plantation: A sustainable ecological approach. Ecological Engineering 71: 479–489. https://doi.org/10.1016/j.ecoleng.2014.07.062.

111. Stern, David I., Michael S. Common and Edward B. Barbier. July 1996. Economic growth and environmental degradation: The environmental Kuznets curve and sustainable development. World Development 24(7): 1151–1160. https://doi.org/10.1016/0305-750X(96)00032-

112. Temperate Desert. (http://environmentalsciencebiomes.weebly.com/temperate-desert.html) and (http://homeguides.sfgate.com/temperate-desert-plants-64858.html).

113. Ten fastest growing trees & plants in the world. April 2014. [http://www.conservationinstitute.org/10-fastest-growing-trees-plants-in-the-world/].

114. Tiwari, K.K., U.N. Rai, S. Dwivedi, H.N. Dutta and N.K. Singh. Nov./Dec. 2005. Fog water harvesting: assessment of quality and its potentiality over northern India (SVI/P-73). Proc. 3rd Int. Conf. on Plants and Environmental Pollution (ICPEP-3). ISEB &NBRI Lucknow, India.

115. Top 10 Fuel Trees for Zone 5 and Above - Permaculture Reflections (https://www.permaculturereflections.com/top-10-fuel-trees-for-zone-5-and-above/.

116. Trees good for fire wood (http://www.ehow.com/list_6523319_trees-good-firewood_.html).

117. Trees and Shrubs for Saline Land (http://agriculture.vic.gov.au/agriculture/farm-management/soil-and-water/salinity/trees-and-shrubs-for-saline-land).

118. Townsend, C.R., M. Begon and J.L. Harper. 2003. Essentials of Ecology No. Ed. 2, pp.xix + 530 pp. Blackwell Science, Oxford.

119. Tripathi, Nimisha, Raj Shekhar Singh. November 2008. Ecological restoration of mined-out areas of dry tropical environment, India. Environmental Monitoring and Assessment. 146(1–3): 325–337.

120. Verma, M., S. Verma and S. Sharma. 2018. Eco-friendly termite management in tropical conditions. pp. 137–164. *In*: Khan, M. and Ahmad, W. (Eds.). Termites and sustainable management. Sustainability in Plant and Crop Protection. Springer, Cham. https://doi.org/10.1007/978-3-319-68726-1_6.

121. Wani, S.P., P. Pathak, and P. Singh. Efficient Management of Rainwater for Increased Crop Productivity and Groundwater Recharge in the SAT. Int. Crops Research Institute for the Semi-Arid Tropics (ICRISAT), Patancheru, Andhra Pradesh, India. [http://www.iwmi.cgiar.org/assessment/files/word/proposals/abstracts/paperwani.doc.].

122. Warriner, Doreen. Economics of Peasant Farming. Routledge, Taylor and Francis Group, Imprint ROUTLEDGE, London.

123. Wikipedia, the free encyclopedia. Phytoremediation, Hyper accumulators. [https://en.wikipedia.org/wiki/List_of_hyperaccumulators].

CHAPTER 3

Alcohols

3.1 Alcohols and their Sources

a. *Alcohols*

Alcohols are the hydroxyl derivatives of alkanes like methane, ethane, propane, butane, etc.; they are formed by replacement of one or more hydrogen atoms (H) by the hydroxyl group (OH). The alcohol containing one OH group is called monohydric alcohol, two OH groups is called dihydric alcohol, and three OH groups is called trihydric alcohol. The examples of these three types of alcohols are given below:

Monohydric Alcohol:

- (i) Methyl alcohol or Methanol (CH_3OH),
- (ii) Ethyl alcohol or Ethanol (C_2H_5OH),
- (iii) Propyl alcohol or Propanol ($CH_3\ CH_2\ CH_2OH$),
- (iv) Butyl alcohol or butanol ($CH_3\ CH_2\ CH_2\ CH_2\ OH$)
  Dihydric Alcohol: Glycol ($CH_2OH\ CH_2OH$)
  Trihydric Alcohol: Glycerol ($CH_2OH\ CHOH\ CH_2OH$) (also called glycerine)

b. *Use of Alcohol as Fuel*

Throughout history, alcohols have been used as fuel. The first four aliphatic alcohols (methanol, ethanol, propanol, and butanol) are of interest as fuel because they can be synthesized chemically or produced biologically.

Various alcohols are used for different purposes, but two alcohols namely *ethyl alcohol* or ethanol **(C_2H_5OH)** and *butyl alcohol* or *butanol* **(C_4H_9OH)** evoke interest as potential sources of fuel for vehicles. These alcohols (derived from biological sources) are called bioethanol and biobutanol, respectively

These alcohol-fuels can be derived from crops containing sugar or starch. They can also be derived from lignocellulosic materials like agricultural wastes, forest wastes, and other biomass. *The alcohol can be formed by the biological fermentation of sugars with the help of enzymes.* (For Butanol see Para 3.11 of this chapter.)

c. *Various Sugars*

1. Sugars are divided into two major groups, namely monosaccharides and oligosaccharides.
2. Monosaccharide sugars, like pentoses and hexoses, cannot be hydrolyzed into smaller molecules.
3. Oligosaccharides can be disaccharide or trisaccharide.
4. The disaccharide group sugars can be hydrolyzed into two molecules of monosaccharide sugars, while trisaccharide group sugars can be hydrolyzed into three molecules of sugars.

5. The disaccharide group of sugars include cane sugar (sucrose), malt sugar (maltose), and milk sugar (lactose) to name a few.

6. An example of trisaccharide sugar is 'raffinose' which is formed by the combination of one molecule of glucose, one molecule of fructose, and one molecule of galactose. Raffinose comprises a fairly substantial amount (5 to 8%) of beetroot and cotton seed, and in small quantities in many foods. It is fermentable by certain bacterial enzymes. (It is not hydrolyzed by enzymes present in the human gastrointestinal track, but is fermented by bacterial enzymes in the colon producing gas (biogas)).

7. *Pentoses $(C_5 H_{10} O_5)$ include xylose, which is also called wood sugar and occurs in bran, straw, wood gum, arabinose (which occurs in 'gum Arabic' from the plant), and ribose (which occurs in the plant nucleic acid).

8. **Hexoses $(C_6 H_{12} O_6)$ include sugars like 'glucose', 'galactose', 'idose', 'altrose' 'mannose', 'fructose', 'sorbose, 'tagatose', 'osazone', etc., but predominantly, glucose and fructose.

9. Glucose and fructose are present in a free state in many fruits.

10. Galactose does not occur in a free state, but is widely distributed in a combined state. Galactose combine together to form lactose, that is, milk sugar. Galactose is a constituent of galactans, which occurs in certain sea algae and seaweeds.

11. Mannose does not occur in a free state, but is a constituent of a polysaccharide named 'mannan', which occurs in ivory nuts/seeds of the tagua palm. It is also a constituent of albumin, globulins/protein, mucoid (mucous-like substances), etc.

d. *Enzymes*

Enzymes are catalysts of biological origin that accelerate various reactions without themselves undergoing any apparent change during the course of action. Chemically, enzymes are globular proteins, but some enzymes are associated with some non-protein groups called prosthetic groups. If this group is a metal group, it is called a cofactor, but if this group is a small organic molecule, it is called a coenzyme. Enzymes are highly specific to a particular reaction and it is difficult to replace them by any other enzyme. They work best at a certain temperature, called optimum temperature, for specific reactions. Enzymes are denatured at very high temperatures and their activity is greatly inhibited at low temperatures.

(1) The *maltase enzyme* in yeast converts malt sugar 'maltose' (obtained from starch) into glucose. Maltose is a disaccharide group of sugar obtained from starch by treatment with hot malt. [See para 3.4.2].

(2) The **xymase enzyme* contained in yeast converts 'monomeric sugars (glucose and fructose' into ethanol.

(3) The ***diastase enzyme* contained in malt (germinated barley) helps to convert starch into maltose sugar by hydrolysis.

e. *Ethanol*

Ethanol or ethyl alcohol (C_2H_5OH) is a very important chemical derived from plants and can also be used as a fuel. The molecular structure of ethanol is as follows.

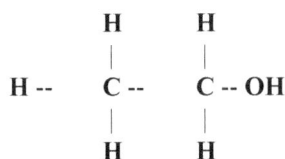

```
      H       H
      |       |
H --  C --  C -- OH
      |       |
      H       H
```

Most conveniently, ethanol is obtained from food crops having sugar content, like sugarcane and sugar beet. The substances that contain fermentable sugars include sugarcane, dates, beets, grapes, etc., and left over by-products from sugar manufacture from sugarcane, like molasses.

The starting material for making alcohol can be starch-containing grains such as barley, wheat, maize, rice, or other materials such as potatoes, etc. Efforts are afoot to make use of cellulose and biomass for the production of ethanol.

3.2 Ethanol from Sugarcane Molasses

Approximate composition of sugarcane is:

(i) Water 70–80 %, (ii) Fiber 12–18 %, (iii) Sugar 12–14 %, (iv) Non-sugar organic and inorganic material 2–3 %. From one tonne of sugarcane, one can get about: 110–114 kg of crystal sugar; 22–24 kg sugar in molasses; 132–138 kg total sugar (sucrose); 145 kg of dried bagasse (the weight of 'undried bagasse' is about 30% of the initial weight of sugarcane). All sugar (132–138 kg) if converted to ethanol, will yield about 72 kg of ethanol.

The energy content of sugarcane is approximately: total sugar 30%; bagasse 35%; and leaves, stems, tips, etc., 35%. Bagasse heat can be used for drying, distillation, and producing low-pressure steam for the process of electricity generation. Bagasse can also be used for fiber- based industries.

The first step in the process of making alcohol from cane sugar, that is, 'sucrose', is hydrolysis (*reaction with water*), which converts sucrose into glucose and fructose.

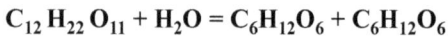

$$C_{12}H_{22}O_{11} + H_2O = C_6H_{12}O_6 + C_6H_{12}O_6$$

Cane sugar + Water = Glucose + Fructose

The hydrolysis is helped by the enzyme 'invertase' found in yeast. The second step, therefore, is the addition of 'yeast', and the maintenance of the temperature of tank containing a solution of glucose, fructose, etc., at about 30°C. The yeast contains various enzymes like 'invertase', 'maltase'*, 'zymase'** etc.

**Zymase*

$$C_6H_{12}O_6 = 2 C_2H_5OH + 2CO_2$$

The ethanol is separated from the broth by distillation.

Molasses usually contains enough nitrogenous matter to act as food for yeast microorganisms during fermentation. If the nitrogen content of the molasses is poor, it may be fortified by the addition of ammonium sulphate or ammonium phosphate. A flow diagram for making ethanol from molasses is shown in Figure 3.1a.

3.3 Ethanol from Food Crops

Apart from sugarcane and sugar beet, alcohol can be produced from any 'starch' contained in corn, seeds, potatoes, sweet potatoes/materials, as already stated.

For preparation of alcohol from grains (such as wheat), the grain is mashed with hot water and heated with 'malt' (germinated barley) at 50°C for 1 hour. The 'malt' contains the enzyme 'diastase'***, which helps in the conversion of starch $[(C_6H_{10}O_5)n]$ into a sugar called maltose $[C_{12}H_{22}O_{11}]$ by hydrolysis. When 'yeast' is added to the maltose solution and allowed to ferment, it is converted into ethanol. Yeast contains an enzyme called 'maltase'*, which converts maltose sugar into glucose and the enzyme 'zymase'** contained in yeast converts glucose into ethanol, which can be extracted by distillation. A flow diagram for making ethanol from potatoes/starch is shown is Figure 3.1b. The chemical reactions are shown below:

***Diastase*

$$2(C_6H_{10}O_5)n + nH_2O = nC_{12}H_{22}O_{11}$$

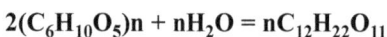

Starch + water = Sugar 'Maltose'

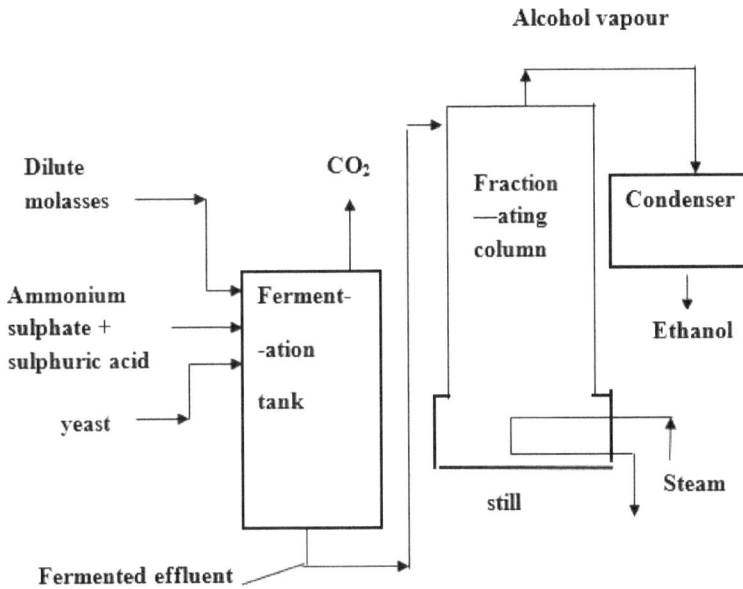

Fig. 3.1a. Making Ethanol from Molasses.

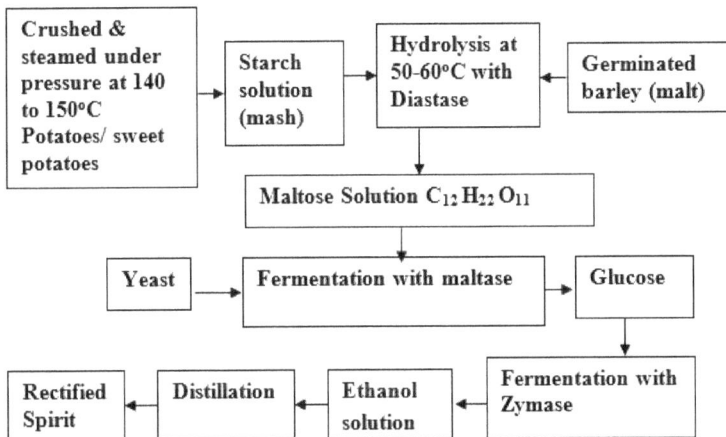

Fig. 3.1b. Making Ethanol from Potatoes.

*** Maltase**

$$C_{12}H_{22}O_{11} + H_2O = 2C_6H_{12}O_6$$

Maltose sugar + Water = Glucose

****Zymase**

$$C_6H_{12}O_6 = 2(C_2H_5OH) + 2CO_2$$

Glucose = Ethanol + Carbon dioxide

   Average yield of Alcohol from various Agricultural Crops: The probable average yield from some edible agricultural crops is given in Table 3.1a.

Note: *Producing alcohol for use as fuel from food crops is not advisable for countries deficient in food as it would require a vast amount of resources, including fertile agricultural land, water, fertilizer, etc. and would adversely affect the food security of the country.*

Table 3.1a. Average yield of Alcohol from Crops.

Crop	Crop yield Tonnes/hectare/year	Alcohol yield litres/ tonne	Alcohol yield liters/hectare/ year
1. Sugarcane	40–120	70	2800–8400
2. Sugar beet	10–40	120	1200–4800
3. Maize	1–4	400	400–1600
4. Sweet sorghum	20–60	55	1100–3300
5. Sweet potato	10–40	125	1250–5000
6. Cassava	10–40	180	1800–7200

3.4 Biomass to Alcohol

3.4.1 Lignocellulosic Materials [7,12,27,38,49,60,65,102]

Efforts are being made world over to produce liquid fuel from cellulosic material. The production of alcohol from lignocellulosic materials does not require extra land and has no adverse effect on food security. The 'lignocellulosic feedstock' includes materials like wheat straw, rice straw, corn stover, sweet sorghum stalks, crop field trashes such as stalks, clipping, corn cobs, grasses like switch grass (*Pancium virgatum*—one of the perennial grasses), hay, etc. Even green seaweed like *Ulva fasciata* can be used for the production of ethanol through enzymatic hydrolysis by using the enzyme 'cellulase' produced from the marine fungus '*Cladosporium sphaerospermum*'.

It may be mentioned that the stalks of sweet sorghum yield 15 to 20% fermentable sugar and therefore, it is a good source of biomass for making alcohol. Sweet sorghum can be grown in semiarid areas as it needs less water. Though commercial and the large-scale production of alcohol from sorghum has not been realized yet to its full potential, a few sugar-producing mills and corporates such as 'Tata Chemicals' in India have started production. Organizations such as ICRISAT and the National Research Centre for Sorghum have developed open pollinated varieties and photo-period sensitive hybrids, enabling year-round production.

The production of cellulosic ethanol is also more energy efficient than the production of grain-based ethanol, which requires extra fuel like natural gas. Any reduction in the fuel requirement for the process of making ethanol results in a decrease in the greenhouse gas emissions. In the production of biomass, the atmospheric carbon dioxide is fixed to form many carbon compounds/complexes. Cellulose produced by plants is composed of both lightly packed regions containing large voids and other irregularities as well as tightly packed crystalline regions. *On a life-cycle basis, ethanol produced from agricultural residues or dedicated cellulosic crops has significantly lower greenhouse gas emissions and higher sustainability.*

Major Components of Biomass

There are three major components that are present in biomass, namely '*cellulose', '**hemicellulose', and '***lignin', whose characteristics are mentioned below:

1. ***Cellulose:** Pure cellulose* has the molecular formula $(C_6 H_{10} O_5)_n$. It is a polysaccharide having a different structure from starch, which is also a polysaccharide with the same molecular formula. A pure form of cellulose is cotton. Cellulose is not acted upon by the enzyme 'amylase' present in the human digestive juice. It is acted upon by the enzyme cellulase, which is produced by some bacterial molds.

2. ****Hemicellulose:** It is also a polysaccharide with a different structure. Hemicellulose can be hydrolyzed by hot dilute acids. (*It is not acted upon by human digestive juices, but can be acted upon by some bacteria present in the large intestine*). Hemicellulose is present along with the 'lignin' in the plant material.

3. ***Lignin:** It is an organic polymer, mainly aromatic in nature, containing reactive groups like carbonyl (=C=O), carboxylic (O=C-OH), hydroxyl (OH), methoxy groups (-O-CH$_3$). Lignin is a part of the plant cell wall and contributes to the structural rigidity of the plant. It is responsible for the resistance of cell wall to microbial degradation. Lignin is a polyphenolic polymer containing phenylpropane structural units. (*It is not a polysaccharide; when fiber containing lignin is eaten it is not acted upon in the human digestive track, but has a property of adsorbing 'cancer-causing hydrocarbons' present in food or produced in the colon by bacterial action.*)

It may be noted that cellulose, hemicellulose, and lignin in the lignocellulosic materials are not present separately, but in the form of complexes.

Uses of Lignin

The 'lignin' is either burnt to generate process steam or converted to value-added products such as dispersing agents, animal feed binders, concrete additives, drilling mud additives, and soil stabilizers. It is a part of dietary fiber in association with cellulose and hemicellulose.

Lignin is a valuable raw material for some valuable products. The lignin can be precipitated from liquor by acidifying it with either carbon dioxide or sulphuric acid. The flue gases from the boiler/furnace contain carbon dioxide and the cleaned flue gas can be used here. The precipitate (*difficult and slow to filter out*) is washed with water and dried. This lignin thus separated can be used for the manufacture of important organic compounds like phenol, vanillin **[CH$_3$CHCH(NH$_2$) CO$_2$H]** and related products, organic sulphur compounds, etc.

As lignin contains phenolic compounds, it could be used as a substitute in phenol-formaldehyde resins/phenol-epoxy resins. (*The properties of extracted crude lignin depend on the method of its extraction from original biomass; for example, if the lignin is extracted by the organosol process (solvent extraction), it can be directly used as a substitute in the formaldehyde resins as it has good curing properties*). To increase the reactivity, lignin can be chemically modified by reactions with methylated phenols. Lignin can be used as an adhesive for fiberboard production.

Composition of Various Lignocellulosic Materials

The contents of cellulose, hemicellulose, and lignin of various lignocellulosic materials are mentioned in Table 3.1b.

Table 3.1b. Percent Cellulose, Hemicellulose, Lignin Contents of Various Lignocellulosic Materials. [79]

Lignocellulosic materials	% Cellulose	% Hemicellulose	% Lignin
1. Hardwood stems	40–55%	24–40 %	18–25 %
2. Soft wood stems	45–50%	25–35%	25–35%
3. Nut shells	25–30%	25–30%	30–40%
4. Corn Cobs	45%	35%	15%
5. Grasses	25–40%	35–50%	10–30%
6. Newspapers	40–55%	25–40%	18–30%
7. Paper	85–99%	0%	0–15%
8. Switch grass	45%	31.4%	12.0%
9. Coastal Bermuda grass	25%	35.7%	6.4%
10. Wheat straw	30%	50%	15%
11. Leaves	15–20 %	80–85%	0%
12. Cotton seed hairs	80–95%	5–20%	0%

3.4.2 Production of Ethanol from Biomass [15,27,28,42,46,99,121]

For converting the biomass to 'ethanol', the biomass would have to be converted to a useable fermentation feedstock (*typically some form of sugar**). This can be achieved by a variety of different process technologies. These different processes for making feedstock for fermentation (*i.e. feedstock containing sugar formed from cellulosic materials*) constitute the critical differences among the bioethanol production technology options. Fermenting the sugary feedstock using biocatalysts (*microorganisms, including yeast and bacteria*) for making ethanol is one of the oldest form of biotechnologies developed by humankind. Further processing the fermented product yields fuel-grade ethanol and by-products that can be used to produce other fuels, chemicals, heat and/or electricity. It may be mentioned here that the ethanol production from one ton of dry wood would be approximately 0.23 ton, while lignin, which is recovered, would be about 0.39 ton and the balance would consist of carbon dioxide evolved and residue, etc.

Generally, the steps in the production of alcohol from lignocellulose/biomass consist of pretreatment (*which may consist of milling of biomass, prehydrolysis, and washing*), hydrolysis by acid/enzyme, and then washing and fermentation, and product purification (which would consist of distillation, recovery of alcohol, and processing/disposal of the balance) as shown in Figures 3.2a and 3.2b.

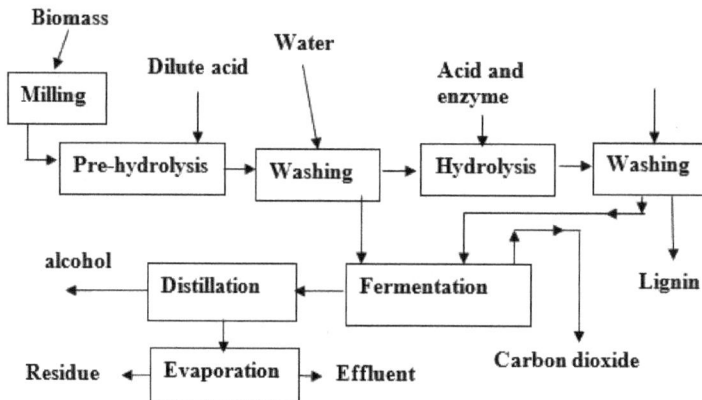

Fig. 3.2a. Production Process of Ethanol from Lignocellulosic Materials.
(Separate processes for pre-treatment, hydrolysis, fermentation and purification of product)

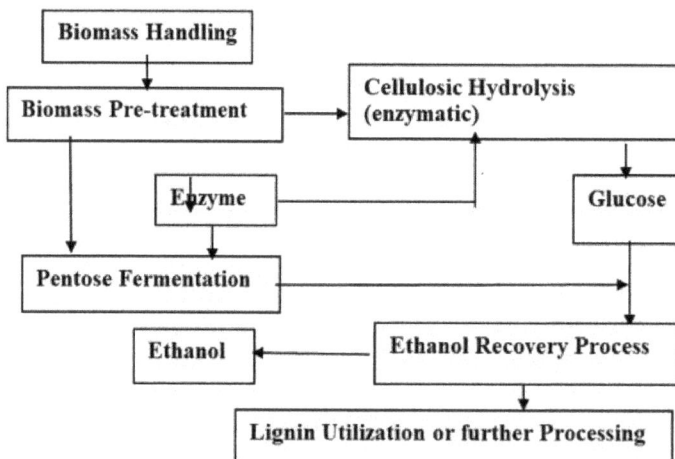

Fig. 3.2b. Biochemical Process for Production of Cellulosic Ethanol.

Main Steps in the Production of Alcohol from Biomass

The main steps in the production of alcohol from biomass are:

- Pretreatment
- Hydrolysis
- Fermentation
- Product purification

a. *Pretreatment of Biomass* [123]

The typical levels of cellulose, hemicellulose, and lignin in the biomass may be taken as: cellulose 30 to 55%, hemicellulose 25 to 50%, and lignin 15 to 30%. The goal of the pretreatment of biomass prior to the manufacture of liquid fuel from it is:

- o Breakdown of the sheathing of 'Hemicellulose and Lignin'.
- o Disrupting the crystalline structure of cellulose to make it more accessible for 'enzymes' to act on it for converting it into 'sugars'.

A *'good pretreatment'* should meet the following requirements:

- o Reduction of biomass to reactive cellulose.
- o Avoidance of the formation of possible inhibitors for hydrolytic enzymes and fermenting microorganisms.
- o Utilization of 'cheap chemicals'.
- o Recovery and recycling of 'chemicals' as far as possible.
- o Reduction of production 'costs'.

The *'reactive cellulose'* may typically include:

- o Cutting and milling
- o Irradiation by gamma (γ) rays, electron beam, microwave, etc.

The thermo-chemical method may include: (i) High-pressure steaming, steam explosion, (ii) extrusion, (iii) pyrolysis, (iv) hydrothermolysis.

In the 'steam explosion process', the cellulosic material, usually in the form of chips, is saturated with water at a pressure of 21–35 kg/cm^2 at a temperature of 215–260°C. When the pressure is released, the water evaporates rapidly and the wood fibers tend to separate, that is, explode, and hence, it is called 'steam explosion process'. In addition to the separation of fibers, acetic acid is formed, which acts as a catalytic agent for the hydrolysis of hemicellulose.

b. *The chemical and physicochemical pretreatments may comprise*

- o Treatment by acids like sulphuric acid, hydrochloric acid, or phosphoric acid
- o Treatment by alkalis like sodium hydroxide or ammonia
- o Treatment by ammonium sulphite
- o Treatment by gases like chlorine dioxide (ClO_2), nitrogen dioxide (NO_2), sulphur dioxide (SO_2), ozone (O_3), or chemicals like hydrogen peroxide (H_2O_2)
- o Explosion by steam, carbon dioxide, or sulphur dioxide
- o Solvent extraction of lignin may be by:
  - Ethanol + Water Extraction
  - Benzene + Water Extraction
  - Ethylene Glycol Extraction
  - Butanol + Water Extraction

Biological Pretreatment

Pretreatment can be affected biologically by fungi and '*actinomycetes'. These microorganisms have enzymes which help in separating lignin from the source substance.

Several enzymes can be employed for pretreatment such as 'esterase'. The phenolic acid released in the process must be continuously removed as it is toxic for the microbes which inhibit the enzymes. Several fungi have two types of enzymatic systems for degradation, namely:

(i) The hydrolytic extracellular system, which produces hydrolases responsible for the degradation of polysaccharides.

(ii) The oxidative extracellular ligninolytic system responsible for the degradation of lignin and the opening of phenyl rings in its structure.

The fungal pretreatment has a low cost and energy requirement, and no need of chemicals, but suffers from the disadvantage of being very slow to act.

Pretreatment of Wheat Straw by '<u>White Rot Fungi</u>' [100]

The pretreatment of wheat straw for bioethanol production by 'white rot fungi namely, *Pleuotus ostreatus*' and chemical pretreatment by acid and alkali leads to the significant removal of lignin. The pretreated straw is then saccharified using *Aspergillus ellipticus* and *Aspergillus fumigates* for further conversion to ethanol.

[***Actinomycetes** are remarkable soil bacteria supplying most of our antibiotics, fixing nitrogen, and decomposing organic matter, cellulose, and lignin. These are a group of gram-positive bacteria. They comprise some of the most common soil life and play a vital part in the organic matter turnover and carbon cycle. They replenish the supply of nutrients in the soil and are an important part of humus formation.]*

Other Methods of Pretreatment: [57,77,123]

(i) Pretreatment by 'supercritical liquid carbon dioxide',

(ii) Pretreatment by 'supercritical hot water',

(iii) Pretreatment by 'wet oxidation by high-pressure water at high temperatures of 170–200°C and oxygen.

A comparison of various pretreatment methods is given in Table 3.1c.

c. Hydrolysis of Pretreated Biomass

The hydrolysis can be done by mineral acids, enzymes, etc. Hydrolysis converts carbohydrates polymers $(C_6 H_{10} O_5)_n$ like cellulose and starch, into monomeric sugars (glucose, fructose).

i. Hydrolysis by Dilute Acids

Hydrolysis occurs in two stages to maximize sugar yields from the hemicellulose and cellulose fractions of biomass. The first stage is operated under milder conditions to hydrolyze hemicellulose, while the second stage is optimized to hydrolyze the more resistant cellulose fraction. *Liquid hydrolyzates* are recovered from each stage, neutralized, and fermented to ethanol.

Hydrolysis by dilute acid is carried out in the following steps:

• Feedstock preparation

• Prehydrolysis of hemicellulose to recover hemicellulose sugars (pentose and hexose) before they are degraded at a higher temperature (> 180°C), during the acid hydrolysis of cellulose

• Hydrolysis of cellulose

• Neutralization

• Clean up/filtration, cooling followed by the fermentation process of sugars

The dilute acid hydrolysis is already a practice in the case of some 'pulp and paper mills'. They utilize a dilute acid hydrolysis process to separate hemicellulose and lignin from wood, and produce specialty cellulose pulp for papermaking. There is some amount of hexose sugars in the spent

Table 3.1c. A Comparison of Various Pretreatment Methods. [65,77,99]

Pre-treatment methods	Main effects	Advantages	Disadvantages and Limitations
Concentrated acid	Hydrolysis of both cellulose and hemicelluloses	High yield of sugar, operation temp. is lower, no need of enzymes	Acid has to be recovered
Dilute acid	Hydrolyzes hemicellulose; cellulose rendered more amenable for enzymatic hydrolysis	Lesser corrosion problems than with concentrated acid	Higher operation temp. and lower sugar in exit stream than with conc. acid, generation of degradation products due to higher temp.
Alkali	Removes lignin & hemicellulose; increases accessible surface area	Good digestibility, better removal of lignin	Long residence time, the salts formed are not recoverable
Ionic liquids	Reduces crystallization of cellulose; removes lignin	Good digestibility, green solvents	Generally not being used
Steam explosion	Causes lignin transformation and hemicellulose solubilization	Cost effective, higher yield of glucose and hemicellulose	Generation of inhibitory compounds; partial degradation of hemicellulose, incomplete disruption of the matrix of lignin and carbohydrates
Biological pretreatment	Degrades lignin and hemicellulose	low cost, and low energy consumption,	Slow rate of hydrolysis
Supercritical fluid technology	Increases the accessible surface area	no formation of inhibitors, enhances enzymatic hydrolysis of cellulose	High energy requirement
Wet oxidation (hot water & oxygen at temp. (170–200°C)	Lignocellulose is broken into solubilized hemicellulose and cellulose-rich fractions (also forms carboxylic acids), removes lignin; it is decomposed into CO_2, H_2O, and Carboxylic acids	Reduces energy requirement as the process is exothermic due to oxygen at a temp. above 170°C; minimum formation of inhibitors	High cost because of the use of oxygen and high temperatures

Note: It may be observed that the pretreatment of wheat straw by steam explosion at about 215°C seems to be the most successful technology, followed by the treatment by hydrogen peroxide at about190°C, supercritical water at about 190°C, and acid at about 190°C.

sulphite liquor, which can be fermented to ethanol. The lignin is either burnt to provide process heat or utilized to make value-added products.

ii. Concentrated Acid Technology

The use of concentrated acid in biomass processing decrystallizes the cellulose. This is followed by dilute acid hydrolysis into sugars. The separation of acid from sugars, acid recovery, and acid reconcentration follow, which is succeeded by the fermentation of sugars.

Two companies in the United States (Arkenol and Masada) are working to commercialize this technology. Arkenol holds a series of patents vis-à-vis the use of concentrated acid to produce ethanol. In view of the advantageous opportunities of obtaining rice straw (a cheap feedstock) to converting into ethanol, Arkenol has located the facility in Sacramento County. This technology further improves the economics of raw straw conversion by allowing for the recovery and purification of 'silica' present in the 'rice straw'.

The Masada Resource Group holds several patents related to the conversion of municipal solid waste (MSW) to ethanol. Their plant located in Middletown, NY, US, processes the lignocellulosic fraction of municipal solid waste into ethanol using technology based on the concentrated sulphuric acid process. The process is well suited to complex and highly variable feedstocks like municipal solid waste.

   iii.   Enzymatic Hydrolysis [10,65,72]

In the separate hydrolysis and fermentation process, the acid hydrolysis is replaced by hydrolysis of woody matter by application of an enzyme. The fermentation is done separately thereafter.

   An important process modification made for the enzymatic hydrolysis of biomass was the introduction of simultaneous saccharification and fermentation (SSF), which has now been improved to include the 'co-fermentation' of multiple sugar substrates. In the SSF process, enzyme and fermenting microbes are combined. As sugars are produced, the fermentative organisms convert them to ethanol. As the present high cost of 'cellulase enzymes' is a prominent barrier to the economical production of bioethanol from lignocellulosic material, ongoing research is focused on achieving greater reduction in the cost of these enzymes. (Enzymatic hydrolysis is used in a plant of Logen Petro Canada, Ottawa.)

   It may be mentioned that the acid, alkaline, or fungal pretreatment of lignocellulosic biomass can be saccharified enzymatically to get fermentable sugars like glucose. Bacteria and fungi are good sources of enzymes like cellulases, xylanases, hemicellulases, and mannanases. There is a group of microorganisms like Clostridium, Cellulomonas, Trichoderma, Penicillum, Neurospora, Fusarium, Aspergillus, etc., which show celluloytic and hemicellulolytic activity and are highly capable of fermenting monosaccharides (like glucose). The enzymic hydrolysis can have glucose yield of 70–85%. The enzymatic hydrolysis may be performed at about 70°C, but takes a much longer time than hydrolysis by acid (more than 24 hours).

Cleaning of the Substrate

During the pretreatment and hydrolysis process, some degradation products of C5 sugar [pentose* $(C_5H_{10}O_5)$] and C6 sugars [**hexose $(C_6H_{12}O_6)$] are formed. These include furfural, hydroxylmethyl furfural (HMF), and weak acids from acid pretreatment and hydrolysis. These degradation products are toxic and inhibit enzymatic hydrolysis and fermentation. Several chemicals and biological methods have been devised to remove these inhibitors and increase the hydrolyzate fermentability. Recycling of the acid is a possible way of reducing costs.

d.   Fermentation of Carbohydrates in Hydrolysed Biomass

Under anaerobic conditions, many microorganisms may ferment carbohydrates in the product obtained after the hydrolysis of biomass to ethanol. The microorganisms may be bacteria, yeast, or fungi.

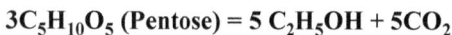

$C_6H_{12}O_6$ (Glucose/Hexose) = 2C_2H_5OH (ethanol) + 2CO_2 (Carbon dioxide)

$3C_5H_{10}O_5$ (Pentose) = 5 C_2H_5OH + 5CO_2

   According to the reactions, the theoretical maximum yield of ethanol can be 0.51 kg, while that of carbon dioxide can be 0.49 kg per kg sugar.

   The alcohols and other products obtained are purified by distillation or converted chemically to the required products.

   It may be mentioned that the waste of furniture/particle boards can be converted to chemicals. The acid-catalysed reaction of cellulosic material with methanol or ethanol at 200°C gives a good yield of 'levulinates' and 'formate esters' as well as other useful by-products. The ethanol or methanol used is recovered. The other useful by-products include 'resin, ethyl formate, furfural, and ethyl levulinates'.

e.   Distillation for the Separation of Ethanol

The purification of ethanol is first done by distillation. The limit to which the ethanol can be separated from water is about 95.6% by volume (89.5 mole %), as is clear from the graph below. This limit is known as the 'azeotropic point' of ethyl alcohol. This mixture is an 'azeotrope' with a boiling point of 78.1°C

   At pressures less than the atmospheric pressure, the composition of the ethanol-water azeotrope shifts to more ethanol-rich mixtures, and at pressures less than 70 torr (9.333 kPa or 0.0733 bar), there is no azeotrope, and it is possible to distill absolute ethanol from an ethanol-water mixture. While the vacuum distillation of ethanol is not economical at present, pressure-swing distillation is a topic of current research. In this technique, a reduced-pressure distillation first yields an ethanol-water mixture of more than 95.6% ethanol. Then, fractional distillation of this mixture at atmospheric pressure distills the 95.6% azeotrope, leaving anhydrous ethanol at the bottom.

Molecular sieves and desiccants can be used to selectively absorb the water from the 95.6% ethanol solution. Synthetic zeolite in pellet form can be used, as well as a variety of plant-derived absorbents, including cornmeal, straw, and sawdust.

The zeolite bed can be regenerated almost an unlimited number of times by drying it with a blast of hot carbon dioxide. Cornmeal and other plant-derived absorbents cannot readily be regenerated, but where ethanol is made from grain, it is often available at a low cost. Absolute ethanol produced this way has no residual benzene. The vapor-liquid equilibrium is shown in Figure 3.3.

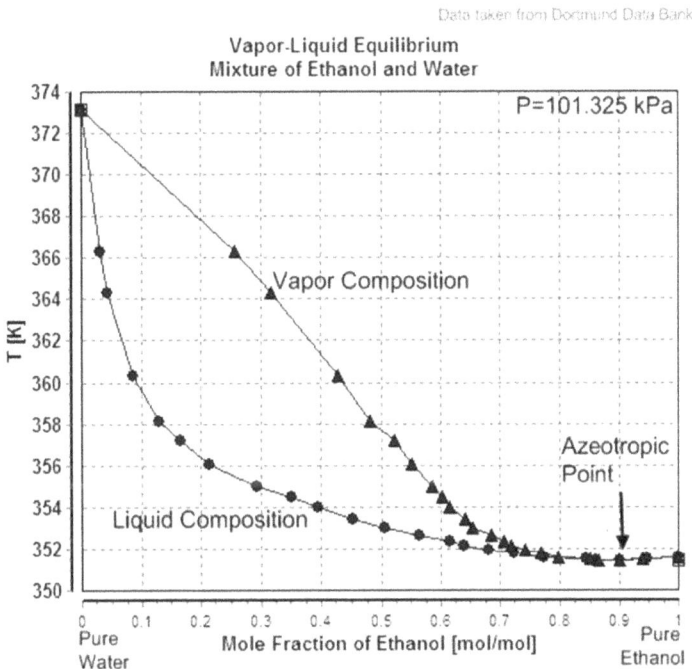

Fig. 3.3. Vapor-liquid Equilibrium of Mixture of Ethanol as 'Mole fractions'.
[Source: https://en.wikipedia.org/wiki/File:Vapor-Liquid_Equilibrium_Mixture_of_Ethanol_and_Water.png]

3.5 Dehydration of Ethanol

3.5.1 Drying of Ethanol Using Lime or a Hygroscopic Material [74]

After distillation, the ethanol can be further purified by *drying* it using lime or a hygroscopic material like rock salt. When lime is mixed with the water in ethanol, calcium hydroxide forms. The calcium hydroxide can then be separated from the ethanol. Similarly, a hygroscopic material will dissolve some of the water content of the ethanol as it passes through, leaving behind purer alcohol.

3.5.2 Membrane Separation, Pervaporation, Vapor Permeation [9]

Membranes can also be used to separate ethanol and water. Membrane-based separations are not subject to the limitations of the water-ethanol azeotrope because their separation is not based on the vapor-liquid equilibrium. Membranes are often used in the so-called hybrid membrane distillation process. This process uses a pre-concentration distillation column as the initial separating step.

Further separation is then accomplished with a membrane operated either in the *vapor permeation* or *pervaporation* mode. Vapor permeation uses a vapor membrane feed and pervaporation uses a liquid membrane feed.

The membrane acts as a selective barrier between the two phases, the liquid phase feed, and the vapor phase permeate. It allows the desired component(s) of the liquid feed to transfer through it by vaporization. The separation of components is based on a difference in the transport rate of individual components through the membrane.

Typically, the upstream side of the membrane is at ambient pressure and the downstream side is under a vacuum to allow for the evaporation of the selective component after permeation through the membrane. The driving force for the separation is the difference in the partial pressure of the components on the two sides and not the volatility difference of the components in the feed.

The driving force for the transport of different components is provided by a chemical potential difference between the liquid feed/retentate and vapor permeate at each side of the membrane. (*The retentate is the remainder of the feed leaving the membrane feed chamber, and which does not permeate the membrane.*)

The separation of components (e.g., water and ethanol) is based on a difference in the transport rate of individual components through the membrane. This transport mechanism can be described using the solution-diffusion model, based on the rate/degree of dissolution of a component into the membrane and its velocity of transport (*expressed in terms of diffusivity*) through the membrane, which will be different for each component and membrane type leading up to the separation.

Zeolite membranes have been developed that can be used for dehydrating ethanol. These membranes should not be used in an acidic medium as they are damaged at low pH values below 6. In the case of bioethanol separation, a combination of distillation and complete dehydration by vapor permeation is better. Mitsui Zosen Machinery (MZM) manufactures zeolite membranes. The dehydration step after distillation may be a fully heat-integrated vapor permeation unit using the zeolite membrane.

3.6 Developments in Fermentation Technologies for Converting Cellulose to Alcohol

3.6.1 Degradation of Cellulosic Materials [14,23,70,84,109,117,119]

Organisms capable of degrading the cellulose and utilizing it as a source of carbon are very important ecologically. In aerobic systems, the cellulose generally degrades into water and carbon dioxide. Most of the cellulose in nature is degraded in aerobic systems. In anaerobic systems, methane (CH_4) and hydrogen are also produced. Some amount of cellulose (about 5 to 10%) is degraded under anaerobic conditions in nature or artificially by a diverse range of bacteria (cellulolytic bacteria). Clostridia, which belong to the cellulolytic class of bacteria, have been closely studied and characterized. These bacteria are found everywhere in soil environment and form endospores. The cellulose is digested by these bacteria, which converts it into several products of metabolism via an exocellular enzymatic complex called 'cellulosome'*. The majority of the gram-positive cellulolytic bacteria' are found within firmicutes and belong to the class Clostridia and the genus *Clostridium.*

The cellulolytic species of bacteria are found within the phyla Thermotogae, Proteobacteria, Actinobacteria, Spirochaetes, Firmicutes, Fibrobacters, and Bacteroids. Most of the cellulolytic bacteria which have been isolated, are found within phyla Firmicutes and Actinobacteria. The microorganisms use several strategies to digest cellulose by secretion of enzymes or multienzyme complexes.

3.6.2 *Fermentation of Cellulosic Material* [6,18,35,70,114,117,118,120]

The 'C. *Thermophilic* bacteria', whose optimum growth is at a temperature of 60°C, can carry out cellulose fermentation by utilizing it as the sole carbon source. The products of fermentation have various proportions of 'lactate, formate, acetate, ethanol, hydrogen, and carbon dioxide' under different growth conditions. The C. *thermocellum* produces the 'cellulosome* enzyme complex' on its surface and degrades cellulose at a high rate.

When lignocellulosic biomass is pretreated with acid and fermented with C. thermophilic bacteria in batches, ethanol and acetate have been observed to be produced in the ratio of 2.3:1 *(mol/ mol)*. In continuous culture, the ethanol production has been observed to be lower (ethanol: acetate mol/mol 1.3:1).

The hydrogen production is observed to be affected similarly in batch culture and continuous culture. The yield of hydrogen is affected by the concentration of the cellulosic substrate. The variation in quantities of end products can be induced by the addition of some substances and by varying pressures in the reactor. [For example, (i) the addition of carbon monoxide in the reactor's head space significantly increases the production of ethanol and decreases the production of hydrogen, carbon dioxide, and acetate, (ii) the addition of ethanol to the growth medium at the initiation of the fermentation process increases the production of hydrogen.]

Notes on* Cellulosome and ++Cellobiose

Cellulosome is a multicomponent cellulolytic exocellular complex of proteins that mediates cellulose binding and degradation. The cellulosome hydrolyzes the biopolymer to its building blocks, 'disaccharide, ++ cellobiose, and cellodextrins'. The non-enzymatic 'polypeptide' binds and supports the enzymatic subunits of the cellulosome to the cell. ++Cellobiose ($C_{12}H_{22}O_{11}$) is derived from cellulose. When acted upon by acetic acid anhydride in the presence of concentrated sulphuric acid, the cellulose produces the octaacetate of cellobiose, which on hydrolysis by potassium hydroxide, produces cellobiose. [5,47]

3.6.3 *Consolidation of the Bioprocessing Steps for Biomass to Ethanol* [52,54,70,72,75,78,95,106,110,115]

The consolidation of the various steps in bioprocessing cellulose can be done to save energy and achieve higher efficiency during its conversion to third-generation fuels (like alcohol/hydrogen from lignocelluloses). The various steps of the process may be integrated in different ways. In all the cases, however, the pretreatment of the lignocellulosic biomass would be required to convert the hemicellulose present in biomass to sugars so that the total cellulose can be made accessible to the enzymes for hydrolysis. When using these approaches, the acid hydrolysis is replaced by enzymatic hydrolysis.

(i) *Separate Hydrolysis and Fermentation (SHF):*

The first approach of consolidation is SHF. In this approach, after hydrolysis, the liquid enters the 'glucose fermentation reactor' The mixture is then distilled in the beer column, where the ethyl alcohol along with some moisture is separated from the substance. The remaining liquid containing unconverted 'xylose (wood sugar)' is then sent to the second reactor where xylose is fermented to ethanol and the effluent is distilled to recover ethanol in a beer column. *Figure 3.2a shows all the processes of pretreatment, hydrolysis, fermentation, and the purification of ethanol separately.*

(ii) *Simultaneous Saccharification and Fermentation (SSF):*

Simultaneous Saccharification and Fermentation (SSF) is the second approach of consolidation. Saccharification means the conversion of cellulose to saccharides (sugars). There is no separate hydrolysis reactor and hydrolysis and fermentation are consolidated. This avoids the problem of the inhibition of enzymatic hydrolysis by the presence of glucose. *The process is shown in* Figure 3.4.

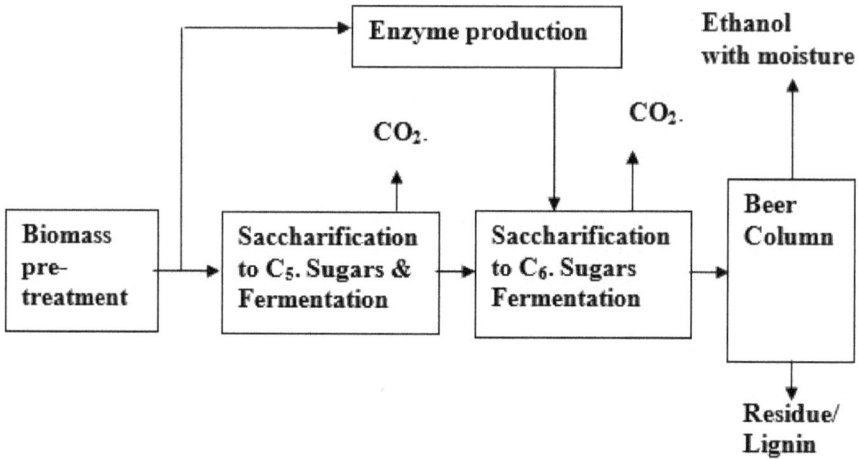

Fig. 3.4. Simultaneous Saccharification and Fermentation (SSF).

(iii) *Simultaneous Saccharification and Co-fermentation (SSCF):*

Simultaneous Saccharification and Co-fermentation (SSCF) is the third approach. It is a modification of SSF, and involves a mixture of enzymes being used for the fermentation of glucose and other sugars like xylose pentose in a single reactor. *The process is shown in* Figure 3.5.

(iv) *Consolidated Bioprocessing (CBP):*

Consolidated bioprocessing (CBP) is the fourth approach in which the enzyme (cellusome) production, substrate hydrolysis, and fermentation by celluloytic microorganisms is achieved in a single step. This process is shown in Figure 3.6.

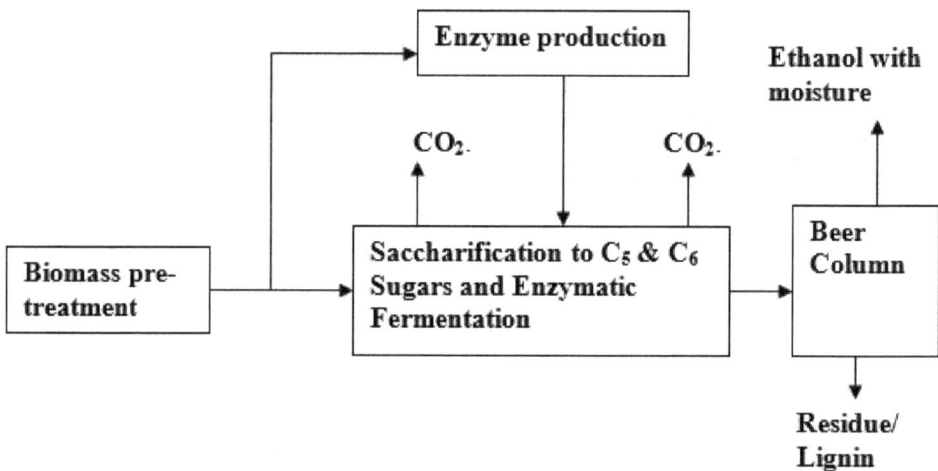

Fig. 3.5. Simultaneous Saccharification and Co-fermentation (SSCF).

3.6.4 *Conversion of Woody Matter to Alcohol by the Enzymes of Some Organisms* [51,58,59,83,112]

Some microorganisms and a marine pest have enzymes that are capable of converting woody mass to sugars, leading to alcohol production. A few examples are given below:

   i. *Clostridia bacteria:* These bacteria may turn woody biomatter into *butanol as* reported by the University of Nottingham, UK. *(Butanol is a better substitute fuel for gasoline than ethanol).*

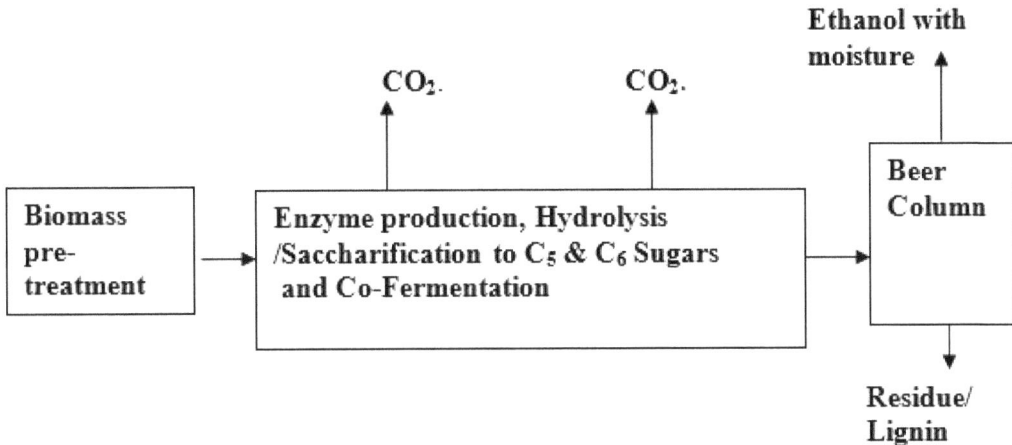

Fig. 3.6. Consolidated Bio-processing (CBP).

ii. *Fungal Enzyme:* There are about twenty-five stable 'fungal enzyme catalysts' which can break down 'Cellulose' into sugars at high temperatures.

iii. *Elephant yeast:* Research has shown that when 'elephant yeast' is slightly modified, it can create ethanol from nonedible sources like straw/grass.

iv. *Gribble:* A marine pest named 'gribble' can be the key to a biofuel production break-through. The researchers at the universities of York and Portsmouth have studied the digestive genes from the gribble species *'Limnoria quadripuncta'* and found that enzymes produced by this tiny marine pest can break down woody cellulose and turn it into sugars. These marine pests trouble seafarers as they bore through (eat away) the wooden planks used in boats and ships. These crustacean (like cactus) pests resemble pink woodlice. [1]

v. *Bacterial Strain Found in Compost Manure Heaps:* TMO Renewables (UK) is reported to have developed bacteria from a bacterial strain found in compost manure heaps, which produces ethanol from cellulose or woody biomass materials. TMO's process demonstration unit was completed in summer 2008 and it exhibited how many cellulosic feedstocks including wheat straw, grasses, residue grains/corn could be converted to beer/ethanol. All these feedstocks and even municipal solid wastes having similar materials may be used on an industrial scale to produce beer to be distilled into ethanol. The process seems very suited to retrofitting in existing corn ethanol plants. [3]

vi. *Zymomonas* sp.: A research and development (R&D) project 'Microbial Production of Alcohol from Cellulose Waste' (sponsored by the Ministry of Renewable Energy, India) was taken up at Murugappa Chettiar Research Institute, Chennai, India. One hundred and two 'cellulolytic and alcohol-producing' microbes were isolated from various sources out of which 42 isolated microbes possessed 'cellulolytic activity'—the best five were selected and identified. All the isolates (microbes) from banana pseudostems were found to be positive. Alcohol-producing microbe isolates have been identified as *'Zymomonas* sp.' A laboratory-scale alcohol production was carried out with enzyme hydrolyzates using eight strains of the *Zymomonas* sp.

It may be mentioned that the useful enzymes as present in 'clostridia bacteria', or 'gribble crustacean sea pest' or 'elephant yeast', when obtained or alternatively produced on the mass scale economically, can be used for converting biomass-containing cellulose to sugars for manufacturing alcohol from cellulosic materials/wastes like stalk, rice straw, and wheat straw.

3.6.5 *Ethanol from Seaweed* [44,45]

One good source of renewable biomass is macroalgae or seaweeds. They can be a competitive biofuel crop as they grow fast and do not compete for land with food crops. They are the source of many essential items such as fertilizers, animal feed, seafood, phycocolloids, etc., and are grown in a number of Asian countries.

Agars and carrageenans are extracted from red seaweed and alginates are extracted from brown seaweed. Phycocolloids are modified polysaccharides that are suitable for fermentation into bioalcohols. The red seaweed's biomass can also be a source of bioethanol in addition to being used for other high-value products.

The floating green seaweed, which has extremely high productivity and is widely distributed across diverse geoclimatic conditions in the world, can be a good resource of biomass for biofuel production. They are proving harmful at some places because of their high growth and therefore, their utilization would be very advantageous.

Though it is easy to release sugar molecules from seaweed, it's not at all easy for microbes such as yeast to ferment all the sugar molecules into ethanol. Brown macroalgae, a seaweed found all over the world, especially in colder seas, mainly contains the sugars glucan, mannitol, and alginate. Yeast can ferment glucan quite efficiently, but struggles with mannitol, while it cannot ferment alginate at all. In order to ferment alginate, scientists had to genetically engineer the microorganism *E. coli*.

Some genes of *E. coli* were deleted and some genes were added to it. While testing this engineered *E. coli* strain on a species of brown algae, *Saccharina japonica*, the scientists found that it was able to ferment alginate into ethanol. Furthermore, this bacterium also proved better at fermenting mannitol than conventional yeast. [29]

[It may be mentioned here that alcohol from the gasification route can be made from any biomass including seaweed, irrespective of their composition].

The Salt and Marine Chemical Research Institute, Bhavnagar (of CSIR, India) has successfully developed bioethanol from the green seaweed *Ulva fasciata* via fermentation using the marine fungus 'Cladosporium sphaerospermum'. *[In their study, the substrate of the green seaweed Ulva fasciata containing inoculated* Cladosporium spaeospermum *with 60% moisture content cultured at 25°C and pH value 4 for four days showed the optimum production of the enzyme cellulase. The enzyme performs hydrolysis, saccharification and fermentation of the cellulosic feed stock/ seaweed, resulting in bioethanol.]*

Extraction of Ethanol and other Products from the Seaweed Ulva fasciata [38,102]

The flow diagram Figure 3.7 shows the process of extracting ethanol from the seaweed *Ulva fasciata* containing about 90% water. It is extracted along with intra-cellular and extra-cellular minerals (by their special process). The extract is called MRLE and contains nutrients (minerals like Na, K, Mg, Ca, Cr, Mn, Fe, Cu, Zn, along with nitrogen, sulphur, boron, and carbon), which are suitable as food supplements for humans and animals.

The residue after water extraction is treated by a solvent to extract lipid, which is rich in essential polyunsaturated fatty acids that are nutritionally important. The lipids comprise about 2.7% of the dry weight of *Ulva fasciata*. The residue from the solvent extraction process is subjected to water extraction, which results in separation of cellulose and ulvan. The latter is a major fraction (about 25%) of the dry weight. It is a water-soluble sulphated polysaccharide with potential applications in the food, pharmaceutical, agricultural, and chemical industries. The cellulose is subjected to hydrolysis and fermentation to yield ethanol. Since there are so many valuable coproducts, the bioethanol from these weeds would be advantageous.

Processing one tonne of fresh *U. fasciata,* can yield 3.8 kg lipid, 34.6 kg Ulvan, and 14.0 kg cellulose on a dry weight basis. Furthermore, 14.0 kg cellulose on hydrolysis and fermentation can produce about 5.85 kg ethanol.

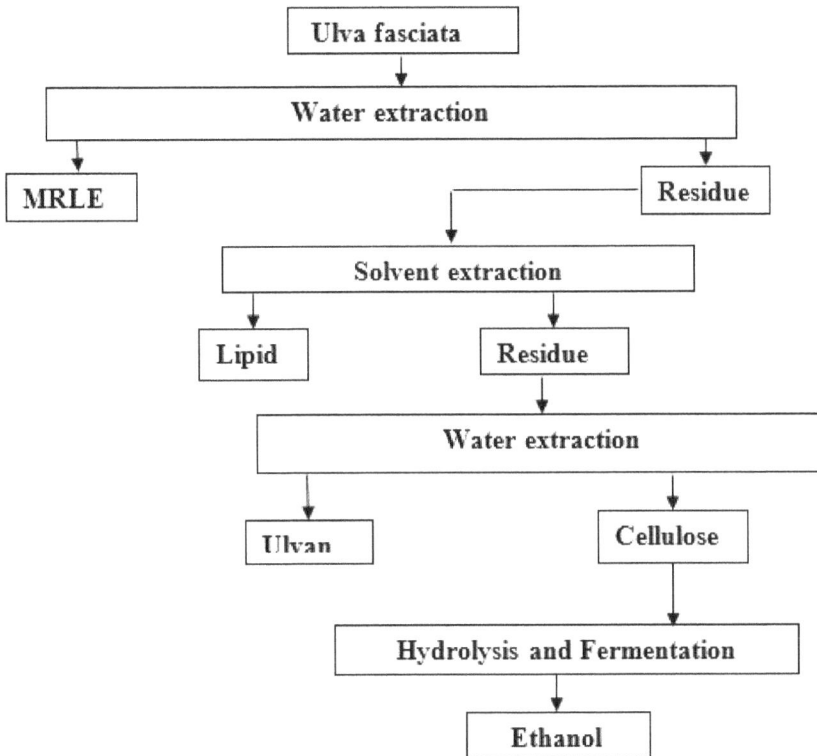

Fig. 3.7. Ethanol and Other Products from Biomass of Seaweed Ulva Fasciata.

The scientists at the same institute in Bhavnagar, have also examined eight different red seaweeds for their potential for biorefinery. The results revealed *Gelidella acerosa* as the best candidate, for which a bench-scale model of the biorefinery was developed. The process so developed enabled the production of bioethanol, and co-products *like lipids, natural pigments, high gel-strength agar, and mineral-rich liquid.* [101,103]

3.7 Bioethanol from Organic Wastes via the Gasification Route [64,66,73,89,94,107]

During the gasification of biomass, all the combustible matter—including 'lignin'—is converted to syngas, which is later processed for making alcohol either by biofermentation or synthesis. A schematic diagram for the typical thermochemical production process of alcohols from lignocellulosic biomass is shown in Figure 3.8. In this process, any kind of sun-dried biomass can be used.

Gasification and Synthesis

The biomass is gasified at a high temperature, preferably using oxygen as the oxidant (not air) to convert the biomass into syngas, which is a mixture of (CO and H_2), and is then converted to the desired product, that is, either alcohols or biodiesel by various synthesis processes. *[Also see Para. 6.15, Chapter 6 regarding the conversion of the gasifier output gases to methanol, methane, or other fuels.]*

Gasification and Fermentation [13]

The process developed by INEOS Co. (US) for producing bioethanol from wastes for its first commercial plant is mentioned below: [79]

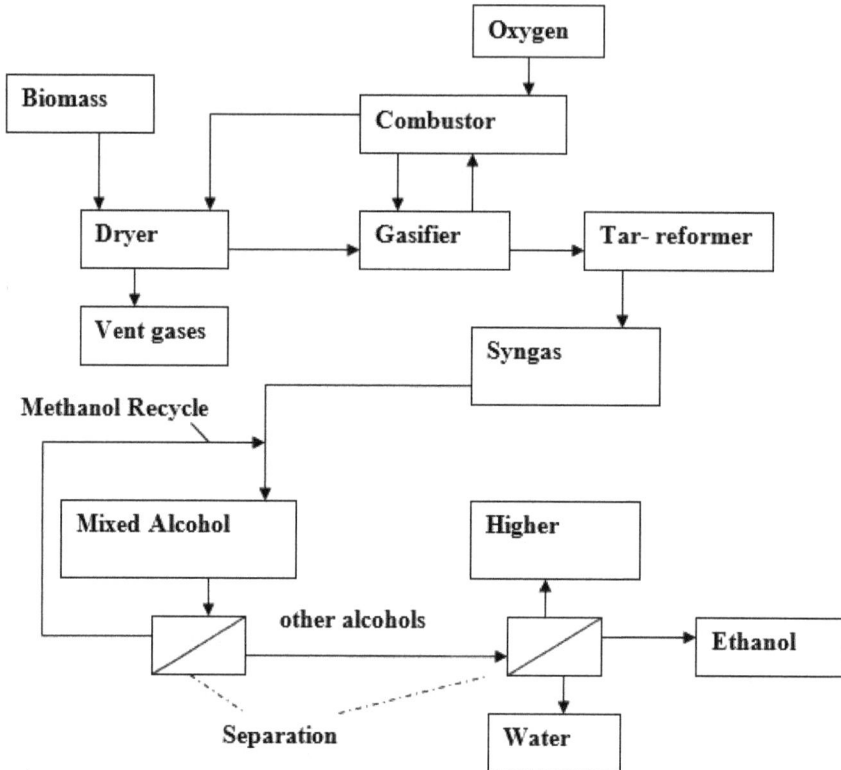

Fig. 3.8. Thermo-Chemical Process for Production of Alcohol from Lignocellulosic Biomass.

- The gasification of organic waste using oxygen to produce synthesis gas' which is cooled and cleaned.

- The Cooled and cleaned 'synthesis gas' is then passed into a reactor for fermentation where the aerobic bacteria convert it into 'ethanol'. A controlled amounts of nutrients are added into the reactor to sustain the process.

- The 'off-gas' from the reactor and the heat recovered in the process of gasification and cooling of gas are used for producing 'steam/power'.

- The ethanol produced is purified by the distillation up to 96%, beyond which other methods like physical absorption process using molecular sieves can be used.

3.8 Use of Ethanol as a Fuel for Motor Vehicles [27,64,97]

It is well known that 'ethanol' is used for blending with 'petrol'; the blending may be 5%, 10%, or more. Brazil embarked on a program for the massive cultivation of sugarcane, production of ethanol, and the substitution of ethanol as fuel in motor vehicles in a very big way. In Brazil, by May 2006, all 'gasoline' meant for motor vehicles was blended with at least 25% ethanol, and it was planned to have three million motor cars run on 100% ethanol, and 60% of the motor vehicles produced in Brazil were of the 'flexi-type', having the ability to run on any mixture of ethanol and gasoline, and also on 100% ethanol. All vehicle manufacturers in Brazil (Fiat, Ford, VW, GM, Toyota, Honda, Peugeot, Citroen, and others), produce 'flex-fuel vehicles' that can run on 100% ethanol or any mix of ethanol and gasoline. These flex-fuel cars represented 90% of the sale of personal vehicles in Brazil, in 2009.

In USA, (by May 2006), 4 million vehicles used a '85% ethanol + 15% gasoline' mixture. A small amount of gasoline content prevents starting trouble in cold weather). Ethanol may also be utilized as a rocket fuel (currently in lightweight rocket-powered racing aircraft).

Australian law limits the use of pure ethanol (sourced from sugarcane waste) to up to 10% in automobiles. It has been recommended that older cars (and vintage cars designed to use a slower burning fuel) have their valves upgraded or replaced.

Air Quality Improvement by the Use of Ethanol or Ethanol-blended Gasoline

Ethanol is an oxygenate, which, when added to gasoline, raises the octane level of gasoline. Gasoline blended with ethanol burns more cleanly and cuts down emissions of carbon monoxide, polycyclic aromatic hydrocarbons [PAH] (including benzene and butadiene), particulate matter, formaldehyde, and other air pollutants. It also reduces the emission of greenhouse gases, which are linked to global warming. If the ethanol is produced from lignocellulosic sources, the greenhouse gas reduction benefit will be obviously more.

(According to an industry advocacy group for promoting ethanol called the 'American Coalition for Ethanol', ethanol as a fuel reduces harmful tailpipe emissions of carbon monoxide, particulate matter, oxides of nitrogen, and other ozone-forming pollutants.)

Due to the cutting down of the generation of hydrocarbons, soot, and carbon monoxide, ground-level ozone and smog generation also get reduced.

The Renewable Fuels Association reports that ethanol can reduce tailpipe soot and particulate emissions by as much as 50% overall, with the greatest reductions being achieved in the highest-emitting vehicles. Given that the American Lung Association links these emissions to cancer, asthma, and heart attacks, ethanol blending can play an important role in improving public health.

The use of ethanol/gasoline blended with ethanol as fuel in vehicles decreases emissions of benzene—a hydrocarbon classified by the EPA as a known human carcinogen. Exposure to benzene can lead to blood disorders, including anemia, and higher instances of leukemia, as well as short-term impacts such as headache and respiratory irritation. Studies have indicated that blending 10% ethanol with gasoline can reduce benzene emissions by as much as 25%. [3]

(Argonne National Laboratory (USA) analyzed the greenhouse gas emissions of many different engine-and-fuel combinations. Comparing ethanol blends with gasoline alone, they showed reductions of 8% with the biodiesel/petrol-diesel blend known as B20, 17% with the conventional E85 ethanol blend, and that using cellulosic ethanol lowers emissions by 64%.)

Other Advantages of Using Ethanol as a Fuel

i. Ethanol has a slightly higher octane rating than petrol. The octane rating of petrol [C_nH_{2n+2} where n = 5 – 12] is 90–100 and that of ethanol is 108–115. (Octane rating indicates the fuel's ability to resist premature detonation (knocking) in the combustion chamber.)

ii. Ethanol can replace anti-knock like benzene, tetraethyl lead, and methyl tertiary butyl ether (MTBE), etc., which are very toxic. MTBE used as an additive causes surface water and groundwater pollution, but ethanol decomposes fast in water and soil and does not cause pollution.

iii. Running Otto cycle engines with pure ethanol allows for increasing the compression ratio of an engine up to 12 from maximum of 10 with gasoline, so more power can be produced from the engine.

The properties of some fuels including ethanol/blended fuels and other fuels are given in Table 3.2a, and Table 3.2b. The detailed property data of basic fuels namely—methanol, ethanol, propane, butane, gasoline (C4–C12), kerosene (C9–C16 and down to C5), and Diesel (C12–C23 and down to C5) is available. [97]

Table 3.2a. Energy Content of Some Fuels Compared to Ethanol. [97]

Fuel type	MJ/L	MJ/kg	Octane number
Dry wood (20% moisture)		~ 19.5	
Methanol	16–17.9	17.5–19.9	108.7
Ethanol	19–21.2	24–26.8	108.6
E85(85% ethanol, 15% gasoline)	25.2	33.2	105
Liquefied natural gas		43~50	
Liquefied Petroleum Gas (LPG) (60% propane + 40% butane)		50	
Aviation gasoline (high-octane gasoline, not jet fuel)	33.5	46.8	100/130 (lean/rich)
Gasohol (90% gasoline + 10% ethanol)	33.7	47.1	93/94
Regular gasoline/petrol	32–34.8	40.8–44.4	min. 91
Premium gasoline/petrol			max. 104
Diesel	38.6–40.3	45.4	25
Charcoal, extruded	50	23	

Table 3.2b. Physical and Chemical Properties of Methanol, Ethanol, and Gasoline. [5,97]

Properties	Methanol	Ethanol	Gasoline
Formula	CH_3OH	C_2H_5OH	C_4–C_{12}
Molecular weight	32	46	About 114
Vapour density relative to air	1.10	1.59	3 to 4
Liquid density g/cm³ at 289°K	0.79	0.79	0.74
Boiling point °K	338	351	300 to 518
Melting point °K	175	129	--
Vapour pressure @ 311°K in Pisa	4.6	2.5	8–10
Viscosity (c p)	0.54	1.20	0.56
Flash point °K	284	287	228
Lower Flammability Limit (LFL)	6.7%	3.3%	1.3%
Upper Flammability Limit (UFL)	36%	19%	7.6%
Autoignition temperature °K	733	636	523–733
Peak flame temperature °K	2143	2193	2303
Minimum ignition energy in air mJ	0.14	--	0.23

Note: Oxygenates additives are oxygenated organic compounds, such as MTBE and ethanol, added to gasoline to increase octane or extend gasoline supplies. The oxygenates commonly used are either alcohols or ethers. They reduce carbon monoxide and soot that is created during the burning of the fuel. Compounds related to soot, such as polyaromatic hydrocarbons (PAHs) and nitrated PAHs, are also reduced. These may affect the evaporative emissions performance and other properties of the fuel as indicated by the tables above.

Otto Cycle

The petrol engines are designed on Otto cycle shown in Figure 3.9 which comprises of:

o 1–2 Adiabatic compression

o 2–3 Constant volume heating (fuel burning)

o 3–4 Adiabatic expansion

o 4–1 Constant volume heat rejection

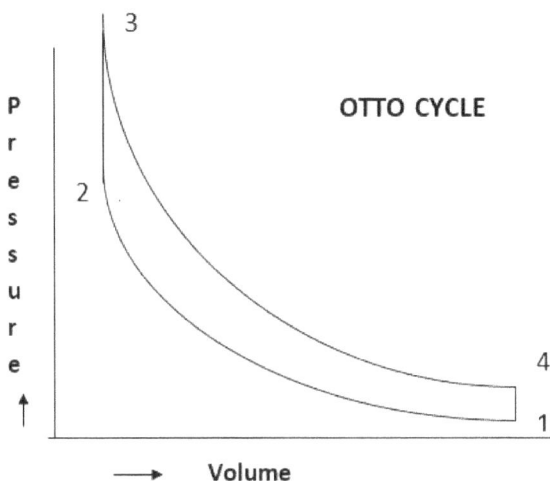

Fig. 3.9. Thermodynamic Processes of Otto Cycle.

Efficiency of Otto Cycle

The efficiency of the otto cycle is given by: $1 - (1/r)^{\gamma-1}$

where γ = ratio of specific heat at constant pressure and constant volume, that is, C_p/C_v.

$\gamma = 1.4$ (for air), and r = Compression ratio

Increase in compression ratio 'r' raises the efficiency as evident from the formula.

Alternate Use of Ethanol [105]

Ethanol can be used for the manufacture of many chemicals and useful products. For example, ethylene glycol *ether (CH$_3$CH$_2$OCH$_2$CH$_2$OH)* also called monoethyl ether of glycol can be produced from alcohol. (*Glycol ether* is the main constituent of the hydraulic brake fluid of vehicles.) A plant for manufacturing this chemical was set up in Kashipur (India) in 2001 (by the plant supplier Sulzer Chemtech). Ethylene glycol ether is produced by the reaction of ethyl alcohol (C$_2$H$_5$OH) with ethylene oxide **(CH$_2$OCH$_2$).CH$_2$OCH$_2$ + C$_2$H$_5$OH = CH$_3$CH$_2$OCH$_2$CH$_2$OH**

3.9 Commercialization of Ethanol Production from Lignocellulosic Materials [2,8,27,36,42]

Several plants were built and operated in various countries during World War II for producing ethanol from cellulosic materials because of the change in priorities and difficult economic and market conditions. These countries included Germany, Russia, China, Korea, Switzerland, the United States, and also other countries. Today, due to competition from ethanol produced in cheaper ways, only a few of these plants are operating (mostly in Russia), but now the technology is being revived.

In Timiskaming, Quebec, there is a paper plant which utilizes sulphite liquor containing 2% fermentable sugars for making alcohol (Tembec Inc.). Tembec and Georgia Pacific are operating sulphite pulp mills in North America, which utilize a dilute acid hydrolysis process to dissolve hemicellulose and lignin from wood, and produce specialty cellulose pulp. The hexose sugars in the spent sulphite liquor are fermented to ethanol.

There is great potential in producing ethanol from widely available cheap raw material containing cellulose. There are a number of technologies for such production, but cost-effective technology has been difficult to achieve. Some success is achieved in the cost-effective production and several plants to manufacture ethanol from these resources are coming up/or are already constructed. Efforts are being made the world over to commercially produce ethanol from cellulosic materials. Some companies like Igoen, Broin, and Abengoa are constructing or have constructed refineries, which can process biomass and make alcohol from it. The enzymes to be used in the process of making alcohol from cellulose are being made by companies like Diversa, Novozymes, Dydic. <u>Nippon Oil Corporation, Japan</u>, and other Japanese manufacturers, including <u>Toyota Motor Corporation</u>, which planned to set up a research body to develop cellulose-derived biofuels. The consortium had planned to produce 250,000 kiloliters (1.6 million barrels) per year of bioethanol by March 2014, and at 40 yen ($ 0.437) per litre (about $70 a barrel) by 2015.

In March 2009, <u>Honda Motor Co. Ltd.</u> announced an agreement for the construction of a new cellulosic ethanol research facility in Japan. The new Kazusa branch facility of the Honda Fundamental Technology Research Center was to be built within the <u>Kazusa Akademia Park</u>, in Kisarazu, Chiba. Construction was scheduled to begin in April 2009, with the aim to begin operations in November 2009.

On 30th October 2015, Dupont celebrated the opening of its cellulosic biofuel facility in Nevada, Iowa (US). This biorefinery is the world's largest cellulosic ethanol plant, with the capacity to produce 30 million gallons per year of clean fuel that offers a 90-percent reduction in greenhouse gas emissions as compared to gasoline. The raw material used to produce the ethanol is corn stover—the stalks, leaves, and cobs left in a field after harvest. [22] A few examples of firms developing alcohol from lignocellulosic resources are given in Table 3.3.

*Note: TIFAC (Technology Information, Forecasting and Assessment Council, India) under its 'Bioprocess and bio-products program' has supported the National Institute for Interdisciplinary Science & Technology (NIIST), Trivandrum to establish a dedicated **center for biofuels** to carry out advanced research in lignocellulosic ethanol production along with the development of other biochemicals.*

As there remains shortage of cattle fodder in India, only non-fodder biomass should be focused upon for conversion to biofuels. Apart from biomass, waste from forest, some non-fodder agricultural biomass like sugarcane tops, plant residue from the cotton plant, chili plant, certain pulse plants and pods waste, nonedible oilseed plants, seaweeds, etc., are of greater interest for the purpose, together with surplus rice and wheat stalk/straw. R&D should aim for improved process economics to address some of the critical technology issues in these residues, which are typically burnt in the fields or used to meet household energy needs by farmers. Sugarcane tops are by far the highest-available surplus crop residue followed by oilseed residue, and cotton.

3.10 Biobutanol as Fuel for Vehicles [11,19,20,43]

Butanol has a formula C_4H_9OH. It may have four isomeric structures *(see Notes). It is primarily used as a solvent, an intermediate in chemical synthesis, and as a fuel. When it is produced biologically, it is called biobutanol. Like many alcohols, butanol is considered to be toxic.

Butanol can also be manufactured from plant resources and therefore it is considered a potential renewable biofuel and can be used in internal combustion engines. It contains more energy for a given volume than ethanol and almost as much as gasoline, so a vehicle using butanol will have fuel consumption more comparable to gasoline than ethanol. A much higher percentage of it can be blended with gasoline than in ethanol, without requiring any modifications in the engine. The relative properties of butanol, ethanol, methanol, and gasoline are mentioned in Table 3.4.

Butanol can be distributed via conventional infrastructure (pipelines, blending facilities, and storage tanks). There's no need for a separate distribution network. Butanol is less corrosive and explosive than ethanol. EPA test results show that biobutanol reduces emissions, namely hydrocarbons, CO, and oxides of nitrogen (NOx). Butanol can also be used as a blended additive to diesel fuel to reduce soot emissions. It is also used as a component of hydraulic fluids and brake fluids.

The fuel in an engine has to be vaporized before it will burn. Insufficient vaporization is a known problem with alcohol fuels during cold starts in very cold weather. As the heat of vaporization of butanol is less than half of that of ethanol, an engine running on butanol should be easier to start in cold weather than the one running on ethanol or methanol.

Table 3.3. Examples of Plants Dedicated to Cellulosic Ethanol Production (under Operation or Construction). [21,22,27,42]

Se. No.	Company	Feed Stock	Remarks
1.	Bioengineering Resources, Fayetteville, AR, US	Wheat straw	Thermochemical gasification with fermentation (Pilot plant operating)
2.	Ethxx International, Aurora, ON, US	Wood	Thermochemical gasification with catalytic conversion (Pilot plant operating)
3.	Fuel Cell Energy, Lakewood, CO, US	Wood	Thermo-chemical gasification with catalytic conversion (Pilot plant operating)
4.	Iogen, Ottawa, ON, Canada	Oathulls, switchgrass, wheat straw, & corn stover	Enzymatic process (experimental plant operating, Capacity 378 M. liters/yr.)
5.	Paszner Technologies, Inc., Surrey, BC	Wood	Acidified aqueous acetone process
6.	PureVision Technology, Ft.Lupton, CO, US	Wood	Enzymatic process
7.	Abengoa Bioenergy Hugoton, KS, US	Wheat straw	
8.	DU Pont Nevada, IA, US	Corn stover	10 MG /year commercial 2014
9.	Colusa Biomass Energy Corp., Sacramento, CA, US	Waste rice straw	
10.	Fulcrum Bioenergy Reno, NV, US	Municipal solid waste	10 MG/year commercial 2013
11.	Gulf Coast Energy, Mossy Head, FL, US	Wood waste	
12.	KL energy Corp. Upton, WY, US	Wood	
13.	Biogasol Bornholm, Denmark	Wheat Straw	5 M liters/year commercial 2012
14.	POET LLC, Emmetsburg, IA, US	Corn cobs	20–25 MG/year commercial 2014
15.	US Envirofuels, Highlands, county, FL, US	Sweet sorghum	
16.	Xethanol, Auburndale, FL, US	Citrus peels	
17.	Masada, Birmingham, AL, US	Municipal solid waste	Process uses concentrated acid, Capacity 3780 million liters/year
18.	Abengoa Bioenergy (AMB_MCE)		Building a 5 MG plant in Spain
19 .	Ethotec, Harwood, New South Wales, Australia	Wood residue	
20.	China resource alcohol Corp., China	Stalks & leaves of corn stover	Plant operating
21.	Abengoa Bioenergy Hugoton, KS	Wheat straw	25–30 MG/year Commercial in 2013
22.	Blue Fire Ethanol Irvine, CA, US	Multiple sources	3.9 million gallons/year commercial
23.	MOs Coma Kinross MI, US	Wood waste	20 million gallons/year commercial
24.	Inbicon owned by Dong Energy Kalunborg Zealand Denmark	Wheat straw	5.4 M liters/year
25.	Sweet Water Energy Rochester NY, US	Multiple sources	
26.	DU Pont Nevada, Iowa US	Corn stover/stalks, leaves, and cobs	30 million gallons/year

Table 3.4. Relative Properties of Butanol, Ethanol, Methanol. [55]
[Source: https://en.wikipedia.org/wiki/Octane_rating Octane rating From Wikipedia, the free encyclopedia]

Fuel	Energy density	Air-fuel ratio	Specific energy	Heat of vaporization	RON*	MON**
Gasoline and biogasoline	32 MJ/L	14.7	2.9 MJ/kg air	0.36 MJ/kg	91–99	81–89
Butanol fuel	29.2 MJ/L	11.1	3.2 MJ/kg air	0.43 MJ/kg	96	78
Anhydrous Ethanol fuel	19.6 MJ/L	9.0	3.0 MJ/kg air	0.92 MJ/kg	107	89
Methanol fuel	16 MJ/L	6.4	3.1 MJ/kg air	1.2 MJ/kg	106	92

It may be mentioned that the octane number of the fuel must be considered with care while choosing the fuel for an engine as it is related to the 'knocking phenomena'.

Butanol (C_4H_9OH) may be stored onboard hydrogen-fueled vehicles having fuel cells using hydrogen. The reforming butanol onboard can provide hydrogen for the fuel cells of the vehicle. Butanol has obviously a higher hydrogen content (because of the long chain structure) than ethanol (C_2H_5OH) or methanol (CH_3OH) and therefore it can better solve the problem of onboard storage of hydrogen on vehicles. One of the biggest challenges facing the development of hydrogen cell vehicles is the storage of enough onboard hydrogen and the lack of hydrogen infrastructure for fueling. This problem can be solved by storing biobutanol instead of hydrogen on the vehicle for conversion to hydrogen onboard.

The energy content of ethanol is just two-thirds that of conventional gasoline. Cars using the most common ethanol blend, E10 (gasoline blended with 10% ethanol), need 1.03 gallons of the fuel to travel the distance they could cover with one gallon of gasoline.

Knocking in Engine

A fuel with a higher-octane rating is less prone to knocking. The control system of any modern car engine can take advantage of this by adjusting the ignition timing. This will improve energy efficiency, leading to a better fuel economy than the comparisons of the energy content in different fuels indicate. By increasing the compression ratio, further gains in fuel economy, power and torque can be achieved.

Conversely, a fuel with a lower octane rating is more prone to knocking and will lower the efficiency. (The fuel with a very low octane rating is subject to extremely rapid and spontaneous combustion of the end charge by compression, which causes knocking.) Knocking can also cause engine damage. Engines designed to run on 87 Octane will not have any additional power/fuel economy from higher octane fuel.

The octane rating of n-Butanol* is lower than that of ethanol and methanol, but similar to that of gasoline. The n-Butanol may be used as an engine fuel.

Notes

(1) Note on the Isomeric Structures of Butanol

There are four isomers of butanol. These are: n-Butanol, s-Butanol, t-Butanol, and Isobutanol.

The n- butanol has a boiling point of 117.4°C and is widely used as a solvent. The s-Butanol has a boiling point of 100°C. It is used for the preparation of butanone, esters, and a lacquer solvent. The t-Butanol has a boiling point of 83°C. It is used as an alkylating agent (alkylation means introducing an alkyl group in a compound). The Isobutanol has a boiling point of 108°C. Isobutanol behaves as a primary alcohol, but it rearranges readily due to the presence of a branched chain near the COH group. Isobutanol, when treated with hydrochloric acid, yields isobutyl chloride and t-Butanol.

n-Butanol has a RON (research octane number) of 96 and a MON (motor octane number) of 78 (with a resulting "(R+M)/2 pump octane number" of 87, as used in North America), while t-butanol has octane ratings of 105 RON and 89 MON. The t-Butanol is used as an additive in gasoline, but cannot be used as a fuel in its pure form because its relatively high melting point of 25.5°C causes it to gel and freeze near room temperature.

Alcohol fuels have less energy per unit weight and unit volume than gasoline. To make it possible to compare the net energy released per cycle—a measure called the fuel's specific energy is sometimes used. It is defined as the energy released per air fuel ratio. The net energy released per cycle is higher for butanol than ethanol or methanol and about 10% higher than that for gasoline.

Butanol has been used as one of the raw materials for the production of butadiene (a raw material for synthetic rubber production). As the butanol has a greater similarity to gasoline, it will work in vehicles without modifications.

(2)   Note on the kinematic viscosity of butanol

The kinematic viscosity of butanol is several times higher than that of gasoline and about as viscous as high-quality diesel fuel. **Kinematics viscosity at 20°C:** *Gasoline 0.4–0.8 cSt; Butanol 3.64 cSt; Ethanol 1.52 cSt; Methanol 0.64 cSt; Diesel >3cSt.*

(3)   Note on (*RON—research octane number; **MON—Motor octane number; and CN—Cetane Number

Gasolines are made up of comparatively lighter hydrocarbons with carbon numbers usually between 4 and 11 while diesel fuels contain heavier, higher boiling-point compounds, typically with carbon numbers between 10 and 21.

Fuel's combustion characteristics are mainly classified by their autoignition quality, which is measured by the research and motor octane numbers, RON and MON, respectively, for gasolines.

Octane number is a measure of a gasoline's ability to resist autoignition, which can cause engine knock and can severely damage engines. Gasolines need to be resistant to autoignition to avoid knock - an abnormal combustion phenomenon which limits SI engine efficiency.

Two laboratory test methods are used to measure octane: one determines the Research Octane Number (RON) and the other determines the Motor Octane Number (MON). RON correlates best with low speed, mild-knocking conditions and MON correlates with high-temperature knocking conditions and with part-throttle operation. RON values are typically higher than MON, and the difference between these values is the sensitivity, which should not exceed 10.

Diesel fuels are characterized by the cetane number (CN). Conventional diesel fuels should auto-ignite easily.

For Gasoline, the higher the RON, the more resistant is the gasoline to autoignition while higher CN indicates that the fuel is more prone to autoignition—there is an inverse correlation between RON and CN.

The RON test is run in a single-cylinder engine at an engine speed of 600 rpm (revolutions per minute) and an intake temperature of 52°C while the MON test is run at 900 rpm and with a higher intake temperature of 149°C. The octane scale is based on two alkanes, n-heptane and iso-octane which are primary reference fuels (PRF). The RON or MON is the volume percent of iso-octane in the PRF. A fuel is assigned the RON (or MON) value of the PRF that matches its knock behavior in the RON (or MON) test. Cetane number (CN) of a fuel is measured by comparing its ignition characteristics with reference fuels in a single cylinder diesel engine. The reference scale is based on normal cetane (n-hexadecane), which is defined to have a CN of 100 and alpha methylnaphthalene of CN = 0. It is the percentage by volume of **cetane** *in a mixture of liquid alpha methylnaphthalene that gives the same ignition characteristics as the oil being tested. Hepta-methyl-nonane, which is a highly branched paraffin is assigned a CN of 15.*

A derived cetane number, DCN, is calculated from the measured ignition delay. Fuels can be said to be in the diesel autoignition range if CN > 30 and fuels with RON > 60 (CN < 30) can be considered to be in the gasoline autoignition range. Practical diesel fuels have CN ranging from 40 to 60 while practical gasolines in most markets have RON between 90 and 100.

A rule of thumb for calculating the lower heating value from the ultimate analysis of a liquid or solid fuel is:

Energy content LHV [MJ/kg] = 32.8 x m(C) + 101.6 x m(H) – 9.8 x m(O)

and two examples of using the formula are: (1) n-Heptane = 26.9 + 18.1 – 0.0 = 45 MJ/kg, and (2) n-Hexanol = 23.3 + 14.2 – 1.5 = 36 MJ/kg

where x denotes the proportion of elements and m denotes the molecular weight of the element

Possible Butanol* Fuel Mixtures

The butanol can be mixed with gasoline in a range from 8% to 16%. Companies are planning to market a fuel that is 85% ethanol and 15% butanol (E85B), so existing E85 internal combustion engines can run on a 100% renewable fuel (without fossil fuels).

Potential Problems with the Use of Butanol as a Fuel

The potential problems with the use of butanol are similar to those of ethanol:

i. The utilization of butanol fuel as a substitute for gasoline requires a slight increase in fuel flow as the butanol has slightly less energy than gasoline. (*The requirement of increase in the fuel flow may be about 10%, compared to 40% for ethanol.*)

ii. Alcohol-based fuels are not compatible with some *fuel system components*.

iii. Alcohol fuels may cause erroneous gas gauge readings in vehicles (if the fuel level gauge is based on capacitance).

iv. Butanol has a lower octane number than ethanol and methanol, which have lower energy densities than butanol. The higher octane numbers of ethanol and methanol allow for greater compression ratios.

v. Butanol is one of many side products produced from the current fermentation technologies; as a consequence, current fermentation technologies allow for very low yields of pure, extracted butanol.

vi. Ethanol can be produced at a much lower cost and with much greater yields than butanol.

vii. Consumer acceptance may be limited due to the potentially offensive 'banana-like' smell of n-Butanol.

3.11 Butanol Production from Plant-Based Materials [49,55,98,111,122]

3.11.1 Process of Production of Biobutanol [17]

Butanol was generally traditionally from fossil fuels, but is now being produced by bacterial fermentation from a variety of feedstocks like food crops (such as sugarcane, sugar beet, corn grain, wheat) and non-food biomass (lignocellulosic materials) like switch grass, agricultural residues (straw, corn stalk bagasse, etc.).

One such bacteria used for fermentation is Clostridium acetobutylicum, *which is also known as the Weizmann organism.* The problem with this type of microbe is that it is poisoned by the very butanol it produces once its (that is, butanol's) concentration rises above 2%. Great strides have been made by researchers in creating "designer microbes" that can tolerate concentrations of butanol without being killed off.*

The ability to withstand harsh, high-concentration alcohol environments, plus the superior metabolism of these genetically enhanced bacteria have fortified them with the endurance necessary to degrade the tough cellulosic fibers of biomass feedstocks/lignocellulosic materials like pulpy woods and switch grass. The cost of biobutanol is coming down due to such processes.

*Note: Chaim Weizmann was the first to use microorganisms for the production of acetone from starch in 1916 and a large quantity of butanol was obtained as a by-product in this process of fermentation. The other byproducts of the process were acetic acid, lactic acid, propionic acid, isopropanol, ethanol, and hydrogen. It may be mentioned that the clostridium bacteria cultures used can cause stomach problems if ingested.

Bioethanol production plants can be modified for the production of biobutanol as the difference between ethanol production and butanol production is primarily in the fermentation of feedstock and minor changes in distillation. (A decreased pressure/vacuum distillation leads to a lower boiling point. At different boiling points, different fractions/products can be separated.)

For converting an ethanol plant to a butanol plant some changes are required to be made such as:

• The yeast in the fermentation step is changed to the bacteria *Clostridium Acetobutylicum.*

• The volume of the fermentation reactor is increased.

• The product produced is to be separated continuously to keep the butanol content low in the broth so that it does not become toxic to the bacterial culture.

- There is a mixture of butanol, ethanol, and acetone in the product obtained from fermentation and more water is included in the process. The distillation part of the plant is expanded for separation of different fractions.

- As with the distillation, the evaporation part of the plant is expanded.

As in the case of ethanol production, the lignocellulosic feedstock (*lignocellulose is a combination of cellulose, hemicellulose, and lignin*) can be used to produce butanol. Cellulose is a polysaccharide consisting of multiple glucoses on a chain; hemicellulose is similar but consists of several different sugars, not only glucose. Lignin is a large macromolecule which acts as a glue binding together all the polysaccharides. These three substances are the main components of trees/plants. These substances have to be separated/broken down before the butanol-producing cells can utilize them as feed. The lignin component is not utilized, while cellulose and hemicellulose are utilized in fermentation, as in case of ethanol production. A flow diagram for the production of butanol from biomass is shown in Figure 3.10.

Fractional distillation can be used for separating the products, but there are other processes also for obtaining butanol from the broth. These methods are described in next paragraph.

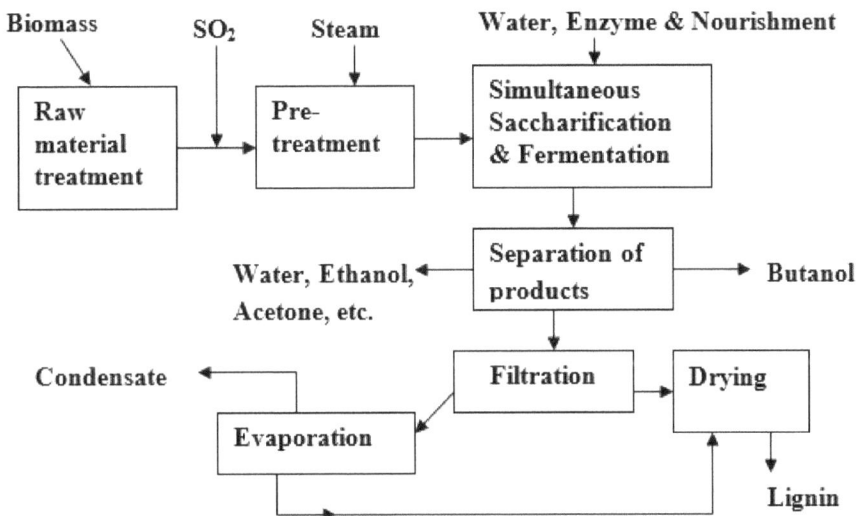

Fig. 3.10. Production of Butanol from Biomass.

3.11.2 Methods for Removal of Butanol from Fermentation Broth [55]

Because the inhibitory effect of butanol is toxic to bacterial culture, it can only be produced as a diluted component, that is, the butanol produced has to be continuously removed from the broth. There are several methods for doing this, such as:

a. Adsorption, b. Gas stripping, c. Pervaporation, d. Perstraction, e. Liquid-Liquid, or Ionic Liquid Extraction, f. Vacuum Distillation. These are briefly described below:

a. Butanol Recovery by Adsorption [39,40,88]

The fermentation broth from a butanol-producing reactor should be first filtered/centrifuged for removal of insoluble material/biomass or can be run through a membrane filter or a centrifuge for separating the biomass. The solids/biomass should be recirculated to the fermentation reactor. The filtrate consisting of water, butanol, ethanol, and acetone is sent to adsorption columns filled with silica-based zeolite-adsorbent hydrophobic material (which does not absorb water). The substances that would bond to the adsorbents are butanol, ethanol, acetone, and some water. The un-adsorbed components are recirculated to the fermentation reactor. A flow diagram of the adsorption process is shown in Figure 3.11.

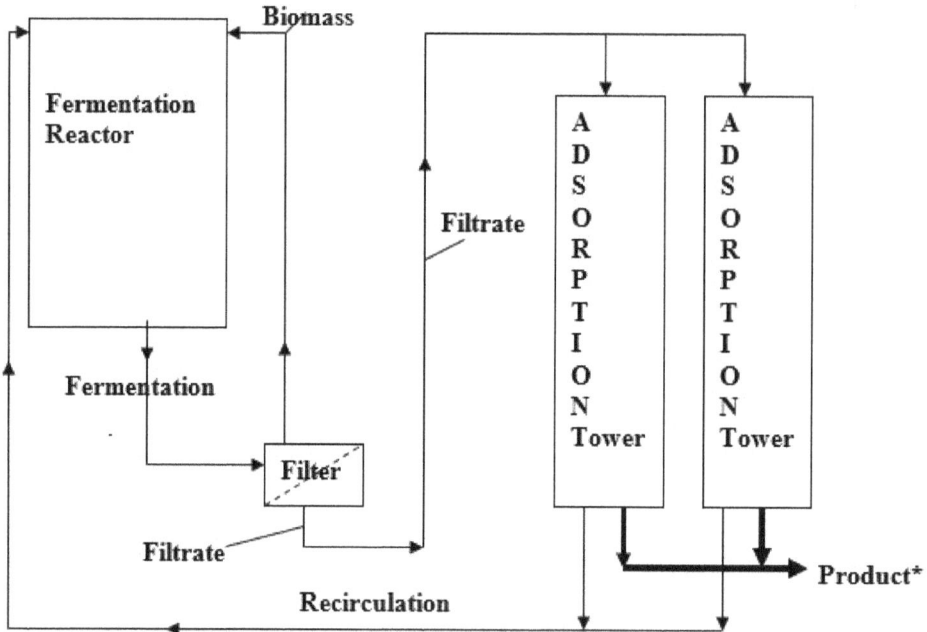

Fig. 3.11. Separation of Fermentation Product by Adsorption.
[*Note: The product obtained may require separation by fractional distillation]

As two adsorption columns are used, they can be used one by one to adsorb and de-adsorb the products. The de-adsorption is done by increasing the temperature to around 200°C so that the products adsorbed are released. Thus the fermentation process can be run continuously. Some zeolite materials have been found which prefer the adsorption of butanol over ethanol and acetone. The process is suitable for small plants.

b. *Butanol Recovery by Gas stripping* [30,86,116]

Gas stripping is a simple and economical method of separating butanol because it does not require expensive equipment and it lowers the butanol concentration. An inert gas such as nitrogen is passed continuously through the fermentation broth during fermentation so that the volatile butanol vaporizes along with ethanol and acetone and goes out with the gas stream from the top of the reactor. The vapors condense when the inert gas along with vapors passes through a condenser and the inert gas is recycled. The condensed product is then put through fractional distillation for obtaining butanol and other fractions separately. The remaining effluent from the distillation would mostly contain lignin and water. Lignin can be recovered for further use or processing. The liquid can also be evaporated, condensed, and recovered.

Gas stripping does not remove the reaction intermediates from the broth and may be considered more advantageous than membrane filtration and adsorption. The bacteria are also not removed by gas stripping.

It may be mentioned here that the butanol has the highest boiling point of all the fractions in the broth and therefore is disfavored when the stream takes up the volatile substances from the fermentation broth. However, most of the butanol, ethanol, and acetone will be taken up by the gas stream and therefore maintain a very low concentration of butanol (not toxic for bacteria) in the broth—as required for efficient production of butanol. Figure 3.12 shows a flow diagram of gas stripping.

c. *Butanol Recovery by Pervaporation Separation* [48,69,72,85,87,113]

In pressure-driven membrane processes (ultra-, micro-, and nano-filtration), the bulk component is purified by passing it through a porous membrane. The minor component is held back and the

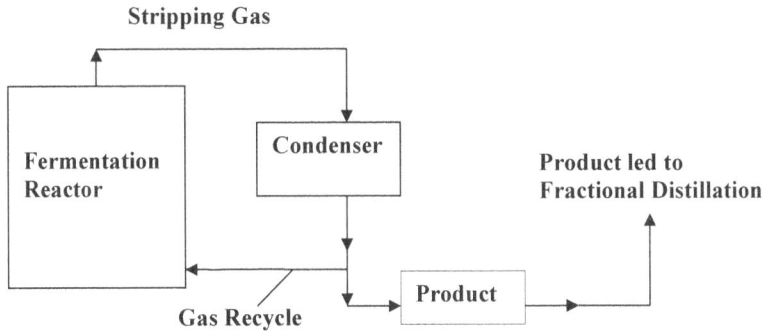

Fig. 3.12. Gas Stripping of Fermentation Product.

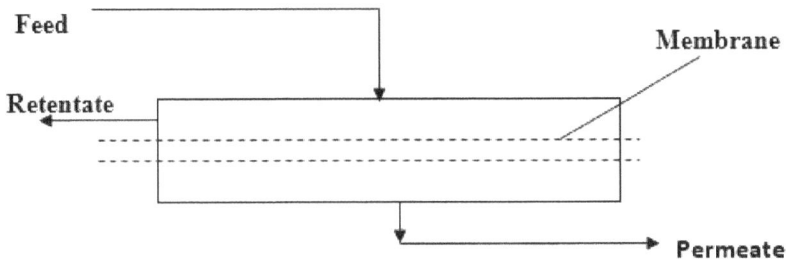

Fig. 3.12a. Schematic of Membrane Separation.

major component selectively permeates the membrane by preferential absorption, diffusion, and desorption. The principle of membrane separation is shown schematically in Figure 3.12a.

Pervaporation is one of the most promising techniques for the removal of toxic substances. This method involves the selective transport by diffusion of some components through a membrane. Vacuum is applied to the side of the permeate. The permeated vapors should be condensed on the low-pressure side.

In the pervaporation during the butanol process, the minor component is preferentially permeated through a membrane. The principle of pervaporation is schematically indicated in Figure 3.12b. The bulk fluid is held back by the membrane. Vacuum is applied to the permeate side of the membrane. The substance that permeates membrane is reasonably volatile and the application of vacuum always causes the permeate to be desorbed in the vapor form (*hence the name 'pervaporation'*).

Pervaporation or vapor permeation* are used where distillation is difficult, costly, or not possible due to the limitations of the water-alcohol azeotrope. (**If the stream of mixture is in the form of vapor/gas, the permeation process is termed as 'vapor permeation', otherwise it is called 'pervaporation'.*)

Pervaporation is a membrane process, which can be employed for:

i. Wastewater treatment having higher levels of soluble organic compounds/material

ii. Recovery of important compounds during the chemical processes

iii. Separation of 99.5% pure ethanol-water solutions

iv. Harvesting of organic substances from the fermentation broth

Pervaporation can be used in batch processes or in a continuous process. Batch pervaporation has great flexibility, but has to maintain a tank, which acts as buffer. Continuous pervaporation consumes very little energy during the whole process and is generally preferred for liquid-vapor separation.

A flow diagram of separation of biobutanol from the fermentation broth by pervaporation is shown in Figure 3.12c. The process involves the simultaneous saccharification and fermentation

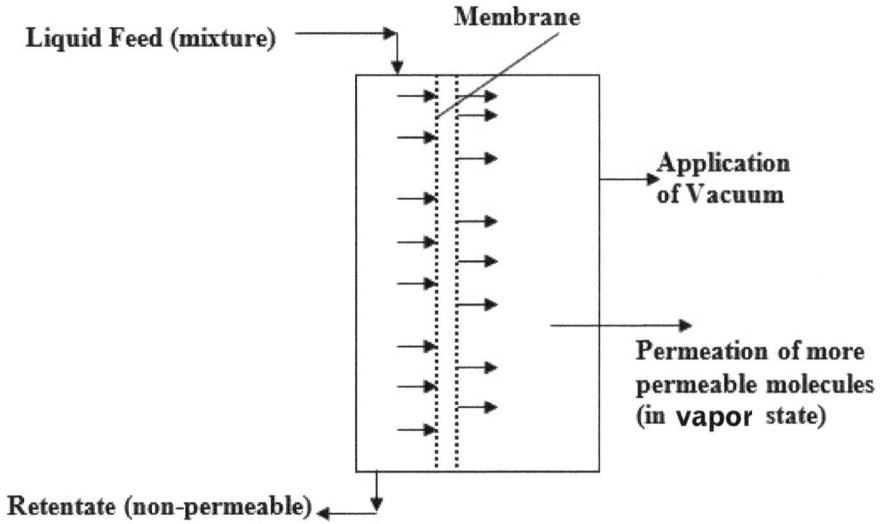

Fig. 3.12b. Schematic of Liquid Pervaporation Through Membrane.

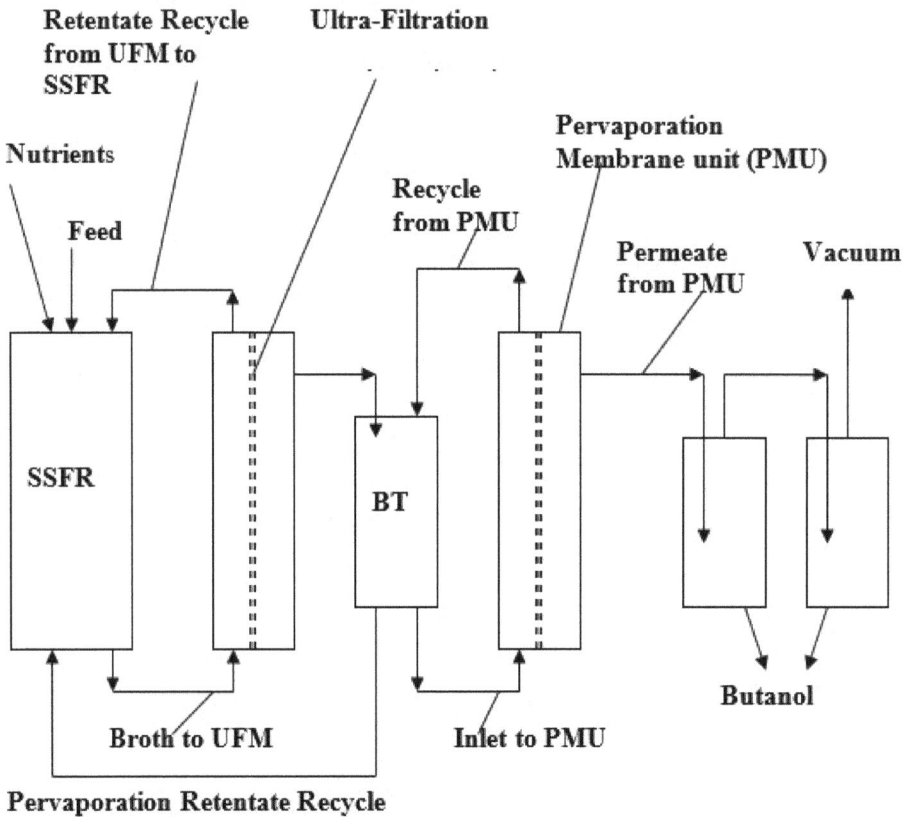

Legend: SSFR—Simultaneous Saccharification Fermentation Reactor; BT—Buffer tank;
UFM—Ultra filtration membrane; PMU—Pervaporation membrane unit; CT—Cold traps

Fig. 3.12c. Seperation of Bio Butanol by Pervaporation.

reactor (SSF Reactor), ultrafiltration membrane (UFM), buffer tank (BT), cold traps (CT), that is, condensers. The reactor temperature may be about 35°C, but the pervaporation temperature is about 78°C. The pervaporation membrane of the silicate-silicone may be used to selectively permeate butanol.

(i) Applicability of Pervaporation Technique

The pervaporation technique requires much less energy than fractional distillation. It should be applied for the purification of volatile organic liquids contaminated with a small amount of water as only the minor component (water) will have to be evaporated and permeated through hydrophilic membrane.

It should always be applied to organic liquids contaminated with a small amount of water/ moisture where the volatility difference is very small (that is, an ethanol water mixture near the azeotropic point). In this situation, pervaporation or vapor permeation is required as distillation fails to separate the liquids. Mixtures of two organic liquids with close volatility differences should also be separated by these techniques.

The pervaporation or vapor permeation techniques are not generally applicable for the separation of high boiling point organics except under special situations like foaming systems where kinetic effects dominate.

(ii) Membranes for Pervaporation or Vapor permeation

Membranes used for pervaporation may be of two types namely: 'hydrophilic/membranes' or 'organophilic membranes'.

Hydrophilic membranes (membranes which selectively filter out water from a mixture): These membranes are used for the removal of water from organic solutions or the concerned feed. These membranes have a glass transition temperature above room temperatures. Such membranes are used for the removal of water from organic compounds like polyvinyl alcohol.

Organophilic Membranes (membranes which selectively filter out an organic compound from a mixture): These membranes are used to recover or extract organic compounds from the mixture. Generally, they are made up of elastomers (elastic materials). The glass transition temperature of these membranes is below the room temperature. Styrene butadiene rubber, nitrile butadiene rubber, polydimethylsiloxane, silicate-silicone, etc., materials are suitable for these types of membranes. Membranes required for selectively passing certain organic compounds should also be 'hydrophobic' (i.e., they do not pass water through them) to keep the water away.

Polydimethylsiloxane (PDMS) membranes are used as organophilic membranes in the pervaporation technique for organic compounds as it is selective to them.

Silicate-silicone membrane has a very high selectivity for butanol and it can be used for butanol separation from a mixture of acetone, ethanol, etc., during the production of biobutanol.

It may be mentioned that the rate of permeation varies from membrane to membrane. The selection of a suitable membrane material is very important as it is the most active part of pervaporation.

Zeolite membranes can be used for dehydrating ethanol. However, these membranes are not suitable for pH values less than 6. Polymer membranes are restricted by a working temperature of less than 110°C and to some extent by the geometry of shape. The industrial-chemical industry uses plate-and-frame modules for pervaporation. These units are sealed with graphite compression gaskets that universally require adhesives, which are not resistant to solvents.

Ceramic porous membranes based on microporous silicon have also been developed. These membranes are not sensitive to acidity and can be operated at high temperatures up to 240°C. They are chemically and mechanically robust as well as hydrophilic and can be used to filter out water from organics. The ceramic membrane can also be prepared into tubular modules, which support and seal the membranes during the pressure difference (which drives the pervaporation).

d. Perstraction-Type Membrane Separation [124]

Perstraction is a type of membrane separation. It is a process where a substance to be separated is made to diffuse through a membrane into a solvent on the other side. The bacteria in the broth are separated from the solvent by the membrane. The design rate of diffusion of butanol from the membrane should always be higher than the rate of production of butanol in the reactor so that there is no chance of accumulation of butanol to a toxic level. Solvents like (i) Oley alcohol, or (ii) 50% Decane + 50% Oleyl alcohol may be used in the above recovery process. Mesitylene is also a possible candidate solvent for the above. A flow diagram of the perstraction process is shown in Figure 3.13.

The product obtained from the perstraction process is a mixture of solvent (about 50% Decane + 50% Oleyl alcohol), butanol, ethanol, and acetone. For efficient separation of these liquids, five different columns operating at different pressures and temperatures and associated condensers may be used. However, the energy consumption in the process is quite high.

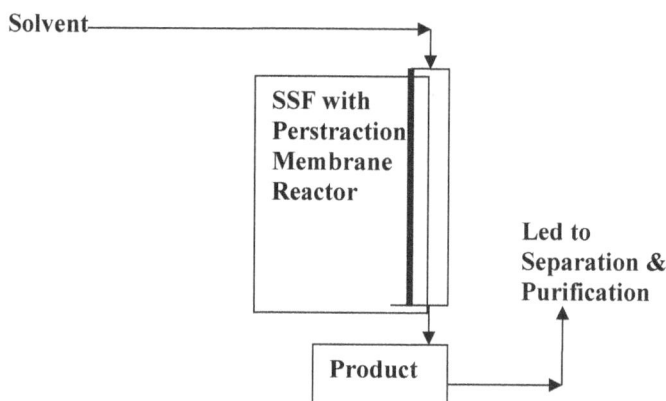

Fig. 3.13. Simultaneous Saccharification and Fermentation (SSF) with Perstraction Membrane Reactor.

e. Butanol Recovery by Ionic Liquid (IL)* [92]

The release of butanol from the fermentation broth is a difficult technical problem. The extraction process using conventional solvents may be useful, but requires solvents that are often volatile, toxic, and dangerous. Nowadays, ionic liquids (IL) are being increasingly used for butanol extraction. The ionic liquid is a mixture of non-volatile, environmentally friendly solvents used for various chemical processes. The use of ionic liquids for the extraction of butanol from the fermentation broth can be realized through the direct application of the liquid in the bioreactor, and the separation of butanol outside the bioreactor. Due to the extremely low solubility of ionic liquid (IL) in water, the fraction of the extract (having ionic liquid + butanol) from the reactor would constitute a separate phase, which has to be taken out from the reactor and led to an evaporator where the ingredients are extracted and distilled. Regenerated ionic liquid after suitable cooling is recycled to the fermentation tank.

Note on Ionic Liquid: *Ionic liquids are organic salts present in the liquid state at room conditions, and have very low vapor pressure and low solubility in water. The combinations of cations and anions give sixteen different ionic liquids (IL). In addition, the substitution of the corresponding radical in the structure of cations allows obtaining several times more IL. This can be taken into account while designing a suitable ionic liquid to test the extraction process.*

3.11.3 Some Researches Regarding Biobutanol Production

A lot of research work is being carried out in the world for efficient and cheaper production of biobutanol. In view of this, increased and a larger scale production of biobutanol may come up in future and it is also expected that in the future some ethanol-manufacturing facilities could be modified to produce butanol. Some of the research regarding biobutanol production are mentioned below.

(i) Clostridiaceae family of Bacteria [1,4,16,17,25,31,33,53,61,76,80,81,82]

The Clostridiaceae family of bacteria is used to produce butanol. The bacteria produce acetone, butanol, and ethanol, along with acetate, butyrate, carbon dioxide, and hydrogen. It is therefore sometimes referred to as 'ABE fermentation'. One strain of this bacteria (called *Clostridium saccharoperbutylaccetonicum*) is a hyper butanol- producing strain. It is somewhat resistant to product inhibition, and can therefore produce a higher concentration of butanol than the commonly used bacteria.

For large-scale butanol production, it is better to use immobilized microbial cells which have advantages like: (i) easier separation from product, (ii) reaching higher cell density, (iii) greatly improved productivity.

(ii) Ralstonia Eutropha H16—Production of butanol from CO_2 [24,32,41,63,67,68]

The lithoautotrophic microorganism known as '*Ralstonia eutropha* H16' was found to produce isobutanol and 3-methyl-1-butanol (3MB) by fermentation using an electro-bioreactor with the input of carbon dioxide as the carbon source and electrical energy. Solar panels were used to convert sunlight to electrical energy, which was then used along with the microorganism for conversion to a chemical intermediate.

The photovoltaic device is relatively efficient in converting sunlight to electricity. The solar cell is linked with biological CO_2 fixation and fuel production. Theoretically, H_2 generated by solar electricity can drive CO_2 fixation in lithoautotrophic microorganisms engineered to synthesize high-energy density liquid fuels. The efficiency and scalability of the process is limited by the low solubility, low mass transfer rate, and safety issues of H_2 in microbial cultures. Formic acid is a better energy carrier compared to H_2. The electrochemical production of formic acid from CO_2 and H_2O can achieve relatively high efficiency. Formate is highly soluble and is readily converted to CO_2 and NADH* in the cells, providing a safe replacement for H_2. However, the high solubility of formate increases the cost of separation. Accumulated formate will decompose at the anode, decreasing the yield of the process. Therefore, simultaneous electrochemical formate production and biological formate conversion to higher alcohols is desirable. Unfortunately, the introduction of electricity to microbial cultures may impede cell growth. Therefore, an integrated process for the reduction of CO_2 to liquid fuel powered by electricity requires (i) the metabolic engineering of a lithoautotrophic organism to produce liquid fuels, (ii) electrochemical generation of formate from CO_2 in fermentation medium, and (iii) enabling microbes to withstand electricity. So, a strain of the microorganism *R. eutropha* was engineered (named as *R. eutropha* LH74D) to do the job.

Electricity powered the electrochemical CO_2 reduction on the cathode to produce formate, which is converted to isobutanol and 3MB by the engineered microorganism *R. eutropha* LH74D. (The microorganism shows healthy growth and produces biofuels in the integrated electro-microbial reactor with electricity and CO_2 as the sole energy and carbon sources, respectively.) [67,68]

Note: Nicotinamide Adenine Dinucleotide, that is, NAD, is a coenzyme, partly made from nicotinamide (vitamin B3). NAD^+ is an oxidizing agent—it accepts electrons from other molecules and becomes reduced. This reaction forms NADH, which can then be used as a reducing agent to donate electrons. These electron transfer reactions are the main function of NAD/NADH.

(iii) Strain of Clostridium named TU-103 [17,71]

In 2011, Tulane University (US) scientists found a unique strain of the bacteria genus Clostridium that could convert any cellulose into butanol even in the presence of oxygen.

(iv) Isolation of Butanol Molecules during Fermentation [34,91]

A method was found by Hao Feng [34] to isolate the butanol molecules during the fermentation process so that the butanol-producing bacteria was not poisoned at the higher concentration of butanol. After the fermentation process, an energy-efficient process called 'cloud point separation' was used to recover the butanol.

(v) Algae/Diatoms [26,42,56,93,96]

Scientists have attempted to engineer reaction pathways that can enable photosynthetic organisms (like blue-green algae) to produce butanol more efficiently.

Some of the microalgae contain a relatively high percentage of sugars in dry matter, such as Chlorella contains about 30–40% of sugars, which greatly increases their usefulness in the production of biobutanol.

The green waste left over from the algae oil extraction can be used to produce butanol. Some macroalgae (seaweed) can be fermented by the bacteria genus 'Clostridia' to produce butanol and other solvents.

(vi) Use of Glycerol for Production of Butanol

The metabolic engineering of microorganisms (for example, *Clostridium pasteurianum* bacterium) allows the production of butanol from glycerol, which is a by-product created during the production of biodiesel. Butanol generated from glycerol by fermentation with '*C. pasteurianum*' is low. A hyper butanol-producing bacterial strain 'MBEL GLY2' has been created using a chemical mutant. It may be used during glycerol fermentation to get butanol.

(vii) Use of Succinate and Ethanol to Produce Butanol

A combination of succinate and ethyl alcohol can be fermented to butyrate, which can then be converted to butanol. Metabolic engineered microorganisms are used for the process.

3.12 Synthesis of Butanol from Bioethanol [62,90,104]

As bioethanol manufacture is much simpler than manufacturing biobutanol, the ethanol obtained from plant sources can be converted to butanol for actual use in vehicles. The chemical reactions involved in conversion/synthesis of butanol from ethanol are as follows:

$$2C_2H_5OH \text{ (By dehydrogenation)} = 2CH_3 CHO + 2 H_2 \tag{1a}$$

Ethanol = acetaldehyde + hydrogen

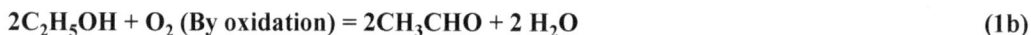

$$2C_2H_5OH + O_2 \text{ (By oxidation)} = 2CH_3CHO + 2 H_2O \tag{1b}$$

Ethanol + oxygen = acetaldehyde + water

$$2CH_3CHO \text{ (in presence of dilute NaOH)} = CH_3 CHO H CH_2 CHO \tag{2}$$

acetaldehyde = Aldol

$$CH_3CHOHCH_2CHO \text{ (on heating)} = CH_3CHCHCHO + H_2O \tag{3}$$

Aldol = Crotonaldehyde water

$$CH_3 CH CH CHO + 2 H_2 = 2 CH_3CH_2CH_2 CH_2OH \text{ (or } C_4H_9OH) \tag{4}$$

Crotonaldehyde+hydrogen = Butanol

The reaction **(1a)** shows the dehydrogenation of ethanol yielding 'acetaldehyde' and 'hydrogen'. The dehydrogenation of ethanol can be done by passing the ethanol vapor over a catalyst at 260–310°C. The catalyst may be nickel supported by tin oxide (SnO_2), nickel supported by alumina (Al_2O_3), or nickel supported by silica (SiO_2).

The acetaldehyde can also be produced through oxidation as per reaction **(1b)**, where water is produced instead of hydrogen. The product of the reaction is put through fractional distillation to separate unreacted ethanol, which is recycled.

The acetaldehyde in presence of dilute sodium hydroxide undergoes condensation to form a syrupy liquid known as aldol as per reaction **(2)**. On heating, aldol forms an unsaturated compound known as 'crotonaldehyde' and separates water from the substance as per reaction **(3)**. 'Crotonaldehyde' is reacted with 'hydrogen' at about 150°C in the presence of a catalyst, copper and aluminum oxide as per reaction **(4)** to form butanol.

The hydrogen (which is recovered in reaction **(1a)**), is used up in the conversion of crotonaldehyde to butanol on reaction **(4)**. In case acetaldehyde is formed by oxidation reaction as per reaction **(1b)**, the hydrogen has to be provided additionally for reaction **(4)**.

3.13 Summary

This chapter is devoted to alcohols–a secondary renewable energy source. Alcohols can be made from sugar or sugar-containing materials, starch or starch-containing materials, and even lignocellulosic biomass. The technology of producing alcohols from lignocellulosic biomass has come as a boon as the production of alcohol from this source would not interfere with food production. Lignocellulosic feedstock includes materials like wheat straw, rice straw, corn stover, sweet sorghum stalks, crop field trashes such as stalks, clipping, corn cobs, grasses like switch grass (Pancium virgatum—one of the perennial grasses), hay, etc. Even green seaweed like Ulva fasciata can be used for the production of ethanol through enzymatic hydrolysis. Many technologies used for the production of alcohols from various materials are discussed in the chapter. The process for producing ethanol from biomass through biological fermentation involves pretreatment, hydrolysis, fermentation, and product purification.

The pretreatment breaks down the sheathing of hemicellulose and lignin in the biomass and disrupts the crystalline structure of cellulose to make it more accessible for enzymes to act on it for its conversion into sugars. Cutting and milling are the first steps in the pretreatment of biomass, after which the chemical and physicochemical pretreatment, biological and enzymatic pretreatments, or other pretreatments like by supercritical liquid carbon dioxide, supercritical hot water, or by wet oxidation by high-pressure water at a high temperature (between 170–200°C) and oxygen may be implemented. The options for physicochemical treatment are: treatment by acids like sulphuric acid/hydrochloric acid/phosphoric acid; treatment by alkalis like sodium hydroxide/ammonia, treatment by ammonium sulphite, treatment by gases like chlorine dioxide (ClO_2)/nitrogen dioxide (NO_2), sulphur dioxide (SO_2), ozone (O_3), or chemicals like hydrogen peroxide (H_2O_2), explosion by steam/CO_2/SO_2 and the solvent extraction of lignin.

The hydrolysis of pretreated biomass can be done by mineral acids, enzymes, etc. The hydrolysis converts carbohydrate polymers ($C_6H_{10}O_5$) like cellulose and starch into monomeric sugars (glucose, fructose).

Under anaerobic conditions, many of the microorganisms may ferment carbohydrates in the product obtained after hydrolysis of biomass to ethanol. The microorganisms may be bacteria, yeast, or fungi.

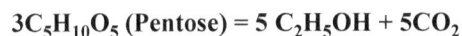

$C_6 H_{12} O_6$ (Glucose/Hexose) = $2C_2H_5OH$ (ethanol) + $2CO_2$ (Carbon dioxide)

$3C_5H_{10}O_5$ (Pentose) = 5 C_2H_5OH + $5CO_2$

Purification of ethanol is first done by distillation, by which ethanol with a purity of 95.63% (maximum by mass) can be obtained. This limit to which the ethanol can be separated from water is known as the 'azeotropic point' of ethyl alcohol. Further drying or purification may be performed by hygroscopic materials like quick lime, rock salt, or using membrane separation/ pervaporation/ vapor permeation.

Consolidation of various steps in bioprocessing of cellulose can be done to save energy and achieve higher efficiency in conversion to fuels like alcohol/hydrogen from lignocelluloses. The various steps of the process may be integrated in different ways. In all cases, however, the pretreatment of lignocellulosic biomass would be required to convert the hemicellulose present in biomass to sugars so that the total cellulose can be made accessible to the enzymes for hydrolysis. In all these approaches, the acid hydrolysis is replaced by enzymatic hydrolysis.

(i) SHF: The first approach of consolidation is 'separate hydrolysis and fermentation (SHF)'. In this approach, after hydrolysis, the liquid enters the 'glucose fermentation reactor'. The mixture is then distilled in the beer column, where the ethyl alcohol along with some moisture is separated out. The remaining liquid containing unconverted 'xylose (wood sugar)' is then sent to a second reactor where xylose is fermented to ethanol and the effluent is distilled to recover ethanol in a beer column.

(ii) SSF: The second approach of consolidation is the 'simultaneous saccharification and fermentation (SSF)'. The saccharification means the conversion of cellulose to saccharides (sugars). There is no separate hydrolysis reactor and hydrolysis and fermentation are consolidated. This avoids the problem of inhibition of hydrolysis by presence of glucose. (The presence of glucose tries to inhibit the reaction of enzymatic hydrolysis.)

(iii) SSCF: The third approach is the modification of SSF where a mixture of enzymes is used for fermentation of glucose and other sugars like xylose and pentose in a single reactor. This is called 'simultaneous saccharification and co-fermentation (SSCF)'.

(iv) CBP: The fourth approach called 'consolidated bioprocessing (CBP)', in which the enzyme (cellusome) production, substrate hydrolysis, and fermentation by celluloytic microorganisms is achieved in a single step.

There is a great potential for producing ethanol from widely available cheap raw material containing cellulose. There are a number of technologies available for such production. Efforts are being made world over to commercially produce ethanol from cellulosic materials. In gasification process, all the combustible matter including `Lignin' is converted to syn-gas which is later processed for making alcohol either by bio-fermentation or synthesis. It is well known that `Ethanol' is used for blending with `Petrol'; the blending may be 5%, 10%, or more. Gasoline blended with ethanol has been used in many countries. Gasoline blended with ethanol cuts down emissions of CO, PAH (including benzene and butadiene), particulate matter, formaldehyde, and other air pollutants. It also reduces the emission of greenhouse gases.

Butanol is primarily used as a solvent, and as a fuel. When it is produced biologically, it is called *bio butanol*. Like many alcohols, butanol is considered to be toxic. Butanol can also be manufactured from plant resources and therefore it is considered as a potential renewable biofuel and can be used in internal combustion engines. It can be blended with gasoline in much higher percentage than ethanol without any modifications in the engine. Bio-butanol reduces emissions, namely hydrocarbons, CO and oxides of nitrogen (NOx). Butanol can also be used as a blended additive to diesel fuel to reduce soot emissions. It is also used as a component of hydraulic fluids and brake fluids. It should be easier to start the vehicles in cold weather than the one running on ethanol or methanol.

Butanol may be stored on board on hydrogen fueled vehicles having fuel-cells using hydrogen. The on-board reforming butanol can provide hydrogen for fuel cells of the vehicle as it has higher hydrogen content than ethanol or methanol and therefore it can serve better to solve problem of on board storage of hydrogen for vehicles. One of the biggest challenges facing the development of hydrogen fuel cell vehicles is the storage of enough on-board hydrogen and the lack of hydrogen infrastructure for fueling. This problem can be solved by storing bio butanol instead of hydrogen on the vehicle for conversion to hydrogen on board.

The butanol can be mixed with gasoline in a range from 8% to 16%. Companies are planning to market a fuel that is 85% Ethanol and 15% Butanol (E85B), so existing E85 internal combustion engines can run on a 100 % renewable fuel (without fossil fuels).

Butanol was generally produced from fossil fuels, but now it is being produced by bacterial fermentation of a variety of feed stocks. *One such bacteria used for fermentation is Clostridium acetobutylicum.* The problem with this type of microbe is that it is poisoned by the very butanol it produces once its (butanol) concentration rises above approximately 2 %. Great strides have been made by researchers in creating "designer microbes" that can tolerate concentrations of butanol without being killed off.

Bio ethanol production plants can be modified for production of *bio-butanol. Fractional Distillation can be used for separating the products, but there are other processes also for obtaining butanol from the broth.* Because of inhibitory effect of butanol as it is toxic to bacterial culture, the butanol can only be produced as a diluted component i.e. the butanol produced has to be continuously removed from the broth. There are several methods of removing the product continuously from the fermentation broth such as: (i) Adsorption, (ii) Gas stripping, (iii) Pervaporation, (iv) Perstraction, (v) Ionic liquid extraction, and (vi) Vacuum distillation.

As the bio-ethanol manufacture is much simpler than manufacturing biobutanol, the ethanol obtained from plant sources can be converted to butanol for actual use in vehicles. Enormous amount of research is being carried out the world over for efficient and cheaper production of bio-butanol. Some specific methods have been described in the text.

QUESTIONS

1. Describe the procedure for obtaining ethanol from sugar/molasses.
2. How is the process of obtaining ethanol from starch different from producing ethanol from molasses? Describe the procedure.
3. What is meant by lignocellulosic biomass? How do the lignin, hemicellulose, and cellulose differ from each other?
4. What is meant by cellulosic ethanol?
5. Which lignocellulosic feed materials can be used for the production of alcohols by biological methods? Why should the production of alcohols from biomasses be preferred over the production of alcohols from food crops?
6. What are the differences in sugars obtained by the pretreatment and hydrolysis of hemicellulose and cellulose?
7. What is the importance of pretreatment in a biological process of producing ethanol?
8. How is the lignocellulosic material/biomass pretreated before it is used to obtain ethanol from it by a biological process? In your opinion, which is the best process for the pretreatment of biomass?
9. Describe the biological procedure of obtaining ethanol from lignocellulosic materials/biomass. What is the fate of the carbon dioxide gas that is produced in the fermentation process?
10. What are enzymes? Give examples of enzymes used in the production of alcohols.
11. What is the biological method of pretreatment of biomass? Which bacteria, enzyme, or fungi can be used in the treatment?
12. What are the functions of the following with regard to conversion of woody biomass to alcohol: (a) White rot fungus, (b) Clostridia bacteria, (c) Elephant yeast, (d) Gribble, (e) Zymomonas sp.?
13. Which sugars are obtained by the pretreatment and hydrolysis of seaweed biomass? Can you ferment them to obtain alcohol? If so, indicate the microorganisms capable of fermenting alginate and mannitol sugars.
14. Describe a method of converting seaweed to alcohol.
15. What are the technologies for purifying alcohols? What is the azeotropic point and what is its importance?
16. Describe the procedure for obtaining alcohols through gasification route.

17. *How can you obtain alcohol from lignin? What are the alternate uses of lignin?*

18. *How would you convert municipal solid waste to alcohols?*

19. *Enumerate the advantages of bioethanol as a fuel for vehicles. Describe the advantages and disadvantages of biobutanol as a fuel for vehicles.*

20. *Briefly describe the biological method for producing butanol. What are the methods used to separate butanol from the fermentation broth? Why is it essential to separate butanol continuously during the biological production of butanol?*

21. *Enumerate the various microorganisms and enzymes used in making bioethanol and biobutanol.*

22. *Describe the process of pervaporation. How does it differ from vapour permeation?*

23. *Write a note on the membranes used for pervaporation process. What do you mean by hydrophilic and organophilic membranes?*

24. *Write short notes on the use of the following processes in biobutanol production: (i) Adsorption, (ii) Gas stripping*

25. *Write short notes on the use of the following processes in biobutanol production:*

    (i) Perstraction, (ii) pervaporation, (iii) Ionic Liquids

26. *Compare the advantages and disadvantages of various processes of manufacturing/separation of biobutanol from the broth.*

27. *Compare the advantages and disadvantages of using the following fuels in the vehicles: (i) Petrol/Gasoline; (ii) Mixture of ethanol and petrol; (iii) Pure ethanol (iv) Butanol*

28. *Give the principle of manufacturing butanol from bioethanol by synthetic route.*

29. *What are the potential problems in using biobutanol as a fuel for engines.*

30. *What do the following numbers indicate with regard to fuels for internal combustion engines? (i) RON (ii) MON (iii) CN*

31. *How does ethanol's energy output compare with that of gasoline?*

32. *What are the effects on air pollution and greenhouse gas emission by replacing gasoline by ethanol or ethanol-blended gasoline as fuel in vehicles? What are the other advantages?*

References

1. Adsul, M.G., M.S. Singhvi, S.A. Gaikaiwari and D.V. Gokhale. March 2011. Development of biocatalysts for production of commodity chemicals from lignocellulosic biomass (Review). Bioresource Technology 102(6): 4304–4312. https://doi.org/10.1016/j.biortech.2011.01.002.

2. Achinas, Spyridon, Gerrit Jan and Willem Euverink. August 2016. Consolidated briefing of biochemical ethanol production from lignocellulosic biomass (Review). Electronic J. of Biotechnology 23: 44–53. open access. https://doi.org/10.1016/j.ejbt.2016.07.006.

3. Air Quality: Ethanol a clear net benefit (Pacific Ethanol Inc.). [http://www.pacificethanol.net/resources/air-quality-ethanol-a-clear-net-benefit].

4. Ánxela, Fernández-Naveira, Haris Nalakath Abubackar, María C. Veiga and Christian Kennes. April 2016. Efficient butanol-ethanol (B-E) production from carbon monoxide fermentation by Clostridium carboxidivorans. Applied Microbiology and Biotechnology 100(7): 3361–3370. Bioenergy and biofuels, Online: 25 January 2016.

5. Badger, P.C. 2002. Ethanol from cellulose: A review. pp. 17–21. *In*: Janick, J. and A. Whipkey (eds.). Trends in new crop and new uses. ASHS Press, Alexandria, VA.

6. Balat, M. and H. Balat. 2009. Recent trends in global production and utilization of bioethanol fuel. Applied Energy 86(11): 2273–2282.

7. Balat, Mustafa. February 2011. Production of bioethanol from lignocellulosic materials via the biochemical pathway: A review. Energy Conversion and Management 52(2): 858–875. https://doi.org/10.1016/j.enconman.2010.08.013.

8. Banerjee, Saumita, Sandeep Mudliar, Ramkrishna Sen, Balendu Giri, Devanand Satpute, Tapan Chakrabarti et al. December 2009. Review: Commercializing lignocellulosic bioethanol, technology bottlenecks and

possible remedies. Published online in Wiley Inter Science (www.interscience.wiley.com). DOI: 10.1002/bbb.188; Biofuels, Bioprod. Bioref. 4: 77–93 (2010).

9. Basile, Angelo, Alberto Figoli and Mohamed Khayet (eds.). 2015. Pervaporation, Vapour Permeation and Membrane Distillation-Principles and Applications. Elsevier Ltd.

10. Bergquist, P.L., M.D. Gibbs, D.D. Morris, V.S. Te'O, D.J. Saul and H.W. Morgan. 1999. Molecular diversity of thermophilic cellulolytic and hemi cellulolytic bacteria. FEMS Microbiology Ecology 28: 99–110.

11. Biobutanol: The Next Big Biofuel? http://biomassmagazine.com/articles/1605/biobutanol-the-next-big-biofuel?

12. Bioconversion of biomass to mixed alcohol fuels From Wikipedia, the free encyclopedia. https://en.wikipedia.org/wiki/Bioconversion_of_biomass_to_mixed_alcohol_fuels.

13. Bioethanol by Gasification of Biomass INEOS group. (http://www.ineos.com/news/ineos-group/ineos-bio-produces-cellulosic-ethanol-at-commercial-scale/). Also see: https://doi.org/10.1002/cite.201900118.

14. Bisaria, Virendra S. and Tarun K. Ghose. April 1981. Biodegradation of cellulosic materials: Substrates, microorganisms, enzymes and products: Review. Enzyme and Microbial Technology 3(2): 90–104. https://doi.org/10.1016/0141-0229(81)90066-1.

15. Bohlmann, Gregory M. March 2006. Process economic considerations for production of ethanol from biomass feedstocks. Industrial Biotechnology 2(1). Published Online: https://doi.org/10.1089/ind.2006.2.14.

16. Boisset, C., H. Chanzy, B. Henrissat, R. Lamed, Y. Shoham and E.A. Bayer. 1999. Digestion of crystalline cellulose substrates by the *Clostridium thermocellum* Cellulosome: Structural and morphological aspects. Biochemical Journal 340: 829–835.

17. Branduardi, Paola, Francesca de Ferra, Valeria Longo, Danilo Porro. January 2014. Review: Microbial n-butanol production from Clostridia to non-Clostridial hosts. Engineering in Life Sciences 14(1): 16–26. https://doi.org/10.1002/elsc.201200146.

18. Brethauer, Simone and Charles E. Wyman. July 2010. Review: Continuous hydrolysis and fermentation for cellulosic ethanol production. Bioresource Technology 101(13): 4862–4874. https://doi.org/10.1016/j.biortech.2009.11.009.

19. Butanol from biomass used as a fuel. Wikipedia, the free encyclopedia. https://en.wikipedia.org/wiki/Butanol_fuel.

20. Butanol fuel From Wikipedia, the free encyclopedia. https://en.wikipedia.org/wiki/Butanol_fuel#Production_of_biobutanol. 6.

21. CBIP Water & Energy Research Digest April – June 2008.

22. Cellulosic ethanol commercialization, From Wikipedia, the free encyclopedia. https://en.wikipedia.org/wiki/Cellulosic_ethanol_commercialization.

23. Cellulosic ethanol From Wikipedia, the free encyclopedia. https://en.wikipedia.org/wiki/Cellulosic_ethanol.

24. Chen, Xiaoli, Yingxiu Cao, Feng Li, Yao Tian and Hao Song. April 2018. Enzyme-assisted microbial electrosynthesis of poly(3 hydroxybutyrate) via CO2 bioreduction by engineered Ralstonia eutropha. ACS Catal. 8(5): 4429–4437. https://doi.org/10.1021/acscatal.8b00226.

25. Desvaux, M. 2005. *Clostridium celluloyticum*: model organism of mesophelic cellulolytic clostridia. FEMS Microbiology Reviews 29: 741–764.

26. Diatom, from Wikipedia, the free encyclopedia. https://en.wikipedia.org/wiki/Diatom.

27. DuPont Celebrates the Opening of the World's Largest Cellulosic Ethanol Plant. (http://www.dupont.com/corporate-functions/media-center/press-releases/dupont-celebrates-open).

28. Ethanol from Biomass. December 2013. Extension, Wood Energy. https://articles.extension.org/pages/70128/ethanol-from-biomass.

29. Evans, Jon. January 2012. New microbe turns sugary seaweed into fuel. (https://www.chemistryworld.com/news/new-microbe-turns-sugary-seaweed-into-fuel/3002849.article)].

30. Ezeji, T.C., N. Qureshi and H.P. Blaschek. August 2003. Production of acetone, butanol and ethanol by Clostridium beijerinckii BA101 and in situ recovery by gas stripping. World J. of Microbiology and Biotechnology 19(6): 595–603.

31. Ezeji, Thaddeus C., Nasib Qureshi and Hans P. Blaschek. December 2007. Production of acetone butanol (AB) from liquefied corn starch, a commercial substrate, using Clostridium beijerinckii coupled with product recovery by gas stripping. J. of Industrial Microbiology & Biotechnology 34(12): 771–777.

32. Fast, Alan G. and Eleftherios T. Papoutsakis. November 2012. Stoichiometric and energetic analyses of non-photosynthetic CO2-fixation pathways to support synthetic biology strategies for production of fuels and chemicals. Current Opinion in Chemical Engineering 1(4): 380–395. https://doi.org/10.1016/j.coche.2012.07.005.

33. Felix, R.C. and L.G. Ljungdahl. 1993. The cellulosome: the exocellular organelle of clostridium. Annual Reviews of Microbiology 47: 791–819.

34. Feng, Hao. August 2012. Scientist Makes Use of Butanol as Biofuel more Appealing. University of Illinois, Illinois, USA.

35. Fujita, Yasuya, Shouji Takahashi, Mitsuyoshi Ueda, Atsuo Tanaka, Hirofumi Okada, Yasushi Morikawa et al. Oct. 2002. Direct and efficient production of ethanol from cellulosic material with a yeast strain displaying cellulolytic enzymes. Appl. Environ. Microbiol. 68(10): 5136–41. DOI: 10.1128/aem.68.10.5136-5141.2002.

36. Galbe, Mats and Guido Zacchi. 2007. Pretreatment of lignocellulosic materials for efficient bioethanol production. Biofuels, pp. 41–65. Part of the Advances in Biochemical Engineering/Biotechnology book series ABE, volume 108.

37. Ge, Leilei, Peng Wang and Haijin Mou. January 2011. Study on saccharification techniques of seaweed wastes for the transformation of ethanol. Renewable Energy 36(1): 84–89. https://doi.org/10.1016/j.renene.2010.06.001.

38. Ghazala, Mustafa. A., Hassan A.H. Ibrahimb, Nayrah A. Shaltouta and Alaa. E. Alic. 2016. Biodiesel and Bioethanol Production from Ulva fasciata Delie Biomass via Enzymatic Pretreatment using Marine-Derived Aspergillus niger. Int. J. Pure App. Biosci. 4 (5): 1–16. DOI: http://dx.doi.org/10.18782/2320-7051.2374 ISSN.

39. Groot, W.J. and K.Ch.A.M. Luyben. October 1986. *In situ* product recovery by adsorption in the butanol/isopropanol batch fermentation. Applied Microbiology and Biotechnology 25(1): 29–31.

40. Groot, W.J., R.G.J.M. vanderLans and K.Ch.A.M. Luyben. March 1992. Review: Technologies for butanol recovery integrated with fermentations. Process Biochemistry. 27(2): 61–75. https://doi.org/10.1016/0032-9592(92)80012-R.

41. Hawkins, Aaron S., Patrick M. McTernan, Hong Lian, Robert M. Kelly and Michael W.W. Adams. June 2013. Biological conversion of carbon dioxide and hydrogen into liquid fuels and industrial chemicals. Current Opinion in Biotechnology 24(3): 376–384. https://doi.org/10.1016/j.copbio.2013.02.017.

42. Hebbale, Deepthi, R. Bhargavi and T.V. Ramachandra. March 2019. Saccharification of macroalgal polysaccharides through prioritized cellulase producing bacteria. Heliyon. 5(3): e01372. doi: 10.1016/j.heliyon.2019.e01372.

43. Hönig, V., M. Kotek and J. Mařík. 2014. Use of butanol as a fuel for internal combustion engines. Agronomy Research 12(2): 333–340.

44. Horn, S.J., I.M. Aasen and K. Østgaard. November 2000. Ethanol production from seaweed extract. J. of Industrial Microbiology and Biotechnology 25(5): 249–254.

45. Horn, S.J., I.M. Aasen and K. Østgaard. January 2000. Production of ethanol from mannitol by Zymobacter palmae. J. of Industrial Microbiology and Biotechnology 24(1): 51–57.

46. http://en..wikipedia.org/wki/cellulosicethanol and http://www.egineer.ucla.edu./newsroom/featured news/archieve/2012/ucla-engineering-research-use-electricity-to-generate-alternative-fuel.

47. https://www.researchgate.net/publication/24238399_Third_Generation_Biofuels_via_Direct_Cellulose_Fermentation.

48. Huang, J. and M.M. Meagher. 2001. Pervaporative recovery of n-butanol from aqueous solutions and ABE fermentation broth using thin-film silicalite-filled silicone composite membranes. J. Membr. Sci. 192: 231–242.

49. Isikgor, Furkan H. and Remzi C. Becer. May 2015. Lignocellulosic biomass: a sustainable platform for production of bio-based chemicals and polymers(Review). Polym. Chem. 6: 4497–4559. Open access. https://doi.org/10.1039/C5PY00263J.

50. Jang, Ji-Suk, YuKyeong Cho, Gwi-Taek Jeong and Sung-Koo Kim. January 2012. Optimization of saccharification and ethanol production by simultaneous saccharification and fermentation (SSF) from seaweed, Saccharina japonica. Bioprocess and Biosystems Engineering 35(1-2): 11–18.

51. Janusz, Grzegorz, Anna Pawlik, Justyna Sulej, Urszula Świderska-Burek, Anna Jarosz-Wilkołazka and Andrzej Paszczyński. November 2017. Lignin degradation: microorganisms, enzymes involved, genomes analysis and evolution. FEMS Microbiol. Rev. 41(6): 941–962. Published online 2017 Oct 27. DOI: 10.1093/femsre/fux049.

52. Jin, M., V. Balan, C. Gunawan and B.E. Dale. June 2011. Consolidated bioprocessing (CBP) performance of Clostridium phytofermentans on AFEX-treated corn stover for ethanol production. Biotechnology and Bioengineering 108(6): 1290–1297. https://doi.org/10.1002/bit.23059.

53. Joshua, T., Ellis Neal, N. Hengge Ronald, C. Sims Charles and D. Miller. May 2012. Acetone, butanol, and ethanol production from wastewater algae. Short Communication. Bio-resource Technology 111: 491–495. Elsevier. https://doi.org/10.1016/j.biortech.2012.02.002.

54. Jouzani, Gholamreza Salehi and Mohammad J. Taherzadeh. 2015. Advances in consolidated bioprocessing systems for bioethanol and butanol production from biomass: a comprehensive review. Biofuel Research Journal. 2(1): Article 4, pp. 152–195. DOI: 10.18331/BRJ2015.2.1.4.

55. Kaminski, W., Elwira Tomczak and Andrzej Górak. Jan. 2011. Biobutanol - Production and purification methods. Ecological Chemistry and Engineering. S = Chemia i Inżynieria Ekologiczna. S 18(1): 31–37. https://www.researchgate.net/publication/279908533_Biobutanol_-_Production_and_purification_methods.

56. Karthick, B., P.B. Hamilton and J.P. Kociolek. 2013. An Illustrated Guide to Common Diatoms of Peninsular India. Ist edition, Gubbi Labs LLP.

57. Kim, H.K. and J. Hing. 2001. Super critical carbon dioxide pre-treatment of lignocellulose enhances enzymatic cellulose hydrolysis. Bio-resource Technology 77(2): 139–144.

58. Kim, S. and M.T. Holtzapple. 2005. Lime pretreatment and enzymatic hydrolysis of corn stover. Bioresour Technol. 96: 1994–2006.

59. Kuhad, Ramesh Chander, Sarika Kuhar, Krishna Kant Sharma and Bhuvnesh Shrivastava. February 2013. Microorganisms and Enzymes Involved in Lignin Degradation Vis-à-vis Production of Nutritionally Rich Animal Feed: An Overview. Biotechnology for Environmental Management and Resource Recovery. pp. 3–44, Chapter, DOI 10.1007/978-81-322-0876-1_1.

60. Kusch-Brandt, Sigrid and Maria V. Morar. August 2009. Integration of Lignocellulosic Biomass Into Renewable fuel Energy Generation Concepts. Pro-Environmen. 2: 32–37. https://www.researchgate.net/publication/47641644_Integration_of_Lignocellulosic_Biomass_into_Renewable_Energy_Generation_Concepts.

61. Lamed, R., E. Setter and E.A. Bayer. 1983. Characterization of a cellulose-binding, cellulose-containing complex in *Clostridium thermocellum*. J. of Bacteriology. 156: 828–836.

62. Lan, Ethan I. and James C. Liao. May 2013. Microbial synthesis of n-butanol, isobutanol, and other higher alcohols from diverse resources. Bioresource Technology 135: 339–349. https://doi.org/10.1016/j.biortech.2012.09.104.

63. Lan, Ethan I. and James C. Liao. July 2011. Metabolic engineering of cyanobacteria for 1-butanol production from carbon dioxide, Metabolic Engineering, Volume 13(4): 353–363. https://doi.org/10.1016/j.ymben.2011.04.004.

64. Larsen, Ulrik, Troels Johansen and Jesper Schramm. May 2009. Ethanol as a Fuel for Road Transportation. Main Report Technical University of Denmark. Denmark.

65. Laureano-Perez, L., Farzaneh Teymouri, Hasan Alizadeh and Bruce E. Dale. 2005. Understanding factors that limit enzymatic hydrolysis of biomass: Characterization of pretreated corn Stover. Applied Biochemistry and Biotechnology 124(1-3): 1081–1099.

66. Li, Demao, Limei Chen, Xiaowen Zhang, Naihao Ye and FuguoXing. May 2011. Pyrolytic characteristics and kinetic studies of three kinds of red algae. Biomass and Bioenergy 35(5): 1765–1772. https://doi.org/10.1016/j.biombioe.2011.01.011.

67. Li, Han, Paul H. Opgenorth, David G Wernick, Steve Rogers, Tung-Yun Wu, Wendy Higashide et al. March 2012. Integrated Electromicrobial Conversion of CO_2 to Higher Alcohols. Science. 335(6076): 1596. doi: 10.1126/science.1217643. (https://www.researchgate.net/publication/223990342_Integrated_Electromicrobial_Conversion_of_CO2_to_Higher_Alcohols).

68. Li, Jiansheng, Yao Tian, Yinuo Zhou and Yongchao Zong. June 2020. Abiotic–biological hybrid systems for CO2 conversion to value-added chemicals and fuels. Transactions of Tianjin University. DOI: 10.1007/s12209-020-00257-5.

69. Liu, F., L. Liu and X. Feng. 2005. Separation of acetone-butanol-ethanol (ABE) from dilute aqueous solution by pervaporation. Separation Purification Technology 42: 273–282.

70. Liu, Zhuo, Kentaro Inokuma, Shih-Hsin Ho, Riaan den Haan, Tomohisa Hasunuma, Willem H. van Zyl and Akihiko Kondo. Sep. 2015. Combined cell-surface display- and secretion-based strategies for production of cellulosic ethanol with Saccharomyces cerevisiae. Biotechnol Biofuels 8: 162. doi: 10.1186/s13068-015-0344-6. eCollection 2015.

71. Maiti, Sampa, Gorka Gallastegui, Satinder Kaur Brar, Yann LeBihan, Gerardo Buelna, Patrick Drogui et al. March 2016. Quest for sustainable bio-production and recovery of butanol as a promising solution to fossil fuel. Review Paper, Int. J. of Energy Research. 40(4): 411–438. https://doi.org/10.1002/er.3458.

72. Martin, C., Mette H. Thomsen, Henrik Hauggaard-Nielsen and Anne Belinda Thomsen. December 2008. Wet oxidation pretreatment, enzymatic hydrolysis and simultaneous Saccharification and fermentation of clover-ryegrass mixtures, Bio resource Technology 99(18): 8777+8782. https://doi.org/10.1016/j.biortech.2008.04.039.

73. Martin, Mariano and Ignacio E. Grossmann. Sept. 2010. Energy Optimization of Bioethanol Production via Gasification of Switchgrass. Department of Chemical Engineering, Carnegie Mellon University, Pittsburgh, PA, USA. http://dx.doi.org/10.1002/aic.12544.

74. Mathewson, S.W. 1980. The Manual for the Home and Farm Production of Alcohol Fuel. Chapter 12. Ten Speed Press, J.A. Diaz Publications.

75. Mbaneme, V. and M. Chinn. 2015. Consolidated bioprocessing for biofuel production: recent advances. Archived Journals: Energy and Emission Control Technologies. 2015(3): 23–44. DOI https://doi.org/10.2147/EECT.S63000.

76. MGreen, Edward. June 2011. Fermentative production of butanol—the industrial perspective. Current Opinion in Biotechnology 22(3): 337–343. https://doi.org/10.1016/j.copbio.2011.02.004.

77. Mosier, Nathan, Richard Hendrickson, Nancy Ho, Miroslav Sedlak and Michael R. Ladisch. Dec. 2005. Optimization of pH controlled liquid hot water pretreatment of corn stover. Bioresour. Technol. 96(18): 1986-93. PMID: 16112486. DOI: 10.1016/j.biortech.2005.01.013.

78. Mosier, Nathan, Charles Wyman, Bruce Dale, Richard Elander, Y.Y. Lee, Mark Holtzapple et al. 2005. Features of promising technologies for pretreatment of lignocellulosic biomass. Bioresource Technology 96: 673–686.

79. Mourato, Miguel P., Inês N. Moreira, Inês Leitão, Filipa R. Pinto, Joana R. Sales and Luisa Louro Martins. August 2015. Review: Effect of heavy metals in plants of the genus *Brassica*. Int. J. Mol. Sci. 16: 17975–17998. doi: 10.3390/ijms160817975.

80. Ndaba, B., I. Chiyanzu and S. Marx. December 2015. n-Butanol derived from biochemical and chemical routes: A review. Biotechnology Reports, 8: 1–9. Open access, Corpus ID: 9745292. https://doi.org/10.1016/j.btre.2015.08.

81. Nielsen, David R., Effendi Leonard, Sang-Hwal Yoon, Hsien-Chung Tseng, Clara Yuan and Kristala L. Jones Prather. July–September 2009. Engineering alternative butanol production platforms in heterologous bacteria. Metabolic Engineering 11(4-5): 262–273. https://doi.org/10.1016/j.ymben.2009.05.003.

82. Oudshoorn, A., L.A.M. van der Wielen and A.J.J. Straatho. 2009. Adsorption equilibria of bio-based butanol solutions using zeolite. Biochemical Engineering Journal 48: 99–103.

83. Pothiraj, C., P. Kanmani and P. Balaji. Dec. 2006. Bioconversion of Lignocellulose Materials. Mycobiology. 34(4): 159–165. Published online 2006 Dec. 31. doi: 10.4489/MYCO.2006.34.4.159.

84. Potthast, Antje, Thomas Rosenau, Herbert Sixta and Paul Kosma. October 2002. Degradation of cellulosic materials by heating in DMAc/LiCl. Tetrahedron Letters 43(43): 7757–7759. https://doi.org/10.1016/S0040-4039(02)01767-7.

85. Qureshi, N. and H.P. Blaschek. 1999. Butanol recovery from model solution/fermentation broth by pervaporation: evaluation of membrane performance, Biomass and Bioener. 17: 175–184.

86. Qureshi, N. and H.P Blaschek. April 2001. Recovery of butanol from fermentation broth by gas stripping. Renewable Energy 22(4): 557–564. https://doi.org/10.1016/S0960-1481(00)00108-7.

87. Qureshi, N., M.M. Meagher, J. Huang and R.W. Hutkins. 2001. Acetone butanol ethanol (ABE) recovery by pervaporation using silicalite–silicone composite membrane from fed-batch reactor of *Clostridium acetobutylicum*. J. of Membrane Science 187: 93–102.

88. Qureshi, N., S. Hughes, I.S. Maddox and M.A. Cotta. July 2005. Energy-efficient recovery of butanol from model solutions and fermentation broth by adsorption. Bioprocess and Biosystems Engineering 27(4): 215–222.

89. Raiser, Thomas. 2012. Eco friendly solutions, Innovative production in a minimum of space, turning wastes into energy. Sulzer Technical Review 3.

90. Riittonen, Toni, Esa Toukoniitty, Dipak Kumar Madnani, Anne-Riikka Leino, Krisztian Kordas, Maria Szabo, et al. 2012. One-pot liquid-phase catalytic conversion of ethanol to 1-Butanol over Aluminium Oxide—The Effect of the Active Metal on the Selectivity. Catalysts 2(1): 68–84. doi: 10.3390/catal2010068.

91. Roffler, S.R., H.W. Blanch and C.R. Wilke. March 1987. *In situ* recovery of butanol during fermentation, Part 1: Batch extractive fermentation. Bioprocess Engineering 2(1): 1–12.

92. Sangoro, J.R., A. Serghei, S. Naumov, P. Galvosas, J. Kärger and C. Wespe et al. May 2008. Charge transport and mass transport in imidazolium-based ionic liquids. Phys. Rev. E.77.051202.

93. Seckbach, J. and J. P. Kociolek (eds.). 2011. Life in Extreme Habitats and Astrobiology. The Diatom World. Vol. 19. Cellular Origin. Dordrecht: Springer Netherlands. http://link.springer.com/10.1007/978-94-007-1327-7.

94. Sikarwar, Vineet Singh, Zhao Ming, Paul S. Fennell, Nilay Shah and Edward J. Anthony. July 2017. Progress in biofuel production from gasification. Progress in Energy and Combustion Science 61: 189–248. https://doi.org/10.1016/j.pecs.2017.04.001.

95. Singh, Nisha, Anshu S. Mathur, Deepak K. Tuli, Ravi. P. Gupta, Colin J. Barrow and Munish Puri. 2017. Cellulosic ethanol production via consolidated bioprocessing by a novel thermophilic anaerobic bacterium isolated from a Himalayan hot spring', Open Access, Biotechnology for Biofuels 10: 1–18. https://doi.org/10.1186/s13068-017-0756-6.

96. Smol, J.P. and E.F. Stoermer. 2010. The Diatoms: Applications for Environmental and Earth Sciences. P 667, Second Edition, Cambridge University Press.

97. Stokes, H. January 2005. Alcohol fuels (Ethanol and Methanol): Safety Presentation at ETHOS. Project Gaia.

98. Sun, Chongran, Shuangfei Zhang, Fengxue Xin, Sabarathinam Shanmugam & Yi-Rui Wu. February 2018. Genomic comparison of *Clostridium* species with the potential of utilizing red algal biomass for biobutanol production. Biotechnology for Biofuels. 11, Article number: 42.

99. Sun, Ye and Jianyang Cheng. Oct. 2001. Hydrolysis of lignocellulosic materials for ethanol production. Review Paper No. BAE 200-08. Journal Series. Dept. of Biology and Agricultural Engineering, North Carolina State University, Raleigh, USA.

100. Thakur, Shilpi, Bhuvnesh Shrivastava, Snehal Ingale, Ramesh C. Kuhad and Akshaya Gupte. 2013. Degradation and selective ligninolysis of wheat straw and banana stem for an efficient bioethanol production using fungal and chemical pretreatment. 3 Biotech. 3(5): 365–372. Published online 2012 Nov 15. doi: 10.1007/s13205-012-0102-4. PMCID: PMC3781266, PMID: 28324332.

101. Trivedi, N., R. Baghel, Bothwell, J. Vishal Gupta, C.R.K. Reddy, Arvind M. Lali and Bhavanath Jha. 2016. An integrated process for the extraction of fuel and chemicals from marine macroalgal biomass. Sci. Rep. 6: 30728. https://doi.org/10.1038/srep30728.

102. Trivedi, Nitin, Vishal Gupta, C.R.K. Reddy and Bhavanath Jha. Dec. 2013. Enzymatic hydrolysis and production of bioethanol from common macrophytic green alga Ulva fasciata Delile. Bioresour. Technol. 150: 106–12. doi: 10.1016/j.biortech.2013.09.103. Epub 2013 Oct 2.

103. Trivedi, Nitin, C.R.K. Reddy, Ricardo Radulovich and Bhavanath Jha. May 2015. Solid state fermentation (SSF)-derived cellulase for sacchirification of the green seaweed Ulva for bioethanol. Algal Research 9: 48–54. DOI:10.1016/j.algal.2015.02.025 [https://www.researchgate.net/profile/CRK_Reddy2/publication/273004044_Solid_state_fermentation_SSFderived_cellulase_for_saccharification_of_the_green_seaweed_Ulva_for_bioethanol_production/links/54f544f50cf2ba6150655d09.pdf?origin=publication_list].

104. Tsuchida, Takashi, Shuji Sakuma, Tatsuya Takeguchi and Wataru Ueda. 2006. Direct synthesis of n-butanol from ethanol over nonstoichiometric hydroxyapatite. Ind. Eng. Chem. Res. 45(25): 8634–8642. DOI: 10.1021/ie0606082.

105. Urs Haller and Daniele Malossa. April 2001. A New Technical Solution for Glycol Ether Production: Brake Fluid from Sugar Cane. Sulzer Chemtech – Sulzer Technical Review, pp. 14–17.

106. Vaid, Surbhi, Parushi Nargotra and Bijender Kumar Bajaj. July 2017. Consolidated bioprocessing for biofuel-ethanol production from pine needle biomass. Environmental Progress and Sustainable Energy AIChE. https://doi.org/10.1002/ep.12691.

107. Valkenburg, C., Y. Zhu, C.W. Walton, B.L. Thompson, M.A. Gerber, S.B. Jones et al. March 2010. Design Case Summary: Production of Mixed Alcohols from Municipal Solid Waste via Gasification. Biomass Program, US Department of Energy.

108. Van der Wal, Hetty, Bram L.H.M. Sperber, Bwee Houweling-Tan, Robert R.C. Bakker, Willem Brandenburg and Ana M.López-Contreras. January 2013. Production of acetone, butanol, and ethanol from biomass of the green seaweed Ulva lactuca. Bioresource Technology 128: 431–437. https://doi.org/10.1016/j.biortech.2012.10.094.

109. Van Loon, L.R., M.A. Glaus, A. Laube and S. Stallone. January 1999. Degradation of cellulosic materials under the alkaline conditions of a cementitious repository for low- and intermediate-level radioactive waste II. degradation kinetics. J. of Environmental Polymer Degradation 7(1): 41–51.

110. Van Zyl, W.H., L.R. Lynd, R. Den Haan and J.E. McBride. 2007. Consolidated bioprocessing for bioethanol production using *Saccharomyces cerevisiae*. Advances in Biochemical Engineering Biotechnology 108: 205–235.

111. Vivek, Narisetty, Lakshmi M. Nair, Binoop Mohan, Salini Chandrasekharan Nair, Raveendran Sindhu, Ashok Pandey et al. September 2019. Bio-butanol production from rice straw – Recent trends, possibilities, and challenges. Bioresource Technology Reports. 7, 100224. https://doi.org/10.1016/j.biteb.2019.100224.

112. Wyman, Charles E., Bruce E Dale, Richard T Elander, Mark Holtzapple, Michael R. Ladisch, Y.Y. Lee et al. Mar.-Apr. 2009. Comparative sugar recovery and fermentation data following pretreatment of poplar wood by leading technologies. Biotechnol. Prog. 25(2): 333–9. doi: 10.1002/btpr.142. PMID: 19294662.

113. Wynn, Nick. Oct. 2001. Reactions and Separation—Pervaporation comes to age. SULZER CHEMTECH, Membrane Systems. pp. 66–72. CEP. wwwcepmagzine.org.

114. Xiros, Charilaos, Evangelos Topakas, Paul Christakopoulos. September 2012. Hydrolysis and fermentation for cellulosic ethanol production. Advanced Review. WIREs Energy and Environment. https://doi.org/10.1002/wene.49.

115. Xu, Q., A. Singh and M.E. Himmel. June 2009. Perspectives and new directions for the production of bioethanol using consolidated bioprocessing of lignocellulose. Current Opinion in Biotechnology 20(3): 364–371.

116. Xue, Chuang, Jingbo Zhao, Fangfang Liu, Congcong Lu, Shang-Tian Yang, Feng-WuBai. May 2013. Two-stage in situ gas stripping for enhanced butanol fermentation and energy-saving product recovery. Bioresource Technology 135: 396–402. https://doi.org/10.1016/j.biortech.2012.07.062.

117. Yamada, Ryosuke, Yuki Nakatani, Chiaki Ogino and Akihiko Kondo. June 2013. Efficient direct ethanol production from cellulose by cellulase- and cellodextrin transporter-co-expressing Saccharomyces cerevisiae. AMB Express, volume 3, Article number: 34. https://amb-express.springeropen.com/articles/10.1186/2191-0855-3-34.

118. Yamada, Ryosuke, Naho Taniguchi, Tsutomu Tanaka, Chiaki Ogino, Hideki Fukuda and Akihiko Kondo. April 2011. Direct ethanol production from cellulosic materials using a diploid strain of *Saccharomyces cerevisiaewith* optimized cellulase expression. Biotechnology for Biofuels. Volume 4, Article number: 8. https://doi.org/10.1186/1754-6834-4-8.

119. Yanjuan, Zhang, Qian Li, Jianmei Su, Ye Lin, Zuqiang Huang, Yinghua Lu et al. February 2015. A green and efficient technology for the degradation of cellulosic materials: Structure changes and enhanced enzymatic hydrolysis of natural cellulose pre-treated by synergistic interaction of mechanical activation and metal salt. Bioresource Technology 177: 176–181. https://doi.org/10.1016/j.biortech.2014.11.085.

120. Ye Sun and Jiayang Cheng. May 2002. Hydrolysis of lignocellulosic materials for ethanol production: a review. Bioresource Technology 83(1): 1–11. https://doi.org/10.1016/S0960-8524(01)00212-7.

121. Zabed, H., J.N. Sahu, A.N. Boyce and G. Faruq. December 2016. Fuel ethanol production from lignocellulosic biomass: An overview on feedstocks and technological approaches. Renewable and Sustainable Energy Reviews 66: 751–774. https://doi.org/10.1016/j.rser.2016.08.038.

122. Zhao, Tao, Yukihiro Tashiro and Kenji Sonomoto. November 2019. Smart fermentation engineering for butanol production: designed biomass and consolidated bioprocessing systems. Applied Microbiology and Biotechnology 103: 9359–9371.

123. Zheng, Y., H.-M. Lin and G.T. Tsao. 1998. Pre-treatment for cellulose hydrolysis by carbon dioxide explosion. Biotechnology Progress 14(6): 890–896.

124. Zheng, Yan-Ning, Liang-Zhi Li, Mo Xian, Yu-Jiu Ma, Jian-Ming Yang, Xin Xu et al. September 2009. Problems with the microbial production of butanol: Review. J. of Industrial Microbiology & Biotechnology. 36(9): 1127–1138.

CHAPTER 4

Biodiesel from Vegetable Oils

4.1 Biodiesel [2,14,15,18,19,52,104,105]

Biodiesel is the name of a clean burning alternative fuel (secondary renewable fuel, sometimes referred to as advanced renewable fuel) produced from renewable resources. It contains no petroleum, but it can be blended in proportions with petroleum diesel to create a biodiesel blend. It can be used in compression ignition (diesel) engines with no major modifications. Biodiesel is biodegradable, nontoxic, and essentially free of sulphur and aromatics. (*Technically, biodiesel is a fuel composed of mono-alkyl esters of long-chain fatty acids derived from vegetable oils or animal fats, designated as B100, and meets the requirements of ASTM (American Society for Testing & Materials) D 6751.*)

Biodiesel blends are denoted as "BXX" with "XX" representing the percentage of biodiesel contained in the blend (i.e., B20 is 20% biodiesel and 80% petroleum diesel). It can also be used exclusively in engines. For the impact of substituting biodiesel for petroleum diesel on emissions refer to Para 1.3.1 and 1.3.2. [123].

'Biodiesel' can be made from various oils from seeds of food crops or non-food crops, or even fats. Oils and fats are simple lipids. These are compounds of glycerol and various 'fatty acids'; that is, they are 'glycerol esters' (termed as 'glycerides/triglycerides'.

The fatty acids present in the glycerides are almost exclusively straight-chain acids, and almost always contain an even number of carbon atoms. The oils are liquid at 20°C and fats are solid/semisolid at 20°C. The fatty acids are divided into two main groups: (a) Saturated fatty acids and (b) unsaturated fatty acids (containing one or more double bonds).

Examples of Saturated Fatty Acids:
- (i) Lauric acid's molecular formula: or $C_{11}H_{23}COOH$
- (ii) Palmitic acid's molecular formula: $C_{15}H_{31}COOH$
- (iii) Stearic acid's molecular formula: $C_{17}H_{35}COOH$
- (iv) Capric acid's molecular formula: $C_9H_{19}COOH$
- (v) Myristic acid's molecular formula: $C_{13}H_{27}CO\ OH$

Examples of Unsaturated Fatty Acids:
- (i) Palmitoleic acid's molecular formula: $C_{15}\ H_{29}\ COOH$
- (ii) Oleic acid's molecular formula: $C_{17}H_{33}COOH$
- (iii) Linolenic acid's molecular formula: $C_{17}H_{29}COOH$
- (iv) Linoleic acid's molecular formula: $C_{17}H_{31}COOH$

(v) Erucic acid's molecular formula: $C_{21}H_{41}COOH$

(vi) Eleostearic acid's molecular formula: $C_{17}H_{29}COOH$

It may be of interest to mention here that the *hydrogenation of oils/glycerides* of unsaturated fatty acids changes them into fats like *vegetable ghee*. Oils and fats when treated with caustic soda form soaps.

Glycerol (also commonly called "glycerine") is a polyhydric alcohol* having three OH groups (that is, it is a *trihydric alcohol*) in the form of $CH_2OH.CHOH.CH_2OH$.

Esters are compounds with the form $R_1O\,R_2$, where R, R_1, R_2, R_3 are the alkyl groups or long carbon chains.

Examples of simple glycerides ($CH_2OCOR.CHOCOR.CH_2OCOR$) are:

- Tristearin ($CH_2OCO\,C_{17}H_{35}.CHOCO\,C_{17}H_{35}.CH_2OCO\,C_{17}H_{35}$)
- Tripalmitin ($CH_2OCO\,C_{15}H_{31}.CHOCO\,C_{15}H_{31}.CH_2OCO\,C_{15}H_{31}$)
- Triolein ($CH_2OCO\,C_{17}H_{33}.CHOCO\,C_{17}H_{33}.CH_2OCO\,C_{17}H_{33}$)

Example of mixed glycerides ($CH_2OCOR_1.CHOCOR_2.CH_2OCOR_3$) are:

- Oleopalmitostearin ($CH_2OCO\,C_{17}H_{35}.CHOCO\,C_{15}H_{31}.CH_2OCO\,C_{17}H_{33}$)
- Palmito diolein ($CH_2OCO\,C_{17}H_{33}.CHOCO\,C_{17}H_{33}.CH_2OCO\,C_{15}H_{31}$)

*Types of Alcohols: Simple alcohols having one OH group of the form ROH are termed as monohydric alcohols, while polyhydric alcohol of the form $CH_2OH.CH_2OH$ having two OH groups are called dihydric alcohols (glycols), and polyhydric alcohol having three OH groups are called trihydric alcohols.

The process of converting vegetable oil/fat to biodiesel (which are methyl esters) essentially consists of the following steps:

- Vegetable oils are converted into methyl esters ($RCOOCH_3$) by 'transesterification'.
- The process of 'transesterification' removes 'glycerol ($CH_2OH.CHOH.CH_2OH$)' from Triglycerides ($CH_2OCOR_2.CHOCOR.CH_2OCOR_2$) and thus reduces the viscosity of the resulting product, which is called biodiesel. Glycerol is also called glycerine.
- A two stage 'integrated pre-esterification' of free fatty acids and 'base catalysed transesterification process' has proved to be an effective method of converting feedstock with a high amount of free fatty acids. However, with the 'acid catalyst for transesterification', it was found that the removal of free fatty acids or the pretreatment of feedstock oils may not be necessary. The heterogeneous catalyst tolerates moisture and free fatty acids.

Note: Biodiesel can also be made from lignocellulosic waste. This subject is dealt with in Chapter 5.

4.2 Transesterification [22,31,51,68,71,79,83,86,96,117]

The transesterification is defined as the splitting of an ester by an alcohol to form a new ester. It is effective in replacing a higher alkyl group by a lower alkyl group in the ester and, in return, getting an ester of a lower alkyl group and higher alcohol as shown in the reaction. In fact, it is the alcoholysis of ester.

The 'alcoholysis' is carried out by refluxing the ester with a large excess of alcohol, preferably in presence of acid or sodium alkoxide (C_2H_5ONa).

$R\,COO\,R_1 + R_2OH = RCOOR_2 + R_1OH$ (Here, R represents alkyl group and $R_1 > R_2$)

Example:

$CH_3COOC_4H_9 + C_2H_5OH \leftrightarrow CH_3COOC_2H_5 + C_4H_9OH$

The mechanism of alcoholysis in basic media may be put as:

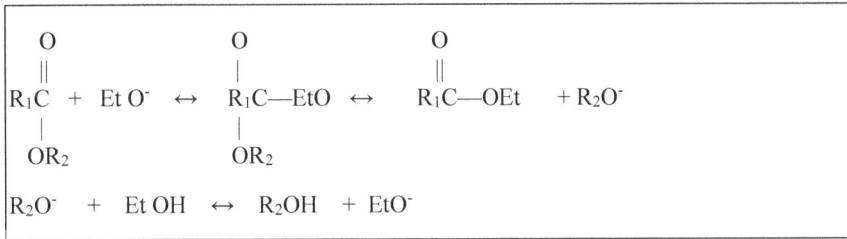

$$R_1\overset{\overset{\displaystyle O}{\|}}{\underset{\underset{\displaystyle OR_2}{|}}{C}} + EtO^- \leftrightarrow R_1\overset{\overset{\displaystyle O}{|}}{\underset{\underset{\displaystyle OR_2}{|}}{C}}{-}EtO \leftrightarrow R_1\overset{\overset{\displaystyle O}{\|}}{C}{-}OEt + R_2O^-$$

$$R_2O^- + EtOH \leftrightarrow R_2OH + EtO^-$$

The scheme of the transesterification of mixed glyceride to methyl ester is as follows:

$$CH_2OCOR_1.CHOCOR_2.CH_2OCOR_3 + 3\ CH_3OH = CH_2OH.CHOH.\ CH_2OH + R_1\ COOCH_3 + R_2COOCH_3 + R_3\ COOCH_3$$

that is, Glycerides + Methyl Alcohol = Glycerol + Methyl Esters

Where, R_1, R_2, and R_3 represent long carbon chains.

'EtOH' denotes Trihydric alcohol, glycerol $(CH_2OH.CHOH.CH_2OH)$.

Animal fats and vegetable oils* are typically made of triglycerides which are esters of free fatty acids with the trihydric alcohol, that is, glycerol. In the transesterification process, the alcohol is deprotonated with a base to make it a stronger nucleophile. Commonly, ethanol or methanol is used. As can be seen, the reaction has no other inputs than the triglyceride and the alcohol.

The transesterification reaction would take place either exceedingly slowly or not at all without an acid or base and heating. It is important to note that the acid or base are not consumed by the transesterification reaction; thus, they are not reactants but catalysts.

Almost all biodiesel is produced in the world from virgin vegetable oils using the base-catalyzed technique as it is the most economical process for treating virgin vegetable oils. It only requires low temperatures and pressures and gives over 98% conversion (*provided the starting oil is low in moisture and free fatty acids*). However, the production of biodiesel from other sources or by other methods may require acid catalysis, which is much slower.

The alcohol reacts with the oils and fats to form the monoalkyl ester (or biodiesel) and crude glycerol. The reaction between the biolipid (*fat or oil*) and the alcohol is a reversible reaction so the alcohol must be added in excess to drive the reaction towards the right and ensure complete conversion. Usually a ratio of about 6:1 of methanol to oil is kept.

The approximate amount of materials required for obtaining *one liter of biodiesel* from waste oil can be put as: (i) waste cooking oil (1.064 liter), (ii) methanol (0.213 liter), and (iii) potassium hydroxide (8.512 g).

4.3 Extraction of Oil from various Seeds [14,15,18,19,30,123]

The first step in the production of biodiesel from vegetable oil extracted from any seed would be to dry the seeds, which are then ground/crushed. Some dry seeds can be directly put in an oil expeller, which crushes and expels oil and the remaining solid material (oil cake). However, for some solvents, the extraction of oil from the seeds is necessary; it is also essential to crush the seeds. Hexane solvent is normally used for the solvent extraction of oil. The solvent is separated (for example, by distillation) and recycled. The extracted oil is purified before proceeding with the transesterification. A flow diagram for solvent extraction of oil from seeds is shown in Figure 4.1.

Oil seeds

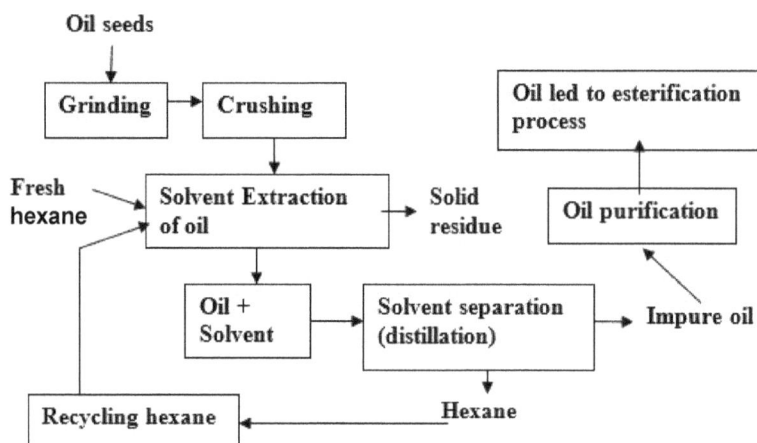

Fig. 4.1. Solvent Extraction of Seed Oil.

4.4 Biodiesel Production from Oils/Fats [2,18,19,30,123]

4.4.1 Base-catalyzed Transesterification [31,51,68,117]

In the base-catalyzed transesterification reaction, any strong base/alkali capable of deprotonating the alcohol can be used, such as sodium hydroxide (NaOH), potassium hydroxide (KOH), sodium methoxide (CH_3ONa), potassium methoxide (CH_3OK), etc.

Usually, the base potassium hydroxide or caustic soda is dissolved in the methyl alcohol (CH_3OH) using a standard agitator or mixer. The dissolution of catalyst in alcohol is a convenient method to disperse the catalyst (which is otherwise solid) in the oil. The alcohol/catalyst mix is then charged in a closed reaction vessel and the biolipid (vegetable or animal oil or fat) is added. The system from here on is totally sealed off to prevent the loss of alcohol.

The reaction that follows replaces the alkyl group in the triglyceride in a series of steps. The carbon in the ester of the triglyceride has a slight positive charge, and the carbonyl oxygen (C = O) has a slight negative charge. This polarization of the C = O bond is what attracts the RO^- to the reaction site.

Empirically, 6.25 grams/liter of sodium hydroxide (NaOH) produces a very usable fuel. About 6 g/L of NaOH is used when the waste vegetable oil (WVO) is light in color and about 7 g/L of NaOH is used when it is dark in color.

(For a large-scale continuous flow production process, alkyl oxide solutions of sodium methoxide or potassium methoxide (instead of sodium hydroxide or potassium hydroxide) in methanol may be used).

The feed oils and fats used for transesterification may contain some moisture and free fatty acids and therefore, it is important to monitor their amount. The free fatty acids and water form soap with alkali (saponification), which may create problems in the separation of glycerin (glycerol), which is a byproduct of the reaction. The presence of water causes hydrolysis of some of the methyl ester produced and then forms soap with the alkali present. The methanol (CH_3OH) also needs to be very dry as any water present in the process promotes the saponification reaction. The *saponification* consumes the alkali, and thus inhibits the transesterification reaction.

The alkali required for the neutralization of free fatty acids can be determined by the titration of oil. The excess alkali would need to be added over and above the amount required for catalyzing the transesterification. Normally, the free fatty acids should be less than 1% for the base-catalyzed process. If the fatty acid content is higher, the acid catalyzed reaction is better.

The oil used for making biodiesel should be filtered and dried either by heating it to 100°C for some time with continuous stirring or by vacuum distillation in a temperature range of 30–40°C.

The boiling point of methanol is about 65°C. The reaction temperature should not exceed the boiling point of methanol, otherwise it will evaporate and form bubbles which will inhibit the reaction. A lower temperature slows down the reaction and increases the reaction completion time. Normally, the reaction temperature is kept in the range of 60–65°C to have an optimum rate of reaction, but sometimes, for safety reasons, a reaction temperature of 55°C is also used.

Excess alcohol is normally used to ensure the total conversion of the fat or oil to its esters. The glycerin/glycerol phase (along with excess catalyst, soaps formed and some alcohol content) is much denser than the biodiesel phase (*i.e., along with some alcohol content*) and the two can be separated by gravity and glycerin can be simply drawn off the bottom of the separating vessel. There is a vast difference in the specific gravity of the biodiesel (about 0.88) and that of glycerol (1.26), and therefore gravity separation works quite well.

In order to quickly separate the two materials, sometimes other methods like centrifuging or coalescence are used. When coalescence is used, the glycerol is forced to coalesce into larger globules which settle quickly. The coalescer forces the mixed liquid to sometimes flow through constrictions in hydrophilic or hydrophobic surfaces.

After the glycerin and biodiesel phases are separated, the excess alcohol in each phase is removed with a flash evaporation process or by distillation.

In some systems, the alcohol is removed first, and the mixture is neutralized before the glycerin and esters are separated. In either case, the alcohol is recovered using distillation equipment and it is reused; however, it should be ensured that no water accumulates in the recovered alcohol stream.

The glycerol separated from biodiesel contains unused catalyst and soaps that are neutralized with an acid and stored as crude glycerin. Water and alcohol are removed later, chiefly using evaporation, to produce 80–88% pure glycerin/glycerol. Once separated from the glycerin, the biodiesel can be purified by washing it gently with warm water to remove any residual catalyst or soap. However, before sending the biodiesel to storage, any water in it should be removed. A schematic flow diagram of the process of the conversion of oils and fat to biodiesel is shown in Figure 4.2.

As mentioned earlier, glycerol is a byproduct of the biodiesel industry. As biodiesel production is increasing exponentially, the crude glycerol (of approximately 80% purity) is also generated from the transesterification of vegetables oils in a large quantity. Despite the wide applications of pure glycerol in food, pharmaceutical, cosmetics, and many other industries, it is too costly to refine the crude glycerol to a high level of purity, especially for medium and small biodiesel producers. Many research projects and studies have been conducted on this subject and the innovative utilization of the crude glycerol is under investigation.

An effective usage or conversion of crude glycerol to specific products will cut down the biodiesel production costs. It can be converted into useful products like 1,3-propanediol, 1,2-propanediol, dihydroxyacetones, hydrogen, polyglycerols, succinic acid, and polyesters. Glycerol, when used in combination with other compounds, yields other useful products. [118]

4.4.2 *Acid Catalysis of Transesterification* [79,83,86]

Base-catalyzed catalysis is sensitive to free fatty acid and moisture, but acid catalysis is insensitive to free fatty acids. Therefore, for any oil having more than 1% free fatty acid content, acid catalysis is beneficial. As there are many impurities in waste oils, acid catalysis may be better suited for them, but it should be borne in mind that the acid catalysis of transesterification suffers from a number of disadvantages such as: (i) the pace of reaction is slow, (ii) the effluent is acidic, (iii) the catalyst is not reusable, and (iv) the equipment cost is high. However, for virgin nonedible oils, the base-catalyzed process is better and economical.

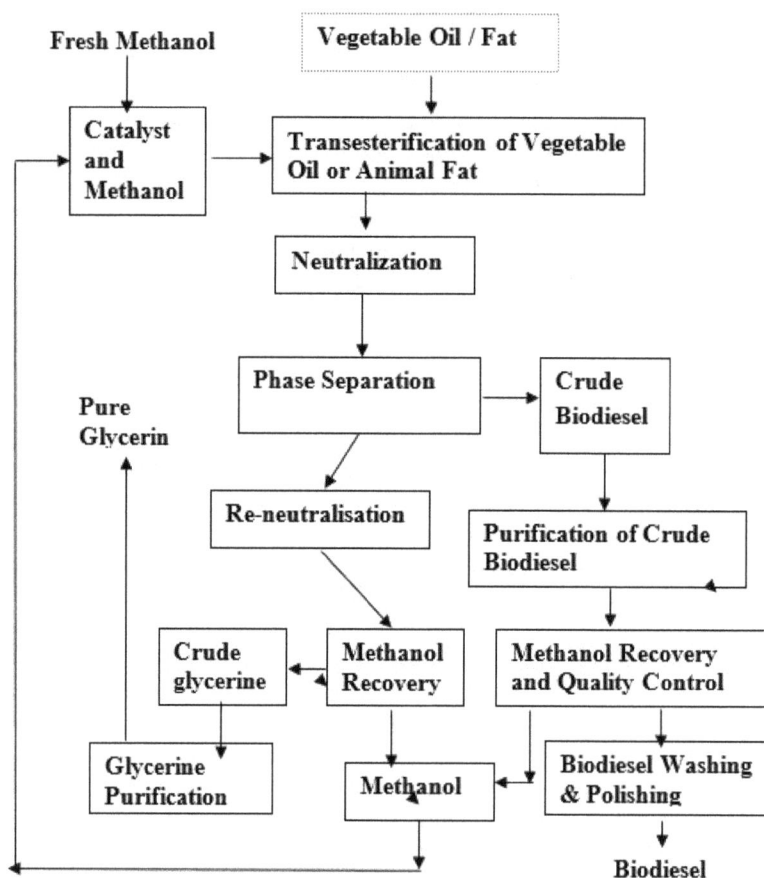

Fig. 4.2. Conversion of Vegetable Oil/Animal Fat to Biodiesel.

4.4.3 Use of 'Ultra and High-shear In-line Mixer' in Batch Reactors [58]

The use of ultra and high-shear mixers can reduce the droplet size of the immiscible liquids like oil/fat and methanol. The smaller droplets of oil/fat and methanol have a larger surface area for the catalyst to act faster and better and therefore, the reaction proceeds faster in 'ultra- and high-shear in-line batch reactors'.

Biodiesel can be produced continuously, semi-continuously, and in batch-mode in reactors with the use of ultra and high-shear in-line mixers. The reduction of production time and increase in production volume is achieved by this method.

4.4.4 Reactors with Ultrasonic Devices [13,36,89]

In the reactors with ultrasonic devices (ultrasonic reactor), the low frequency ultrasonic waves (about 20 kHz) produce and collapse the bubbles of reaction mixtures constantly, providing simultaneous mixing and heating for carrying out the transesterification process. The use of an ultrasonic reactor to produce biodiesel drastically reduces the reaction time, reaction temperatures, and energy input. This process of transesterification can therefore run continuously and is better than using the time-consuming batch process. Industrial-scale ultrasonic devices are available for large-scale processing.

4.4.5 Use of Microwave Heating in the Transesterification Reaction [9,45,57,71,80,81]

Microwaves can provide process heat for the transesterification of oils/fats. Research in this direction is ongoing. Intense localized heating can be provided by microwaves, but it may be kept in mind that local temperatures may be higher than the recorded temperature of reaction. The energy consumption is quite low when microwave heating is provided. The development of the microwave-heated reactors hold great potential for efficiency and cost cutting in the commercial production of biodiesel.

4.4.6 Enzymatic Transesterification Using Lipases* as a Catalyst [5,43,49,53,65,103,113,129,130]

The enzymatic transesterification uses the biocatalysts, 'lipases'. The lipases catalyze 'the hydrolysis of vegetable oils or fats and help in the formation of linear chain esters resulting in release of glycerol. Very high yields can be obtained from used oils and fats by using lipases as catalysts as they make the reaction less sensitive to high free fatty acids (FFA) content, which is a problem faced in the standard biodiesel process. One problem with the lipase reaction is that methanol cannot be used because it inactivates the lipase catalyst after one batch.

This problem can be surmounted by using methyl acetate instead of methanol as the lipase is not inactivated by this chemical and can be used for several batches, making the lipase system more cost effective. Alternatively, stepwise addition of methanol can prevent the inactivation of lipase and allow the continued usability

The problem can be overcome with the use of immobilized lipase, which, on hydrotalcite, has been found to catalyze the formation of biodiesel (methyl esters) from waste vegetable oil and methanol. With immobilized lipase as well, the stepwise addition of methanol is considered better.

It may be mentioned that a minimum water content is needed for lipase activity and that the lipase enzyme is inactivated when there is a rise in temperature. Table 4.1 shows a comparison of alkali catalysis and lipase catalysis.

Note: Use of Lipase Candida Antarctica for conversion of vegetable oil to biodiesel [113]

Experimentally continuous methanolysis of vegetable oil by an enzymatic process was carried out for which immobilized Candida antarctica lipase was found to be the most effective among the several lipases tested. The enzyme was inactivated by shaking it in a mixture containing more than 1.5 molar equivalents of methanol in proportion to oil. To fully convert the oil to its corresponding methyl esters, at least three molar equivalents of methanol are needed. Thus, the reaction was conducted by adding methanol stepwise to avoid lipase inactivation. The first step of the reaction was conducted at 30°C for 10 h in a mixture of oil/methanol (1:1, mol./mol.) and 4% immobilized lipase with shaking at 130 oscillations/min. After more than 95% methanol was consumed in the formation of ester, a second molar equivalent of methanol was added, and the reaction continued for 14 h. The third molar equivalent of methanol was finally added, and the reaction continued for 24 h (total reaction time, 48 h). This three-step process converted 98.4% of the oil to its corresponding methyl

Table 4.1. Comparison of Alkali Ccatalysis with Lipase Catalysis for Biodiesel Production. [49,113]

	Alkali catalysis	**Lipase catalysis**
Reaction temperature K	333–343	303–313
Free fatty acid in raw material	Form Saponified Products	No interference with reaction
Water in raw material	Interference in reaction	No influence
Yield of methyl esters	Normal	Higher
Recovery of glycerol	Difficult	Easy
Purification of methyl esters	Repeated washing required	None
Production of catalyst cost	Cheap	Relatively expensive

esters. To investigate the stability of the lipase, the three-step methanolysis process was repeated by transferring the immobilized lipase to a fresh substrate mixture. As a result, more than 95% of the ester conversion was maintained even after 50 cycles of the reaction.

4.4.7 Use of a Heterogeneous Catalyst in Transesterification [21,27,42,50,69,74,88,90,92,98,103,112,115,132]

Some solid catalysts, called heterogeneous catalysts, can be used for the transesterification of oils for production of biodiesel. This solid catalyst may be of two kinds, namely acidic and basic. Because of their higher stability even when the content of free fatty acids is higher in feed oil, they can be used in the production of biodiesel, although their catalytic activity is lower than the basic catalysts. The acidic catalysts are acid zeolites, heteropolacids, immobilized sulfonic acids, sulphated zirconia, and mixed oxides. Some carbon-based solid catalysts have also been reported to have been investigated.

Note: *The chemists at Brown University were successful in converting the waste vegetable oil-to-biodiesel in a single reaction vessel using environmentally friendly catalysts and making the conversion six times faster than the current methods. Bismuth triflate and scandium triflate were used as catalysts in the process, which was performed in a microwave reactor instead of conventional thermal heater. It was found that these catalysts could convert the oil to biodiesel in about 20 minutes in the microwave reactor, whereas the reactions without catalysts using a conventional heater generally took two hours. Although the microwave method needs a higher temperature (150°C) for the reaction (compared to 60–65°C under base- or-acid catalyzed transesterification), overall, it uses less energy because of the very short reaction time. [59]*

Figure 4.3 shows the flow diagram for heterogeneously catalyzed biodiesel production.

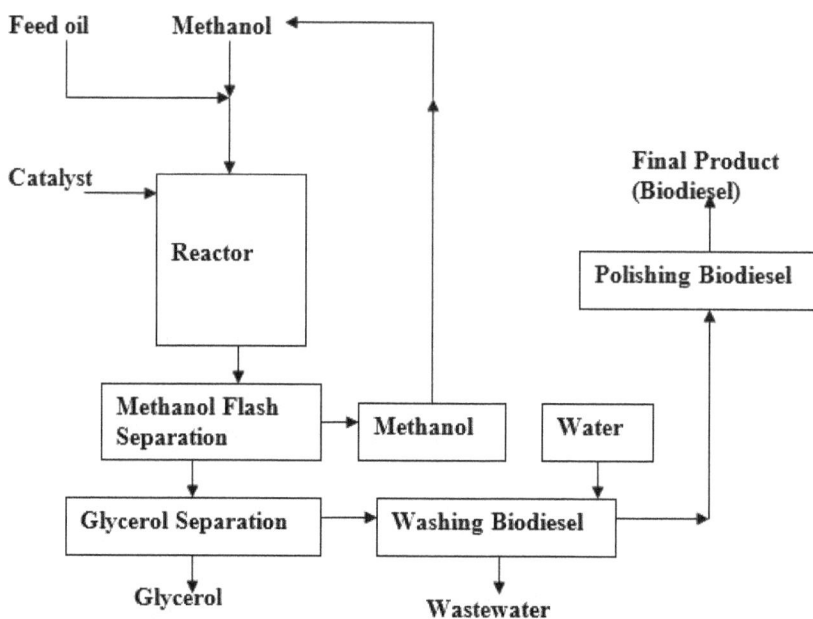

Fig. 4.3. Heterogeneously Catalyzed Biodiesel Production.

4.4.8 Reactive Extraction of Biodiesel from Oilseed [73,84,85,116,120]

A new process of reactive extraction has been invented in which instead of oil, the ground oilseeds are treated with methanol and catalysts directly for transesterification, to produce diesel oil. In this process, a very large amount of excess methanol is required, which increases the costs and the

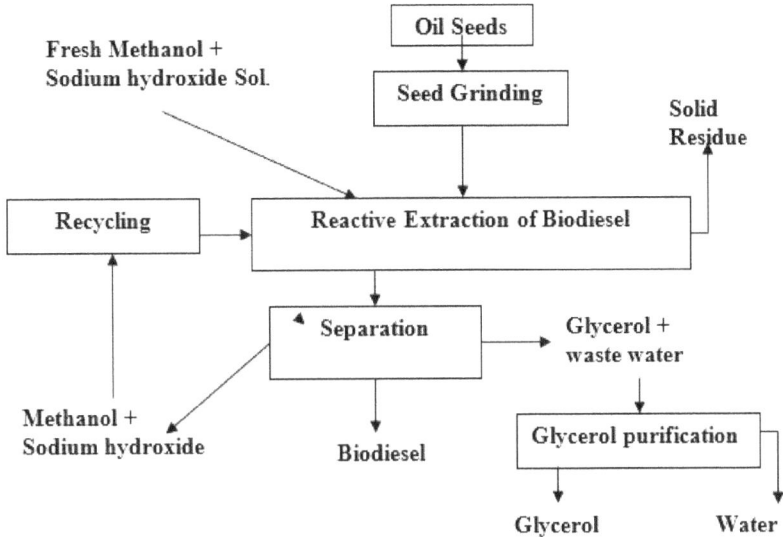

Fig. 4.4. Reactive Extraction of Biodiesel from Oil Seeds.

process is still not yet competitive with the process using oil for transesterification. A flow diagram for this process is shown in Figure 4.4.

4.4.9 Catalyst-free Supercritical Process [25,75,91,96,99,108,109,110,111,122, 125,126]

A catalyst-free method that uses supercritical methanol at high temperatures and pressures in a continuous process has been developed. In the supercritical state, the oil and methanol remain in a single phase and the reaction occurs spontaneously and rapidly. This process can tolerate water in the feedstock. The free fatty acids in the oil are converted to methyl esters instead of soap and therefore it is possible to use a wide variety of feedstocks. As no catalyst is used, the catalyst removal step is eliminated. The drawback is the use of high temperatures and pressures, which increases the capital costs of the plant. However, the energy costs of production may be approximately similar to those of the catalytic production routes.

The transesterification of triglycerides by the non-catalyst supercritical method works at a temperature of 350°C and a pressure of 43 MPa in the presence of excess methanol.

To achieve moderate working conditions, a 2-phase process can be used. Oils and fats are hydrolyzed at subcritical conditions for producing fatty acids and glycerol. The upper portion of the product of hydrolysis is the oil phase containing mostly of fatty acids and the lower portion mainly consists of glycerol and water. The fatty acids separated by decantation or otherwise are then mixed with methanol and treated at supercritical conditions to produce methyl esters (biodiesel). The advantages of the supercritical process may be summed up as follows:

- The process tolerates a high moisture content in the feed stock
- No removal of the catalyst is involved
- Glycerides and fatty acids react with equivalent rates
- If high methanol/oil ratios are used, the total conversion of oil can be achieved in a short time.

4.5 Properties of Biodiesel [12,17]

Biodiesel has a number of standards for its quality including DIN EN 14214, DIN EN 590, ASTM International D6751, and others.

Biodiesel properties are beneficial vis-à-vis the flash point and emission level of polluting gases as evident from Table 4.2, which provides a comparison of the properties of petroleum diesel and biodiesel. The Table 4.3 gives the energy contents per liter of some fuels. It may be noted that biodiesel and butanol are near to gasoline in per liter energy content, which means that the volume occupied onboard any vehicle by these fuels would be near to that occupied by gasoline.

From the above table, it may be seen that the properties of biodiesel are quite suitable for diesel engines. Biodiesel can be used exclusively or in combination with petroleum diesel.

Studies have shown that the increase in the proportion of biodiesel in petroleum diesel will reduce both the carbon monoxide and hydrocarbon emission from diesel engines as per studies. The impact of 100 % biodiesel substitution for petroleum diesel in an urban bus can be seen in the report from NERL (US) given in Para 1.3.2.

Table 4.2. Comparison of Properties of Petroleum Diesel and Biodiesel.

Se. No.	Properties	Petroleum Diesel	Biodiesel
1.	Fuel composition	C10-C21 hydrocarbons	C12-C22 Methyl esters
2(a). 2(b).	Kinematic viscosity@40°C (cSt) Density (20–40°C) (kg/L)	1.3–4.1 0.82–0.86	1.9–6.0 0.875 (Jatropha biodiesel)
3.	Boiling point	188–343°C	182–338°C
4.	Flash point	32°C	> 160°C
5.	Percent composition by weight	87% C, 13%H 0.05 %S, O = Nil	C = 77%, H = 12% O = 11%, S = 0.00–0.0024%
6.	Heating value	40.3 MJ/liter	33–39 MJ/litre (*depending on source*)
7.	Emission Level	Higher	Lower particulate, SO_x, CO, and unburnt hydrocarbons
8.	Cetane No.	42	60–65 (biodiesel from Jatropha/palm oil) 45 (biodiesel from soybean oil)
9	Cloud point K	258–278	270–285
10.	Pour point K	238–258	258–289
11.	Standard	ASTM D975	ASTM D6751 and EN14214
12.	Production Process	Reaction + Fractionation	Chemical Reaction
13.	Lubricity	Low Sulphur fuel has low lubricity factor	Much greater than Diesel, comparable to all lubricants
14.	Biodegradability	Poor	Good
15.	Toxicity	Highly toxic	Essentially non-toxic
16.	Oxygen	Very low	Up to 11 % free oxygen
17.	Aromatics		No aromatic compounds
18.	Spill hazard	High	None
19.	Material Compatibility	No effect on natural butyl rubber	Degrades natural, butyl rubber
20.	Shipping	Hazardous	Shipped as a nonhazardous and nonflammable material
21.	Alternative fuel	No	Yes

Table 4.3. Energy Content Comparison of Different Fuels with Petroleum Gasoline.

Liquid Fuel	Energy Content (Approx.) MJ/L	Gasoline Equivalent (approx.) of one Liter
Petroleum Gasoline	32	1
Bioethanol	19.6	0.61
Biobutanol	29.2	0.91
Petroleum Diesel	40.3	1.25
Biodiesel	33–39 (depending on source)	1.04 or more

4.6 Problems to be Addressed before the use of Biodiesel [6,18,48,51,60,62,90,91]

4.6.1 Compatibility of Materials with Biodiesel [47,56,119]

It should be ensured that the materials used in handling and storing the biodiesel do not react with it or do not deteriorate when coming into contact with it. The biodiesel itself should also not get deteriorated by such materials. A few examples are given below:

- *Plastics:* High-density polyethylene (HDPE) is compatible but polyvinyl chloride (PVC) is slowly degraded. Polystyrenes dissolve on contact with biodiesel.
- *Metals*: Biodiesel has an effect on copper-based materials (e.g., brass), and it also affects zinc, tin, lead, and cast iron but stainless steels (316 and 304) and aluminum are unaffected.
- *Rubber:* Biodiesel also affects types of natural rubbers found in some older engine components. Studies have also found that fluorinated elastomers (FKM)* cured with peroxide and base-metal oxides can be degraded when biodiesel loses its stability caused by oxidation. However, testing with FKM- GBL-S* and FKM- GF-S* were found to be the toughest elastomers, which could handle biodiesel in all conditions. *(*Commercial names)*

4.6.2 Gelling Property of Biodiesel [44,67,114]

On cooling of biodiesel below a certain point, the molecules of biodiesel start aggregating and forming crystals. It appears cloudy when the crystals become larger than one quarter of the wavelength of visible light (this point is referred to as 'cloud point'). When biodiesel cools further, these crystals become larger. The lowest temperature at which fuel can pass through a 45 micron filter is the 'cold filter plugging point' (CFPP). As biodiesel cools further, it will gel and then solidify. There are differences in the CFPP requirements between countries as different national standards are followed. The temperature at which pure biodiesel (B100 that is, 100% biodiesel) starts to gel varies significantly and depends upon the mix of esters and therefore, on the feedstock oil used to produce the biodiesel. For example, biodiesel produced from low 'erucic acid'* varieties of canola seed starts to gel at approximately $-10°C$ (14°F). Biodiesel produced from 'tallow' tends to gel at around $+16°C$ (6°F). There are a number of commercially available additives that will significantly lower the pour point and the cold filter plugging point of pure biodiesel. Biodiesel can be utilised in winter operation by blending biodiesel with other fuel oils, including low sulphur diesel/kerosene.

Note: Erucic acid occurs as 'glycerol esters' in various oils like rapeseed oil, canola oil, cod liver oil, etc. Erucic acid's melting point is 34°C and has the following molecular formula:

$CH_3 (CH_2)_7 CH = CH (CH_2)_{11} COOH$

4.6.3 *Contamination of Biodiesel by Water* [3]

Biodiesel may contain small but problematic quantities of water. Although it is not miscible with water, it is hygroscopic like ethanol, that is, it absorbs atmospheric moisture. One of the reasons biodiesel absorbs water is the persistence of monoglycerides and diglycerides left over from an incomplete reaction as these molecules can act as an emulsifier and allow water to mix with the biodiesel. In addition, there may be residual water from processing or resulting from storage tank condensation. The water content of the biodiesel can be measured by the use of 'water-in-oil sensors'.

Problems Created by Presence of Water in Biodiesel:

a. Water reduces the heat of combustion of the bulk fuel. This means more smoke, more difficulty in starting the engine, and less power.

b. Water causes corrosion of vital fuel system components such as fuel pumps, injector pumps, fuel lines, etc.

c. Water and microbes cause the paper element filters in the system to fail (rot), which in turn results in premature failure of the fuel pump due to the ingestion of larger particles.

d. Water freezes to form ice crystals near 0°C (32°F) and these crystals provide sites for nucleation and accelerate the gelling of the residual fuel.

e. Water accelerates the growth of microbe colonies, which can plug a fuel system. Biodiesel users who do not have heated fuel tanks therefore face a year-round microbe problem.

f. Water can cause pitting and corrosion in the pistons and other parts of a diesel engine.

g. Water contamination is also a potential problem when using certain chemical catalysts involved in the production process as it may substantially reduce the catalytic efficiency of base (high pH) catalysts such as potassium hydroxide.

However, the *'supercritical methanol production process'* in which the transesterification process of oil feedstock by methanol takes place under high temperature and pressure, is largely unaffected by the presence of water contamination during the production phase. (See Para 4.4.9 above).

4.6.4 *Storage of Biodiesel* [20,23,28,35,40,82]

In general, the standard storage and handling procedures used for petroleum diesel can be used for biodiesel. The fuel should be stored in a clean, dry, dark environment. Acceptable storage tank materials include aluminum, stainless steels (304 and 316), and high-density polyethylene (HDPE). Rubbers/elastomers FKM- GBL-S* and FKM- GF-S* were found to be the toughest elastomer capable of handling biodiesel in all conditions.

Copper, brass, lead, tin, and zinc should be avoided as they are affected by biodiesel. Polyvinyl chloride (PVC) is slowly degraded. Polystyrenes are dissolved on contact with biodiesel.

4.6.5 *Operating Existing Diesel Engines with Biodiesel* [76,121,127]

Biodiesel can be used in any diesel engine with little or no modification to the engine or the fuel system. *However, biodiesel has a solvent effect that may release deposits accumulated on tank walls and pipes from previous diesel fuel storage. The release of deposits may clog filters initially and precautions should be taken.* The continued use of biodiesel will not cause an increased frequency of filter changes.

Biodiesel can actually extend the life of the engine. Biodiesel has superior lubricating properties and there is no sulphur dioxide in its exhaust. There can be a reduction in the wear of vital engine parts.

4.7 Tree-Borne Oilseeds (TBO) [4,37,38,46,70,87,107]

There are several trees which produce seeds that contain oils, which may be edible or non-edible. The advantage of tree-borne oilseed production on non-agricultural land in comparison with oilseed production on agricultural lands is obvious. It is always preferable to use non-edible oil by growing oilseed trees on non-agricultural lands/unutilized lands anywhere, or wastelands for making biodiesel.

- Oilseed-producing trees for edible oil for Indian climatic condition include Oil Palm, Coconut, *Pongamia pinnata* (Karanja/Dahur Karanja/Pungu/Punnu/Kranuga, etc.)
- Oilseed-producing trees for non-edible oil include *Azadirachta indica* (Neem), *Madhuca* (Mahua), *Calophyllum inophyllum* (Sultana chumpa/Punnaga/Undi), *Jatropha curcas*, Soapnut (*S. mukorossi* and *S. trifoliatus*) etc.

In agroforestry, the choice of oil-producing trees should take into account the local needs of the community living in area. Normally, the choice of trees is dictated by the faster growth of trees and economic considerations, including the need for fuelwood, fruits, timber, oil, etc.

The large-scale production of nonedible oil from seeds of various trees has not been practiced in general, although local communities extract nonedible oil from some seeds for medicinal purposes. In traditional agroforestry, the planting of trees bearing nonedible oil has no intended purpose. However, jatropha is planted by farmers for providing a live fence. This requires planting them in a row at close spacing. Other species like *Pongamia pinnata*, *Madhuca* species, and *Calophyllum inophyllum* probably came up as wildlings and were retained by farmers. Their presence is usually limited to a few trees in the middle of crop fields or along the borders with the intention of providing shade/fence/traditional medicines, etc., but not for large-scale production of oil. Now some innovative farmers are developing or modifying existing agroforestry systems to suit local conditions for growing such trees. Biodegradable natural surfactants obtained from plants can be an attractive alternative to synthetic surfactants for the remediation of contaminated soils.

Wastelands belonging to farmers can be brought under the 'Tree-borne Oilseed' (TBOS)' program. The present agricultural production levels of some marginal farmlands are so low that the farmers will be happy to try out alternative crops. Therefore, land availability for oilseed production is unlikely to be a constraint. However, the vital resource that will determine the area eventually brought under oilseed production is irrigation water. Although most TBOS are hardy, they will still require adequate irrigation to produce satisfactory yields. As the priority allocation of available water is for food production (including edible oil), the oilseed production of non-edible oil may be done on land where competition for water does not occur. In fact, irrigation facilities have to be established in dry lands by constructing canals systems, watershed management, and rainwater harvesting so that adequate land can be brought under cultivation properly.

4.8 Use of Nonedible Oils for Manufacturing Biodiesel [8,16,26,32,33,34,54,55,77,78,101,102,103]

Any edible or non-edible (used or fresh) oil/fat can be used for making biodiesel. In some countries, there is a perpetual shortage of edible oils and therefore, it may not be possible for them to spare any edible oil for manufacturing biodiesel. Additionally, the use of resources which are essential to grow food crops cannot be spared to grow oilseeds just for making biodiesel. However, the choice of oils which can be used for production of biodiesel are the non-edible oils obtained from seeds of *Madhuca indica* (Mahua), *Azadirachta indica* (Neem), *Calophyllum inophyllum* (Sultana champa/Punnaga/Undi), *Jatropha curcas* (Bagbhereda/Kananeranda), Soapnut, that is, Ritha (*S. mukorossi* and *S. trifoliatus*), etc., grown intentionally for the purpose or grown traditionally for other purposes. Also, any old/used or surplus vegetable oils or fat rendered inedible can be used for making biodiesel. The characteristics of the abovementioned trees are mentioned below:

a. Madhuca indica (Mahua)

The Madhuca indica or Mahua tree is a slow-growing tree with an average height of 18 meters and a diameter of 80 cm. It grows well in subtropical/tropical regions with temperature in the range of 2–46°C and rainfall 550–1500 mm, altitude up to 1200 meters, and in deep loamy or sandy soils with good drainage. The wood is hard, durable timber, the flowers are sweet, yield distilled spirit, and the seeds contain 20% to 50% oil, which is not considered edible although tribal communities often use it for cooking. It is locally classified as a medicinal tree.

b. Azadirachta indica (Neem)

The Azadirachta indica (Neem) is found all over India in all type of climates and soils. It prefers a rainfall range of 750–1000 mm and a temperature range of 15–45°C and an altitude up to 1200 meters. The tree can survive in hot arid/semiarid zones and acidic soils and alkaline soils with a pH value of 10. It is largely a medicinal tree (all its parts have medicinal properties), and it grows to a height of 20 meter. The wood of the tree is hard and provides good timber and its seeds contain 30–40% oil, which is not edible, but medicinal.

All parts of the Neem tree are medicinal and have many uses. It provides timber resistant to decay. Neem based biopesticides are very useful. The propagation of this tree is highly advisable both from the point of view of making biodiesel from its oil and other uses. The plant grows well in subtropical/tropical regions on almost all types of soils, plains, highlands. It can sustain itself in hot semiarid lands as well. It is not affected by insects and plant diseases because of the presence of a protective compound in it called 'Azadirachtin' (See Note in Para 2.1.4 on Biopesticide).

c. Calophyllum inophyllum (Punnaga/Sultana champa/Undi)

The Calophyllum inophyllum is a slow-growing tree found in coastal areas or in areas up to an altitude of 200 meters in a rainfall range of 750–5000 mm and a temperature range of 7–48°C. It is sensitive to frost and fire. It prefers deep, sandy, loamy soils, but can survive on highly sandy soils if required. The wood of the tree is hard and durable and is used for boat building. The leaves, flowers, seeds, and seed oil, are used locally as medicine. It is often grown as ornamental tree and its seeds contain 50–70% of oil.

d. Jatropha curcus

Jatropha curcus grows in subtropical parts of the world—Africa, South America, India, etc. It is a shrub about 3–4 meters high and can be propagated by seeds as well as branch cuttings. Its seeds have an oil content of about 30%, which can be converted to biodiesel.

Jatropha plant normally starts bearing seeds after 3 years of sowing, but full fruition begins only after 5 years and lasts for 30 years. The Jatropha plant (in irrigated area) yields about three times more seeds per Jatropha plant than in an unirrigated area.

Hybrid variety of Jatropha has been developed by Chinese scientists which is superior to the ordinary variety in respect of: (i) shorter maturing period of plants, (ii) higher yield per hectare. (This variety of Jatropha is grown in large tracts of Sichuan, Henan, and Inner Mongolia in China.)

Jatropha curcus can be grown in different agroclimates both on irrigated and nonirrigated land. Jatropha is a drought-resistant shrub especially suited to semiarid and arid conditions. Normally, it grows well in areas with rainfall of 480 mm–2,400 mm and at an altitude between 0 (sea level) to 1000 meter. It can grow in soil with stones or gravel and even on calcareous or shallow soils. It has a low demand for fertiliser and moisture. However, the growth of jatropha is not good in low-fertility alkaline soils and eroded soils.

The lower calorific value of biodiesel/methyl ester made from jatropha seed oil is about 39 MJ/liter, which is higher than the calorific value of biodiesel made from other crops like peanut oil, soybean, sunflower, palm oil for which the values are 33.6, 33.5, 33.5, 33.3 MJ/liter (lower calorific value) respectively. Jatropha cake, after extraction of oil, is rich in protein (it contains 35% protein), but has some toxic substances and therefore, cannot be used in foods. However, it can be used as organic manure because it has 2.09 % phosphorous, 1.69% potassium, and 4.44 % nitrogen. Jatropha leaves and bark yield blue dye, which is good for colouring cloth, fishing nets, etc., Jatropha curcus is of particular interest in India because of the above favourable characteristics.

The oil content in Jatropha is influenced by the condition of the soil and the age of the plant. The jatropha plant has an average life with effective yield of up to 50 years. The economic yield on the average can be considered as 1–2 kg/plant and 4–6 MT/hectare/year depending on the agroclimatic zone and agricultural practices. One hectare of plantation on average soil will give about 1.6 MT of oil. Methyl ester (biodiesel) from Jatropha may have the following content of fatty acids:

Oleic acid about 45–46%, Palmitic acid about 13–14%,

Linoleic acid about 32%, Stearic acid about 5.5%.

Note: There are three genes that have been identified in jatropha, which are involved in the production of oil. These genes show high homologies with three of the genes present in the genome of the algae, 'Chlamydomonas reinharddtii' These findings may help in providing a better understanding of lipid synthesis in plants/algae and exploring the mechanism of economic oil production from algae.

The yield from jatropha plantation for biodiesel production in irrigated and nonirrigated areas is illustrated in Table 4.4.

Table 4.4. Jatropha Plantation Vis-à-Vis the Production of Biodiesel.

Se. No.	Particulars	Under Irrigated Conditions	Under Rainfed Conditions
1.	Spacing of plants	2x2	1.5x1.5
2.	No. of plants/hectare	2500	4444
3.	Tentative yield of seeds/plant after six years of sowing	5 kg.	1.5 kg.
4.	Tentative total yield of seeds/ha after six years	12,500 kg	6666 kg
5.	of sowing	25 to 30 %	25 to 30 %
6.	Oil content of seeds Cake	75 to 70 %	75 to 70 %
7.	Tentative yield of oil per hectare	3125 to 3750 kg	1666 to 2000 kg
8.	Tentative yield of cake per hectare	9375 to 7500 kg	5000 to 4666 kg

e. Soapnut Tree (Ritha Tree)

Soapnut (Ritha) is a fruit of the soapnut tree, generally found in tropical and subtropical climate areas in various parts of the world, including Asia, America, and Europe. The two main varieties (*S. mukorossi* and *S. trifoliatus*) are widely available in India, Nepal, Bangladesh, Pakistan, and many other countries.

The oil content in *S. trifoliatus* (which is very similar to that in *S. mukorossi* seed kernels) is on average 51.8% of the seed weight. The oil from soapnut is non-edible oil and has significant potential for biodiesel production; it is extracted from the nut kernel which is waste material.

Soapnut has several applications, ranging from medicinal treatments to soap and surfactant. Soapnut fruit shells have been in use from ancient times as natural laundry detergents for washing fabrics, for bathing, and as traditional medicines.

Soapnut trees can be integrated in the community's agroforestry plan. Planting soapnut trees in community forests and in barren lands provides a sink for carbon sequestration as well as feedstock for biodiesel production.

Soapnut oil is found to have 9.1% free FA, 84.43% triglycerides, 4.88% sterol, and 1.59% others. Methyl ester (biodiesel) from soapnut may have the following content of fatty acids:

Oleic acid about 52%, palmitic acid about 6–9%, linoleic acid about 4%, stearic acid about 1.5%, and eicosenic acid 23–24%.

The glycerol, which is a byproduct of biodiesel production, can be used to produce organic acids such as succinic acids by bacterial fermentation. This, therefore, would boost the economics of biodiesel production.

f. Seeds of Salicornia brachiata for biofuel production

The seeds of Salicornia brachiata, a salt- and drought-resistant group of plants occurring in coastal estuarine regions of tropical/subtropical/temperate zones are also a source of edible oil and a potential source of biodiesel [see Para 2.2.6 (e)].

Note: Raising plants on arsenic-polluted land for biofuel production and simultaneous phytoremediation of the soils

It is hoped that a new phytoremediation technology can enable the detoxification of arsenic- polluted soils. Using this technology, the high-biomass, fast-growing 'Crambe abyssinica and Brassica juncea' were subjected to genetic engineering by transferring ArsC and γ-ECS genes to them. These plants, when put in arsenic-polluted soil, will accumulate the arsenic after absorbing it from the soil. In due course of time, the soil will be detoxified. In the meantime, the seeds and the biomass of these plants can be used for production of biofuels.

4.9 Biodiesel and Jatropha Plantations in Developing Countries [1,11,24,29,72,100,101,102,124,128]

Example of Tanzania

As an example of jatropha-based biodiesel development, let us take the case of Tanzania (Africa). Jatropha grows wild in Africa. The plant is widely believed to have potential to help combat the greenhouse effect, stop local soil erosion, create additional income for the rural poor, and be a major source of energy. The production per hectare of jatropha seeds widely differs in different regions in the world depending on climate, soil condition, and water availability. In Tanzania, the expected seed production may be 10–20 ton/ha/year. The seed contains about 30% oil, which may be extracted by manually operated presses or power-operated screw expellers. The manual oil extraction is less frequent than the oil extraction by power-operated expellers. The seeds are harvested in the dry season, when the labour is not engaged in farm activities. In Tanzania, jatropha oil can be used for cooking, lighting lamps (in especially developed stoves/lamps), or for making soap, besides being used in the production of biodiesel. The oilseed cake can be used for making briquettes, which can be used as cooking fuel. It can also be used to make fertilizer/manure and aid in the production of biogas. Tanzania can spare enough land for growing jatropha.

Example of India

In India, biodiesel was initially utilized as fuel by three wheelers and state transport buses. It was experimentally used to run state transport corporation buses in Karnataka. A research and development project for biodiesel production and field trials on car using biodiesel took place at Indian Institute of Technology, Delhi. The trials were carried out on a diesel car, which blended 20% of biodiesel with diesel 'successfully'.

There are plans to cultivate plants on barren land in India and to use the oil from jatropha seeds for biodiesel production. In order to organize the industry, the Biodiesel Association of India was formed to encourage energy plantations for increasing feedstock supplies. The University of Agriculture Sciences at Bangalore has identified many elite lines of *Jatropha curcas* and Pongamia pinnata, that could be planted in unused areas through government initiatives.

Large-scale plantations were initiated in Northeast India and Jharkhand by D1 Williamson Magor Bio Fuel Limited, a joint venture between D1 Oils of U.K. and Williamson Magor of India. The hilly areas of the Northeast are ideal for growing this hardy, low-maintenance plant.

Indian Oil Corporation tied up with Indian Railways to introduce biodiesel crops over one (1) million square kilometers. Also, Jharkhand and Madhya Pradesh tied up with Indian Oil Corporation to cultivate large tracts of land with jatropha, which is favored choice for producing biodiesel in India.

A model biofuel park with 116,300 *Jatropha curcas* plants has been established at Defence Institute of Bio-energy Research (DIBER) at Harsola Military farm, Mhow. The park was established in 2007 and has started fruition in 2012. For maximum utilization of land, *Camelina sativa* (another high oil-yielding crop) is being grown via intercropping in jatropha plantation during winters, when the jatropha plants shed leaves and do not hinder sunlight penetration.

4.10 Advantages of Jatropha Plantations for Biodiesel and the Resulting Carbon Sinks [39,41,106,131]

There are several advantages of jatropha plantations for biodiesel production and the resulting carbon sinks such as:

a. *Reclamation of land: Jatropha curcas* can be grown on any type of soil, including wasteland. Once planted, the wasteland is virtually converted to fertile land. This is an excellent plant for India to reclaim vast tracts of over 130 million hectares of wasteland.

b. *High rate of survival: Jatropha curcas* are strong, drought-resistant plants suitable for extreme climatic conditions.

c. *Forest development: Jatropha curcas is* extremely useful for the accelerated development of green belts/forest cover.

d. *Saving in greenhouse gas (GHG) net emissions:* Twelve thousand trees of Jatropha curcas can create an annual absorption of over 10 tonnes of CO_2 (carbon dioxide). It is considered a useful plant to combat the rising menace of global warming.

e. *Grow-diesel carbon sinks:* The large-scale development of grow-diesel carbon sinks' on waste/forest land can result in the greatly increased production of biodiesel, which can be an import substitute for imported fossil fuel to some extent.

f. *Employment generation:* The large-scale creation of grow-diesel carbon sinks will help in generating employment.

g. *Payback of investment:* Since the Jatropha seeds shall be used for the manufacture of biodiesel, the investment in the creation of grow-diesel carbon sinks is recoverable.

h. *Land cost appreciation:* Grow diesel energy plantation is a suitable investment in land. Normally the lands near cities are converted to commercial and residential purposes over a span of one or two decades. Such lands can be ideally sown with jatropha plantation to get a regular annual income till the land is ready for commercial use.

i. The leaf extract of *jatropha curcas* is found to be effective in killing termites*.

*Note: The water extract from jatropha leaves may be prepared by boiling the leaves in water. Ether, butanol, and the hexane extracts of jatropha leaves may be obtained by using standard extraction methods like Soxhlet**. It is reported that in an experiment (at 28°C/80% RH), the termite mortality by these extracts was found as indicated below:*

(a) *For 0.35 g/ml ether extract, the mortality rate of termites after 48 hours was 73%.*

(b) *For 0.39 g/ml butanol extract, the mortality rate of termites after 48 hours was 67%.*

(c) *For 0.62 g/ml hexane extract, the mortality rate of termites after 48 hours was 60%.*

(d) *For 10% hot water extract, the mortality rate of termites after 48 hours was 27%.*

It has been suggested that the mortality of termites may be enhanced with some other chemical/ biochemical additives.

*[** The Soxhlet extraction process is used to extract oil from algae. In this process, oil from algae is extracted through the repeated washing or percolation with an organic solvent such as hexane or petroleum ether. After recovery, the solvent is recycled. (See Para 5.5.2 of Chapter 5)].*

4.11 Biodiesel from Edible Oils [7,10,46,60,61,62,63,93,97,101,102,103]

As already stated, biodiesel can be derived from edible oils like palm, soybean, corn, and sunflower, non-edible oils (*Jatropha curcas, Calophyllum inophyllum, Ceiba pentandra,* and *Mahua*), algae, and waste vegetable oils and animal fats (tallow, lard, yellow grease, and chicken fat). The countries where there is surplus of any kinds of vegetable oils or fats can be involved in the production of biodiesel. The waste frying oil (used/waste vegetable oil or fat) is already being converted to biodiesel in many countries. An example of palm oil, which is an edible oil, is considered here.

As of today, the biodiesel demand keeps increasing as the biofuel has been introduced to more than 60 countries worldwide. Palm oil plays a significant role in expanding the world's production and the consumption of oils and fats, surpassing soybean, rapeseed, and sunflower oil. The global production of palm oil is predominantly dominated by Indonesia and Malaysia, which together supply about 84% of the total palm oil worldwide. This is followed by Thailand.

As mentioned earlier, compared to the combustion of fossil fuels, the use of biodiesel, as an alternative fuel, results in a significant reduction of harmful emissions. The extensive utilization of food-based crops for biofuel production has been subject to lot of criticisms. The use of palm biodiesel can cause global imbalance in the food industries. Therefore, the sustainability of the palm biodiesel industry is very crucial.

Oil palm plantation was first introduced in Southeast Asia. Palm oil is edible and is produced in the southern part of India. Table 4.5 shows the various uses of oil palm. The oil is obtained from both the flesh and kernel of oil palm. The following are its characteristics:

- Plum-sized fruits obtained from oil palm consist of flesh and hard nut (kernel).
- A yellow-red liquid extracted from its flesh is called 'crude palm oil' (CPO).
- The oil is extracted from the kernel and is called 'palm kernel oil' (PKO).
- Crude palm oil (CPO) contains 5% free fatty acid, which must be removed during refining, otherwise it will decompose.
- CPO (crude palm oil) contains about 50% saturated fats. The sludge from the fruit flesh after extracting crude palm oil (CPO) contains fiber (which is good material for paper making), and protein.
- PKO (palm kernel oil) contains about 80% saturated fats. The cake of the kernel after extracting palm kernel oil (PKO) also contains protein.

Thus, several standards and policies have been implemented to ensure the sustainable development of this industry, which includes forest conservation, environmental management, biological control of pests, and reduction of greenhouse gases. Both Malaysia and Colombia are equipped with advanced technologies, facilities, and expertise in this particular field in ensuring that the palm oil biodiesel can bring many more benefits to the global market, especially for transportation and energy ecosystem.

The palm biodiesel industry is facing many challenges in implementing the biodiesel program in Malaysia. Palm oil is the backbone of Malaysia's economy, comprising more than 5% of its gross domestic product (GDP). Till now, B5 and B7 biodiesel programs have been successfully implemented in Malaysia. They have just started to implement B10. Implementation of B30 in Indonesia has been postponed.

Table 4.5. Uses of Palm Fruit and Nut for Biodiesel, Food, and Other Products. [46]

Oil Palm Fruit			Oil Palm Nut	
Crude Palm Oil (CPO)	**Fiber**	**Sludge**	**Kernel**	**Palm Cake, that is, Palm Kernel Meal (PKM)**
Food: Frying oil, pastries, margarine, mayonnaise, and others	Pulp & paper	Animal feed, soap, fertilizer	Food: frying oil, pastries, margarine, mayonnaise, ice-cream, & others	Animal feed
Soap, detergent, biodiesel, oleochemicals, cosmetics			Oleochemicals-cosmetics, soap, detergent	

Note: Fresh palm oil, which is a good edible oil, should not be used for making biodiesel in countries with edible oil shortage. However, used palm oil and palm oil wastes can be used for it.

The high oil yield of the trees has encouraged wider cultivation, leading to the clearing of forests in parts of Indonesia and Malaysia to make space for oil palm monoculture. Palm plantations have caused a large amount of deforestation in Southeast Asia. France banned palm oil from the country's biofuel scheme in January 2020. This was hailed as a positive step in the fight against wide-scale deforestation with the European Union (EU) tightening its net around a controversial substance and its member countries beginning to define legislation to prevent environmental damage. Palm has become controversial issue. Some other regions will benefit, if not the West.

Besides producing palm oil, this industry generates a huge quantity of dry and wet residues, which can be processed to produce biofuel. Many technologies have been developed to process oil palm and palm oil wastes into biofuel and many more are in process.

4.12 Summary

This chapter is devoted to producing biodiesel from vegetable oils, which is a secondary renewable source. Biodiesel can be made from various oils from the seeds of food crops, non-food crops, or even fats. Oils and fats are simple lipids. These are compounds of glycerol and various fatty acids, that is, they are glycerol esters or termed as 'glycerides/triglycerides.

The fatty acids present in the glycerides are almost exclusively straight-chain acids, and almost always contain an even number of carbon atoms. The oils are liquid at 20°C and fats are solid/semisolid at 20°C. The fatty acids are divided into two main groups: (a) saturated fatty acids and (b) unsaturated fatty acids (containing one or more double bonds)

The examples of saturated fatty acids are: capric acid ($C_9H_{19}COOH$), myristic acid ($C_{13}H_{27}COOH$), palmitic acid ($C_{15}H_{31}COOH$), stearic acid ($C_{17}H_{35}COOH$), etc.

The examples of unsaturated fatty acids are: oleic acid ($C_{17}H_{33}COOH$), linoleic acid ($C_{17}H_{31}COOH$), linolenic acid ($C_{17}H_{29}COOH$), erucic acid ($C_{21}H_{41}COOH$), etc.

The glycerol (also commonly called 'glycerine') is polyhydric alcohol* having three OH groups (i.e., trihydric alcohol) with the formula $CH_2OH.CHOH.CH_2OH$.

Animal fats and vegetable oils are typically made of triglycerides, which are esters of free fatty acids with the trihydric-alcohol, that is, glycerol.

Esters are compounds of the form $R_1O R_2$, where R, R_1, R_2, R_3 are the alkyl groups or long carbon chains. Example of simple glycerides ($CH_2OCOR.CHOCOR.CH_2OCOR$) are:

Tristearin—[$CH_2OCO C_{17}H_{35}.CHOCO C_{17}H_{35}.CH_2OCO C_{17}H_{35}$]

Tripalmitin—[$CH_2OCO C_{15}H_{31}.CHOCO C_{15}H_{31}.CH_2OCO C_{15}H_{31}$]

Triolein—[$CH_2OCO C_{17}H_{33}.CHOCO C_{17}H_{33}.CH_2OCO C_{17}H_{33}$]

Examples of mixed glyceride [$CH_2OCOR_1.CHOCOR_2 .CH_2OCOR_3$] are:

Oleopalmitostearin—($CH_2OCO C_{17}H_{35}.CHOCO C_{15}H_{31} .CH_2OCO C_{17}H_{33}$)

Palmito diolein—($CH_2OCO C_{17}H_{33}.CHOCO C_{17}H_{33} .CH_2OCO C_{15}H_{31}$)

The process of converting oil to biodiesel essentially consists of the following steps:

- Vegetable oils are converted into methyl Esters ($RCOOCH_3$) by transesterification.

- The process of transesterification removes glycerol ($CH_2OH.CHOH.CH_2OH$) from Triglycerides ($CH_2OCOR_2.CHOCOR.CH_2OCOR_2$) and thus reduces the viscosity of the resulting product, which is called 'biodiesel'. Glycerol is also called glycerine.

- A two stage integrated pre-esterification of free fatty acids and base-catalyzed transesterification process proved to be an effective method in converting feedstock with a high amount of free fatty acids. However, with acid catalysts for transesterification, it was found that the removal of free fatty acids or the pretreatment of feedstock oils may not be necessary. The heterogeneous catalyst tolerates moisture and free fatty acids.

- The scheme of the transesterification of mixed glyceride to methyl ester is as follows:

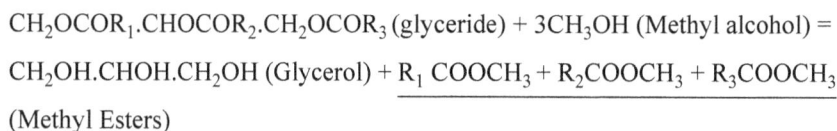

$CH_2OCOR_1.CHOCOR_2.CH_2OCOR_3$ (glyceride) + $3CH_3OH$ (Methyl alcohol) =

$CH_2OH.CHOH.CH_2OH$ (Glycerol) + $\underline{R_1\ COOCH_3 + R_2COOCH_3 + R_3COOCH_3}$

(Methyl Esters)

(Where; R_1, R_2, and R_3 represent long carbon chains)

The transesterification reaction takes place either exceedingly slowly or not at all without acid or base and heating. It is important to note that the acid or base are not consumed by the transesterification reaction, thus, they are not reactants but catalysts. The catalyst may be alkali, acids, or enzymes. Some solid *heterogeneous catalysts also can be used for Transesterification*. The alcohol reacts with the oils and fats to form the monoalkyl ester (or biodiesel) and crude glycerol. The reaction between the biolipid (fat or oil) and the alcohol is a reversible reaction so the alcohol must be added in excess to drive the reaction towards the right and ensure complete conversion. Usually, the ratio of methanol to oil is kept at about 6:1.

The various processes of making biodiesel have been dealt with along with their merits and demerits. Briefly, the processes dealt with are:

i. Base Transesterification: The transesterification may be catalyzed by alkalis like sodium hydroxide, potassium hydroxide, sodium methoxide, or potassium methoxide. The alcohol/ catalyst mix is charged in a closed reaction vessel (totally sealed off) and the vegetable or animal oil/ fat is added. The problem of soap formation is encountered when there are free fatty acids and moisture/water present in the materials.

ii. Acid transesterification: The transesterification reaction can be catalyzed by acid. The acid catalysis is not sensitive to moisture and free fatty acids and is therefore, advantageous in the case of waste/used oils. However, the disadvantages are: slow pace of reaction, acidity of the effluent, non-usability of catalyst, and the high cost of equipment.

iii. Enzymatic transesterification: Very good yields can be obtained from used oils and fats by using lipases as catalysts and methyl acetate instead of methanol. The process is less sensitive to free fatty acids.

iv. Reactive extraction of biodiesel: The reactive extraction of biodiesel from oilseed directly is also possible by treating the ground oilseeds with methanol. As a very large amount of excess methanol is required, the process is very costly and economically unviable.

v. Catalyst-free transesterification supercritical process: The catalyst-free method uses supercritical methanol at high temperatures and pressures in a continuous process where the oil and methanol remain in a single phase and the reaction occurs spontaneously and rapidly. This process can tolerate water and free fatty acids in the feedstock.

The various types of reactors used for making biodiesel have been highlighted. These include batch reactors with ultra- and high-shear inline mixers, reactors with ultrasonic devices, and reactors using microwave heating for the transesterification reaction.

The comparison of the properties of biodiesel with other liquid fuels and petroleum diesel have been given. Problems with use of biodiesel such as compatibility of the materials, gelling property, and water contamination are dealt with.

Nonedible vegetable oils and waste vegetable oils should be used for making biodiesel oils so that it does not interfere with food. Oilseed-producing trees (for non-edible oil) include *Azadirachta indica*, *Madhuca indica*, *Calophyllum inophyllum*, *Jatropha curcas*, soapnut ((*S. mukorossi* and *S. trifoliatus*), etc. Jatropha can grow on unirrigated wasteland as well. Details about its production and advantages of its plantations are highlighted.

QUESTIONS

1. *What is a vegetable oil chemically? Illustrate giving the chemical formulas.*
2. *What is biodiesel chemically? Illustrate by giving chemical formulas.*
3. *What are the advantages of using biodiesel over petroleum diesel?*
4. *Is biodiesel used as a pure fuel or is it blended with petroleum diesel?*
5. *Do you need special facilities to store biodiesel other than those usually required for petroleum diesel? What are the requirements of storage?*
6. *Are there any problems in using biodiesel in existing diesel engines?*
7. *Is biodiesel good for your engine?*
8. *Is biodiesel the same thing as raw vegetable oil? What are the differences?*
9. *What is transesterification? How is this process used for making biodiesel?*
10. *How is the biodiesel obtained from various vegetable oils/ fats. Illustrate the process using a flow diagram.*
11. *What is the solvent extraction of oil from oilseeds?*
12. *Which catalysts are used in the transesterification reaction of oils/ fats?*
13. *Which catalyst is better in case you want to make biodiesel from used vegetable oil or fat and why?*
14. *How does the use of the ultra and high-shear inline mixer help in manufacturing biodiesel?*
15. *How does the ultrasonic device in reactors help in the production of biodiesel?*
16. *How does it help to use microwave heating for the transesterification reaction?*
17. *State the merits and demerits of enzymatic transesterification using lipases as a biocatalyst.*
18. *What are heterogeneous catalysts for transesterification?*
19. *Describe the catalyst-free supercritical process of making biodiesel.*
20. *Compare the properties of biodiesel with those of petroleum diesel.*
21. *What problems need to be addressed in the use of biodiesel?*
22. *What is gelling property with regard to biodiesel?*
23. *What are the problems created if the biodiesel gets contaminated with moisture/water?*
24. *What is reactive extraction of biodiesel directly from oilseeds?*
25. *What are the sources of edible oils and tree-borne non-edible oils?*
26. *Enumerate the sources of non-edible oils that can be used for obtaining biodiesel.*
27. *What are the respective advantages of planting various trees which bear oilseeds? Give examples of trees which produce oilseeds.*
28. *Write a short note on jatropha plantation for obtaining biodiesel. Why is this tree being given so much importance?*

29. *What are the advantages of jatropha plantations apart from oilseed production, which can then be converted to biodiesel?*

30. *Write a note on the production of biodiesel from oil palm. What is the difference between oil from 'fruit of oil palm' and that from the 'oil palm nut kernel'?*

References

1. Achten, Wouter M.J., Erik Mathijs, Louis Verchot, Virendra P. Singh, Raf Aerts and Bart Muys. November 2007. Jatropha biodiesel fuelling sustainability. Biofuels, Bioproducts and Biorefining. https://doi.org/10.1002/bbb.39.

2. Alalwan, Hayder A., Alaa H. Alminshid and Haydar A.S. Aljaafari. March 2019. Promising evolution of biofuel generations. Subject review. Renewable Energy Focus 28: 127–139.

3. Atadashi, I.M., M.K. Aroua, A.R. Abdul Aziz and N.M.N. Sulaiman. June 2012. The effects of water on biodiesel production and refining technologies: A review. Renewable and Sustainable Energy Reviews 16(5): 3456–3470. https://doi.org/10.1016/j.rser.2012.03.004.

4. BAIF. 2004. Jatropha and other perennial oilseed species. Proc. of National Workshop. Hegde, N.G., J.N. Daniel and S. Dhar (eds.). 160 pages. BAIF Development Research Foundation, Pune, India.

5. Bajaj, Akhil, Purva Lohan, Prabhat N. Jha and Rajesh Mehrotra. January 2010. Biodiesel production through lipase catalyzed transesterification: An overview. J. of Molecular Catalysis B: Enzymatic 62(1): 9–14. https://doi.org/10.1016/j.molcatb.2009.09.018Get rights and content.

6. Balan, Venkatesh. 2014. Current Challenges in Commercially Producing Biofuels from Lignocellulosic Biomass. Review Article, ISRN Biotechnology Volume. Article ID 463074, 31 pages. http://dx.doi.org/10.1155/2014/463074.

7. Balat, Mustafa. February 2011. Potential alternatives to edible oils for biodiesel production—A review of current work. Energy Conversion and Management 52(2): 1479–1492. https://doi.org/10.1016/j.enconman.2010.10.011.

8. Banković-Ilić, Ivana B., Olivera S.Stamenković and Vlada B.Veljković. August 2012. Biodiesel production from non-edible plant oils. Renewable and Sustainable Energy Reviews 16(6): 3621–3647. https://doi.org/10.1016/j.rser.2012.03.002.

9. Barnard, T. Michael, Nicholas E. Leadbeater, Matthew B. Boucher, Lauren M. Stencel and Benjamin A. Wilhite. 2007. Continuous-flow preparation of biodiesel using microwave heating. Energy Fuels 21(3): 1777–1781. DOI: 10.1021/ef0606207.

10. Barnwal, B.K. and M.P. Sharma. August 2005. Prospects of biodiesel production from vegetable oils in India. Renewable and Sustainable Energy Reviews 9(4): 363–378. https://doi.org/10.1016/j.rser.2004.05.007.

11. Baur, H., V. Meadu, M. van Noordwijk et al. 2007. Biofuel from Jatropha curcas: Opportunities, Challenges and Development Perspectives; World Agroforestry Centre: Nairobi.

12. Benjumea, Pedro, John Agudelo and Andrés Agudelo. August 2008. Basic properties of palm oil biodiesel–diesel blends. Fuel 87(10-11): 2069–2075. https://doi.org/10.1016/j.fuel.2007.11.004.

13. Berger, Harald, Norbert Dragesser, Rudolf Heumueller, Erich Schaetzer, and Manfred Wagner (Inventers). US5484573A, United States assigned to Hoechst Aktiengesellschaft. Reactor for carrying out chemical reactions. https://patents.google.com/?inventor=Harald+Berger.

14. Best uses of biomass. Sept/Oct 2004. Martim Tampier of Enviro. Chem. Services Inc. Canada. Refocus ISES p. 22–25.

15. Biofuels from agricultural wastes. January 2019. DOI: 10.1016/B978-0-12-815162-4.00005-7.

16. Biodiesel Technocrat's Reports on Biodiesel Industry. http://www.biodieseltechnocrats.in/general_information.html.

17. Biodiesel Handling and Use Guide (Fifth Edition) November 2016. DOE/GO-102016-4875. (Lead Authors: Teresa L. Alleman and Robert L. McCormick). https://afdc.energy.gov/files/u/publication/biodiesel_handling_use_guide.pdf

18. Biodiesel from Wikipedia, the free encyclopedia https://en.wikipedia.org/wiki/Biodiesel#Contamination_by_water

19. Biofuelling Brazil. May/June 2006. Refocus (ISES).

20. Bondioli, Paolo, Ada Gasparoli, Armando Lanzani, Enzo Fedeli, Sergio Veronese and Maura Sala. June 1995. Storage stability of biodiesel. J. of the American Oil Chemist's Society 72(6): 699–702.

21. Borges, M.E. and L. Díaz. June 2012. Recent developments on heterogeneous catalysts for biodiesel production by oil esterification and transesterification reactions: A review. Renewable and Sustainable Energy Reviews. 16(5): 2839–2849. https://doi.org/10.1016/j.rser.2012.01.071.

22. Boulal, Ahmed, Mostefa Khelafi and Cherif Khelifi. 2018. Biodiesel from restaurant cooking oils using trans-esterification process. Int. J. of Applied Engineering Research (ISSN 0973-4562) 13(6): 4157–4161 © Research India Publications. http://www.ripublication.com 4157.

23. Bouaid, Abderrahim, Mercedes Martinez and José Aracil. November 2007. Long storage stability of biodiesel from vegetable and used frying oils. Fuel. 86(16): 2596–2602. https://doi.org/10.1016/j.fuel.2007.02.014.

24. Brittaine, R. and N. Lutaladio. 2010. Jatropha: A smallholder bioenergy crop—The potential for pro-poor development. In Integrated Crop Management. Vol. 8. FAO: Rome.

25. Bunkyakiat, Kunchana et al. 2006. Continuous Production of Biodiesel via Transesterification from Vegetable Oils in Supercritical Methanol. Energy and Fuels (American Chemical Society) 20: 812–817. doi:10.1021/ef050329b.

26. Cadenas, Alfredo and Sara Cabezudo. May–June 1998. Biofuels as Sustainable Technologies: Perspectives for Less Developed Countries. Technological Forecasting and Social Change, ELSEVIER. Volume 58(1-2): 83–103. https://doi.org/10.1016/S0040-1625(97)00083-8.

27. Cao, F., Y. Chen, F. Zia et al. Sept. 2008. Biodiesel from high acid value waste frying oil catalyzed by superacid heteropolyacid. Biotechnology and Bioengineering 101(1): 93–100.

28. Carioca, J.O.B. 2010. Biofuels: problems, challenges and perspectives. Biotechnology Journal 5(3): 260–273.

29. Case Study Jatropha Curcas: Global Facilitation Unit for Underutilized Species. 2004. Deutsche Gesellschaft fuer Technische Zusammenarbeit (GTZ), Frankfurt. www.underutilized species.org/Documents/PUBLICATIONS/jatropha_curcas_india.pdf.

30. Cervero, J.M., J. Coca and S. Luque. 2008. Production of Biodiesel from vegetable oil. Grasas Y Aceites. International of Fats and Oils 59(l): 76–83.

31. Çetinkaya, Merve and Filiz Karaosmanoğlu. 2004. Optimization of base-catalyzed transesterification reaction of used cooking oil. Energy Fuels 18(6): 1888–1895. DOI: 10.1021/ef049891c.

32. Chhetri, Arjun B., Martin S. Tango, Suzanne M. Budge, K. Chris Watts and M. Rafiqul Islam. 2008. Non-edible plant oils as new sources for biodiesel production. Int. J. Mol. Sci. 9(2): 169–180. https://doi.org/10.3390/ijms9020169.

33. Chhetri, A.B., Y.R. Pokharel, H. Mann and M.R. Islam. 2007. Characterization of soapnut and its use as natural additives. Int. J. of Material and Products Technology.

34. Chhetri, A.B., K.C. Watts, M.S. Rahman and M.R. Islam. 2009. Soapnut extract as a natural surfactant for enhanced oil recovery. J. Energy Sources, Part A: Recovery, Utilization, and Environmental Effects 31(20): 1893–1903. https://doi.org/10.1080/15567030802462622.

35. Christensen, Earl and Robert L. McCormick. December 2014. Long-term storage stability of biodiesel and biodiesel blends. Fuel Processing Technology 128: 339–348. https://doi.org/10.1016/j.fuproc.2014.07.045.

36. Colucci, J.A., E.E. Borrero, F. Alape. 2005. Biodiesel from an alkaline transesterification reaction of soya bean oil using ultrasound mixing. J. of American oil Chemists Society 82(7): 525–530.

37. Daniel, J.N. 1997. Pongamia pinnata: a nitrogen fixing tree for oilseed. FACT 97-03, Forest Farm and Community Tree Network, USA.

38. Daniel, J.N. and N.G. Hegde. 2007. Tree-Borne Oil Seeds in Agroforestry. Proc. of the National Seminar on Changing Global Vegetable Oils Scenario: Issues and Challenges before India. pp. 263–276. Hegde, D.M. (ed.). Indian Society of Oilseeds Research, Hyderabad, India.

39. Danvan, der Horst and Saskia Vermeylen. June 2011. Spatial scale and social impacts of biofuel production. Biomass and Bioenergy 35(6): 2435–2443. https://doi.org/10.1016/j.biombioe.2010.11.029.

40. Das, L.M., Dilip Kumar Bora, Subhalaxmi Pradhan, Malaya K. Naik and S.N. Naik. November 2009. Long-term storage stability of biodiesel produced from Karanja oil. Fuel 88(11): 2315–2318. https://doi.org/10.1016/j.fuel.2009.05.005.

41. Deng, Xiangzheng, Jianzhi Han and Fang Yin. 2012. Net energy, CO2 emission and land-based cost-benefit analyses of jatropha biodiesel: a case study of the Panzhihua Region of Sichuan Province in China. Energies. 5(7): 2150–2164. https://doi.org/10.3390/en5072150.

42. Dhawane, Sumit H., Tarkeshwar Kumar and Gopinath Halder. August 2015. Central composite design approach towards optimization of flamboyant pods derived steam activated carbon for its use as heterogeneous catalyst in transesterification of *Hevea brasiliensis* oil. Energy Conversion and Management 100: 277–287. https://doi.org/10.1016/j.enconman.2015.04.083.

43. Du, Wei et al. 2004. Comparative study on lipase-catalyzed transformation of soybean oil for biodiesel production with different acyl acceptors. J. of Molecular Catalysis B: Enzymatic 30: 125–129.

44. Dunn, Robert O. Jul 2015. Cold flow properties of biodiesel: a guide to getting an accurate analysis. Published online:https://www.researchgate.net/deref/http%3A%2F%2Fwww.tandfonline.com%2Floi%2Ftbfu20.

45. Encinar, J.M., J.F. González, G. Martínez, N. Sánchez and A. Pardal. May 2012. Soybean oil transesterification by the use of a microwave flow system. FUEL 95: 386–393. https://doi.org/10.1016/j.fuel.2011.11.010.

46. Fassler, Peter. Jan. 2006. Bio diesel fuel from Palm too good to be used as fuel. Sulzer Technical Review. 88:14-16. Sulzer Chemtech. https://www.researchgate.net/publication/291137844_Too_good_to_be_used_as_fuel/citation/download.

47. Fazal, M.A., A.S.M.A. Haseeb and H.H. Masjuki. February 2011. Biodiesel feasibility study: An evaluation of material compatibility; performance; emission and engine durability. Renewable and Sustainable Energy Reviews 15(2): 1314–1324. https://doi.org/10.1016/j.rser.2010.10.004.

48. Ferrari, Roseli Ap., Anna Leticia M. Turtelli Pighinelli and Kil Jin Park. 2011. Biodiesel Production and Quality. Biofuel's Engineering Process Technology. Dr. Marco Aurelio Dos Santos Bernardes (Ed.). ISBN: 978-953-307-480-1. InTech. Available from: http://www.intechopen.com/books/biofuel-s-engineering-processtechnology/biodiesel-production-and-quality.

49. Fukuda, Hideki, Akihiko Kondo and Hideo Noda. 2001. Biodiesel fuel production by transesterification of oils. Journal of Bioscience and Bioengineering 92(5): 405–416. https://doi.org/10.1016/S1389-1723(01)80288-7.

50. Georgogianni, K.G., A.P. Katsoulidis, P.J. Pomonis and M.G. Kontominas. May 2009. Transesterification of soybean frying oil to biodiesel using heterogeneous catalysts. Fuel Processing Technology 90(5): 671–676. https://doi.org/10.1016/j.fuproc.2008.12.004.

51. Gholami, Zahra, Ahmad Zuhairi Abdullah and Keat-Teong Lee. November 2014. Dealing with the surplus of glycerol production from biodiesel industry through catalytic upgrading to polyglycerols and other value-added products. Renewable and Sustainable Energy Reviews, ELSEVIER 39: 327–341. https://doi.org/10.1016/j.rser.2014.07.092.

52. Global Agriculture - Agrofuels and Bioenergy. https://www.globalagriculture.org/report-topics/agrofuels-and-bioenergy.html.

53. Gog, Adriana, Marius Roman, Monica Toşa, Csaba Paizs and Florin DanIrimie. March 2012. Biodiesel production using enzymatic transesterification-current state and perspectives- Review. Renewable Energy. 39(1): 10–16. https://doi.org/10.1016/j.renene.2011.08.007.

54. Guik, M.M., K.T. Lee and S. Bhatia. November 2008. Feasibility of edible oil vs. non-edible oil vs. waste edible oil as biodiesel feedstock-Review. Energy 33(11): 1646–1653. https://doi.org/10.1016/j.energy.2008.06.002.

55. Hegde, Krishnamoorthy, Niharika Chandra and Saurabh Jyoti Sarma. April 2015. Genetic engineering strategies for enhanced biodiesel production. Molecular Biotechnology 57(7). DOI: 10.1007/s12033-015-9869-y

56. Haseeb, A.S.M.A., M.A. Fazal, M.I. Jahirul and H.H. Masjuki. March 2011. Compatibility of automotive materials in biodiesel: A review. Fuel. 90(3): 922–931. https://doi.org/10.1016/j.fuel.2010.10.042.

57. Hayes, B.L. 2004. Recent advances in microwave- assisted synthesis. Aldrichimica Acta 37(2): 66–77.

58. High-shear mixer From Wikipedia, the free encyclopedia. https://en.wikipedia.org/wiki/High-shear_mixer.

59. https://news.brown.edu/articles/2010/10/biodiesel.

60. https://link.springer. com/article/10.1007/s12155-020-10165-0.

61. https://res.mdpi. com/d_attachment/processes/processes-08-01244/article_deploy/processes-08-01244.pdf.

62. https://www.power-technology.com/features/does-biofuel-have-a-palm-oil-problem/.

63. https://www.researchgate.net/publication/304781931_Advances_in_biofuel_production_from_oil_palm_and_palm_oil_processing_wastes_A_review.

64. Huang, Daming, Haining Zhou and Lin Lin. 2012. Biodiesel: an alternative to conventional fuel. Energy Procedia 16: 1874–1885. Int. Conf. on Future Energy, Environment, and Materials. doi:10.1016/j.egypro.2012.01.287. Available online at www.sciencedirect.com Energy P.

65. Iso, Mamoru, Baoxue Chen, Masashi Eguchi, Takashi Kudo and Surekha Shrestha. November 2001. Production of biodiesel fuel from triglycerides and alcohol using immobilized lipase. J. of Molecular Catalysis B: Enzymatic 16(1): 53–58. https://doi.org/10.1016/S1381-1177(01)00045-5.

66. Issues relating to biofuels, From Wikipedia, the free encyclopedia.https://en.wikipedia.org/wiki/Issues_relating_to_biofuels.   https://en.wikipedia.org/wiki/Issues_relating_to_biofuels#Social_and_economic_effects.

67. István Barabás and Ioan-Adrian Todoruţ. 2011. Biodiesel Quality, Standards and Properties, Biodiesel Quality, Emissions and By-Products. Dr. Gisela Montero (Ed.). ISBN: 978-953-307-784-0, InTech. Available from: http://www.intechopen.com/books/biodiesel-quality-emissions-and-by products/biodiesel-qualitystandards-and-properties.

68. Jain, Siddharth and M.P. Sharma. October 2010. Kinetics of acid base catalyzed transesterification of Jatropha curcas oil. Bioresource Technology 101(20): 7701–7706. https://doi.org/10.1016/j.biortech.2010.05.034.

69. Jitputti, Jaturong, Boonyarach Kitiyanan, Pramoch Rangsunvigit, Kunchana Bunyakiat, Lalita Attanatho and Peesamai Jenvanitpanjakul. February 2006. Transesterification of crude palm kernel oil and crude coconut oil by different solid catalysts. Chemical Engineering Journal. 116(1): 61–66. https://doi.org/10.1016/j.cej.2005.09.025.

70. Joshi, S. and S. Joshi. 2004. The oil tree—Simarouba glauca DC. *In*: Hegde, N.G., J.N. Daniel and S. Dhar (eds.). Jatropha and Other Perennial Oilseed Species. Proc. of National Workshop: 133–137. BAIF Development Research Foundation, Pune, India.

71. Kang-Shin, Chen, Yuan-Chung Lin, Kuo-Hsiang Hsu and Hsin-KaiWang. February 2012. Improving biodiesel yields from waste cooking oil by using sodium methoxide and a microwave heating system. Energy 38(1): 151–156. https://doi.org/10.1016/j.energy.2011.12.020.

72. Kant, Promode and Shuirong Wu. 2011. The extraordinary collapse of jatropha as a global biofuel. Environmental Science & Technology 45: 7114–7115. dx.doi.org/10.1021/es201943v | Environ. Sci. Technol.

73. Kasim, Farizul H. and Adam P. Harvey. July 2011. Influence of various parameters on reactive extraction of *Jatropha curcas* L. for biodiesel production. Chemical Engineering Journal 171(3): 1373–1378. https://doi.org/10.1016/j.cej.2011.05.050.

74. Kho, T.S. and K.H. Chung. 2008. Production of Biodiesel from waste frying oil by transesterification on Zeolite catalyst with different acidity. Journal of Korean Industrial and Engineering Chemistry 19(2): 214–221.

75. Kok, Tat Tan, Keat Teong Lee and Abdul Rahman Mohamed. February 2010. A glycerol-free process to produce biodiesel by supercritical methyl acetate technology: An optimization study via Response Surface Methodology. Bioresource Technology 101(3): 965–969. https://doi.org/10.1016/j.biortech.2009.09.004.

76. Kumar, Vijaya et al. Feb. 2015. Evaluation of performance and characteristics of biodiesel blends with diesel in a single cylinder D1 diesel engine. 5th Int. Conf. on plant and environmental pollution ICPEP-5, NBRI, Lucknow, India.

77. Kumar, Ashwani and Satyawati Sharma. May 2011. Potential non-edible oil resources as biodiesel feedstock: An Indian perspective. Renewable and Sustainable Energy Reviews 15(4): 1791–1800. https://doi.org/10.1016/j.rser.2010.11.020.

78. Kumar, Pravin, Dinkar Patil and Masoomraja Zakir Mulla. July 2017. Review on different sources for the production of biodiesel. Int. J. of Latest Technology in Engineering, Management & Applied Science (IJLTEMAS). VI(VII): 2278–2540. www.ijltemas.in Page 21. https://www.researchgate.net/publication/318588870_Review_on_Different_Sources_for_the_Production_of_Biodiesel.

79. Lam, Man Kee, Keat Teong Lee and Abdul Rahman Mohamed. July–August 2010. Homogeneous, heterogeneous and enzymatic catalysis for transesterification of high free fatty acid oil (waste cooking oil) to biodiesel: A review. Biotechnology Advances 28(4): 500–518. https://doi.org/10.1016/j.biotechadv.2010.03.002.

80. Leadbetter, Nicholas E., T. Michael Barnard and Lauren M. Stencel. 2006. Fast, easy preparation of biodiesel using microwave heating. Energy & Fuels 20(5): 2281–2283. [J. of Physical Chemistry A 2015, 119(34): 8971–8980]. DOI: 10.1021/acs.energyfuels.6b00017.

81. Leadbetter, Nicholas E., T. Michael Barnard, Matthew B. Boucher, Lauren M. Stencel. and Benjamin A. Wilhite. 2007. Continuous-flow preparation of biodiesel using microwave heating. Energy & Fuels 21(3): 1777–1781. DOI: 10.1021/acs.energyfuels.5b03024.

82. Leung, D.Y.C., B.C.P. Koo and Y. Guo. January 2006. Degradation of biodiesel under different storage conditions. Bioresource Technology 97(2): 250–256. https://doi.org/10.1016/j.biortech.2005.02.006.

83. Leung, Dennis Y.C., Xuan Wu and M.K.H. Leung. April 2010. A review on biodiesel production using catalyzed transesterification. Applied Energy 87(4): 1083–1095. https://doi.org/10.1016/j.apenergy.2009.10.006.

84. Lim, Steven and Keat Teong Lee. March 2013. Process intensification for biodiesel production from *Jatropha curcas* L. seeds: Supercritical reactive extraction process parameters study. Applied Energy. 103: 712–720. https://doi.org/10.1016/j.apenergy.2012.11.024.

85. Lim, Steven, Shuit Siew Hoong, Lee Keat Teong and Subhash Bhatia. September 2010. Supercritical fluid reactive extraction of *Jatropha curcas* L. seeds with methanol: A novel biodiesel production method. Bioresource Technology 101(18): 7169–7172. https://doi.org/10.1016/j.biortech.2010.03.134.

86. Lotero, Edgar, Yijun Liu, Dora E. Lopez, Kaewta Suwannakarn, David A. Bruce and James G. Goodwin. 2005. Synthesis of biodiesel via acid catalysis. Ind. Eng. Chem. Res. 44(14): 5353–5363. DOI: 10.1021/ie049157g, Publication Date (Web): January 25, 2005.

87. Mandava, S.S. 1994. Application of a natural surfactant from sapindus emerginatus to in-situ flushing of soils contaminated with hydrophobic organic compounds. M.S. Thesis in Civil and Environmental Engineering, Faculty of Louisiana State University and Agricultural and Mechanical College.

88. Mardhiah, H. Haziratul, Hwai Chyuan Ong, H.H. Masjuki, Steven Lim and H.V. Lee. January 2017. A review on latest developments and future prospects of heterogeneous catalyst in biodiesel production from non-edible oils. Renewable and Sustainable Energy Reviews 67: 1225–1236. https://doi.org/10.1016/j.rser.2016.09.036.

89. Marhasin, Evgeny, Marina Grintzova, Vicktor Pekker and Yuri Melnik. High power ultrasonic reactor for sonochemical applications. US7157058B2, Google Patent. https://patents.google.com/patent/US7157058B2/en.

90. Matsuhashi, H., S. Fruta and K. Arata. 2004. Biodiesel fuel production with solid super-acid catalysis in fixed bed reactor under atmospheric pressure. Catalyst Communications 5(12): 721–723.

91. Minami, Eiji and Shiro Saka. December 2006. Kinetics of hydrolysis and methyl esterification for biodiesel production in two-step supercritical methanol process. Fuel. 85(17-18): 2479–2483. https://doi.org/10.1016/j.fuel.2006.04.017.

92. OguzhanIlgen. March 2011 . Dolomite as a heterogeneous catalyst for transesterification of canola oil. Fuel Processing Technology 92(3): 452–455. https://doi.org/10.1016/j.fuproc.2010.10.009.

93. Ong, Yee Kang and Subhash Bhatia. January 2010. The current status and perspectives of biofuel production via catalytic cracking of edible and non-edible oils. Energy 35(1): 111–119. https://doi.org/10.1016/j.energy.2009.09.001.

94. Ottinger, Richard L. August 2007. Biofuels – Potential, Problems & Solutions. Biofuels Conf. Sponsored by Pace University School of Law Pontoficia Universidade Catolica Do Rio De Janeiro, National Energy-Environment Law and Policy Institute The University of Tulsa College of Law.

95. Pachauri, Naresh and Brian He. July 2006. Value-added Utilization of Crude Glycerol from Biodiesel Production: A Survey of Current Research Activities. An ASABE Meeting Presentation Paper Number: 066223, ASABE Annual Int. Meeting, Portland, Oregon.

96. Patil, Prafulla, Shuguang Deng, J. Isaac Rhodes and Peter J. Lammers. February 2010. Conversion of waste cooking oil to biodiesel using ferric sulphate and supercritical methanol processes. Fuel. 89(2): 360–364. https://doi.org/10.1016/j.fuel.2009.05.024.

97. Patil, Prafulla D. and Shuguang Deng. July 2009. Optimization of biodiesel production from edible and non-edible vegetable oils. Fuel. 88(7): 1302–1306. https://doi.org/10.1016/j.fuel.2009.01.016.

98. Peng, B.X., Q. Shu et al. 2008. Biodiesel production from waste feed stock by solid acid catalysts. Process safety and environmental Protection 86(6): 441–447.

99. Pilar Olivares-Carrillo and Joaquín Quesada-Medina. October 2011. Synthesis of biodiesel from soybean oil using supercritical methanol in a one-step catalyst-free process in batch reactor. J. of Supercritical Fluids. 58(3): 378–384. https://doi.org/10.1016/j.supflu.2011.07.011.

100. Pramanik, Krishna, February 2003. Properties and use of *Jatropha curcas* oil and diesel fuel blends in compression ignition engine. Renewable Energy 28(2): 239–248. DOI: 10.1016/S0960-1481(02)00027-7.

101. Rathore, Vivek and Giridhar Madras. December 2007. Synthesis of biodiesel from edible and non-edible oils in supercritical alcohols and enzymatic synthesis in supercritical carbon dioxide. Fuel. 86(17-18): 2650–2659. https://doi.org/10.1016/j.fuel.2007.03.014.

102. Rathore V., S. Tyagi, B. Newalkar and R.P. Badoni. 2014. Glycerin-free synthesis of Jatropha and Pongamia biodiesel in supercritical dimethyl and diethyl carbonate. Ind. Eng. Chem. Res. 53: 10525–10533. 10.1021/ie5011614 .

103. Refat, A.A. and S.T. Sheltawy. 2008. Comparing three options for biodiesel production from waste vegetable oil. WIT Transactions on ecology and environment, Waste management and the Environment IV, 109: 133–140. Billerica MAMA: WIT Press.

104. Refocus Jan./Feb. 2000, p. 48. Renewable Resources and Renewable Energy-A Global Challenge. Mauro Graziani and Paolo Fornasiero (Eds.). 2006. CRC Press.

105. Renewable Energy Focus. January 2012. Gail Rajgor. How viable are biofuels? http://www.renewableenergyfocus.com/view/23378/how-viable-are-biofuels/.

106. Romijn, Henny A. October 2011. Land clearing and greenhouse gas emissions from Jatropha biofuels on African Miombo Woodlands. Energy Policy 39(10): 5751–5762. https://doi.org/10.1016/j.enpol.2010.07.041.

107. Roy, D., R.R. Kommalapati, S. Mandava, K.T. Valsaraj and W.D. Constant. 1997. Soil washing potential of a natural surfactant. Environ. Sci. Technol. 31(3): 670–675.

108. Saka, S. and D. Kusdiana. January 2001. Biodiesel fuel from rapeseed oil as prepared in supercritical methanol. Fuel. 80(2): 225–231. https://doi.org/10.1016/S0016-2361(00)00083-1.m.

109. Saka, S. and E. Minami. November 2006. A novel non-catalytic biodiesel production process by supercritical methanol as NEDO 'high efficiency bioenergy conversion project'. 2nd joint Int. Conf. on sustainable energy and environment (SEE), Bangkok, Thailand.

110. Saka, Shiro and Yohei Isayama. July 2009. A new process for catalyst-free production of biodiesel using supercritical methyl acetate. Fuel. 88(7): 1307–1313. https://doi.org/10.1016/j.fuel.2008.12.028.

111. Saka, Shiro, Yohei Isayama, Zul Ilham and Xin Jiayu. July 2010. New process for catalyst-free biodiesel production using subcritical acetic acid and supercritical methanol. Fuel 89(7): 1442–1446. https://doi.org/10.1016/j.fuel.2009.10.018.

112. Serio, M. Di, M. Ledda, M. Cozzolino, G. Minutillo, R. Tesser and E. Santacesaria. 2006. Transesterification of soybean oil to biodiesel by using heterogeneous basic catalysts. Ind. Eng. Chem. Res. 45(9): 3009–3014. DOI: 10.1021/ie051402o; Publication Date (Web): March 23, 2006.

113. Shimada, Yuji. et al. 1999. Conversion of vegetable oil to biodiesel using immobilized Candida Antarctica lipase. J. of American Oil Chemists Society 76(7): 789–793.

114. Shrestha, D.S., J. Van Gerpen, J. Thompson and A. Zawadzki. July 2005. Cold flow properties of biodiesel and effect of commercial additives. Paper Number: 056121, ASAE Annual International Meeting Tampa, Florida.

115. Shu, Q., J. Gao, Z. Nawar et al. 2010. Synthesis of biodiesel from waste vegetable oil with large amounts of free fatty acids using a carbon based solid catalyst. Applied Energy 87(8): 2589–2596.

116. Shuit, Siew Hoong, Keat Teong Lee, Azlina Harun Kamaruddin and Suzana Yusup. 2010. Reactive extraction of *Jatropha curcas* L. seed for production of biodiesel: process optimization study. Environ. Sci. Technol. 44(11): 4361–4367. DOI: 10.1021/es902608v, Publication Date (Web): May 10, 2010.

117. Singh, Alok Kumar, Sandun D. Fernando and Rafael Hernandez. February 2007. Base-catalyzed fast transesterification of soybean oil using Ultrasonication. Energy Fuels 21(2): 1161–1164. https://doi. org/10.1021/ef060507g.

118. Song, H. and S.Y. Lee. 2006. Production of succinic acid by bacterial fermentation. Enzyme and Microbial Technology 39: 352–361.

119. Sorate, Kamalesh A. and Purnanand V. Bhale. January 2015. Biodiesel properties and automotive system compatibility issues. Renewable and Sustainable Energy Reviews 41: 777–798. https://doi.org/10.1016/j. rser.2014.08.079.

120. Su, Erzheng, Pengyong You and Dongzhi Wei. December 2009. *In situ* lipase-catalyzed reactive extraction of oilseeds with short-chained dialkyl carbonates for biodiesel production. Bioresource Technology 100(23): 5813–5817. https://doi.org/10.1016/j.biortech.2009.06.077.

121. Sundarapandian, S. 2007. Performance and Emission Analysis of Bio Diesel Operated C I Engine. J. of Engineering, Computing and Architecture l(2): 1–22.

122. Tan, Kok Tat, Keat Teong Lee and Abdul Rahman Mohamed. February 2010. A glycerol-free process to produce biodiesel by supercritical methyl acetate technology: An optimization study via Response Surface Methodology. Bioresource Technology 101(3): 965–969. https://doi.org/10.1016/j.biortech.2009.09.004. Error! Hyperlink reference not valid.

123. Ucciani, E., J.F. Mallet and J.P. Zahra. 1994. Cyanolipids and fatty acids of *Sapindus trifoliatus* L. (Sapindaceae) Seed Oil. Fat Science Technology 96(2): 69–71.

124. Van, J.A.J. Eijck and H.A. Romijn. 2008. Prospects for Jatropha biofuels in developing countries: an analysis for Tanzania with strategic niche management. J. Energy Policy 36(1): 311–325.

125. Van Kasteren, J.M.N. and A.P. Nisworo. June 2007. A process model to estimate the cost of industrial scale biodiesel production from waste cooking oil by supercritical transesterification. Resources, Conservation and Recycling 50(4): 442–458. https://doi.org/10.1016/j.resconrec.2006.07.005.

126. Vera, C.R., S.A. D'Ippolito, C.L. Pieck and J.M. Parera. August 2005. Production of biodiesel by a two-step supercritical reaction process with adsorption refining. (PDF). 2nd Mercosur Congress on Chemical Engineering, 4th Mercosur Congress on Process Systems Engineering. Rio de Janeiro. (http://wwwenpromer2005.eq.ufrj. br/nukleo/pdfs/o818papu818.pdf).

127. Viswanath, S. Abinav and V. Dinesh. 2012 . Modeling and Analysis of Performance, Combustion and Emission Characteristics of Jatropha Methyl Ester Blend Diesel for Cl Engine with Variable Compression Ratio. Int. J. of Engineering Science and Technology 4(7): 3457–3471.

128. Wahl, N., R. Jamnadass, H. Baur et al. 2009. Economic Viability of *Jatropha curcas* L. Plantations in Northern Tanzania assessing Farmers Prospects via Cost-Benefit Analysis, ICRAF Working Paper no. 97. World Agroforestry Centre: Nairobi,

129. Zong, M.H., Z.Q. Duan et al. 2007. Preparation of a sugar catalyst and its use for highly efficient Watanabe, Y, Y. Shimada, A. Sugihara and Y. Tominaga. 2001. Enzymatic conversion of waste edible oil to biodiesel fuel in a fixed bed bioreactor. J. of American Oil Chemist Society 78(7): 703–707.

130. Yagiz, F., D. Kazan and A. Akin. 2007. Biodiesel production from waste oils by using lipase immobilized on Hydrotalcite and Zeolites. Chemical Engineering Journal 134(1-3): 262–267.

131. Zhou, Adrian and Elspeth Thomson. November 2009. The development of biofuels in Asia. Applied Energy. 86(Supplement 1): S11–S20. https://doi.org/10.1016/j.apenergy.2009.04.028.

132. Zong, M.H., Z.Q. Duan, W.Y. Lou, T.J. Smith and H. Wu. 2007. Preparation of a Sugar Catalyst and its Use for Highly Efficient Production of Biodiesel. Green Chemistry 9(5): 434–437. doi:10.1039/b615447f.

CHAPTER 5

Oil/Biodiesel from Algae, Fungi, and Lignocellulosic Biomass and Emerging Alternate Fuels

5.1 Oil from Algae

5.1.1 *Importance of Algae* [111,132,150,175]

Algae are a very diverse group of organisms, but historically, they have often been grouped together with plants because algae are organisms that use *photosynthesis** for their growth, like plants. Similar to plants, algae absorb carbon dioxide and release oxygen. They use sunlight to convert water and carbon dioxide into sugars and oxygen. *(Also see Annexure A of Chapter 1.)* These organisms range in size from being microscopic (microalgae) to many meters in lengths (macroalgae). Algae are primary producers and provide the basis of energy and fixed carbon in almost every ecosystem in which they are present.

Sea algae is a major source of global oxygen and absorption/sequestration of carbon dioxide. Estimates vary, depending on the types of algae considered, but based on the stoichiometry (the accounting, or math, behind chemistry) of photosynthesis, 1 ton of algae may sequester about 1.5 tonne of CO_2. Some microalgae have a large percentage of *'lipids'* which can be exploited to make 'biodiesel'. Like other plants, the algae culture may be a step towards *carbon-neutral** fuel.

Algae have been incredibly important for our global ecology and bio-geochemistry from very ancient times. More than two billion years ago, the production of oxygen by algae photosynthesis dramatically changed the Earth's atmosphere and enabled the evolution and diversification of complex organisms that breathe oxygen (like us). Although they are not as conspicuous as plants in our everyday lives, microalgae in the ocean (*phytoplankton*)** currently do about as much photosynthesis each year as all the plants on land. Some of the phytoplanktons end up sinking to the bottom of the ocean, taking carbon with them. So, without algae, the carbon dioxide concentration in the atmosphere would have been even higher than it already is.

Some countries with a diverse coast line like UK, Spain, USA, Canada, Japan, Malaysia, Indonesia, India, etc., offer many advantages for the utilization of marine biological resources. Marine algae of many kinds are found attached to the bottom of the sea in relatively shallow coastal waters. Algae are also found in rocky seashore areas, lagoons, and reed areas of many countries.

In nature, there are many species of algae—about 900 green species, 4,000 red species, and 1,500 brown species. Some species are commercially utilized. About 145 are used for food and 110 are used for the production of phycocolloids like *agar agar,**** algin, alginates, carrageen, etc. Algae are source of proteins, lipids (oils), polysaccharides, minerals, vitamins, pigments, biomedicines, chemicals, food additives, enzymes, feedstock for pharmaceuticals, etc. Some algae are consumed in Asia as food. Certain types of microalgae are very rich in lipids, which are a very good source of biofuel. The continuous cultivation of 'microalgae' in brackish/sea waters in estuaries, gulfs, and saltwater/brackish water lakes is important for the production and harnessing of '*algal oil*' apart from the advantage of reduction of atmospheric carbon dioxide. [35,87,112,176]

Algae can be grown in freshwater also, but normally, freshwater has many other important uses for mankind/animals/agriculture. Algae can absorb toxic and heavy metals from water and therefore, reduce water pollution. Algae can also be a source of alcohol.

It may be mentioned that petroleum is widely believed to have its origins in 'kerogen', which is easily converted into an oily substance under conditions of high pressure and temperature. 'Kerogen' is formed from algae, biodegraded organic compounds, plankton, bacteria, plant material, etc., by biochemical and/or chemical reactions.

There are mainly three metabolic pathways present in microalgae:

(a) Autotrophy

(b) Heterotrophy

(c) Mixotrophy

The photoautotrophic algae uptake carbon dioxide under light, in simple inorganic media like water (*Photosynthesis******). They mainly store starch and lipids amongst other components.

A few microalgae have the ability to grow fast even without light taking carbon from organic sources, like sugars, in the medium, and accumulate lipids. These may be called heterotrophic.

If organic carbon, such as glucose, is present in water, the growth of some algal species is higher; they take carbon both from the atmospheric carbon dioxide and glucose in the medium. This is called '*mixotrophy*' or '*photoheterotrophy*'. In this case, the lipid content may also be higher.

These metabolic pathways might be switched from one to the other in certain algal species after the acclimatizing of many generations.

Because we are more interested in producing biodiesel and at the same time, reducing atmospheric carbon dioxide and global warming, the photoautotrophic method is the best. Moreover, providing glucose/sugars to the medium for growth of the algae would be too uneconomical and a wastage of resources.

Notes:

*(1) Carbon Neutral: *A system is 'carbon neutral' if the amount of carbon it releases into the atmosphere, during its fuel-burning cycle, is equal to the amount of carbon that is consumed (sequestered) to produce that equivalent amount of fuel. It can also be stated that if the carbon released during its burning is sequestered on the same timescale, the operation can be designated as 'carbon neutral'.*

**(2) Phytoplankton: *The term `phytoplankton' is another word for the small (often microscopic) algae that float freely in the waters of oceans and lakes. By definition, plankton is unable to move or swim faster than the currents. Phytoplanktons are considered to be a rich source of food, biofertilizers, biofuel and feed. Phytoplanktons are like plants and take up CO_2 from atmosphere by photosynthesis. They are a diverse group of photoautotrophic microorganisms which include bacteria, algae and some archaebacterial prokaryotes. There are approximately around 5000 species of phytoplanktons discovered till now. Some examples are -*

a. *Blue-green algae - Synechococcus, Prochlorococcus, Spirulina,*

b. *Green algae - Chlorella vulgaris, Dunaliella salina;*

c. *Diatom - Odontella aurita, Phaeodactylum tricornutum;*

d. *Dinoflagellate - Protoperidinium depressum*

Based on cell wall arrangement and cell structure, the phytoplanktons are classified into five types—Diatoms, Dinoflagellates, Blue-Green Algae, Green Algae and Coccolithophores. Based upon their size they are classified as

   i. *Ultraplankton - < 2 μm;*

   ii. *Nanoplankton - 2 to 20 μm;*

   iii. *Microplankton - 2 to 200 μm*

   iv. *Macroplankton - 200 to 2000 μm;*

   v. *Megaplankton - > 2000 μm*

Phytoplanktons are mostly used as food supplements and play a vital role in both animal and human nutrition. Phytoplanktons are also used in the manufacture of drugs. Also, they have proven to be suitable for synthesising vaccines. Some species of Phytoplanktons have a high tolerance to CO_2 and can significantly capture carbon. some diatoms can absorb around 10 to 20 billion tonnes of CO_2 every year. Some phytoplanktons can also sequester pollutants and are used for bioremediation processes. This is called phycoremediation. Source: https://study.com/academy/lesson/phytoplankton-definition-types-facts.html; https://oceanservice.noaa.gov/facts/phyto.html

****(3) Agar-agar:** It is a well-known polygalactan, insoluble in cold water, but soluble in hot water at a concentration of 1–4% and its solution sets to jelly on cooling. Agar agar is used in media for culturing bacteria. It is also used as a medicine for relieving constipation in children. Galactans occur in large amounts in some seaweed and algae. Galactans are polysaccharides composed of galactose molecules (galactose is a hexose sugar like glucose, mannose, and fructose). In addition, galactans contain D-mannonic acid and sulphuric acid in combination.*

*****(4) Photosynthesis:** It is a biochemical process, during which plants and certain organisms utilize solar energy to 'fix' CO_2 from the atmosphere into carbohydrates and release oxygen (O_2).*

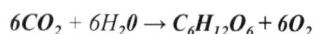

$$6CO_2 + 6H_2O \rightarrow C_6H_{12}O_6 + 6O_2$$

These carbohydrates can then be utilized for energy or as the building blocks of biomass. Chlorophyll is a pigment that is the principle 'photoreceptor' in photosynthesis. It is contained in an organelle called the 'chloroplast' along with other components, which allow the conversion of sunlight energy into carbohydrates.

5.1.2 *Oil Content of Algae* [48,123,159,178,181,205,222,236]

Some microalgae have a high content of oil (> 50 %). The algal oils have the advantage of being *sulphur-free, non-toxic, biodegradable, and from a renewable* source. Their energy content is comparable to that of fossil fuel oils. Algal oil can be converted to biodiesel, but oil production from algae needs to be commercialized. Some strains of microalgae that produce more carbohydrates than oils can be fermented to make bioethanol and biobutanol.

Although algae appear to have a lot of advantages in terms of land and water use, it is essential to analyse the sustainability, environmental impacts, and possible unintended consequences of large-scale algal biofuel production.

Algae could be potentially grown in ponds on marginal lands with brackish water/saltwater, but there will still be a need for some freshwater (for example, to replace water lost due to evaporation). This may sometimes impact other demands for freshwater in marginal environments.

Corn is not considered to be a major oil crop. It produces around 18 gallon (68 litres) of oil per acre per year, which is less than half as much as soybeans. The production of oil from microalgae is still in infancy and needs to be carried out on a large scale. Taking an optimistic view, which may not be unrealistic, an annual yield of about 3,000 to 5000 gallons (11356 to 18927 litres) of oil per acre from microalgae or much more may be possible. Microalgae could be hundred times more productive than corn.

Microalgae have much faster growth-rates than terrestrial crops. The next best crop, palm oil, produces 635 gallons (2,404 litres) per acre. Microalgae has the potential of producing many times more energy per acre than land crops. The use of corn for biodiesel oil production competes with its use as a food crop, whereas this is not an issue for algae. In the majority of microalgae species, the lipids that make the oil come from the various components of the cell, such as the cell membrane or the membrane around organelles. Some algae, such as diatoms, use lipids as energy-storage compounds [48,123]. Table 5.1 gives the comparative yields of oil per acre from various oilseeds and microalgae.

Algal oil can be processed into biodiesel as easily as oil derived from land-based crops. The difficulties in efficient biodiesel production from algae lie not in the extraction of the oil, but in finding a microalgal strain with a high lipid content and fast growth rate that isn't too difficult to harvest, and a cost-effective cultivation system (that is, type of photobioreactor) that is best suited to that strain. [178] The oil content of some microalgae species is given in Table 5.2.
Table 5.3 shows theoretical maximum yield of some algae grown in open ponds.

The following species may be present in the algal culture: Spyrogyra, *Chorococcus* sp., *Scenedesmus* sp., *Spirulina* sp., Hydrodictyon colony, *Urenoma* sp., *Navicula* sp., *Pinullaria* sp., *Frustulia* sp., *Gomphonema* sp., *Euglena* sp., *Closterium* sp., *Ocillatoria* sp., *Zygnema* sp., etc.

[None of the algal species have oil which is edible except spirulina.] Algae yields of various species in open ponds for most blue-green alga are in the range of 10–35 g m^{-2} day^{-1}.

Microalgae to be Grown for Biofuels

When choosing an algal species for biofuels, there are many factors to be considered. The most important is the 'algae strain or strains' to be grown in culture. There is no limit to the types of algae that can be grown for biofuels. *The best species likely depend on the locations of the facility, climate, the type of water source, and quality of water that is available. Which algae will be best, cannot be stated in a simple way and requires trials and estimation. The best algae may still elude.*

Depending on the type of algae grown, the location and the system used, it is possible for wild invasive algae to out-perform the target species. However, if a large enough quantity of algae is used to inoculate the system and/or a closed or semi-closed system is used, the target algae can be grown in sufficient quantities to prevent wild algae from taking over from invasive wild species.

Table 5.1. Yields of Oil Per Acre Per Year in Gallons. [181]

Se. No.	Substance	Oil Yield in Gallons Per Acre Per Year
1.	Corn	18 (68 L)
2.	Soybeans	48 (182 L)
3.	Safflower	83 (314 L)
4.	Sunflower	102 (386 L)
5.	Rapeseed	127 (481 L)
6.	Oil palm	635 (2404 L)
7.	Algae	3,000–5,000 (11356 to 18927 litres L)

5.1.3 *Algae Growth Parameters* [213,237]

The required for growth of algae are:

- *Daylight*: The more the hours of light in the day, the better is the growth.

- *Night Illumination*: Artificial white light during night, if provided, also helps.

- *pH Range:* Most of the cultured algal species are comfortable in a pH range of 7 to 9, the optimum value being 8.2 to 8.7.

- *Mixing and aeration*: Mixing is essential to prevent sedimentation of algae, to ensure that all the cells of algae are equally exposed to the light and nutrients, to avoid stratification (that is, the formation of layers (in outdoor culture)) and to improve the gas exchange between the culture medium and the air. The aeration is important because the atmospheric CO_2 is the carbon source for the growth of algae by photosynthesis.

Table 5.2. Oil Content of Some Species of Microalgae. [159,181,205,222,236]

Se. No.	Species of Algae	Oil Content as % Dry Weight
1.	*Ankistrodesmus TR-87*	28–40
2.	*Botryococcus braunii*	**See Note**
3.	*Chlorella protothecoides fautorophic/heterotrophic*	15–55
4.	*Chlorella species*	28–32
5.	*Cyclotella DI-35*	42
6.	*Dunaliella tertiolecta*	36–42
7.	*Hantischia D- 160*	66
8.	*Nannochloropsis sp.*	31–68
9.	*Nitzschia sp.*	45–47
10.	*Nitzschia TR-114*	28–50
11.	*Phaeodactylum tricornutum*	31
12.	*Neochloris sp,*	14.60
13.	*Stichococcus*	33 (9–59)
14.	*Tetraselmis suecica*	15–32
15.	*Thalassiosira pseudonana*	21–31
16.	*Crpthecodinium cohnii*	20
17.	*Neochloris oleoabundans*	35–54
18.	*Schizochytrium*	50–77
19.	*Scenedesmus TR-84*	45
20.	*Scenedesmus dimorphus*	28–31

Note on Botryococcus: Botryococcus is a well-known hydrocarbon producer. Under unfavourable conditions of growth, botryococcus's un-saponifiable lipid content (which cannot be converted to soap) increases to a level of 90%. The bulk of the Botryococcus hydrocarbon (about 95%) is extracellular (located in the colony matrix and in occluded globules and not in cells). However, the high hydrocarbon concentrations exist only in non-growing, senescent, and even decaying cultures; the hydrocarbon content of growing algae is fairly low. Botryococcus has been extensively investigated for its hydrocarbon-producing ability but the information collected is based only on indoor algal cultivation. The mass accumulation in botryococcus is more than seven days, which is less than that for typical unicellular green algae. However, increased hydrocarbon productivity may be feasible by modifying culture conditions, and/or the better strains used; modifications, which according to the literature, result in doubling times of two to three days. It may be worthwhile to examine a variety of strains for better growth performance and productivity.

Table 5.3. Theoretical Maximum Yields of a few Algae in Open Ponds. [181]

Se. No.	Species of Algae	Yield in Grams/m²/day
1.	*Marine nannochloropsis*	20 (30% lipids)
2.	*Spiriluna plantesis*	10.3
3.	*Dunaliella salina*	12.0
4.	*Scenedesmus species*	13.4
5.	*Ankistrodesmus*	18
6.	*Haematococcus pluvialis*	3.8

- *Temperature:* The optimum temperature for the growth of algae is between 20 and 24°C, while most algal species tolerate temperatures between 16–27°C. The temperatures lower than 16°C would very much slow down the growth, *while the temperature higher than 35°C would be lethal for the algae.*

- *Salinity:* Marine species are extremely tolerant of changes in salinity. Most of the 'cultured algae' are comfortable in a salinity range of 12–40 g/liter, but the optimum value may be 20–24 g/liter.

Area Requirement for the Growth of Algae and Algae oil

- The weight of algae that can be grown on an acre of land is limited by the amount of energy that area receives from the sun. Theoretically, under ideal conditions, algae growth at best may be in the range of 100 to 200 tonne per acre per year; such growth is unlikely to be achieved in practice. More realistic estimates are 30–40 tonne per acre per year.
- Depending on the lipid content of the algae (say @ 34%), 1 tonne of dry algae may produce about 100 gallons of oil.

Limitations of Growth of Algae Using Sunlight

A finite amount of light reaches any given area of the Earth during the day. Any system using the sun as its energy source is limited by the amount of sunlight it receives. Simply increasing the volume of a system within a given area may not translate to more biomass production. It is possible to increase the total production of the algae system by increasing the horizontal area it occupies but increase in production per unit area beyond sunlight availability may not be possible for the *photoautotrophic algae.*

5.1.4 *Algae Growth Dynamics* [50,99,103,194]

The growth of algae culture has five phases:

- *Lag or induction phase*, during which little increase in cell density takes place. *(This is the adaptation period for the cell metabolism to physiological factors, such as the level of nutrients, enzymes, metabolites involved in carbon fixation and cell division, etc. The period is long when the culture is transferred from plate to liquid culture.)*
- *Exponential growth period*, during which the cell density increases exponentially as a function of time according to the formula:

$$C_t = C_0 \cdot e^{mt}$$

where, C_0 and C_t denote the initial cell concentration and at time t respectively, m is the specific growth rate (mainly dependent on algal species, light intensity, and temperature).
- *Phase of declining growth rate*, when the cell division slows down, which may be happening due to a decline in nutrients, light intensity, carbon dioxide level, or other physical and chemical factors.
- *Stationary phase*, when the growth rate attains a balance with the physiological factors and other factors, and the growth rate is more or less constant.
- *Crash phase or death phase*, where, in practice, the algae culture crash (death of algae) can be caused by a variety of reasons such as depletion of nutrients, deficiency of dissolved oxygen in water, overheating, changes in pH value beyond the acceptable range, high pollution in water, development of parasite organisms, harmful bacteria, etc. In this phase, the condition in the environment/water is incapable of sustaining the growth and the cell density starts declining at a rapid rate.

5.1.5 *Growing Algae for Oil* [1,12,71,86]

Efforts are being made to grow microalgae on commercial scales to produce algal oil for conversion to biodiesel. Feasibility studies have been conducted to arrive at the yield estimates as already mentioned in Tables 5.1, 5.2, and 5.3. In addition to its projected high yield, alga-culture (unlike crop-based biofuels) does not entail a decrease in food production, since it requires neither farmland nor freshwater. Many companies are pursuing the development of algae bioreactors for various purposes, including biodiesel production and capturing CO_2.

UV light can be harmful to algae, but they have evolved mechanisms to protect themselves and to repair any damage that occurs. As an example, many algae produce screening pigments that absorb UV light. However, pests, pathogens, and weeds could be problems. They, sometimes, affect the algae cultures, as in the case of agricultural plants.

Features of Ideal Strain of Microalgae for Biofuel Production:
The key factors when selecting algae for biofuel production are:

(i) Higher lipid and carbohydrate content

(ii) High growth rate

(iii) Better photosynthetic efficiency

(iv) High ability of capturing CO_2 and sequestering it

(v) Ability to provide valuable co-products

(vi) Possessing self-flocculating characteristics

(vii) Ability to dominate wild and invasive strains (for growing open ponds)

(viii) Tolerance to wide range of temperatures

(ix) Limited requirement of nutrients

(x) For growing it in photo bioreactors (PBR), it should have the robustness and ability to survive shear stresses common in PBR. *[See Para 5.3 ahead.]*

(xi) Lipid and carbohydrate content, the foundation of biodiesel and ethanol production, must be balanced with growth rate (the amount of biomass produced in a given amount of time). *Environmental conditions should be considered to determine which species would thrive in the algae system locations*.

(xii) Biodiesel obtained from algae and terrestrial plants is similar. It may be possible to select certain species with *unique lipids to develop biodiesel with a lower gel point or even jet fuel.*

A large number of scientists are working on modifying algae using traditional plant breeding methods as well as the latest molecular biology techniques. The research may be able to improve the photosynthetic efficiency, temperature tolerance, cell wall composition favourable for oil production, oil biosynthesis pathways, etc.

Microalgae can be harvested several times in a short time frame and it can grow 20 to 30 times faster than food crops, but the harvesting of microalgae from large volumes of water is a messy and laborious job. There are a number of possible solutions to the problem. Some of the proven approaches include flocculation, flotation, gravity sedimentation, filtration, ultrasonic aggregation, and centrifuging depending on the specific species of algae being grown. Minimizing the cost of production and harvesting the algae is a critical factor for algal oil projects.

Phycologists,** other scientists, and societies have been studying and/or using algae for millennia. However, with the advent of new technologies, materials, and the rising cost of petroleum, it is now becoming feasible to use algae in other applications, such as biofuels, that were not possible even a decade ago.

It is possible to produce biofuels from algae economically. This, however, will require minimizing the cost of the system as well as reducing input costs (nutrients, water, etc.), while at the same time maximizing the systems' energy efficiency as well as the growth rate of the algae. There should be the production of co-products along with biofuels as it would improve the economics of the process. After recovering the lipids, the remaining biomass may be used to produce alcohols. *[**Phycology is the study of algae; phycologist—scientist studying algae.]*

5.1.6 *Main Obstacles in Producing Oil from Algae* [4,8,26,211]

The main problems of algal oil production are:

i. The oil-rich algae are difficult to protect from invading organisms when algae are grown in open ponds.

ii. Algae produce oil best within a narrow range of temperature. Low temperatures at night and excessive heat during the day is a problem in open ponds.

iii. Harvesting of microalgae is difficult with mesh net from the immense size of ocean or large lakes. The energy spent on harvesting may be too high.

iv. It is possible to harvest macroalgae like kelp, which is used in many parts of the world as a source of food and fertilizer. However, these ecosystems can be very fragile. They serve as nursery grounds for many species of fish and shellfish. It is unclear at present if the ecosystem having oil-bearing microalgae could withstand large-scale commercial algal harvesting for the production of biofuels. The potential impact of removing large quantities of the very food that the entire ecosystem depends on may be adverse. There is also an issue of the bycatch of larvae of fish and shellfish that may also be present, many of which could be endangered or threatened.

5.1.7 *Processing of Algal Oil into Biodiesel* [1,4,8,25,26,24,30,39,60,71,72,149]

Algal oil is highly viscous; the viscosity ranges between 10–20 times to that of diesel. The high viscosity is due to large molecular weight and chemical structure of the oil. The high viscosity may lead to problem in pumping, atomization in injectors, etc., and therefore its viscosity needs to be reduced. The most common method of reducing viscosity is 'transesterification' (*discussed in Chapter 4*). Other methods of reducing viscosity may be pyrolysis, microemulsion, blending, and thermal depolarization.

After preparation of biodiesel by transesterification, the impurities, moisture, and excess amounts of additives, if any, are removed. (*Refer to Figure 5.2*).

5.1.8 *Advantages of Biodiesel from Algal Oil* [20,89,90,96,216]

Producing biodiesel from microalgae is being seen as the most efficient way. Briefly, the main advantages of this method are:

• Rapid growth rates of algae.

• A high per-acre yield of algae.

• Algae does not compete with food crops.

• Algae can be grown in saltwater, brackish water, or even polluted water. Although algae can be grown in freshwater, it would be a wastage of an important resource and therefore, it is not practiced.

• Algae can be grown in ponds/tanks created on waste suboptimal land, or in estuaries and saltwater lakes.

• There is almost an insignificant fresh water requirement for growing algae (only for making up the evaporation loss, which may sometimes be needed).

• Algae can be used to remediate polluted water from agricultural runoff and treatment plants. (They absorb many toxic metals and other pollutants from water).

• Certain species of algae can be harvested daily.

• Biofuel from algae does not contain any 'sulphur'. Also, the biodiesel from algae does not contain aromatic compounds. In conventional diesel, the sulphur leads to the formation of sulphur oxide and sulphuric acid, while the aromatic compounds also increase particulate emissions and are considered to be carcinogens.

- Algal biofuel is 'non-toxic'.
- Algal biofuel is highly 'biodegradable'. The biodegradability of biodiesel is particularly advantageous in environmentally sensitive areas.
- Algae consume/absorb atmospheric CO_2 as they grow and therefore the use and production of algal oil is an efficient means of reducing CO_2 in the atmosphere and controlling global warming.
- A system could be designed to use algae to capture CO_2 from power stations and other industries that would otherwise be released into the atmosphere.
- Climate of tropical, subtropical, and monsoon regions is conducive to grow various species of microalgae (for the production of algal biodiesel), which can have natural benefits for other regions of the world.
- In addition to the reduction of emission of CO_2, SOx, and particulate matter, the use of biodiesel confers additional advantages like a higher flashpoint, faster biodegradation, and greater lubricity. The higher flashpoint helps in safer handling and storage. The lubricity of biodiesel is greater than that of diesel fuel and blending it with conventional diesel fuel increases its lubricity. [119]
- Besides lipids and carbohydrates for the production of biofuels, many species of algae can be grown to produce nutraceuticals, such as astaxanthin, vitamins, food supplements, fish and cattle feed, and pharmaceuticals.
- Ethanol can be made from algal biomass like any other biomass. A significant portion of algal biomass is made of carbohydrates, which can be fermented into ethanol. It is possible to derive both lipids for oil and carbohydrates for ethanol from the same batch of algal biomass.

Environmental Effects of Biodiesel [54,113,153,224]

It may be mentioned here that lot of energy is spent on the process of harvesting and processing microalgae to obtain biodiesel. This factor is important in the life-cycle assessment of algal production for biofuels. It may be mentioned here that use of biodiesel is better for the environment as compared to petroleum diesel. Apart from carbon dioxide sequestration, biodiesel is estimated to produce lesser pollution as follows:

- 67% fewer unburned hydrocarbons than petroleum diesel
- 48% less CO than petroleum diesel
- 47% less particulate matter than petroleum diesel
- 80% less PAH (polycyclic aromatic hydrocarbons) than petroleum diesel
- 90% less nPAH (nitrated PAHs) than petroleum diesel
- No generation of SOx and sulphates

However, in all fairness to petroleum diesel, biodiesel does produce 10% more NOx than petroleum diesel. But this may be reduced by various NOx reducing measures. There are some invasive species of algae that may be harmful to the environment and may required to be dominated by species being cultivated. The majority of the species are harmless and even beneficial to the environment.

5.2 Growing Algae in Open Waters/Ponds [3,124,163,179,187,217]

The production of algae in open waters in the sea will have the advantage of not using land and freshwater as it would be cultivated in saltwater in coastal areas. Algae can also be cultivated in raceway ponds.

A raceway pond is made of a closed loop circulation channel that is typically 0.3 m deep. In raceway pond, the temperature fluctuates within a diurnal/daily cycle and seasonally, and the evaporative loss of water may be significant. The temperature and light insolation depends on the climate and weather conditions of a place. The raceways may be plastic lined. Paddle wheels are used to provide the necessary mixing to maintain the algae culture in suspension. Aerators may also be used to supplement CO_2 for maximising growth. The open ponds are subject to contamination by other species of algae and bacteria. The production suffers when it is very cold as no temperature control is possible. To minimize the contamination of culture, strategies like maintaining high salinity and a high pH value may be good for some monocultures of species like spirulina.

New Zealand's 'Aquaflow Bionomics' and USA's Live fuels Incorporated are using open ponds for growing algae for biofuel production. They are using innovative techniques for preventing invasions by unwanted organisms in the system. Live Fuels Inc. has been working with scientists of 'Sandia National Lab. (US)' for developing a 'Green crude product' at a competitive price.

A Tel Aviv Company, Seambiotic Ltd., grows a high yield oil-rich algal strain in open ponds using waste CO_2 emissions. It is working with a company called 'Inventure Capital' of Seattle (US) to utilise the advanced conversion process developed by them for producing biodiesel and ethanol. A company 'Petro Sun Inc. (US)' planned to grow algae in large salt water pond systems in the coastal areas of the 'Texas Gulf (US)'.

Note on cultivating microalgae in sewage domestic wastewater

The National Institute for Water Research (NIWR), in Pretoria, investigated the feasibility of growing algae in domestic wastewater for several years with positive results. They carried out pilot tests to grow algae in raw sewage water in shallow ponds (for example, 400 mm deep). Harvesting algae comprised of three separate, but interdependent process stages that is, separation, dewatering, and drying.

Dissolved air flotation was used along with flocculation by the addition of a flocculant like aluminium sulphate. Slurry of 2 to 3% of dry solids was obtained by this method. It was found that if cationic polyelectrolyte was used instead of aluminium sulphate, slurry of 8% dry solids could be obtained because of easy water separation with polyelectrolyte. Further, separation of water from sludge/screening and drying of algal mass would be required. The liquid effluent would be sent to a wastewater treatment plant for required treatment before discharge.

The separation of algae from water and the drying process of algae account for the major portion of the cost and hence, the main emphasis should now be on improving the efficiency of water/moisture removal. [3]

5.3 Photobioreactors (PBR) for Algae [45,52,79,80,86,93,138,164, 168,186,241]

Because the growth of scum layer on the surface of pond tends to block the passage of light to microalgae lower down in the pond, some pioneers have designed fabricated transparent enclosures called 'photobioreactors' (PBR) to grow microalgae in them instead of in ponds. A photobioreactor is a closed or semi-closed system in which light and nutrients are supplied to the system in an attempt to maximize algal biomass. There are many types of photobioreactors, including tubular, flat panel, airlift, or even plastic bags. Light can be provided by direct sunlight, electric light, or a combination of both.

In such photobioreactors, the microalgae along with nutrients are slowly circulated with the aim to maximise photosynthesis by providing better exposure of algae to light, nutrients, and CO_2 (from the atmosphere), thereby maximising the production of algae. Tubular PBR are sometimes used. They consist of an array of straight transparent tubes which are usually made of plastic/glass generally placed outdoors for illumination by natural light. The cultivation vessels have a large surface area-to-volume ratio. The tubes are generally less than 10 centimeters in diameter to maximize sunlight penetration. This array captures the sunlight for photosynthesis. The medium broth is circulated through the tubes by a pump, where it is exposed to light for photosynthesis, and then back to a reservoir. The algal biomass is prevented from settling by maintaining a highly turbulent flow within the reactor, using a mechanical pump. A portion of the algae is usually harvested after passing through the solar collection tubes. In this way, continuous algal culture is possible.

The photosynthesis process generates oxygen. In an open raceway system, this is not a problem as the oxygen is simply returned to the atmosphere. However, in the closed photobioreactor, the oxygen levels build up until they inhibit and poison the algae. The culture must periodically be returned to a degassing zone—an area where the algal broth is bubbled with air to remove the excess oxygen. Also, the algae use carbon dioxide, which can cause carbon starvation and an increase in pH. Therefore, carbon dioxide must be fed into the system in order to successfully cultivate the microalgae on a large scale. A schematic diagram of a typical PBR for producing/growing algae is shown in Figure 5.1.

Unlike open raceways ponds, the PBR essentially permit the cultivation of a single species culture of microalgae for a prolonged duration. PBR may require cooling during the daylight hours for maintaining optimum temperature, while it would also be advantageous to maintain optimum temperature during night hours by heating if necessary. Heat exchangers can be used for cooling/heating photobioreactors, which may be located either in the tubes themselves or in the degassing column. The optimum temperature range is 20–24°C. *However, it is important to consider that the use of external energy consumption in photobioreactors (PBR) is more than in pond systems growing microalgae. Therefore, the economics and environmental consequences of the process have to be closely examined.*

Light Energy for Algae

The sunlight energy received by the cultivation system can be represented by the hemispherical incident light flux density, that is, photon flux density (PFD). The whole solar spectrum at ground level covers a range of 0.26 to 3 μm. 'Photosynthetic active radiation' (PAR) is in the band width of usually 0.4 to 0.7μm. The PAR range is about 43% of the full solar energy spectrum.

The photosynthetic growth in autotrophic conditions under solar light (or artificial light of a suitable spectrum) would depend on the assimilation of carbon from CO_2 and mineral nutrients in water. Only the photosynthetic active radiation (PAR) part of total light spectrum plays a part in this.

In most artificial light cultivation systems, the normal incidence of light is usually chosen, because it is the most effective way to transfer light into the culture volume of microalgae as it would ensure lesser reflection and better light penetration in the bulk of the culture. This would be different in solar light as the displacement of the sun with respect to earth takes place continuously

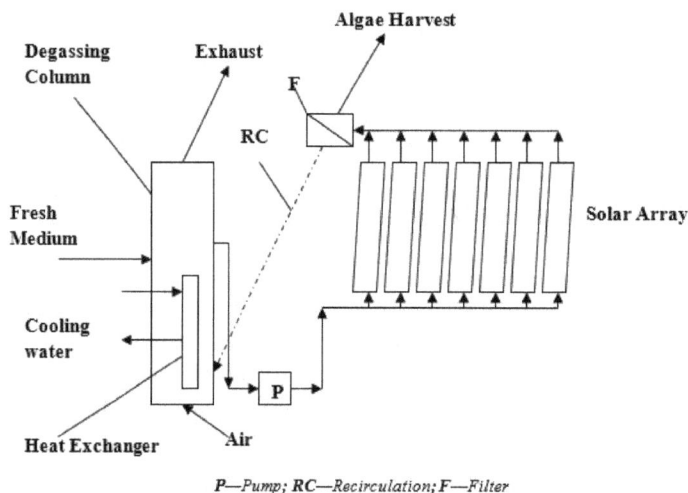

P—Pump; RC—Recirculation; F—Filter

Fig. 5.1. Tubular Photo Bio-reactor with Parallel Run Horizontal Tubes for Production of Microalgae.

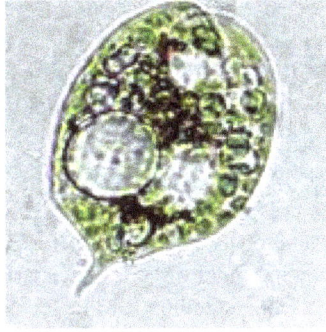

Fig. 5.1a. Image of Algae
[*source: http://www.extension.org/algae-for-biofuel-production*]

during day and a non-normal incident light condition would exist. The sunlight would have a large proportion of diffused radiation because of scattering and reflection from various surfaces.

Here it is relevant to mention that very strong light/an excess of light can create the problem of *photoinhibition, which is the reduction in the rate of photosynthesis due to an excess of light. The phenomena of photoinhibition can lead to the damage of the photosynthetic apparatus of the cell at higher respiration rates.*

Fermenting Equipment (Fermenter)

As stated earlier, a few types of microalgae that have the ability to grow fast and accumulate lipids on organic sources are called heterotrophic; with higher production of biomass and lipid growth. *(Also see Para 5.1.1.)* These types of microalgae can be grown in fermenters, which are different from photobioreactors. Fermenters are similar to bioreactors in that they are closed or semi-closed systems used for the production of biomass. However, fermenters utilize an organic source of carbon (e.g., sugar) as the source of energy instead of carbon from atmospheric CO_2 through light and photosynthesis. These fermenters can usually achieve much higher biomass than a photobioreactor, but it is uneconomical due to the cost of supplying the fixed carbon source like glucose and additionally, it is unhelpful in CO_2 sequestration. (For a gram of algal biomass produced, at least 1 gram of sugar would need to be added.)

5.4 Technological Developments in Algae Production—Examples: [3,9,83,91,101,156,185]

(i) M/s Valcent Products of US in a joint venture with Global Green Solutions of Canada grow microalgae in long rows of suspended moving plastic bags in a patented system called 'Vertigrow'. By growing it in a vertical fashion, much more surface area of the cells of microalgae get exposed to light and a moving system was designed to keep the algae exposed to sunlight long enough to absorb the solar energy needed for photosynthesis. Such a pilot system (a large high-density greenhouse) was created near El Paso, Texas, USA. The firms claim they can produce much more algal oil per year per acre as compared to production from oilseeds.

(ii) Racked glass photobioreactors are used by microalgae-culturist firms like the Massachusetts-based 'Green Fuel Technologies Corporation, USA'. They utilise CO_2 from flue gases of thermal power stations, cement factory, etc., for growing microalgae.

(iii) A2BE Carbon Capture LLC has patented 450 ft. long 50 ft. wide (137.2 m × 15.25 m) photobioreactor consisting of twin transparent plastic algal water beds (i.e., having parallel redundancy to cater for the case when one algal bed is shut down for any reason). Counter-rotating currents are induced within the algal beds to ensure the maximum exposure of microalgae to sunlight as they pass through the phototrophic zone. The internal temperature is controlled. For harvesting, a biological agent aggregates the algal cells into larger and more separable entities, which can be extracted relatively easily. Internal rollers operating in both directions serve to clean the internal surfaces of the waterbed tubes, and re-suspend microalgae.

(iv) Petro Algae, a Florida (US)-based green fuel company, have developed bioreactors and harvesting methods for converting microalgae grown in open freshwater ponds into biodiesel. They have also developed a technology for producing protein as a by-product from the residue (after extracting algal oil) for use as animal feed. Indian Oil Corporation has signed a memorandum of understanding for this technology.

(v) A company named 'Solix Biofuels' worked with Colorado State University's energy conversion laboratory. They designed a 20 meter-long photobioreactor that can utilise CO_2 emission from a 'brewing factory' in the US.

(vi) Scientists at Biofuel Systems in Alicante, Spain produce biofuel from marine phytoplankton* (a microscopic algae). At Biofuel Systems, a small amount of this microscopic algae is collected from sea and then put in a photosynthesis machine, where these organisms reproduce by cell division, or mitosis (a process of cell division involving the arrangement of protoplasmic fibers in definite figures), resulting in a very fast growth rate. Depending on the species of microalgae, the cell division process may take 8–24 hours. When sufficient biological mass is accumulated, it is removed, dried, and pressed into easily transportable bricks. The microalgae biomass bricks are transported to a factory where the biomass is treated for extracting algal oil, which after extraction is converted to biofuels and separated from the byproducts. These by-products can be marketed separately. The residue can also be used for making liquid fuels through the gasification route, like any other biomass. At Biofuels Systems, the microalgae are grown in vertical towers, which occupy one square meter of surface area and can be harvested every 24 hours. The microalgae grown in just one such tower is claimed to produce an energy equivalent to that produced from a1000 square meter sunflower plantation. [*Phytoplankton in high concentrations makes the water look greenish because of the chlorophyll in it. They produce more than half of the planet's oxygen by photosynthesis.]

(vii) The cyanobacterial strain, Nostoc muscurum TISTR8871, shows good results vis-à-vis biomass production and has starch accumulation of up to 32.9% during the photosynthetic growth of green algae.

(viii) *Neochloris* sp., *Nannochloropsis* sp., and *Chlorella* sp. can be selected for enhanced biomass production both in open ponds and photobioreactors.

(ix) The fully automated plant at the Fraunhofer Centre for Chemical-Biotechnological Processes (CBP) in Leuna was designed to produce microalgae on an industrial scale under strictly controlled conditions. It was built by Subitec GmbH hand in hand with researchers from the Fraunhofer Institute for Interfacial Engineering and Biotechnology (IGB) and Fraunhofer CBP. They were able to achieve concentrations of microalgae that are up to five times higher than those produced in other types of closed reactors. The main reason for this is light—a prime growing condition for algae. To ensure that sufficient light reaches all of the algae, the researchers chose to build their pilot plant using flat-panel reactors. These consist of arrays of panels measuring three or five centimeters in thickness, installed vertically on the ground. Over a slotted tube at ground level, a mixture of air and CO_2 is blown into the nutrient medium in which the algae are suspended. This has two effects: Firstly, the gas bubbles rise upward, supplying the algae with the carbon dioxide they need to grow, and secondly the flow of gas stirs up the algae and causes them to move about. In this way, each algae is propelled at regular intervals to the surface of the reactor, where it comes into contact with the light it needs to replenish its store of energy. The reactors are installed either outdoors or in a greenhouse and sunlight is utilized for the job. [3]

(x) *The Sonoran Desert (Mexico) Company's* goal is to produce fuel directly from the algae without killing the microorganisms, allowing for a shorter turnaround time to make fuel. The company claims that its process can produce around 6000 gallons per acre per year. [156]

(xi) *Seambiotic (Israel)* grows microalgal cultures in open ponds at their five-hectare commercial plant. They use flue gases like carbon dioxide and nitrogen from a nearby coal fired plant as feedstock. Its 1000-m^2 facility produces roughly 23 tonnes of algae per day—three tonnes of algal biomass yielding around 100 to 200 gallons of biofuel. [156]

(xii) *Solazyme, South San Francisco* genetically engineered algal cultures using DNA from different strains to maximize oil and biomass production. They grow these algal cultures in large fermentation vessels before harvesting their oil. It first tested its jet fuel in late 2008. They make fuel to supply to the US Navy. [156]

(xiii) *Solix Biofuels, Coyote Gulch, Colo.* uses specialized photobioreactors in which batches of microalgal cultures are grown in large, closed-growth chambers under controlled light-and-temperature conditions. Once the cultures are fully grown, their oil is extracted through the use of chemical solvents like benzene or ether. The solvents are mixed into the chambers to separate the oil from the algae, and it is then collected from the surface.

(xiv) Sapphire Energy, after experimenting with the production of various algae fuels beginning in 2007, now focuses on producing what it calls "green crude" from algae in open raceway ponds. The green crude has the same composition as crude oil, and is therefore compatible with existing refineries. The company has already shown that its fuel can be used in cars and even jets. Sapphire has a 100-acre pilot facility near Las Cruces, N.M. They first commercially demonstrated their algae fuel facility in New Mexico and have continuously produced biofuel since the completion of the facility in that year. In 2013, Sapphire began commercial sales of algal biofuel to Tesoro, making it one of the first companies, along with Solazyme, to sell algae fuel in the market. [83,91,156]

(xv) Solazyme Inc., South San Francisco, California, US has set up a plant in Daphne in Alabama for algal biofuels. It uses disinfected wastewater as a nutrient for algae. Algae are subjected to a hydrothermal liquefaction system to produce biofuel. [91]

(xvi) Muradel, Private Ltd, South Australia produces fuel from halophytic microalgae. Its trademark process produces 'green crude', which upon fractionation yields substitutes of fossil fuels. [91]

(xvii) The *genetic engineering* of algae has been used to increase its lipid content, biomass production, and growth rates. Current research in genetic engineering includes either the introduction or removal of enzymes.

In 2007 Oswald et al. introduced a monoterpene synthase from sweet basil into *Saccharomyces cerevisiae*, a strain of yeast. This particular monoterpene synthase causes the synthesis of large amounts of geraniol, while also secreting it into the medium. Geraniol is a primary component in rose oil, palmarosa oil, and citronella oil as well as essential oils, making it a viable source of triglycerides for biodiesel production.

The enzyme ADP-glucose pyrophosphorylase is vital in starch production, but has no connection to lipid synthesis. The removal of this enzyme resulted in the muted version, which showed an increased lipid content. After 18 hours of growth in a nitrogen-deficient medium, this version had on average 17 ng triglycerides/1000 cells, compared to 10 ng/1000 cells in the non-muted version. This increase in lipid production was attributed to the reallocation of intracellular resources, as the algae diverted energy from starch production. [83]

In 2013, researchers used a "knock-down" of fat-reducing enzymes (multifunctional lipase/phospholipase/acyltransferase) to increase lipids (oils) without compromising growth. The study also introduced an efficient screening process. Antisense-expressing knockdown strains 1A6 and 1B1 contained 2.4- and 3.3-fold higher lipid content during exponential growth, and 4.1- and 3.2-fold higher lipid content after 40 h of silicon starvation. [9]

[When genes are inserted in the reverse orientation into a strand of DNA and used in genetic engineering, it is called 'antisense expressing'. The removal of a gene is called a 'gene knockdown'.]

Sapphire Energy and Bio Solar Cells are using genetic engineering to make algae fuel production more efficient. Genetic engineering could vastly improve algae fuel efficiency as algae can be modified to only build short carbon chains instead of long chains of carbohydrates. Sapphire Energy also uses chemically induced mutations to produce algae suitable for use as a crop. [83]

Algenol Biofuels's (Florida, US) technology produces high yields and relies on patented photobioreactors and proprietary downstream techniques for low-cost fuel production using carbon dioxide from industrial sources. [83]

(xviii) Qeshm Microalgae Biorefinery Co. (QMAB) is an Iran-based biofuels company operating solely on the Iranian island of Qeshm in the Strait of Hormuz. QMAB's original pilot plant has been operating since 2009, and has a 25,000 liter capacity. In 2014, QMAB released BAYA Biofuel, a biofuel derived from the algae Nannochloropsis, and has since specified that its unique strain has up to 68% lipids by dry weight volume. The development of the farm mainly focuses on two phases: the production of nutraceutical products and the conversion of green crude oil to produce biofuel. The main product of their microalgae culture is crude oil, which can be fractioned into various kinds of fuels and chemical compounds. [83]

5.5 Extraction of Oil and other Products from Algae [7,21,78,102,107,127,136,141,166,220]

5.5.1 Mechanical Extraction of Oil from Algae

Oil can be extracted from the dried algal biomass by mechanical presses or centrifuging. This method is being used by some companies, for example, Petro Algae LLC (now acquired by Petro Tech Holdings Corporation of US). They employ a natural strain of microalgae developed by the Arizona State University (US) by cultivating them in photobioreactors. The algal biomass is harvested and dried and the oil is extracted by centrifugation. The leftover is a high protein meal that can be used as animal feed.

5.5.2 Solvent Extraction of Oil from Algae [78,107]

Oil from microalgae can also be obtained by a solvent extraction process. For this, the cell membranes of microalgae have to be first broken down by pretreatment with enzymes or by chemicals. The pretreated algae are treated with solvent to extract oil from it. The dissolved oil can be obtained by separating the solvent by distillation, which is reused. Possible solvents for use in this process are hexane, benzene, ether, etc. Hexane, which is widely used in the food industry, is a common choice. The organic solvent used is recovered and recycled. As benzene is classified as carcinogenic substance, it is not used.

In the Soxhlet extraction process, the oil from microalgae is extracted through repeated washing or percolation with an organic solvent such as hexane or petroleum ether. The solvent after recovery is recycled.

5.5.3 *Gasification Method of Oil Extraction from Algae* [7]

The gasification method can be used to obtain oil from algae. For this, the algae are dried and heated in the absence of oxygen to vaporise the volatile contents. The resultant vapours are passed through a catalyst system and oil products are obtained. Different catalysts encourage the assembly of different organic molecules so that products ranging from crude oil to diesel, kerosene, petrol, etc., can be formed, depending on catalyst used. This catalytic approach avoids the more usual oil extraction, transesterification, and refinement/cracking and cleaning process.

The 'Solena Group' based in Washington uses gasification technology developed by 'NASA'. A plasma gasifier is used to heat biomass to about 5000°C to produce synthetic gas (syngas). The diesel obtained from it is like petroleum diesel; it is called 'renewable diesel and not biodiesel'. *(see Para 5.11.4.)*

5.5.4 *Enzymatic Extraction of Oil from Algae* [29,70,210]

In this method, enzymes are used to degrade the cell walls with water acting as the solvent. This makes the fractionation of oil much easier. The enzymatic extraction can be supported by ultrasonic waves which break down the cell walls faster. The combination method, the Sono-enzymatic treatment, is faster and gives higher yields.

5.5.5 *Use of Supercritical CO_2 for Oil Extraction from Algae* [11,21,38,127,130,184,197,229]

The tendency of supercritical CO_2 to dissolve oil can be harnessed for taking out oil from algae. The CO_2 for this purpose is liquefied under pressure and then its temperature is brought up to a supercritical point.

5.5.6 *Production of Co-Products from Algae for the Economic Viability of Algal Oil* [2,81,92]

Scientists are investigating all sorts of potential co-products from algae. After extracting the oil from algae, around half of the biomass is left. Any company which develops valuable co-products having a sizeable market could help to make algal biofuels economically more viable. Several high-value products are already made from algae, such as:

a. Long-chain polyunsaturated fatty acids that are added to infant formula

b. Natural pigments which are used in cosmetics

c. Algal biomass residue is high in protein, so it is converted to animal feed. It can be a good soil additive and would act as a fertilizer.

d. Vitamins

e. Agar and other polygalactans

f. Mineral rich liquid

Alternatively, the leftover biomass could be subjected to anaerobic digestion to produce methane ('biogas') as an additional form of fuel. The leftover biomass can also produce liquid fuels like any other biomass through the gasification route. *(See Para 5.9 and 5.10).*

However, the market for these products is relatively small at present compared to the energy market and there is a need to develop the market for these products aggressively.

5.6 New Technologies for Obtaining Algal Oil [18,42,44,46,82,144,146, 171,202,209]

5.6.1 *Technology for Breaking down Algae Cell Walls for Oil Extraction* [82,146,154,171]

Another new method based on 'the technology used in biogas generation' can be applied for breaking down cell walls of algae for extracting oil from it. The principle of this technology used in biogas generation from organic biomass sludge is as follows:

'*It is already known that pre-treatment of biomass to break it down to smallest particles prior to anaerobic digestion for generation of biogas greatly increases the yield of the gas. The smaller the particle size of the feed stock to the digester, more efficient is the digestion. The biogas normally contains 40% CO_2. If the biomass is mixed with sludge from biomass reactor under pressure prior to digestion process, the CO_2 permeates through all the microbial cells. Now if the pressure is dropped suddenly, the cell walls are unable to withstand this rapid expansion of CO_2 which had permeated into the cells. Therefore, the cells rupture and enhance the biogas production in the digester.*'

A similar 'Cell rupture technology' (using CO_2 under pressure for the pretreatment of algae in conjunction with suddenly dropping its pressure, that is, pressure swing) for extracting oil from algae is used by producers.

Ultrasonic Extraction of Oil from Algae

The ultrasonic treatment of microalgae improves the extraction of oil from the algal cells and its conversion to biodiesel. Intense ultrasonic treatment of liquids generates sound waves that propagate into the liquid media, resulting in alternating high-pressure and low-pressure cycles. During the low-pressure cycle, high-intensity small vacuum bubbles are created in the liquid. When the bubbles attain a certain size, they collapse violently during a high-pressure cycle. This is called 'cavitation'. During the implosion, very high pressures and high-speed liquid jets are produced locally. The resulting shear forces break the cell structure mechanically and improve the material transfer. This effect supports the extraction of lipids from algae.

The ultrasonic reactor system can be easily retrofitted into existing facilities, improving algae extraction.

Use of Ultrasonic Technique in Cold Pressing: By ultrasonic treatment of microalgae for cold pressing, good control over the cell disruption can be exercised, to avoid an unhindered release of all intracellular products, including cell debris, or denaturing of the product. By breaking the cell structure, more lipids stored inside the cells can be released by the application of outside pressure.

Use of the Ultrasonic Technique in Solvent Extraction: The high-pressure cycles of the ultrasonic waves support the diffusion of solvents, such as cyclohexane, into the cell structure. As ultrasound breaks the cell wall mechanically by the cavitation shear forces, it facilitates the transfer of lipids from the cell into the solvent. After the oil is dissolved in the cyclohexane, the pulp/tissue is filtered out. The solution is distilled to separate the oil from the cyclohexane. [For the ultrasonic treatment of flammable liquids or solvents in hazardous environments, Hielscher offers certified ultrasonic treatment systems.] [83]

Use of Ultrasonic Treatment in Enzymatic Extraction: Ultrasonic treatment infuses high synergetic effects when it is combined with the enzymatic treatment. The cavitation assists the enzymes in the penetration of the tissue, resulting in faster extraction and higher yields. In this case, water acts as a solvent and the enzymes degrade the cell walls. [154]

New Technique for Fractionation of Algae: New techniques for fractionation of algae have been developed, which also provide the highest recovery of lipids for biofuel production, and have separate by-product streams for protein and carbohydrate fractions for being processed into feed or other biofuels. The development of the SRS Algafrac™ technology was reported to be provided for 'a commercially scalable process which extracts polar and non-polar lipids, whether contained in the algae vacuoles or membrane bonds'.

5.6.2 Technology Developed by Originoil Inc. for Algal Oil [198]

A new technology is reported to have been developed by a Los Angeles-based firm, Originoil Inc. They have addressed some basic issues like those mentioned below:

i. Introducing carbon dioxide and the nutrients needed for growth of microalgae without agitating the water because the preferred strains of algae thrive best in a calm environment.

ii. Distributing light within algal culture.

iii. Maximizing oil yield/extraction by cracking the cell walls of as many algal cells as possible using the smallest amount of energy.

A slurry of micron-sized nutrition bubbles is created and is channelled to the algal culture. The increased contact between micron-sized nutrients and the algae ensures optimum absorption without fluid disruption. The process of creating micron-sized nutrition bubbles has been named Quantum fracturing' by the firm. Even the distribution of light has been addressed by the firm in the design of their 'Helix Bioreactor'. In this reactor, the culture medium is contained in a rotating vertical shaft around which lights are arranged helically. The lighting elements produce light of specific frequencies. For large-scale production, multiple helix bioreactors could be stacked to form an integrated network of automated, remotely monitored growth units. Each reactor group could be connected to a single extraction subsystem to form a networked production facility.

[As mentioned earlier, fracturing the cell walls of microalgae is essential for better extraction of oil. The algal biomass is subjected to low wattage microwaves which weaken the cell walls. Special catalysts and the ultrasonic technique are used for fracturing cell walls. The energy required by this process is lower than that by other mechanical processes for cracking the cell walls].

Milking Oil from Microalgae without Killing it: Harnessing algal oil from algae without killing these organisms is one of the directions of current research. This may include bioengineering the algae to secrete the oil directly into the culture or breaking open algal cells on demand.

OriginOil, based in Los Angeles, announced that a new technology, which promises a better way to "milk" algae to extract their natural oils. They use electrical pulses to get to the oil from inside algae without killing them, leaving them alive to produce more oil. It is a more low-cost solution than genetic engineering the algae for oil production. However, the live lipid extraction is especially beneficial when used with algae that have been genetically engineered for a faster growth rate or higher lipid yield. *[198]*

[The use of solvent extraction by solvents biocompatible with microalgae to extract oil and leave the microalgae alive may be a step in this direction.]

5.6.3 New Technique for Harvesting Self-flocculating Microalgae [18,42,76,75,144,182,209]

The scientists at the Institute for Water and Wastewater Technology, India, and Durban University of Technology, South Africa, have evaluated the applicability of polypropylene non-woven fabric membrane (PNM) for harvesting self-flocculating microalgae, *Scenedesmus obliquus,* by filtration. The self-flocculation property of this strain of algae can be utilized as a preconcentration tool. The filtration through PNM was found to be a low-cost method in the case of self-flocculating algae. Its success can be attributed to the formation of the 'algal bio-filter layer' facilitated by its self-flocculation property.

5.6.4 *Development of Polymeric Composites for Harvesting Algae* [69,81,126,191,207,219,234]

Microalgal harvesting by coagulation and flocculation methods require chemical agents. IIT Delhi's Institute of Water and Wastewater Technology, India, and Durban University of Technology, South Africa, have, in collaboration, developed a high molecular weight highly cationic polymeric composite by polycondensation of Epichlorohydrin, N,N-Di-isopropyl-amine and Ethylenediamine. This polymeric composite when used in coagulation and flocculation studies for the recovery of microalgae (*Scenedesmus* sp.) at a very low dose of 10 ppm, gave a very good recovery of biomass (of 95%). It is hoped that the use of this composite would be very cost effective in harvesting microalgae for biodiesel production. [145]

5.6.5 *Enhancing Lipid Content of Algae by Nutrient Limitations* [114,173,192,225]

Scientists of Gujarat University (India) have found that the lipid content of *Desertifilum tharense Msak01* and *Leptolyngbya* algal strains can be greatly enhanced under conditions of limitation of nitrogen or phosphorus.

In case of the '*Desertifilum tharense* Msak01' algal strain, its lipid content under limited nitrogen content could be enhanced to 198%, while under limited phosphorus conditions it could be enhanced to 110%.

In case of the Leptolyngbya algal strain, its lipid content under limited nitrogen content could be enhanced to 210%, while under limited phosphorus conditions, it could be enhanced to 165%. The presence and percentage content of various chemical nutrients affect the biomass growth and lipid content. [114]

5.6.6 *Effect of pH, Light Intensity, and Phosphate on Algal Biomass and Lipid Content* [128,131,170,223]

Other than light and water, phosphorus, nitrogen, and certain micronutrients are also useful and essential in growing algae. Nitrogen (N) and phosphorus (P) are the two most significant nutrients required for algal productivity, but other nutrients such as carbon (C) and silica (Si) are additionally required. Of the nutrients required, phosphorus is one of the most essential ones as it is used in numerous metabolic processes. The other micronutrients include iron (Fe), cobalt (Co), zinc (Zn), manganese (Mn) and molybdenum (Mo), magnesium (Mg), and calcium (Ca).

The variations in light intensities, pH values, and phosphorus content have an effect on the biomass production and lipid content of algae. Experiments were performed on the *Scenedesmus abundans* sp. under different autotrophic growth conditions using the BG-11 culture medium. It was found that the highest growth of biomass was obtained at a pH value of 5, with 6000 lux illumination and 60 mg/liter of phosphate concentration. On the other hand, the highest lipid content was found at a pH value of 6, with illumination of 6000 lux, and a reduced phosphate content to 20 mg/liter. Overall, it was found in the study, that 6000 lux illumination was the best, both for algal biomass production and its lipid content. *This knowledge can be used for optimizing oil production from algae.* [131]

5.7 Fungi as a Source of Oil [34,115,174,203]

5.7.1 *Fungi oil* [5,36,56,135,140,148,204,215]

Oils accumulated by oleaginous fungi have emerged as a potential alternative feedstock for biodiesel production. Some fungi from mangroves have been known for the production of several lignocellulosic enzymes, but their ability of producing oil has remained unexplored. However, the oleaginous fungi from the mangroves can be used for production of oil.

The accumulation of lipids by some of the fungi species have been investigated for their lipid content with regards to biodiesel production.

The fungal cell oil has a high content of saturated and monounsaturated fatty acids, that is, palmitic (C16:0), stearic (C18:0), and oleic (C18:1) acids, etc. This is very similar to vegetable oils, which are used for the production of biodiesel. Fungal isolates like *Aspergillus terreus* are very promising with regards to high oil yield. The *Aspergillus terreus* IBB M1 strain of fungi was found to accumulate high lipid content by using cheap lignocellulosic materials like bagasse, grape stalks, groundnut shells, etc. There are many wet mangroves which can facilitate biodiesel production with *Aspergillus terreus IBB M1* as a promising fungi candidate. The Mortierella species (*M. alpina* and *M. isabellina* ATHUM 2935), and *Cunninghamella echinulata* are also able to accumulate a high content of lipids. It may be noted that all fungi are not suitable for oil production.

The utilization of oleaginous fungi for the production of fungi oil for making biodiesel has many advantages over other plant sources such as:

• Fungi can be grown in bioreactors and do not need agricultural land.

• Fungi display rapid growth rates and have short life cycle.

• The growth of fungi is not dependent on light/illumination.

• Fungi can utilize any space, and can be grown in any geographical zone by artificially providing favorable growth conditions.

• Fungi production can be easily scaled up.

• Fungi can utilize any type of inexpensive carbon source like lignocellulosic biomass, agroindustrial residues, etc. Hydrolysates of lignocellulose, that is, hydrolysed biomass, sewage, and whey are very well utilized (because of their reduction to pentose and hexose sugars). Even glycerol which is a byproduct of biodiesel production may be used as carbon source by oil-producing fungi.

A. Fungus Gliocladium roseum [145]

It has been discovered that fungus named 'Gliocladium roseum' has the unique property of converting cellulose into the medium-length hydrocarbon typically found in diesel. The fungus was discovered on an Ulmo tree (also called Eucryphia cordifolia) in the rainforest of northern Patagonia (South America).

Researchers at the College of Engineering, Montana State University, USA (MSU) have reported that the *Gliocladium roseum* fungus produces gases. Under limited oxygen conditions, it produces a number of compounds associated with diesel fuel derived from crude oil. These are the first fungi organisms that have been found to produce many ingredients of diesel, including hexadecane, octane, 2 methyl heptane, and 1-octene. The oil to be produced from the fungus is identified as 'mycodiesel'. (This is a topic of further research taken up at the College of Engineering (MSU) and at Yale University.)

Gliocladium roseum shows incredible promise, but more research still needs to be conducted on it before this source of hydrocarbons can be exploited. This is because the fungus only produces hydrocarbons when exposed to certain conditions, which are difficult to create outside the lab. [36,140]

B. Fungus M. circinelloides

It is a potential feedstock for biodiesel production. Extracted microbial lipids showed a high content (> 85%) of saponifiable matter and a suitable fatty acid profile for biodiesel production. Researchers in Spain at Universidad Rey Juan Carlos (URJC), have demonstrated the direct transformation of biomass consisting of the fungus *M. circinelloides* into biodiesel, which complies with the US. ASTM D6751 and the EU's EN14213 and 14214 standards. [56]

C. Camellia japonica

Sources from 'Russian Academy of Sciences' have mentioned that a large amount of lipids can be isolated from some single-cell fungi and turned into biodiesel. Fungal species like *C. Japonica* (known as 'common camellia' or 'Japanese camellia') are interesting in this regard. *Camellia japonica* was employed as a feedstock for the production of biodiesel by trans-esterification with methanol on alkali catalysts.

D. Oil from Fungi found in the Tropical Mangroves Wetlands of India's Western Coast

According to an article published by BioMed Central Ltd. 2012, single cell oil (SCO) of oleaginous fungi from the tropical mangrove wetlands may be a potential feedstock for biodiesel. [135] The accumulation of lipids being species-specific, strain selection is critical and therefore, it is very important to evaluate the fungal diversity of mangrove wetlands. All the cells of these fungi were investigated with respect to their oleaginous property, cell mass, lipid content, fatty acid methyl ester profiles, and physicochemical properties of transesterified SCOs in order to explore their potential for biodiesel production.

In one study, 14 yeasts and filamentous fungi were isolated from the detritus- based mangrove wetlands along India's western coast. The *'Nile red staining technique' revealed that the lipid bodies were present in 5 of the 14 fungal isolates. Lipid extraction showed that these fungi were able to accumulate > 20 % (w/w) of their dry cell mass (4.14–6.44 g L^{-1}) as lipids with a neutral lipid as the major fraction.

The presence of higher quantities of saturated and monounsaturated C_{16} and C_{18} fatty acids and the absence of long chain polyunsaturated fatty acids (PUFAs) are the major features of the single cell oils of these tropical mangrove fungi. The experimentally determined and predicted biodiesel properties based on fatty acid methyl ester (FAME) composition of the fungal SCOs of three isolates are found to lie within the range specified by international biodiesel standard specifications.

Isolate IBB M1, with the highest SCO yield and containing high amounts of saturated and monounsaturated fatty acid (MUFA), was identified as *Aspergillus terreus* through a morpho-taxonomic study and 18 SrRNA gene-sequencing. Batch flask cultures with varying initial glucose concentrations revealed that the maximal cell biomass and lipid content were obtained at 30 gL^{-1}. The strain was able to utilize cheap renewable substrates like, sugarcane bagasse, grape stalk, groundnut shells, and cheese whey for SCO production. Further studies on the utilization of agricultural residues as a carbon source are ongoing, with a focus on optimizing the productivity of cell mass rich in single cell oil, but it is clear that the single cell oil of oleaginous fungi from the mangrove wetlands of India's western coast can be used as a potential feedstock for biodiesel production. The *Aspergillus terreus IBB M1* is a promising candidate fungus for this purpose.

(Though many fungi other than oleaginous fungi from the mangrove ecosystem produce several lignocellulolytic enzymes, they remain unexplored for their SCO-producing ability.)

The lipid content and fatty acid composition of single cell oils varies in response to environmental factors such as the type of carbon source, pH, temperature and according to the nature of the microorganism, that is, it is species- and strain-specific. Since the accumulation of lipids by oleaginous fungi varies, not all oleaginous fungal cells can be used as a feedstock for biodiesel production and a careful selection of the oleaginous strains of the fungal species and a characterization of their lipid composition needs to be performed to ascertain their suitability for biodiesel production *[135, 204]*.

*Note on Nile red staining technique

The dye Nile red, 9-diethylamino-5H-benzo[a]phenoxazine-5-one, is an excellent vital stain for the detection of intracellular lipid droplets by fluorescence microscopy and flow cytofluorometry. Nile red-stained, lipid droplet-filled macrophages exhibit greater fluorescence intensity than Nile red-stained control macrophages, and the two cell populations can be differentiated and analysed by flow cytofluorometry. The yellow-gold fluorescence is more discriminating. The dye is very soluble in the lipids it is intended to show, and it does not interact with any tissue constituent except by a solution.

E. Harnessing algae and fungi to create a new biofuel system

Michigan State University's scientists have found a solution to enhance oil production and harvest using what many consider sea sludge. This platform uses two species of marine algae and soil fungi. It lowers the cultivation and harvesting costs and increases productivity. The species of alga, *Nannochloropsis oceanica*, and fungus, *Mortierella elongata*, both produce oils that can be harvested for human use. When scientists place the two organisms in the same environment, the tiny algae attach to the fungi to form big masses that are visible to the naked eye. This aggregation method is called bioflocculation. When harvested together, the organisms yield more oil than if they were cultivated and harvested on their own. Biofuels systems tend to rely on one species, such as algae, but they are held back by productivity and cost problems. Bio-flocculation is a relatively new approach and has the advantages of (i) sustainablity (the fungi grow on sewage or food waste, while the algae grow in seawater) and (ii) cost-savings (with bio-flocculation, the aggregates of fungi and algae are easy to harvest with simple and cheap tools). (https://msutoday.msu.edu/news/2018/harnessing-algae-and-fungi-to-create-new-biofuel-system)

5.7.2 *Conditions for Increase in the Oil Content of Fungi* [109]

It may be mentioned that the lipid content of fungi cell oil varies according to nutrients (i.e., type of carbon and nitrogen sources), pH value, temperature, and the fungi species and specific strain. There are considerable variations in the lipid content of different species of the same genera when grown under identical culture conditions. The fatty acids composition of fungi oil and content can be altered by altering culture conditions.

For example:

(i) Under lower temperature conditions, there is an increase in the number of unsaturated fatty acids and their content in fungi. The unsaturated fatty acid content lowers the melting point of lipid and viscosity. But there are exceptions to this generalization as in the case of '*Fusarium oxysporum*,' where lowering the temperature increases saturated fatty acids.

(ii) *Aspergillus fischeri* produces a high yield of biomass and oil at neutral or slightly alkaline pH. In the case of *Aspergillus nidulans*, the maximum biomass and oil are formed at a pH of 6.8. But *Fusarium oxysporum* yields the maximum biomass and oil at a pH of 5.9.

The source of nitrogen and ratio of carbon to nitrogen (C:N ratio) plays an important role in the rate of accumulation and quantity of fat by fungi. The temperature plays a role in quality of the fat produced and the degree of unsaturation. The pH value of the medium is also very important.

The accumulation of oil occurs only when the nitrogen concentration in the growth medium falls below a certain critical value. In general, some fungal species produce a large amount of biomass when organic nitrogen is present in the medium, while the inorganic nitrogen helps to synthesize fat. There are certain exceptions to this generalisation, as in case of *Aspergillus nidulans* and *Penicillum lilacinum*, which produced more lipid when organic nitrogen source like asparagines and glycine [$CH_2(NH_2)CO_2H$] were used (in comparison with an inorganic nitrogen source). *Aspergillus nidulans* is reported to more readily utilize ammonium nitrate than sodium nitrate. Generally, at lower temperatures, more unsaturated fats are produced, but there are exceptions, such as *Fusarium oxysporum*.

The pH range of '6.0 to 7.0' is generally more conducive to the growth of several fungi, but it may vary slightly from fungi to fungi. *Aspergillus fischeri* gives a high yield of biomass as well as oil at a pH of 7 or at a slightly alkaline level. In the case of *Aspergillus nidulans*, the maximum biomass and oil may be formed at a pH value of 6.8 using 'glucose as carbon source', while in the case of *Fusarium oxysporum*, the maximum growth and oil formation may be at a pH of 5.9 using 'glucose and ammonium nitrate as nutrients'. It may be pointed out that variations in the carbon source may result in a high variation of lipid content, as in case of '*Penicillum sponulosum*', which accumulates 36.6% lipid content with glucose, 12.2% with sucrose, and 63% with molasses as carbon source.

Fusarium sp. and *Penicillum* sp. have been found to be of special interest as potential fat producers. They can be grown particularly well on natural waste material (for example, cellulose), agricultural waste, and molasses as a carbon source. [109]

5.7.3 Lipid Content of Some Fungi of the Ascomycetes Class [31,109]

The total lipid content of selected fungi is given in Table 5.4.

Table 5.4. Lipid Content of Selected Fungi from the Class Ascomycetes - Percent Dry Weight (under favourable conditions). [159]

Se. No.	Name of Fungi	Lipid as % Dry Weight
1.	*Aspergillus fischeri*	10.5–37
2.	*Aspergillus flavipes*	39.7
3.	*Aspergillus flavus*	5.7–35.5
4.	*Aspergillus minutes*	12.8–34.9
5.	*Aspergillus nidulans*	7.6–20.2
6.	*Chaetomium globosum*	54.1
7.	*Fusarium* sp.	38.6
8.	*Fusarium bulbigenum*	45.6
9.	*Fusarium graminearum*	12.6–31
10.	*Fusarium lycopersicum*	41.1
11.	*Fusarium oxysporum*	25.2–33.9
12.	*Penicillum* sp.	29.7–34.8
13.	*Penicillum roquefortii Thom*	22.9
14.	*Penicillum oxalicum*	15.7–23.2
15.	*Penicillum citrinum Thom*	18.1
16.	*Stillbella thermophile*	38.1
17.	*Trichosporum cutaneum*	45–56
18.	*Malbranchea pulchelta*	26.5
19.	*Mortierella* sp.	34.9

5.7.4 Fungal Oil Extraction and Conversion to Biodiesel [41,63,74,133,172,183,189,215]

For extraction of oil from fungi, the drying of the fungal biomass is the first requirement. Smashing the fungi cell can be done by a pre-reatment for ease of oil extraction. The pretreatment may be an ultrasonic-assisted chemical/enzymatic treatment, as in the case of algae.

The oil extraction from fungi may be carried out as it is in algae. The extracted oil can then be converted to biodiesel by the same method as used in the case of vegetable oil or fat. A schematic flow diagram of the process is shown in the Figure 5.2.

a. Solvent Extraction of Oil from Fungi

The fungal oil can be extracted from fungi by the solvent extraction method as well. Chloroform/methanol (2:1) is a good solvent for the extraction of fungi oil.

The efficacy of three extraction methods for determining the lipid and fatty acid composition of six fungal cultures have been studied. The extraction methods were:

(1) chloroform/methanol (2:1)

(2) hexane/isopropanol (3:2)

(3) soxhlet extraction by using hexane

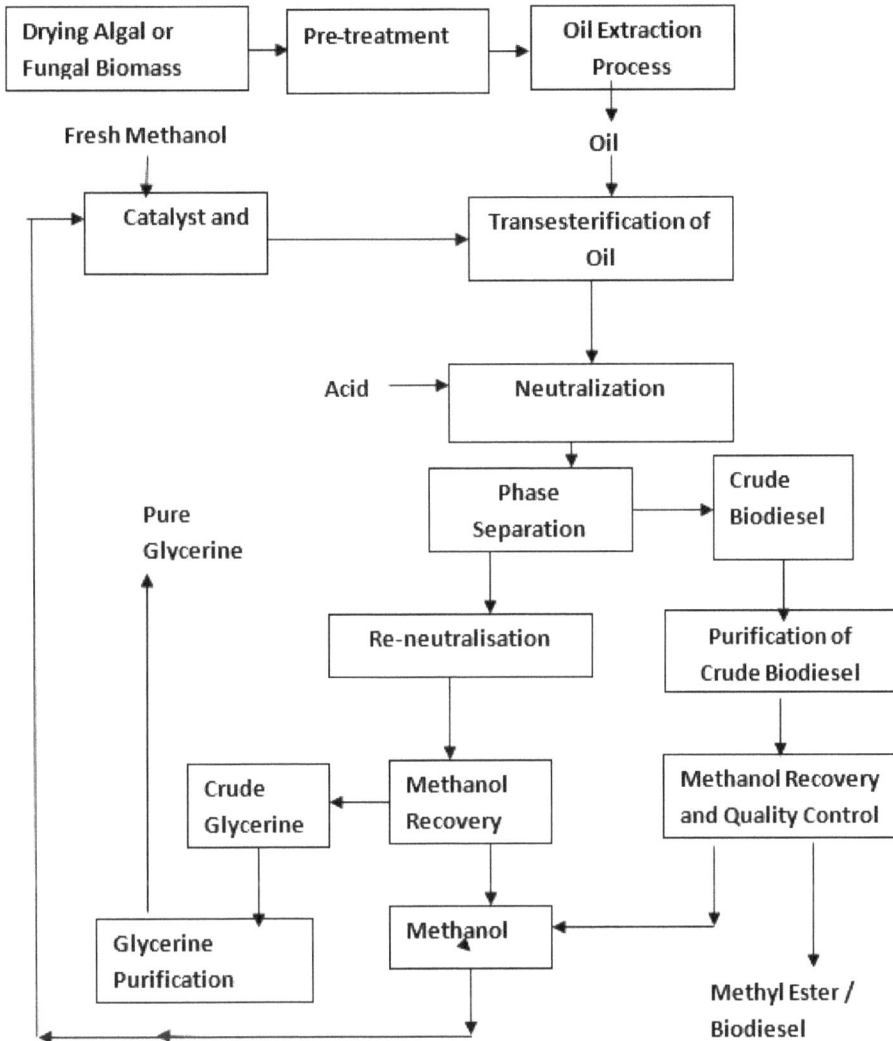

Fig. 5.2. Production of Biodiesel from Algae or Fungi.

It was found that the total lipid and fatty acid composition varied in fungal cultures, depending on the extraction conditions. Out of the above three methods, the chloroform/methanol (2:1) was found to be the best for extraction of lipid and fatty acids from fungal cultures. [183]

b. Extraction of Oil from Fungi at Supercritical Conditions of Solvent

At supercritical conditions, certain substances like CO_2, nitrous oxide (N_2O), trifluoromethane (CHF_3), and sulphur hexafluoride (SF_6) have a high dissolving power and they can extract oil from fungi. (These gases are strong greenhouse gases and their release into the atmosphere must be prevented during processing.)

c. A Particular Study

A Particular Study carried out for the extraction of oil from fungi (*Mortierella* ramannianavar. *angulispora*) using carbon dioxide (CO_2), nitrous oxide (N_2O), trifluoromethane (CHF_3), and sulphur hexafluoride (SF_6) under supercritical (SC) conditions yielded the following result:

'The oil solubility was highest in SC-N_2O (super critical nitrous oxide) followed by SC-CO_2, while both SC-CHF_3 and SC-SF_6 showed poorer solvent power.

The recorded oil solubilities at 333 K and 24.5 MPa were 2.3 wt.% in N_2O, 0.48 wt.% in CO_2, 0.0099 wt.% in CHF_3 and 0.0012 wt.% in SF_6.

The oil solubilities in SC-N_2O and SC-CO_2 were measured over the pressure range 15.7–29.4 MPa and at temperatures ranging from 313–353 K. The N_2O always showed greater solvent power than did CO_2 at the same temperature and pressure. The solvent power of a supercritical fluid increases with density at a given temperature, and increases with temperature at constant density. The change in neutral lipid composition of the extracted oil with the extraction ratio was measured. Free fatty acids or diglycerides were extracted more easily than triglycerides or '*Sterol esters*'.* The change in fatty acid composition was also measured. The proportion of γ-linolenic acid in the extract remained constant throughout the extraction'. [172]

(It may be noted that the greenhouse effect of SF_6 or N_2O is much greater than that of CO_2 and any accidental release of such strong greenhouse gases cannot be accepted. Therefore, it is preferable to use supercritical CO_2 and not SF_6 or N_2O.)

(*Sterol Esters: These are a heterogeneous group of chemical compounds. They are created when the hydroxyl group of a sterol and a fatty acid undergo an esterification reaction. They can be found in trace amounts in every type of cell, but are highly enriched in foam cells and are common components of human skin oil. Plant sterol esters have been shown to reduce the level of low-density lipoprotein (LDL) cholesterol in blood when ingested. Plant sterol esters are used in dietary supplements made from phytosterols and fatty acids, also derived from plants. They are added to certain oil-containing products like margarine, milk, or yogurt to make functional foods for controlling cholesterol levels.)

5.8 Liquid Fuel from Lignocellulosic Material by the Biochemical Route [9]

Some lignocellulosic enzymes are produced by selected fungi, which can be used to produce oil as mentioned in the previous paragraph.

Gribble, a small shrimp-like crustacean, has enzymes in its guts that can break down cellulosic materials into sugar; thus, it has the potential to convert woody biomass into fuel. Using advanced biochemical analysis and X-ray imaging techniques, scientists have determined the structure and function of a key enzyme used by gribble to digest wood. Some fungi have similar cellulase (with the characteristic of being able to degrade wood), but the enzyme from the gribble shows some important differences. The gribble cellulase is extremely resistant to aggressive chemical environments, is very robust, and has an extremely acidic surface. If this enzyme is produced on a large scale, it would be effective in converting lignocellulose to sugars. In view of its robustness, a smaller quantity of enzyme will be used in the process.

Being robust in difficult environments means that the enzymes can last much longer when working under industrial conditions and so, fewer enzymes will be needed.

The research and development of suitable and viable technology at several universities and institutions for using gribble for the conversion of woody biomass to liquid fuel through the biochemical route is underway. Also see *Chapter 3, Para 3.6.4 (iv)*. [9,132]

5.9 Liquid Fuel from Lignocellulosic Biomass through the Gasification Route [15,37,43,55,61,77,132,134,139,147,167,180,214,239]

5.9.1 Fischer-Tropsch (FT) Process for Converting Syngas to Liquid Fuels

The lignocellulosic biomass is gasified* in a gasifier to produce syngas which is a mixture of carbon monoxide and hydrogen with some impurities. Oxygen is used as an oxidant for gasification instead of air to keep the product free from nitrogen. Other impurities in syngas such as sulphur dioxide, chlorine and its compounds, etc., have to be removed before the syngas can be sent for further

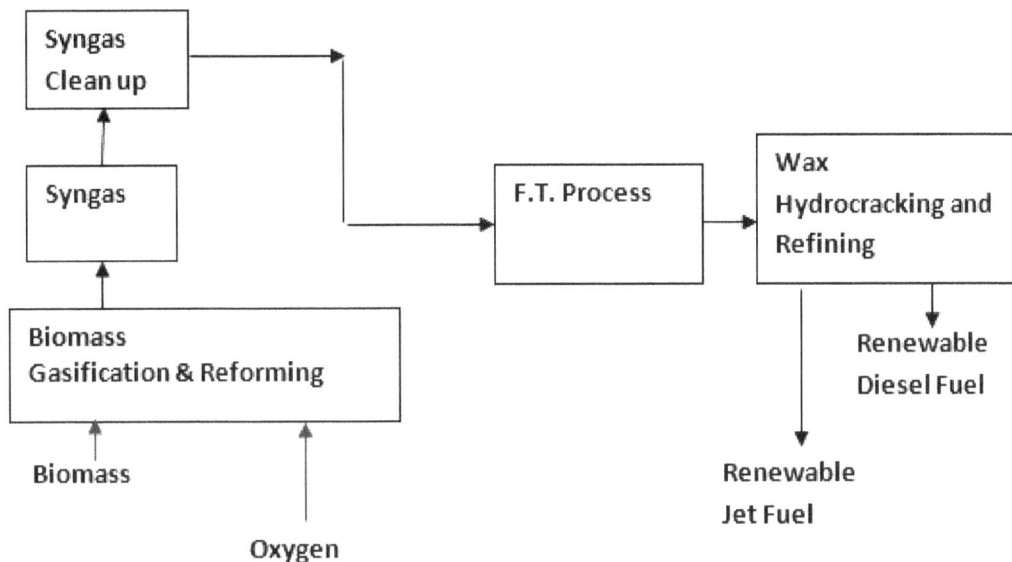

Fig. 5.3. Conversion of Biomass into Liquid Fuels by F T Process.

processing into liquid fuels. The cleaned syngas, which is a mixture of CO and H_2 is sent to a Fischer-Tropsch synthesizer, where a number of hydrocarbons are produced. Catalysts are used in the FT synthesizer. The conversion of CO to alkanes involves the net hydrogenation of CO, the hydrogenolysis of C-O bonds, and the formation of C-C bonds. Such reactions are assumed to proceed via the initial formation of surface-bound metal carbonyls. Many intermediate chemicals form in the reaction before the final product is formed. The process can use any type of biomass, such as tree foliage, grasses, algal biomass, agricultural biomass/crop residues, halophytes like Salicornia brachiata,** etc.

A flow diagram of the process of production of renewable fuels like renewable jet fuel and renewable diesel is shown in Figure 5.3.

The liquid fuel can be obtained from biomass, that is, BTL (Biomass to Liquid Fuel). Carbon/coal/lignite can also be converted to liquid fuel, that is, CTL (Carbon to Liquid Fuel). Also, for the convenience of storage and use, sometimes methane/natural gas or biogas is required to be converted to liquid fuel for ease of transport and use, that is, Gas to Liquid Fuel (GTL).

[*For details about gasification and the pyrolysis of biomass, see Para 6.13 of Chapter 6.

** Regarding the halophyte Salicornia brachiata,' see Para 2.2.6 of Chapter 2].

5.9.2 Chemistry of the Fischer-Tropsch (FT) Process

The FT process involves a series of chemical reactions that lead to the production of a variety of hydrocarbons. The following type of chemical reactions gives 'Alkanes' (saturated paraffin hydrocarbons).

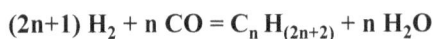

$(2n+1) H_2 + n\, CO = C_n H_{(2n+2)} + n\, H_2O$

where 'n' is a positive integer. The formation of methane (n = 1) is generally unwanted.

Most of the alkanes produced tend to be straight-chain alkanes, although some branched alkanes are also formed. In addition to alkanes, competing reactions result in the formation of alkenes (unsaturated hydrocarbons), as well as alcohols and other oxygenated hydrocarbons. Usually, only relatively small quantities of these non-alkanes products are formed, although catalysts favoring some of these products have been developed.

a. Other Reactions Relevant to the F-T (Fischer-Tropsch) Process

Several reactions are required to obtain the gaseous reactants needed for the F-T catalysis. First of all, reactant gases entering a F-T reactor must first be desulphurized to protect the catalysts that are readily poisoned. The other major reactions employed to adjust the hydrogen to carbon monoxide ratio (H_2/CO) are:

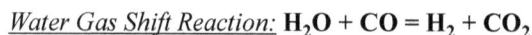

Water Gas Shift Reaction: $H_2O + CO = H_2 + CO_2$

(This reaction provides a source of Hydrogen)

Steam Reforming: $H_2O + CH_4$ **(Methane)** $= CO + 3\ H_2$

(This is an important reaction, which converts methane into CO and H_2 for F-T plants that start with methane.)

b. Process Conditions for the F-T Process

Generally, the FT process is operated in the temperature range of 150–300°C (302–572°F). Higher temperatures lead to faster reactions and higher conversion rates, but also tend to favor methane production. Therefore, the temperature is usually maintained at the low to middle part of the range. Increasing the pressure leads to higher conversion rates and also favors the formation of long-chain alkanes, both of which are desirable. The typical pressures used range from one to several tens of atmospheres. Even higher pressures would be favorable, but the benefits may not justify the additional costs of high-pressure equipment, while the higher pressures may sometimes lead to catalyst deactivation via coke formation.

A variety of synthesis gas compositions can be used. For cobalt-based catalysts,** the optimal 'H_2: CO ratio' is around 1.8 to 2.1. Iron-based catalysts promote the water-gas-shift reaction and thus, can tolerate significantly lower ratios. This reactivity can be important for synthesis gas derived from coal or biomass, which tend to have a relatively low ratio of hydrogen to carbon monoxide ('H_2:CO' < 1).

c. Catalysts Used in Fischer-Tropsch Process

Many catalysts can be used in the FT process, but commonly, cobalt and iron are used for the production of liquid fuels. Nickel as a catalyst favors methane formation and therefore is not used when the aim is to produce liquid fuels.

Cobalt*-based catalysts are very active. The catalysts are supported on high-surface-area binders/supports such as silica, alumina, or zeolites. In addition to the active metal, the catalysts typically contain a number of "promoters," including potassium and copper.

Cobalt catalysts have a tendency of getting *poisoned* by Group-I alkali metals like lithium (Li), sodium (Na), potassium (K), rubidium (Rb), cesium (Cs) and therefore it may not be used in the case of syngas from biomass or other solid fuels, which may contain these metals as impurities.

For the syngas obtained from biomass, the iron catalyst is suitable. The presence of potassium in the syngas acts as promoter when the iron catalyst is used.

It may be mentioned that for FT synthesis, when the feedstock is natural gas, cobalt catalysts are more active and better. When using 'natural gas feedstock' using 'autothermal reforming', a water gas shift reaction takes place to boost the production of hydrogen and the lowering of CO. In this case, the nickel-rhodium catalyst is better. (See Para 6.18.2, Chapter 6. for autothermal reforming.)

d. Wax Hydrocracking Step

The output from the FT synthesizer is sent up to the 'hydrocracker', for cracking the heavy molecules in the gas/oil. They have a higher boiling range than distillate fuel oil. It cracks the heavy molecules into distillate and gasoline in the presence of hydrogen and a catalyst. The hydrocracker upgrades low-quality heavy gas oils into high -quality, clean-burning jet fuel, diesel, and gasoline.

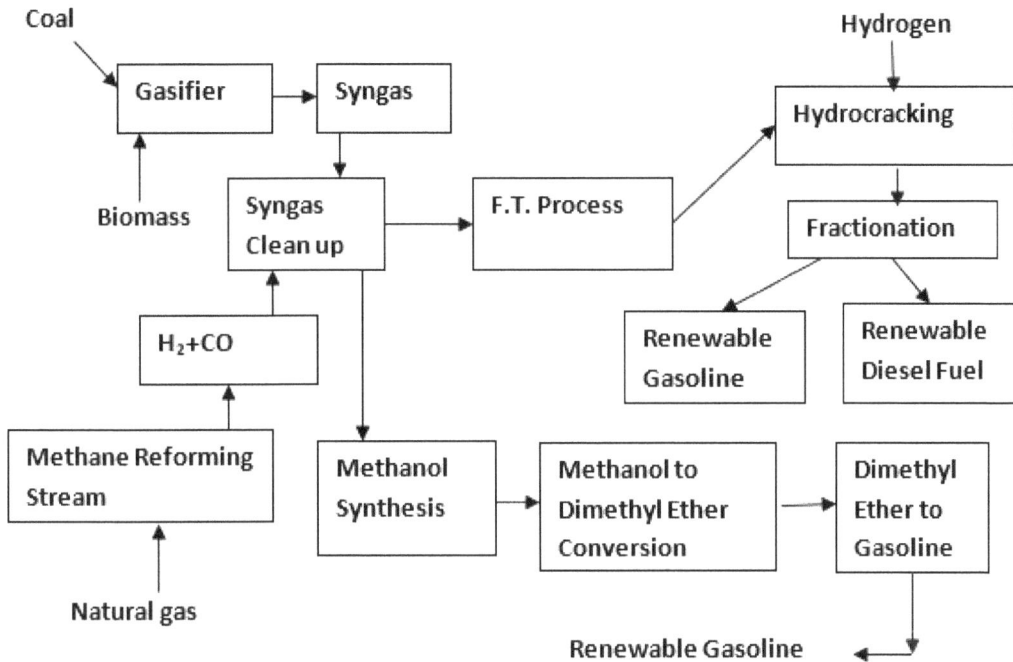

Fig. 5.4. Conversion of Biomass/Natural Gas/Coal into Synthetic Liquid Fuels by FT Process.

Two main chemical reactions occur in the hydrocracker: (i) catalytic cracking of heavy hydrocarbons into lighter unsaturated hydrocarbons and (ii) the saturation of these newly formed hydrocarbons with hydrogen. The former reaction, that is, catalytic cracking of the heavier hydrocarbons uses heat and causes the feed to be cooled as it progresses through the reactor. The later reaction of saturation of the lighter hydrocarbons releases heat and causes the feed and products to be heated up as they go through the reactor. Hydrogen is used to control the temperature of the reactor by feeding it into the reactor at different points. This process of hydrocracking yields valuable gasoline, jet and renewable diesel fuels having higher hydrogen content and excellent burning properties. The 'renewable diesel' formed through the gasification route from biomass is different from biodiesel *(See Para 5.12.3 ahead)*.

e. Conversion of any Feedstock like Biomass/Methane/Coal to Liquid Fuel [37]

It may be stated that any liquid fuels like renewable gasoline, renewable jet fuel, and renewable diesel can be obtained from biomass or from biogas. Instead of biomass or biogas, if we use natural gas or coal, similar liquid fuels obtained would not be called 'renewable' (as natural gas or coal is not designated as 'renewable'). The process from any feedstock like biomass, natural gas and/or coal is shown in Figure 5.4. Also see *Para 5.10* ahead.

5.9.3 Liquid Fuel from Biomass by Fast Pyrolysis [134]

Fast pyrolysis is a special type of pyrolysis process in which organic materials like tyres, plastics, paper, agricultural wastes, crop residues, refuse, food-processing wastes, sewage sludge, municipal solid waste, etc., are rapidly heated in the absence of air. The process breaks down the long chain of organic polymers into simple compounds and is completed in a few seconds in a reactor. Under these conditions, organic vapors, pyrolysis gases, and charcoal are produced. Typically, 60–75 wt.% of the feedstock is converted into oil. The essential features of this process are mentioned below:

- Finely grounded dry biomass is used (size less than 2 mm is used and moisture is around 10% only).

- Strong heating and high heat transfer rates at the reaction interface of the finely ground biomass are required.
- The time exposure to lower temperature is drastically minimized.
- The temperature of pyrolysis is controlled at around 500°C and the vapor phase temperature is controlled in the range of 400–500°C.
- The residence time for vapor is controlled to less than 2 sec in the reactor and the vapors are cooled very rapidly to get the bio oil.

Pyrolysis is one of the several means of producing liquid fuel from biomass. The maximum yield of organic liquid (pyrolytic oil or bio oil) from thermal decomposition may be increased to as high as 80% (dry weight) through the proper choice of heating rate and pyrolysis temperature, and the evacuation of the product from the reaction zone. Pyrolysis offers the possibility of easy handling of the liquids and a more consistent quality compared to any solid biomass. With fast pyrolysis, a clean liquid is produced as an intermediate suitable for a wide variety of applications.

Typical By-product Yields from Different Modes of Pyrolysis of Wood

Different quantities of liquid, solid, and gaseous byproducts are obtained from the pyrolysis of dry wood with about 10% moisture under different conditions of reactions as mentioned in Table 5.5. (*Also see Para.6.13, Chapter 6 for 'Pyrolysis'.)*

Table 5.5. Wood Pyrolysis Under Different Process Conditions & By-products.

Pyrolysis Process	Conditions in Pyrolysis Process	Liquid/ Bio oil	Gases/ Biogas	Solid/ Char
Fast pyrolysis	Moderate temperature around 500°C; high heat transfer rates; Very short residence time particularly for vapor; finely ground dry biomass; rapid cooling of vapors	75%	13%	12%
Gasification	High temperature; long residence time; moderate heat transfer rates	5%	85%	10%
Carbonization	Lower temperature; low heat transfer rates; very long residence time	30%	35%	35%

5.9.4 *Direct Thermochemical Liquefaction of Algal Biomass*

Petroleum like liquid fuel can be formed through the pyrolysis route from algal biomass, like any other biomass. The pyrolysis usually requires a drying procedure in which large amounts of energy are required to vaporize water. An alternative technique involving the direct thermo-chemical liquefaction of algal biomass of high moisture content may be applied for the production of fuel oils from microalgae. This technique uses nitrogen pressure to purge air and raises nitrogen pressure to prevent water from vaporizing during heating the biomass to a desired temperature, maintaining the temperature for a specified time, and then cooling. The oil fraction can be separated with dichloromethane.

5.9.5 *Algal Biomass to Liquid Fuel Using Catalytic Agents (Algal Hydrogenation)*

Algae biomass can be converted to liquid hydrocarbons at temperatures between 400–430°C, and operating hydrogen pressures of 73–159 kg/cm^2 (1025–2250 psig), in the presence of a cobalt molybdate catalyst or other catalysts. In addition, liquid products, hydrocarbon-rich gases are obtained. In general, higher temperatures and longer reaction times increase the degree of conversion and oil yield, and this decreases the yield of tar/asphalt-like compounds in the hydrogenation of algae. The oil yield and the degree of conversion also increase with hydrogen pressure up to a certain limit and then level off. The algal hydrogenation process is a useful means of producing liquid hydrocarbons for use as fuels, or feedstocks for making various chemicals.

5.9.6 Renewable Diesel from Lignocellulosic Materials through Gasification Route by Choren Industries, Germany [61]

The process was developed for making renewable diesel through biomass gasification by the FT process after extensive tests were performed on a 1 MW pilot plant. This process can produce tar-free gas without the use of a catalyst. The dried biomass (for example, with a moisture content of not more than 15–20%) is gasified in a low temperature gasifier (partial combustion/carbonizing temperature 400 and 500°C), producing a gas with tar and coke containing ash. The gas (with tar) from a low-temperature gasifier is led to a Carbo-V patented gasifier.

The heated oxygen from the heat exchanger (HE) is used in Carbo-V gasifier, where the tar is broken down/gasified. There is partial combustion of gases in the Carbo-V gasifier and the temperature is kept at about 1300–1500°C, at which all coke, tar, and hydrocarbons (including methane) break down to give CO, H_2, CO_2, and steam. The temperature, however lowers down to 800–900°C due to the heat of absorption during gasification of coke. This chemical quenching process produces tar-free gas *with a low methane* content and with high proportions of combustible CO and H_2.

The tar-free output gas from the Carbo-V gasifier is sent to the heat exchanger (HE), from where the cooled gas goes to the De-duster/De-dusting equipment. In the HE, the oxygen for use in the Carbo-V gasifier is heated by absorbing the heat from the gases coming to it from the gasifier. After the de-dusting stage, the gas goes through the 'Multi-stage washing/water scrubbing stage' for the removal of undesirable substances like chlorides, sulphides, or soluble gases, etc. From there, the cleaned and cooled syngas is led to the FT synthesizer for producing the desired fuel. The synthesis takes place at 200°C and 20 bar pressure with the help of a (cobalt) catalyst.

The solids removed by the de-duster (which may also contain some residual coke) are also blown in the Carbo-V gasifier. The solid residue melts and flows down the inside wall of the Carbo-V gasifier and is disposed into a water bath at the bottom. The flow diagram is shown in Figure 5.5.

The world's first industrial commercial biomass to liquid fuel production plant is located at Freiberg, Germany, and manufactures 'Sun Diesel' by Choren Industries (annual capacity: 15,000 tonne biodiesel). As per information from 2008, such plants with a capacity of 200,000 tonnes per annum were also found in Lubmin, near Greifswald.

5.10 Conversion of Lignocellulosic Biomass into Alcohols and Furans/Other Chemicals/Fuels [84,125,142,177,208,212,218]

After gasification of biomass, it is possible to convert it to a variety of useful liquid chemicals and fuels such as methanol, ethanol, dimethyl ether, mixed alcohols (a mixture of methanol, ethanol, propanol, butanol, etc.), furans, renewable gasoline, renewable diesel, etc., through the FT process. The sulphur, halides, and other impurities should be removed from the syngas obtained from the biomass before its synthesis to these chemical/fuels.

The syngas can be converted to dimethyl ether, popularly known as 'DME,' which is envisaged as a future fuel for vehicles. The next paragraph (Para 5.10) has been exclusively devoted to **DME**.

5.10.1 Methanol and Mixed Alcohols from Biomass

Methanol

Methanol is methane with one hydrogen atom replaced by a hydroxyl radical (OH). As an engine fuel, methanol has chemical and physical properties similar to ethanol. Methanol can be converted to ether or MTBE (which is blended with gasoline to enhance octane and to create oxygenated gasoline). Methanol can be produced from a variety of feedstock, including natural gas, coal,

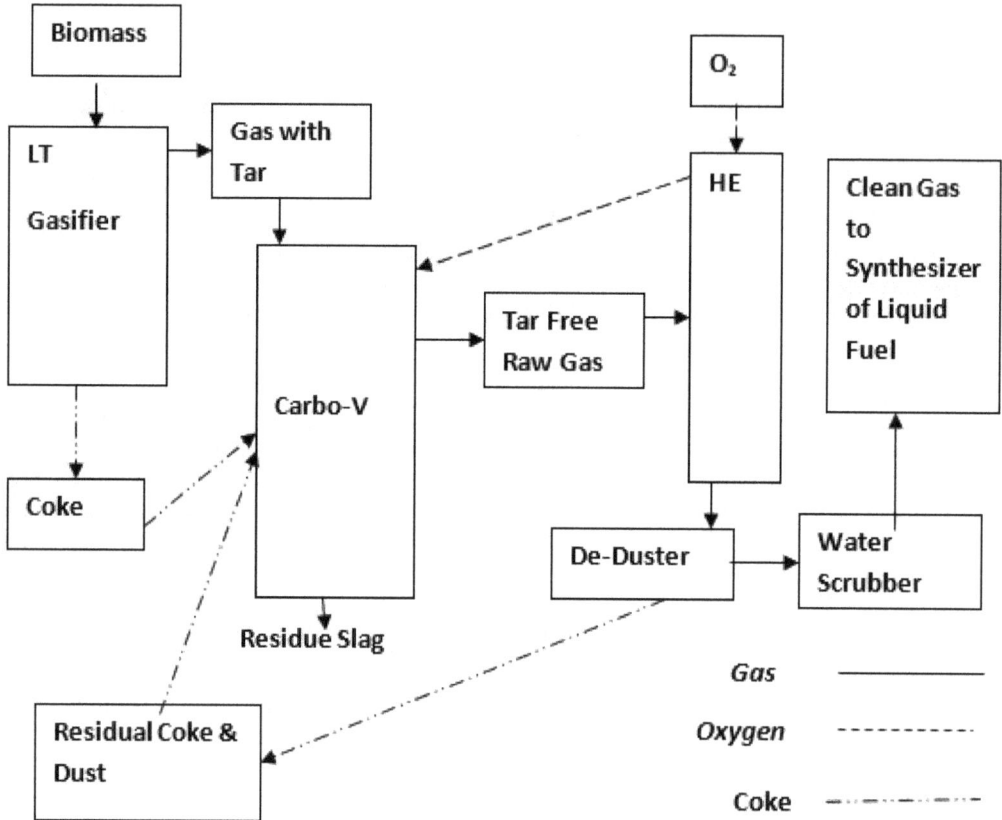

Legend: *LT Gasifier— Low Temperature Gasifier; HE—Heat Exchanger*

Fig. 5.5. Production of Renewable Diesel from Lignocellulosic Materials.

biomass, and cellulose. The syngas is fed into a reactor with a catalyst to produce methanol. The conversion of syngas to methanol can be achieved using a copper/zinc-based catalyst (copper oxide/zinc oxide).

However, at present, it is generally produced by steam-reforming natural gas to create a synthesis gas. Feeding this synthesis gas into a reactor with a catalyst produces methanol and water vapor. Although other feedstocks can produce methanol, at present, natural gas seems to be the most economical.

It is essential that the purity of syngas used for methanol synthesis is as per the quality requirement mentioned in Table 6.3 *(Para 6.14.5 of Chapter 6)*. Process conditions vary with the catalyst used. With the use of a copper oxide catalyst, a temperature range of 220–275°C and pressure range of 50–100 bars are used, while with zinc oxide catalyst temperature of about 350°C and pressure range of 250–350 bars are used. For methanol synthesis, the ratio of (H_2+CO_2) to $(CO+CO_2)$ should be greater than 2 in the syngas, and if not, it is to be adjusted. The synthesis reaction can be written as:

$CO + 2H_2 = CH_3OH$ *(Methanol/Methyl alcohol)*

(The methanol can be blended in gasoline at the rate of 15% for fuelling internal combustion engine light vehicles.)

Advantages and Disadvantages of Methanol as a fuel:

Advantages:

- Methanol has very low ozone-forming potential.
- Emissions of sulphur and sulphur compounds are virtually negligible.
- Very low evaporative emissions due to its low vapor pressure.
- Easy refuelling.
- Methanol has a high-octane quality that is an important factor for engines. The power, acceleration, and payload are comparable to those of equivalent internal combustion engines.
- Methanol is the most practical carrier of hydrogen to run fuel cell vehicles.
- *Pure methanol is also itself a very suitable fuel for certain types of fuel cells that can power vehicles.*
- Methanol and gasoline blends have been used in small vehicles.

Disadvantages :

- High formaldehyde emissions.
- Acute toxicity.
- Low energy content compared to gasoline and so the mileage will be slightly lower.
- Demands special lubricants and spare parts for the engine.
- Neat methanol (M-100) also presents a special safety hazard as it burns without a visible flame and even methanol-water wastes may be flammable.

Mixed Alcohols

The conversion of syngas to mixed alcohols is achieved using a catalyst similar to the FT process and methanol production using alkali metals as catalysts (like alkali molybdenum sulphide) to promote the production of mixed alcohols. The ratio of H_2 to CO required in syngas is 1.0–1.2. *The presence of a small amount of sulphur in syngas may be beneficial when alkali molybdenum sulphide is used, but when other catalysts are used, sulphur oxides or its other compounds must be removed because of their poisoning effect.)* Ethanol and methanol are the main products in the mixed alcohol obtained, while higher alcohol forms a small percentage. *(Also see Para 3.7).*

Alcohols through Fermentation of Syngas

The syngas can be fermented to alcohols under atmospheric conditions with anaerobic bacteria such as *'Clostridium ljungdahlii'*. The process of transforming biomass into ethanol has already been described in *Para. 3.4,* which may also be referred.

5.10.2 Renewable Gasoline from Biomass [27,235]

Gasoline can be produced from syngas by first converting it to methanol, then to dimethyl ether (DME), which is then further dehydrated with a zeolite catalyst (ZSM-5). This catalyst has high selectivity for molecules in the gasoline range (C_4–C_{10}).

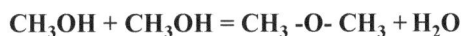

$$CO + 2H_2 = CH_3OH$$

$$CH_3OH + CH_3OH = CH_3\text{-}O\text{-}CH_3 + H_2O$$

By further dehydration of DME

$$n(CH_3\text{-}O\text{-}CH_3) \rightarrow \text{Gasoline } (C_nH_{2n+2}) + \text{water}$$

The conversion of biomass to gasoline is shown in Figure 5.4.

This gasoline is a renewable hydrocarbon biofuel and is called 'biogasoline,' as it is produced from plant-based material. (See Para 5.11.6 ahead).

5.10.3 Furans from Biomass

Heterocyclic compounds/chemicals known as 'furans' such as 5-Hydroxy methyl furfural (HMF) can be produced from 'untreated lignocellulosic biomass' with the help of the compound known as N, N-Dimethyl acetamide (DMA) containing lithium chloride (LiCl) as the solvent. This is a special 'solvent' that enables the synthesis of 'HMF' in a 'Single Step'. In this process, there is high yield from lignocellulosic biomass (untreated) as well as from purified cellulose. The conversion of cellulose into HMF is not affected by the presence of other components of biomass like lignin and protein.

HMF is a 'platform chemical' from which many other fuels and chemicals can be made. HMF can be converted to other products like '2.5-Dimethylefuran' (DMF). The DMF is a fuel with an energy content of 31.5 MJ per liter and its boiling point is 92–94°C, which is higher than that of ethanol (78°C). The compound is not miscible with water.

5.10.4 Non-Enzymatic Conversion Process of Lignocellulosic Biomass to Furans [23]

It is a newly discovered process which uses cheap catalysts, while not using enzymes. The process takes place at temperatures < 250°C and the catalyst used is chromium chloride ($CrCl_2$ or $CrCl_3$). The recycling and recovery of chemicals is done, which is very advantageous both from the point of view of the environment and economics.

The synthesis of 5- Hydroxy Methyl Furfural (HMF) from cellulose and lignocellulosic biomass uses the materials mentioned below:

- Biomass: Corn stover, say with cellulose content of about 34%.
- Solvent: N,N-dimethylacetamide (DMA) and lithium chloride (LiCl) plus hydrochloric acid (HCl) mol. 10% (a solvent from Cologne, Germany).
- Catalyst: $CrCl_3$ and HCl mol. 10%.
- Additives: 1-ethyl-3-methylimdazolium chloride (EMIM)Cl 20 to 60% of the weight of the solvent.

The time of reaction of synthesis is 1 to 3 hours at a reaction temperature of 140°C. Based on the above, the yield of 'HMF' is 23 to 48%. Under the best conditions, 42% of dry weight of the cellulose is converted to HMF and 19% of the dry weight of 'corn stover' is converted into HMF and Furfural in one step. HMF and Furfural products contain 43% of combustion energy available from 'cellulose and xylan' in 'corn stover' which is the starting material.

The biomass components like 'lignin' which cannot be converted to HMF but can be reformed to produce 'hydrogen' or can be burnt to provide process heat. [23]

5.11 Dimethyl Ether (DME) [17,28,53,57,58,59,62,65,66,68,88,100,108,110, 116,117,119,143,158,162,190,206,231,230,232,233,238]

Dimethyl ether is a synthetic fuel. It is gaseous at ambient conditions, but can be liquefied at moderate pressure and can be handled as LPG. Dimethyl ether is also called by alternative names such as: 'methoxymethane', 'wood ether', 'dimethyl oxide', or 'methyl ether'. It is a very simple ether, and is a colourless, slightly narcotic, non-toxic, highly flammable gas at ambient conditions. It is a substitute fuel for the internal combustion engine.

5.11.1 *Properties of DME* [57,62,68,190,206]

The properties of dimethyl ether are given in Table 5.6, along with how they compare to petroleum diesel. DME has a carbon to hydrogen ratio of 1:3, and therefore can be a good potential carrier of hydrogen onboard for vehicles. Dimethyl ether has half the energy of petroleum diesel and requires engine modification for its use as fuel.

Table 5.6. Comparison of Physical and Chemical Properties of DME and Diesel.

Property	DME	Diesel
Preferred IUPAC name and other names	*Wood ether, methoxymethane, dimethyl ether, dimethyl oxide*	Petroleum diesel *standard EN590 petroleum*
Chemical Formula	C_2H_6O or $CH_3 - O - CH_3$	$(C_{10} - C_{21})$ **Hydrocarbons**
Molecular weight	46.07	148.6
Carbon content mass%	52.2	86
Oxygen content mass%	34	80
Hydrogen content mass%	13	14
Carbon-to-hydrogen ratio	0.337	0.516
Appearance	Colorless	Brownish
Odor	Typical	Typical
Boiling point	−24.9 °C	71–193°C
Critical temperature	127°C	-
Critical pressure	53.7 bar	-
Liquid density at 20°C	0.67 kg/L	0.83 kg/L
Surface tension (at 298 K)	0.012 N/m	0.027
Flash point	−41°C	32°C
Autoignition temperature	350°C	210°C
Explosive limits	LEL 3.4%, UEL 27%	LEL 0.6%, UEL 7.5%
Melting point	−141 ; °C	
Boiling point	−24°C	188–343°C
Vapor pressure	>100 kPa	
Specific heat capacity	65.57 J K^{-1} mol^{-1}	
Standard enthalpy of formation ($\Delta_f H^\circ_{298}$)	−184.1 kJ mol^{-1}	Negligible
Standard enthalpy of combustion ($\Delta_c H^\circ_{298}$)	−1.4604 MJ mol^{-1}	-
Liquid viscosity	0.15 cP	2–4 cP
Kinematic viscosity of liquid	< 1 cSt	3
Ignition temperature at 1 atm.	235°C	250°C
Vapor Pressure at 20°C	-	< 0.098 Pa
Cetane number	55–60	40–55
Enthalpy of vaporization	467.13 kJ/kg	300 kJ/kg
Lower heating value	28430 kJ/kg	42500 kJ/kg
Stoichiometric A/F mass ratio	9.0 kg/kg	14.6 kg/kg
Explosion limit in air-vol.%	3.4–17	1.0–6.0
Bulk modulus	6.37 E+08 N/m²	1.49 E+09 N/m²
Flash point	−41°C	52°–96°C

Dimethyl ether (DME) is degradable in the atmosphere and is not a greenhouse gas. DME ($CH_3 - O - CH_3$) has only C–H and C–O bonds and no C–C bond with high oxygen content (34.8%), which results in low smoke and PM emissions. Cetane number is a measure of the fuel's ability to autoignite. The higher cetane number of DME (than diesel fuel) leads to faster ignition and shorter ignition delay, which in turn lowers the premixed burning of the fuel, resulting in lower emission of oxides of nitrogen (NOx) and also lower combustion noise than diesel.

DME vaporizes easily in the cylinder when injected in the liquid phase, due to its low boiling point leading to better atomization, improved combustion, and easier starting of the engine during cold weather conditions. The viscosity of DME is very low (about 10% of diesel fuel) and therefore, it is easier to inject fuel in the engine cylinder during cold weather conditions.

At low concentrations in air (a few percent by volume), DME has hardly any odor and doesn't cause any negative health effects. Even at high mole fractions (> 10% by volume), it has no effect on human health except narcotic effects after a long exposure, but at this concentration, it can be recognized by its odor. DME displays a visible blue flame similar to natural gas when burning over a wide range of air-fuel ratios, which is an important safety characteristic. The exhaust from the combustion of DME has low reactivity.

Some engine modifications are necessary for its use; primarily relating to the injection pump and the installation of a pressure tank, similar to that for LPG. The fuel line must also be adapted with specific elastomers.

It is anticipated that DME will require a lubricity additive due to its low inherent lubricity. Further, it is anticipated that original equipment manufacturers (OEMs) may specify an engine-cleaning additive. It would also require an odorant such as ethyl mercaptan. There is also a possibility that an additive to increase flame luminescence may be needed.

DME's handling requirements are similar to those of propane—both must be kept in pressurized storage tanks at ambient temperature. The use of DME in vehicles requires a compression ignition engine with a fuel system specifically developed to operate on DME. There have been a number of DME vehicle demonstrations in Europe and North America, including the one in which a customer operated 10 vehicles for 750,000 miles.

5.11.2 *Characteristics of DME as a Fuel* [57,62,65,68,190,206]

a. Diemethyl ether (DME) is a colorless clear liquid under 6 atm, but it is a gas at ambient temperature and atmospheric pressure.

b. It can be transported and stored as a liquid at low temperatures.

c. There are no 'sulphur or nitrogen compounds' in DME.

d. DME vapor causes eye, nose, and throat irritation.

e. Prolonged exposure to DME can cause headache, dizziness, and even a loss of consciousness

f. On contact with skin, DME liquid causes symptoms of frostbite.

g. DME is highly flammable liquid or gas and a dangerous fire hazard. It can form explosive peroxides under the influence of light and air. On combustion, it forms irritating fumes. It reacts with oxidants.

h. At a low concentration in air (a few percent by volume), it has hardly any odor.

i. It's oxygen content is about 34% and therefore, promotes smokeless combustion.

j. It can be used in conventional diesel engines with a modified fuel injection system.

k. It has a high cetane number.

l. DME vaporizes easily in the cylinder when injected in the liquid phase due to its low boiling point leading to better atomization, improved combustion, and easier starting of the engine during cold weather conditions.

m. The viscosity of DME is very low (about 10% of that of diesel fuel) and therefore, it is easier to inject fuel in the engine cylinder during cold weather conditions.

n. It provides 100% reduction of SOx.

o. DME displays a visible blue flame similar to natural gas when burning over a wide range of air-fuel ratios. This is an important safety characteristic as it helps in flame detection.

p. DME has quite a high amount of hydrogen in its composition and therefore, can be a good potential carrier of hydrogen onboard for vehicles.

5.11.3 Advantages and Limitations Regarding DME Application as a Fuel [33,57,62,68,157,188,190,199,206]

Advantages of DME

a. Higher energy density than compressed natural gas (CNG).

b. Will not leak due to evaporation over time. It is heavier than air in gaseous form.

c. Does not need cryogenic storage; however, it should be kept under pressure in liquid form.

d. Ease of refuelling (no venting, no compressors).

e. Ultra-low emissions with simple emissions control.

f. It has a very high cetane number, which is a measure of the fuel's ignitibility in compression ignition engines. The energy efficiency and power ratings of DME and diesel engines are virtually the same.

g. Because of its lack of carbon-to-carbon bonds, using DME as an alternative to diesel can virtually eliminate particulate emissions and potentially negate the need for costly diesel particulate filters.

Limitations Regarding DME Application as a Fuel [57, 62, 68, 188, 190, 206]

Despite the advantages of DME as a fuel, there are certain limitations which should be taken into consideration for its use as a substitute fuel. These limitations are outlined below:

a. Due to its lower calorific value and lower density than petroleum diesel, a higher volumetric flow rate of DME would be required than the diesel fuel for the energy input.

b. DME has half the energy density of diesel fuel, requiring a fuel tank twice as large as that needed for diesel.

c. Due to its low vapor pressure and low boiling point, a pressurized fuel system is required to maintain it in the liquid state.

d. Due to the low viscosity of DME, leakages have to be more precisely managed.

e. DME can cause severe wear of moving parts within the fuel injection system due to its low lubricating property. The conventional fuel delivery/fuel injection systems are therefore not compatible with dimethyl ether. DME is also corrosive.

f. Due to its wide flammability limits, the operation of the DME combustion system needs the adoption of rigorous procedures for its safe operation.

g. Comprehensive research would be required for DME as an alternative fuel for internal combustion engines, especially in the areas of: fuel storage, fuel supply, and injection systems to overcome high internal leakage problems, keeping DME in its liquid state, and high volume flow rate requirement.

h. The DME engine is costlier.

5.11.4 *Environmental Friendliness of DME* [10,62,116,158,231,230,232,238]

Important concerns related to any fuel used for transportation or cooking and heating are the potential environmental and human health impacts of the use of the fuel. In the case of DME, there are no adverse issues with regard to human or animal exposure; it is an environmentally friendly fuel. On burning, DME does not emit any sulphur dioxide. The NOx content of exhaust gases is low and the content of particulate matter is ultra-low.

DME was first used as an aerosol propellant because of its environmentally benign characteristics. It is not harmful for the ozone layer. DME producer DuPont Fluorochemicals (which markets DME under the product name "Dymel A"), provides a technical bulletin that gives a good overview of the physical and chemical properties of DME, and the results of their own health and safety studies.

'A two-year inhalation study and carcinogenicity bioassay at exposure levels of up to 20,000 ppm showed no compound-related effects, no signs of carcinogenicity, and no evidence of mutagenicity or teratogenicity in separate reproductive studies. Based on all these studies, the product has been approved by the Dupont Company for general aerosol use, including in personal products.'

DME is one of the most promising alternative automotive fuel solutions among the various ultra-clean, renewable, and low-carbon fuels under consideration worldwide. It is being projected as future alternative fuel and may potentially be used as:

- Substitute for diesel fuel in IC engines
- Transportation fuel
- Power generation fuel
- Domestic gas

Use of LPG Blended with DME for Cooking and Heating

DME can be blended in LPG and used for domestic cooking. Heating-blends containing up to 20% volume DME generally require no modifications to equipment or distribution networks. The blended gas for cooking and heating will be a safer, cleaner, and more environmentally benign fuel for the consumers.

As portable, bottled LPG is being used in developing countries as a cooking fuel and the LPG market is sharply growing, blending DME in LPG will bring about fast growth in DME use and will provide a ready market for DME so that it can be produced at a large scale at a fairly consistent lower cost. In China, facilities that make methanol from coal gasification are being converted to produce DME by adding a methanol dehydration step to the methanol plant's process.

5.11.5 *Blending of DME in Diesel for Use in Vehicles* [17,32,62,108,110,116,158,193,231,230,232,238]

The exclusive use of DME in vehicles may be difficult at present because of the limitations already pointed out (in *Para 5.10.3*), unless these problems are sorted out. But in order to use this clean fuel (DME), it can be blended in diesel in certain proportion for use in vehicles. The blending of DME in diesel fuel changes its physicochemical properties. The increase of the DME mass fraction decreases the density, calorific value, kinematic viscosity, and aromatic fractions of the blends, while the C/H ratio, cetane number, and oxygen content of the blends are enhanced, which improve the performance and emission characteristics of the engine. The blended fuel for the vehicles will be cleaner than pure diesel and would retain the desirable physical properties of diesel. The relative benefits of both fuels would be realized to some extent without much modification to the engine. The emission of smoke as well as NOx and oxides of sulphur would be lower.

Studies are being conducted on engine performance with blended fuel of different proportions such as 10% DME, 20% DME, 30% DME (by mass) in diesel. Recent research and development shows that blending DME with diesel as an alternative fuel for CI engine and its application in automotive vehicles is a promising solution for the future.

Only modest modifications are required to convert a diesel engine to run on DME, and engine and vehicle manufacturers, including Volvo, Mack, Isuzu, Nissan, and Shanghai Diesel, have developed heavy vehicles running on diesel engines fuelled with DME. As a replacement for diesel fuel, DME is particularly advantageous when compared to the environmental problems created by the use of petroleum diesel.

DME-blended fuels can be used as fuel in diesel engines, gasoline engines and gas turbines.

Note: R & D Efforts—In 2013, Pennsylvania State University, Volvo, and Oak Ridge National Laboratory completed the field testing of a prototype DME truck. The heavy-duty truck performed well under real-world driving conditions, achieving comparable efficiency to a conventional diesel truck. Test results indicated that particulate matter emission standards could be met without the use of a diesel particulate filter. As with conventional diesel vehicles, oxides of nitrogen (NO_x) emissions reductions can be handled with standard NO_x aftertreatment systems. Alternatively, the engine can be calibrated in such a way to negate the need for such a system, but this reduces efficiency. There have been a number of DME vehicle demonstrations in Europe and North America, including one in which a customer operated 10 vehicles for 750,000 miles. [46,62,116,172, 223,229]

5.11.6 *Principle of Production of DME from Lignocellulosic Materials* [13,16,19,47,49,51,67,104,129,137,161,165,221,227,228,240]

DME can be produced from organic waste or biomass, natural gas or coal/lignite, which are gasified and converted to clean synthesis gas (syngas). The DME (CH_3-O-CH_3) can then be synthesised from syngas in two ways, that is, indirect synthesis or direct synthesis. Both of these processes have been commercialized.

The principle of conversion of biomass to DME is shown in Figure 5.6.

(a) *Indirect synthesis*—In this method, the syngas is first converted to an intermediate product '*methanol*' in the presence of a catalyst (for example, *copper-based or others*), which then gets converted to dimethyl ether via dehydration. This requires the stoichiometric ratio in syngas of **(H_2-CO_2) to (CO + CO_2)** to be more than 2, and the use of silica-aluminium-based catalyst. Since methanol is produced as an intermediate product, this method has the flexibility of varying liquid products (methanol/dimethyl ether) as required.

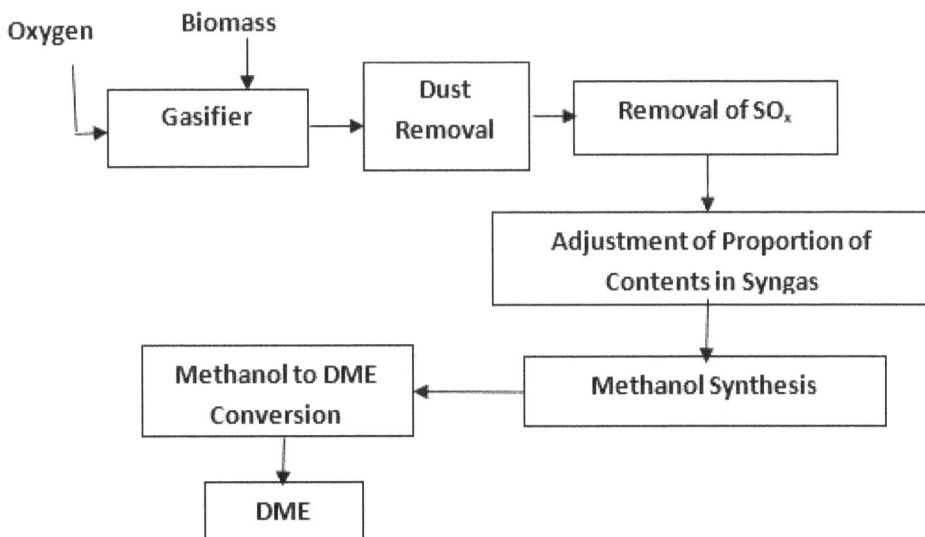

Fig. 5.6. Conversion Process of Biomass into Dimethyl Ether (DME).

The *dehydration reaction* for the production of dimethyl ether would be as follows:

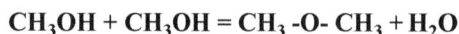

$$CH_3OH + CH_3OH = CH_3 \text{-O-} CH_3 + H_2O$$

Figure 5.6 shows the flow diagram of the process for conversion of biomass to dimethyl ether.

(b) Direct synthesis—This type of synthesis is a newer development. This method requires the stoichiometric ratio in syngas of (H_2-CO_2) to $(CO + CO_2)$ as 1 and the use of a mixed catalyst (dual catalyst system), which permits both methanol synthesis and dehydration in the same process unit without the separation of methanol synthesis stage. This process has higher conversion efficiency. [161,165,221]

Note on DME Production

The majority of global DME production is currently in China. Japan has production facilities, and major new capacity additions are being planned or are under construction in Trinidad and Tobago, North America, Indonesia, and Uzbekistan. The world's first bio-DME plant is in Sweden.

Although dimethyl ether can be produced from biomass, natural gas and syngas from coal may be chosen by most countries, at present. The likely feedstock of choice for large-scale DME production in the United States is natural gas. DME is not commercially available in the United States at this time, but one company is developing small-scale plants to produce DME from natural gas in quantities ranging from 4,500 to 10,000 gallons per day.

Further information regarding DME is available on links like:

- *Process Design and Economics for the Conversion of Lignocellulosic Biomass to Hydrocarbons via Indirect Liquefaction: Thermochemical Research Pathway to High-Octane Gasoline Blend-stock Through Methanol/Dimethyl Ether Intermediates(PDF).*

- *Emissions and Performance Benchmarking of a Prototype Dimethyl Ether-Fuelled Heavy-Duty Truck(PDF).*

- *California Dimethyl Ether Multimedia Evaluation: Final Tier I Report(PDF)*
  The AFDC also provides 'publications search' for more information.

5.12 Emerging Alternative Fuels [27,40,64,155,235]

The emerging alternative fuels for vehicles include:

- Biobutanol
- Methanol
- Renewable hydrocarbon biofuels
- Dimethyl ether

These fuels may increase energy security, reduce emissions, and improve vehicle performance. Some other fuels, such as ammonia, may also meet the criteria for alternative fuels when used in limited quantities. More research is needed to characterize the impacts of these fuels, such as necessary vehicle modifications, required fuelling infrastructure, human health impacts, greenhouse gas emissions, and tailpipe emissions. The Biofuel Policy and Renewable Fuel Standard (US) has been briefly mentioned in Para 1.3.5 and in the Note under it.

5.12.1 Biobutanol

Biobutanol is dealt with in detail in *Para 3.10 to para 3.12* in *Chapter 3*. As mentioned previously, butanol, when produced from plant-based material/ biomass, is termed as 'biobutanol'. At present, butanol is being commonly produced using fossil fuels. Biobutanol is a renewable fuel and qualifies under the Renewable Fuel Standard; the category depends on the feedstock used for production. Though the exclusive use of biobutanol is yet to be established, it can be blended with gasoline and used as a transportation fuel. Butanol can be produced from the same feedstock as ethanol.

Biobutanol manufacturers produce a range of high-value products, including transportation fuel. The primary co-products of biobutanol plants may include solvents/coatings, plastics, and fibers. Production of these co-products helps bio butanol manufacturers improve their economic performance through the diversification of product offerings. A challenge that needs to be met is the production of more biobutanol than ethanol from the same feedstock.

The energy content of butanol is much higher than that of ethanol or methanol and is just less than that of gasoline. The heat of vaporization of butanol is lower (about half) than that of ethanol; it would be easier to start the engine in cold weather conditions with butanol as fuel than ethanol or methanol. Butanol has a lower '*Reid vapor pressure' when compared with ethanol, resulting in a reduction in evaporative emissions.

Research and Development

The US Department of Agriculture's Agricultural Research Service and the National Centre for Agricultural Utilization Research/Bioenergy has taken up research project #417041 with the long-term goal of developing lignocellulosic materials as a feedstock for producing sugars that can be converted to biofuels. The research focuses more specifically on the following objectives:

Objective 1: Develop commercially viable analytical tools that producers and biorefiners of lignocellulosic feedstocks can use to evaluate the quality of harvested biomass for enzymatic and fermentative conversion to ethanol or butanol and that plant breeders can use to select superior cultivars (seedlings for producing superior plants) for biorefineries.

Objective 2: Develop new, commercially viable enzymes and/or protein systems that increase the efficiency of lignocellulosic saccharification.

Objective 3: Identify components in lignocellulosic hydrolysates that reduce saccharification or fermentation efficiencies and develop commercially viable mitigation strategies.

Objective 4: Develop new technologies that enable commercially viable pretreatment processes for lignocellulosic biomass, inhibitor-abatement strategies, and enzyme preparations that are optimized for saccharification, particularly of lignocellulosic feedstocks.

(1) *Note on Reid vapor pressure (RVP)

RVP is a common measure of the volatility of gasoline. It is defined as the absolute vapor pressure exerted by a liquid at 37.8°C (100°F) as determined by the test method ASTM-D-323. The test method measures the vapor pressure of gasoline, volatile crude oil, and other volatile petroleum products, except for liquefied petroleum gases. RVP is stated in kPa and represents a pressure relative to the atmospheric pressure because ASTM-D-323 measures the gauge pressure of the sample in a non-evacuated chamber.

The matter of vapor pressure is important relating to the function and operation of gasoline-powered, especially carbureted, vehicles. High levels of vaporization are desirable for starting and operating vehicles in winter and lower levels are desirable for avoiding vapor lock during hot summers. Fuel cannot be pumped when there is vapor in the fuel line (summer) and starting vehicles in winter will be more difficult when liquid gasoline in the combustion chambers has not vaporized. Thus, oil refineries manipulate the Reid vapor pressure seasonally specifically to maintain gasoline engine reliability.

The Reid vapor pressure (RVP) differs slightly from the true vapor pressure (TVP) of a liquid due to small sample vaporization and the presence of water vapor and air in the confined space of the test equipment. That is, the RVP is the absolute vapor pressure and the TVP is the partial vapor pressure.

(2) Further Note on Biobutanol

As already mentioned earlier, there are four isomers of butanol, but most of the commercialization work focuses on blending isobutanol with gasoline. In the US, there are two Clean Air Act provisions that allow for the blending of up to 12.5% biobutanol with gasoline. The fuel quality standard for biobutanol—ASTM D7862—also allows for a biobutanol blend of up to 12.5% with gasoline. The Octamix waiver allows for a 16% biobutanol blend to be used as a legal fuel equivalent to E10 (10% ethanol, 90% gasoline); however, the testing for effects on human health for the fuel is ongoing. It is important to ensure that biobutanol blended with ethanol gasoline combinations do not result in an oxygen content exceeding the US Environmental Protection Agency's limit of 3.7%.

Currently, major refiners cannot sell biobutanol blends for on road use in US because it is not registered under 40 CFR (US Code of Federal Regulations, Part 79) for companies with revenues exceeding $50 million, as required by the Clean Air Act. However, an additive manufacturer with sales of less than $50 million has sold biobutanol in recent years to specialty markets including marine, jet fuel, and also for conversion to para-xylene, a precursor to a plastic product.

The Oak Ridge National Laboratory has researched the compatibility of fuelling equipment materials with biobutanol and found that equipment compatible with ethanol blends is compatible with biobutanol. Underwriters Laboratories announced in 2013 that equipment certified under testing subject 87A (for blends above E10) could also retain their certification if used with biobutanol. It is anticipated that biobutanol would be distributed by tanker truck and rail, with the potential for transportation in pipelines once research demonstrates its safety.

5.12.2 *Methanol and Fuels Blended with Alcohols* [40,64,106,105,152,94]

As an engine fuel, methanol has chemical and physical fuel properties similar to ethanol. Methanol is a very suitable fuel for fuel cell powered vehicles (it has already been dealt with in Para 5.9.1 (i)). The toxicity and solubility of methanol in water raises concerns about its safe storage and distribution. Although methanol-fuelled vehicles have the advantage of refuelling in a similar way to gasoline or diesel vehicles, adequate training is required by the staff in handling storage of methanol, and maintenance and operation of methanol-fuelled vehicles. It may be mentioned here that methanol (CH_3OH) is considered an alternative fuel under the US Energy Policy Act of 1992.

A. Examples of Some Field Trials and Results for Methanol-Blended Gasoline Fuel. [94]

(i) *Indian Institute of Petroleum (IIP), Dehradun,* in 1983–1986, conducted a fleet trail with M12 under the UNDP/UNIDO-assisted programme on alternate fuel. The M12 gasoline blend (12% methanol with gasoline) was used as the fuel in this study on 14 two-wheelers of various makes. IIP also carried out some experimental studies on the two-wheelers with methanol gasoline blends up to 20%.

The findings of the studies are summarized below:

- Operation beyond or over 15% methanol-gasoline blends were erratic and the engine started misfiring and hunting.
- With M15, it showed marginally better output and a 3–4% improvement in fuel consumption.
- It established that up to 15% blends can be used without any engine modifications.
- A substantial reduction in carbon monoxide was recorded.
- No hot or cold driveability problems with 12% methanol gasoline blend (M12 blends).
- The engine performance of M12 vehicles was found comparable to that of gasoline vehicles.
- An increase of wear with the use of cast iron rings was observed. In the year 1992, the BIS standard was amended to facilitate the use of methanol in gasoline.

(ii) Indian oil companies initiated a pilot scale project in November 1993 to market a methanol-petrol blend of 3% methanol called 'Petrol-M'. This product was supplied from 10 selected retail outlets in the city of Baroda in Gujarat. Initially, this project was taken up for a period of one year.

The key findings of this project were:

- Blending, transportation, and quality wise Petrol-M trail marketing was successful.
- A total quantity of 376 kL of the product was sold.
- It was comfortably used by cars, two-wheelers, and three-wheelers.
- To tackle the apprehended problem of the corrosive effect of methanol on engine parts, the oil industry used corrosion inhibitors.

(iii) Central Pollution Control Board (CPCB), India also commissioned a study to evaluate the emission performance of methanol-gasoline blends through IIP sometime in 1995–1996. The CPCB estimated that if all the petrol-driven vehicles in Delhi use a methanol-gasoline blend of 3% methanol and 97% gasoline, it may be possible to have an 11% reduction in hydrocarbons emissions, 7% in CO, and 30% in NOx compared to pure gasoline-driven vehicles.

B. Examples of Field Trials and Results of Alcohol-Blended Diesel Fuel (Diesohol)

Apart from ethanol/methanol-gasoline blends, an ethanol/methanol-diesel blend is also another alternative option. The blended diesel with alcohols is termed as 'diesohol'. Unlike the ethanol-gasoline blend, the ethanol-diesel blend has some concerns regarding lubricity, reduced flash point, and startability problems.

Ethanol-diesel blend projects are under trial in Brazil and Sweden. Several technologies are currently under trial in different parts of the world. An Australian non-profit organization, 'APACE', has developed an ethanol-diesel emulsion agent which is under trial in Australia. This diesohol technology claims successful blends up to 15% of ethanol in diesel. APACE claims that this emulsion, which allows the use of hydrated ethanol with diesel, not only reduces the emission of NOx, particulate matter (PM) (including PM2.5), and hydrocarbons, but also increases the thermal efficiency of the engine. The vehicles can use diesohol and diesel fuels interchangeably. The fire point of the diesohol is higher than the flash point and the magnitude of the difference depends upon the composition of the diesel fuel. The Australian Government is currently in the process of developing fuel quality and operability standards for diesohol.

C. Issues to be Resolved for the Commercial Use of Diesohol as Fuel

There are still several issues that need to be resolved before diesohol can be introduced commercially. Some of the important ones are:

1. Vapor lock: The use of ethanol changes the vapor lock characteristics of the fuel.

2. The material compatibility of pump seals, timing, belts, and some nitrile rubber seals used in the fuel injection systems of a vehicle are still a concern.

3. Toxicity of the emulsifier.

4. Dosage quantities of the emulsifier, which may vary from base fuel to base fuel. Finding the optimum mixtures of emulsifier and proportions of alcohol and diesel.

5. Blend stability, especially at low temperature.

6. Establishment of diesohol test standards.

7. Stability of fuel with water addition.

Indian Initiative on Use of Diesohol: [94]

It is interesting that India was one of the earliest countries to recognize the merits of burning ethanol in diesel engines. The biofuel system developed by German Professor H.A. Havemann and his colleagues at the Indian Institute of Science (IISc), Bangalore, in the early 1950s is the earliest original published work in technical literature regarding alcohol diesel.

Ministry of Petroleum & Natural Gas, India, is working on introducing a 5% ethanol-diesel blend. For this purpose, Indian Oil Corporation (IOC) has selected some vehicle models to carry out the trial tests. IOC (R&D) has also made a recommendation to the Society of Indian Automobile Manufacturers (SIAM) to carry out the tests with a 5% ethanol-diesel blend to be supplied by the IOC. However, the Bureau of Indian Standards (BIS) specification does not permit the blending of ethanol and diesel at present. (*The Bureau of Indian Standards are total adoption of ISO standards.*) Sufficient field trials in Indian conditions need to be carried out and its benefits on emissions, material compatibility, and drivability, etc., may be assessed before trying this fuel.

There are some field trials data on methanol-diesel fuel. The Indian Institute of Petroleum (IIP) has developed a retrofit kit for the dual-fuel operation of diesel vehicles with alcohols. IIP used the fumigation concept for this purpose and has successfully demonstrated this system on Maharashtra state SRTC and DTC diesel buses under actual commercial passenger service. The fleet consisted of 25 & 35 buses of Tata and Ashok Leyland, respectively, with a total cumulative operation of 42,00,000 km (Methanol used consisted of 10% gasoline to impart flame luminosity due to safety concerns as the methanol flame is almost invisible in sunlight).

The findings of the study of IIP fleet on methanol-diesel fuel with 60 buses with cumulative total operation 42,00,000 km were as follows:

1. Diesel Replacement--------15–20%,

2. Fuel Consumption and Energy Efficiency----- Comparable

3. Better Smoke Reduction -------25–40%,

4. Drivability -----------Good-to-Better

5. Oil Consumption & Degradation --------Comparable

6. Engine Wear------- Lower-to-Comparable

7. Engine Deposits------ Lower-to-Comparable

8. Sludge Deposits------- Marginally Lower

9. Material Compatibility--------Adequate but Rubber components partially hardened

From the above trials, it was concluded that 15–20% substitution of diesel by alcohols is possible by a simple retrofit fumigation system, but more field trials are required on new- generation diesel engines to assess the technical feasibility of the fuel. [169]

5.12.3 *Renewable Hydrocarbon Biofuels and Dimethyl Ether* [40,64,106,105,160,22,98,169,95,196,120,6,201,200,118,14,121,122,195,226]

Renewable hydrocarbon biofuels (also called by various names like green hydrocarbons, 'biohydrocarbons drop in biofuels,' and 'sustainable or advanced hydrocarbon biofuels') are fuels produced from biomass sources through a variety of biological and thermochemical processes. These products are similar to petroleum gasoline, diesel, or jet fuel in the chemical makeup and are therefore, considered infrastructure-compatible fuels. They can be used in vehicles without engine modifications and can utilize existing petroleum distribution systems.

The renewable hydrocarbon biofuels include:

1. **Renewable gasoline**: It is also known as biogasoline or 'green' gasoline. It is a biomass-derived transportation fuel suitable for use in spark-ignition engines. It should abide by the specifications of countries where produced, such as the ASTM D4814 specification in United States, EN 228 in Europe, etc.

2. **Renewable diesel**: It is also called 'green' diesel. It is a biomass-derived transportation fuel suitable for use in diesel engines. It should abide by the specification of country where produced, such as ASTM D975 specification in United States, EN 590 in Europe, etc. *In this connection, it is noteworthy that* **renewable diesel** *is distinct from biodiesel. While renewable diesel is chemically similar to petroleum diesel, biodiesel is a monoalkyl ester, which has different physical properties and hence, different fuel specifications (ASTM D6751 and EN 14214). The production processes of biodiesel and renewable diesel are different. Biodiesel is produced via transesterification of vegetable oils/fats or algal oil. Renewable diesel is produced through various processes such as hydrotreating (isomerization), gasification, pyrolysis, and other thermochemical and biochemical means.*

   A process of production of renewable diesel from lignocellulosic materials through gasification route by **Choren Industries, Germany** *has been described in Para 5.9.6.*

3. **Renewable jet fuel**: It is also called 'biojet' or aviation biofuel. It is a biomass-derived fuel that can be used interchangeably with petroleum-based aviation fuel if it meets the requirement of the relevant specification like ASTM D7566 specification.

Relevant Pathways for the Production of Renewable Hydrocarbon Biofuels:

(i) Pyrolysis or liquefaction of biomass to bio oil with hydroprocessing.

(ii) Production of syngas, up gradation, and cleaning of syngas and the Fischer-Tropsch (FT) process, hydrogenation are important pathways.

(iii) Hydro-processing of lipid feedstock

The commercial plants for the production of renewable diesel have been installed all over the world. Currently, over 5.5 billion liters of renewable diesel is produced globally and is forecasted to grow up to 13 billion liters in 2024. The global regulations and policies are playing a vital role in the development of the renewable diesel production. Additionally, several planned renewable diesel production capacities are coming up in the next five years to the global market. The principles of producing the renewable hydrocarbon fuels are already described earlier in *Para 5.9/ 5.9.1, Para 5.10/5.10.1/5.10.2* and *Figure 5.3*.

As renewable diesel is chemically the same as petroleum diesel, it may be used in its pure form (called '*R100*') or mixed/blended with petroleum diesel similar to biodiesel blending. A blend of 20% renewable diesel and 80% petroleum diesel is called R20, and a blend of 5% renewable diesel and 95% of petroleum diesel is called R5. California uses nearly all of the US-produced renewable diesel and imported renewable diesel mainly because of the economic benefits for its use under California's Low Carbon Fuel Standard.

5.12.4 Advantages of Renewable Hydrocarbon Biofuels [98]

There are many advantages of renewable hydrocarbon biofuels, which include the following:

- Renewable hydrocarbon biofuels are similar to their petroleum counterpart fuels and there are no problems of compatibility with the existing infrastructure built for petroleum-based fuels. The engine, which has been designed for petroleum-based fuels, can accept the renewable hydrocarbon fuel without any problem.

- Renewable hydrocarbon biofuels are replacements for conventional diesel, jet fuel, and gasoline. They provide more flexibility as multiple products can be made from various feedstocks and production technologies.

- Renewable hydrocarbon biofuels may be produced domestically by any country from a variety of feedstocks locally available and can increase employment opportunities. The production would also add to the energy security of the country of production.

- Carbon dioxide captured by growing feedstocks reduces overall greenhouse gas emissions by balancing CO_2 released from burning renewable hydrocarbon biofuels.

- It can also be used neat (100% renewable diesel) or can be blended with petroleum diesel.

- Renewable diesel has a higher cetane number, that is, between 70 and 95 that means, fuel ensures quick and complete ignition with improved efficiency, reduces NOx and PM emissions, and allows the smooth start and less noise.

- Renewable diesel can be stored for a long time with no deterioration in quality.

- Renewable diesel is odorless. It does not contain any aromatics or impurities.

5.12.5 Comparison of Fuel Properties: Renewable Diesel vs. Other Diesels [22,97]

The properties of renewable diesel are very similar to high-quality sulfur-free petroleum diesel. Earlier, the synthetic gas-to-liquids (GTL) diesel was viewed as the best choice for engines and exhaust emissions. Renewable diesel can offer the same compositional advantages and has a lot of similarities as GTL diesel, in addition to that, it is completely renewable. The low sulfur

Table 5.7. Typical properties of petroleum diesel, biodiesel, renewable diesel and Fischer-Tropsch diesel. [22,97]

Se. No.	Property	No. 1 Petroleum Diesel	Biodiesel FAME	Renewable Diesel	Fischer-Tropsch Diesel
1.	Carbon (wt %)	86.8	76.2	84.1	
2.	Hydrogen (wt %)	13.2	12.6	15.1	
3.	Oxygen (wt %)	0.0	11.2	0.0	
4.	Cetane Number (CN)	44.5–67	45–55	> 70	74–80
5.	Lower Heating Value(LHV) MJ/kg	43.1	37.2	43.7–44.5	43.5
6.	Density at 15	796–841	880	770–790	774–782
7.	Sulfur content mg/kg			< 5	< 5
8.	Flash point (°C)	54–148	100–180	> 59	71
9.	Cloud Point (°C)		10.5 – (–13)	–5– (–34)	
10.	Ash (wt %)	0.01	-	< 0.001	
11.	Viscosity at 40 (°C) (mm²/s)	1.9–4.1	2.9–11	2–4	2–4.5
12.	Oxidation stability		2–15 h	< 25 g/m³	0.8

Notes:

1. Standards and typical properties of petroleum Diesel #2, biodiesel, and Green Diesel Data are given in Energies 2019, 12, 809; doi:10.3390/en12050809.

2. Diesel (No. 2) fuel has a sulfur level between 15 ppm and 500 ppm. It is used primarily in motor vehicle diesel engines for on-highway use. It costs less at the pump. Diesel 2 has a lower volatility than Diesel 1 and has increased viscosity, which is a benefit in warmer weather as well as in engines that would typically run very hot.

3. No. 1 Petroleum Diesel: Diesel #1 has shorter ignition delay, added lubricants, added detergents, and other beneficial fuel additives. It is more volatile and flows more easily than Diesel #2, so it's more efficient at lower temperatures. It is free of paraffin wax and offers the best operability during the coldest winter. It has a higher cetane rating, which reduces maintenance requirements. Diesel 1 is the premium version of diesel fuel while Diesel 2 is just the standard version.

4. FAME (Fatty acid methyl ester) biodiesel is produced from the transesterification of vegetable oils and fats.

content, metal-free, and ash-free characteristics of renewable diesel make it a safer fuel for various applications. It possesses higher heating value as compared to other biofuels and higher energy content than its biodiesel form. Renewable diesel also shows excellent cold properties, that is, cloud point below –34.4 to –40°C (–30 to –40°F), which gives a high bio-mandate for blending ratios possible all over the year even in winter. Its density also remains nearly constant even at a low cloud point. The high cetane number and low-density values make it more suitable for compression ignition engines. Moreover, the flashpoint is also higher compared to biodiesel for safe storage and transportation. The typical properties of petroleum diesel, biodiesel, and renewable diesel are given in Table 5.7.

Overall, renewable diesel (RD) is a promising substitute for diesel. Hydroprocessing is a technique used in producing RD. The triglycerides in vegetable oil are transformed into a fuel having a molecular structure similar to that of petro-diesel. RD is used as a "drop-in" fuel, with a high cetane index and good cold flow properties. It can be developed as a sustainable alternative to petro-diesel in a long-term scenario. This may result in reducing the greenhouse gas (GHG) emissions.

Dimethyl Ether

Dimethyl ether (DME) has been dealt with in detail covering all aspects in ***Para 5.11*** of this chapter.

5.13 Summary

This chapter has been devoted to oil/biodiesel from algae, fungi, and lignocellulosic biomass. Biodiesel is a secondary renewable fuel. Oil-producing algae production in estuaries, saltwater lakes, backwaters, etc., may be taken for harnessing algae oil, which can be converted to biodiesel like any other vegetable oil. Some algae are very rich in lipids, which are a very good source of biofuel.

Algae can also be resource for alcohol. Oil yield per acre from microalgae may be thousands of times of oil yield per acre from soybeans. Algae can be harvested several times in a short timeframe and it can grow 20 to 30 times faster than food crops, but the harvesting of algae from large volumes of water is a messy and laborious job. Algal oil does not contain sulphur.

Growth parameters for algae are: white light (natural or artificial), pH value between 7 to 9 (best 8.2 to 8.7), mixing and aeration, temperature between 16 to 27°C (best 20 to 24°C), salinity range 12–40 g/liter (optimum range 20–24 g/liter). The effect of pH, light intensity, and phosphate on algal biomass & lipid content have been highlighted.

The growth of algae has five phases, namely: (i) lag or induction phase, (ii) exponential growth period, (iii) phase of declining growth rate, (iv) stationary phase, (v) crash phase or death phase.

Problems in growing algae in open ponds are: (i) It is difficult to protect oil-rich algae from invading organisms, (ii) algae produce oil best within a narrow range of temperature band-low temperature during night and excessive heat during the day are problems.

Algae can be produced very efficiently in photo bioreactors (PBR) under controlled conditions. A number of organizations are producing algae for oil. A number of algal strains have good growth and also oil content. Enhancing the lipid content of algae by nutrient limitations has been highlighted.

Several methods of harvesting algae have been described such as: algae from open pond, algae from photo bioreactors, applying polypropylene non-woven fabric membrane (PNM) for harvesting of self-flocculating microalgae *Scenedesmus obliquus* by filtration, harvesting by coagulation and flocculation methods using polymeric composites.

Several methods of algal oil extraction have been described such as: mechanical extraction, solvent extraction, enzymatic extraction, extraction by gasification, ultrasonic extraction (by cell wall rupture), and oil extraction by using supercritical CO_2 (for cell rupture by pressure swing).

Fungi as a source of oil has been described in detail, giving the advantages, lipid content of some strains, conditions for the increase in the lipid content of fungi, occurrence of oil-bearing fungi, oil extraction from fungi by solvent extraction, and oil extraction under supercritical conditions.

Conversion of lignocellulosic material to liquid fuel has been dealt with in detail. Several methods have been described namely: (i) oil by the biochemical route from lignocellulosic material; (ii) liquid fuel from biomass by fast pyrolysis; (iii) direct thermo-chemical liquefaction of algal biomass; (iv) liquid fuel from algal biomass using catalytic agents (algal hydrogenation); (v) Fischer-Tropsch process for converting syngas to liquid fuels.

The Fischer-Tropsch process involves a series of chemical reactions that led to the production of a variety of hydrocarbons from purified syngas, which is a mixture of hydrogen and carbon monoxide. The following type of chemical reaction gives 'alkanes'. [Alkanes are saturated paraffin hydrocarbons.]

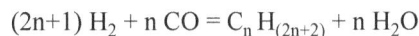

$$(2n+1) \, H_2 + n \, CO = C_n H_{(2n+2)} + n \, H_2O$$

Where 'n' is a positive integer. The formation of methane (n = 1) is generally unwanted.

Many catalysts can be used for FT (Fischer-Tropsch) process, but commonly, cobalt and iron are used for the production of liquid fuels.

The method of producing Renewable Diesel from lignocellulosic materials through the gasification route by Choren industries, Germany has also been described.

Dimethyl Ether (CH_3-O-CH_3)—a synthetic fuel, which may be the future choice as the fuel for vehicles has been dealt with in detail. The properties of dimethyl ether (DME), methods of its production from lignocellulosic biomass including direct and indirect synthesis, the use of diesel blended with DME in vehicles, and present limitations in its exclusive use have been highlighted.

Renewable hydrocarbon biofuels, their relevant pathways for production, and their advantages are described. The comparison of the properties of renewable diesel with other diesel fuels is also given. Renewable diesel is a promising substitute for diesel. It can be developed as a sustainable alternative to petro-diesel in a long-term scenario. This may result in reducing the greenhouse gas (GHG) emissions.

QUESTIONS

1. *What is photosynthesis?*

2. *What are algae?*

3. *What is a photobioreactor/bioreactor?*

4. *How can algae help reduce greenhouse gases such as CO_2?*

5. *How much CO_2 will be consumed in a process of growing one tonne of algae?*

6. *How is fermenting equipment different from photo bioreactors?*

7. *How much algae can be grown on one acre of land?*

8. *If you want to grow more algae using sunlight, can't you increase the volume of the growing system over the same area?*

9. *Which method of algae for biofuels production (plastic bags, open ponds, tubular and flat panel bioreactors) is the most promising?*

10. *Can algae be grown in the dark using bioreactors?*

11. *Is it possible to harvest wild algae for biofuels?*

12. *Why use algae as a biofuel feedstock?*

13. *How much oil can be made from one tonne of algae?*

14. *What types of algae can be used for biofuels? What should be the characteristics of an ideal algal species for biofuel?*

15. *What factors need to be considered in selecting algae for biofuels?*

16. *What are the differences between petroleum diesel and biodiesel?*

17. *Is biodiesel from algae the same as biodiesel from terrestrial plants?*

18. *What are the advantages of using algae over terrestrial crops for biofuels?*

19. *Do you need freshwater to grow algae?*

20. *What are some of the useful products that can be made from algae?*

21. *Can't algae be harmful for our environment? If so, why do we want to grow more of it?*

22. *Can artificial lights be used to grow algae for biofuels?*

23. *What does carbon neutral mean?*

24. *Are biofuels bad for the environment? Does this include biofuels made from algae?*

25. *Can biofuels be produced from algae economically?*

26. *Is it true that farmed algae grown are often overtaken by "wild" algae after just a few days or weeks?*

27. *It seems scientists have known the "promise" of algae for decades. Why now are algae the 'next best thing'?*

28. *Can algae be utilized to make ethanol?*

29. *What is photolimitation?*

30. *What is photoinhibition?*

31. *What is PAR?*

32. *What is chlorophyll?*

33. *What is phytoplankton?*

34. *What is phycology?*

35. *Enumerate advantages of producing biodiesel from algae. Give examples of algae that have high oil content.*

36. *What are the metabolic pathways present in microalgae?*

37. *What are the growth parameters for algae?*

38. *What are the different phases in the growth of algae?*

39. *What are the main problems in growing algae?*

40. *Describe the harvesting of microalgae.*

41. *What are the characteristics of algal oil?*

42. *Describe the methods of oil extraction from algae.*

43. *How would the cell rupture pretreatment of microalgae help in oil extraction? What is the role of: (a) high pressure CO_2 and pressure swing in the operation; (b) ultrasonic waves?*

44. *What are the differences among photoautotrophic, photoheterotrophic (mixotrophic), and heterotrophic algae?*

45. *Write a short note on the Botryococcus species of algae.*

46. *Write a note on the cultivation of microalgae in domestic wastewater.*

47. *Give three relevant examples of technological developments in algae production.*

48. *What are the differences between the 'solvent extraction' and 'gasification' methods of oil extraction from algae?*

49. *Write short note on the 'Production and harvesting of Algae as a source of oil'.*

50. *Describe two technologies for extraction of oil from algae.*

51. *What are the additional products other than oil, which can be obtained from algae?*

52. *What are the main co-products from algae apart from oil?*

53. *Write a short note on the use of ultrasonic technique for algal oil extraction.*

54. *What are the specialities of technology developed by 'Originoil' for algal oil.*

55. *Write a note about harvesting of algae using polymeric composites.*

56. *How can you enhance the lipid content of microalgae?*

57. *What are the effects of pH value, light intensity, and phosphate content on algal biomass and lipid content?*

58. *What are the effects of pH and temperature on the fungi?*

59. *What are the reactions involved in Fischer-Tropsch (FT) process?*

60. *Write a note on process conditions and use of catalyst in FT process.*

61. *Describe the gasification method of obtaining liquid biofuel from lignocellulosic biomass.*

62. *Comment on the statement that 'the wood pyrolysis products contents can be changed by changing process conditions.'*

63. *What do you understand by 'direct thermo-chemical liquefaction of algal biomass'?*

64. *What is 'algal hydrogenation' with reference to production of liquid fuel from algal biomass through gasification?*

65. *What are the different products that can be obtained from lignocellulosic biomass through the gasification route?*

66. *What are the differences between biodiesel and renewable hydrocarbon diesel fuel?*

67. *Enumerate fungi from which oil can be obtained. What are the conditions under which oil content increases? Where has such fungi been found?.*

68. *Write short notes on the (a) pyrolysis of biomass to obtain liquid fuel, (b) gribble, (c) biomass conversion to furans.*

69. *Describe the Fischer-Tropsch process for converting biomass into liquid fuel.*

70. Write short notes on production of liquid fuels by (i) direct thermo-chemical liquefaction of algal biomass and (ii) algal hydrogenation of algal biomass.

71. What are DME's basic physical characteristics?

72. In view of the properties of dimethyl ether, how do you think it can be the 'vehicle fuel' in times to come.

73. Describe the principles of manufacturing DME from biomass.

74. Where is DME being produced?

75. What benefits does DME offer when blended with LPG?

76. What are the benefits of DME use as a transportation fuel?

77. What health and safety considerations are there with DME?

78. What are the advantages and limitations regarding the use of DME as a fuel?

79. What do you know about environmental friendliness of DME?

80. Can you use LPG blended with DME for cooking and heating?

81. Are there any advantages of blending DME with diesel for use in vehicles?

82. Describe the principle of production of DME from lignocellulosic materials.

83. What are the emerging alternative fuels?

84. Write short notes on: (a) methanol as a fuel for vehicles (b) biobutanol as a fuel for vehicles

85. What are the differences between petroleum gasoline and renewable gasoline?

86. Describe how to obtain renewable hydrocarbon fuels from biomass.

87. What are the advantages of renewable fuels?

88. Compare the properties of petroleum diesel with those of other diesels.

89. What are the different pathways for the production of renewable hydrocarbon biofuels?

References

1. A Look Back at the US Department of Energy's Aquatic Species Program: Biodiesel from Algae by the National Renewable Energy Laboratory. June 2007. http://www1.eere.energy.gov/biomass/pdfs/biodiesel_from_algae.pdf.

2. Abishek, Monford Paul, Jay Patel and Anand Prem Rajan. May 2014. Algae Oil: A Sustainable Renewable Fuel of Future', Triantafyllos Roukas (Academic Editor), Biotechnology Research International. Review Article. Open Access. Volume 2014. Article ID 272814, 8 pages. https://doi.org/10.1155/2014/272814.

3. ACHEMA 2015, Press Release, Microalgae-produced on commercial scale. https://www.fraunhofer.de/en/press/research-news/2015/June/microalgae-produced-on-a-commercial-sc.

4. Algaculture - Wikipedia, the free encyclopaedia. June 2007. http://en.wikipedia.org/wiki/Algaculture.

5. Amaretti, A., S. Raimondi, M. Sala, L. Roncaglia, M. Lucia, A. Leonardi et al. 2010. Single cell oils of the cold-adapted oleaginous yeast *Rhodotorula glacialis* DBVPG 4785. Microbe Cell Fact. 9: 73.

6. Amin, Ashraf. December 2019. Review of diesel production from renewable resources: Catalysis, process kinetics and technologies. Ain Shams Engineering Journal 10(4): 821–839. https://doi.org/10.1016/j.asej.2019.08.001.

7. Amin, Sarmidi. July 2009. Review on biofuel oil and gas production processes from microalgae. Energy Conversion and Management 50(7): 1834–1840. https://doi.org/10.1016/j.enconman.2009.03.001.

8. An in-depth look at biofuels from algae. June 2007. http://biopact.com/2007/01/in-depth-look-at-biofuels-from-algae.html.

9. Animals - University of York [Ask 'gribbles' how to turn wood into liquid fuel] Posted by David Garner-York. June 2013. (Research by University of York). DOI: 10.1073/pnas.0914228107. [http://www.futurity.org/ask-gribbles-how-to-turn-wood-into-liquid-fuel/].

10. Arcoumanis, Constantine, Bae Choongsik, Roy Crookes and Eiji Kinoshita. June 2008. The potential of dimethyl ether (DME) as an alternative fuel for compression-ignition engines: A review. Fuel. 87(7): 1014–1030. https://doi.org/10.1016/j.fuel.2007.06.007.

11. Aresta, Michele, Angela Dibenedetto, Maria Carone, Teresa Colonna and Carlo Fragale. December 2005. Production of biodiesel from macroalgae by supercritical CO_2 extraction and thermochemical liquefaction. Environmental Chemistry Letters 3(3): 136–139.

12. Ariede, Maíra Bueno, Thalita Marcílio Candido, Ana Lucia Morocho Jacome, Maria Valéria Robles Velasco, João Carlos M.de Carvalho and André RolimBaby. July 2017. Cosmetic attributes of algae - A review. Algal Research. 25: 483–487. https://doi.org/10.1016/j.algal.2017.05.019.

13. Azizi, Z., M. Rezaeimanesh, T. Tohidian and M.R. Rahimpour. 2014. Dimethyl ether: A review of technologies and production challenges. Chem. Eng. and Proc. 82: 150–172. http://www.toyoeng.com/jp/en/products/energy/dme/.

14. Bailey, B., J. Eberhardt, S. Goguen and J. Erwin. 1997. Diethyl Ether (DEE) as a Renewable Diesel Fuel. SAE Technical Paper 972978. https://doi.org/10.4271/972978. Also in: Alternative Fuels: Technology and Developments-SP-1298, SAE 1997. Transactions - Journal of Fuels and Lubricants-V106-4.

15. Baitalow, F., R. Stahlschmidt, P. Seifert, R. Pardemann, B. Meyer and J. Engelmann. 2016. Production of gasoline from coal based on the stf process. *In*: Litvinenko, V. (ed.). XVIII Int. Coal Preparation Congress. Springer, Cham.

16. Bakhtyari, Ali, Mostafa Mohammadi and Mohammad Reza Rahimpour. September 2015. Simultaneous production of dimethyl ether (DME), methyl formate (MF) and hydrogen from methanol in an integrated thermally coupled membrane reactor. Journal of Natural Gas Science and Engineering 26: 595–607. https://doi.org/10.1016/j.jngse.2015.06.052.

17. Bang, Seung Hwan and Chang Sik Lee. March 2010. Fuel injection characteristics and spray behavior of DME blended with methyl ester derived from soybean oil. Fuel 89(3): 797–800. DOI: 10.1016/j.fuel.2009.10.009.

18. Barros, Ana I. Gonçalves, Ana L. Simões and José C.M. Pires. January 2015. Harvesting techniques applied to microalgae: A review. Renewable and Sustainable Energy Reviews 41: 1489–1500. https://doi.org/10.1016/j.rser.2014.09.037.

19. Bateni, Hamed and Chad M. Able. January 2019. Development of heterogeneous catalysts for dehydration of methanol to dimethyl ether: a review. In Catalysis in Industry 11(1): 7–33. DOI: 10.1134/S2070050419010045.

20. Behera, Shuvashish, Richa Singh, Richa Arora, Nilesh Kumar Sharma, Madhulika Shukla and Sachin Kumar. February 2014. Scope of algae as third generation biofuels. Review Article, Frontiers in Bioengineering and Biotechnology 2: 90. Published online Feb. 2015. https://doi.org/10.3389/fbioe.2014.00090.

21. Bellou, S., M.N. Baeshen, A.M. Elazzazy, D. Aggeli, F. Sayegh and G. Aggelis. 2014. Microalgal lipids biochemistry and biotechnological perspectives. Biotechnol. Adv. 32(8): 1476–1493. https://doi: 10.1016/j.biotechadv.2014.10.003.

22. Bezergianni, Stella and Athanasios Dimitriadis, May 2013. Comparison between different types of renewable diesel. Renewable and Sustainable Energy Reviews 21: 110–116. https://doi.org/10.1016/j.rser.2012.12.042.

23. Binder, Joseph B. and Ronald T. Raines. 2009. Simple chemical transformation of lignocellulosic biomass into furanics for fuels and chemicals. Journal of the American Chemical Society 131: 1979–1985.

24. Biodesel production – Biodiesel from Algae. June 2007. http://www.biodieselnow.com.

25. Biodiesel from Algae Oil – Info, Resources, News & Links. June 2007. http://www.oilgae.com.

26. Biodiesel from Algae Oil – PESWiki, Directory. June 2007. http://peswiki.com/index.php/Directory:Biodiesel_from_Algae_Oil.

27. Biofuels in the European Union. From Wikipedia, the free encyclopedia https://en.wikipedia.org/wiki/Biofuel_in_the_European_Union.

28. Boehman, André L. December 2008. Dimethyl Ether Special Section. Fuel Processing Technology 89(12): 1243–1478.

29. Brasil, Bruno dos Santos Alves Figueiredo, Félix Gonçalves de Siqueira, Thaís Fabiana Chan Salum, Cristina Maria Zanette, and Michele Rigon Spier. July 2017. Microalgae and cyanobacteria as enzyme biofactories, Review Article. Algal Research. 25: 76–89. ScienceDirect.com. https://www.sciencedirect.com/science/journal/22119264/25.

30. Briggs, Michael. 2004. Wide scale Biodiesel Production from Algae. University of New Hampshire. June 2007 from http://www.unh.edu.

31. Bucher, V.V.C., K.D. Hyde, S.B. Pointing and C.A. Reddy. 2004. Production of wood decay enzymes, mass loss and lignin solubilisation in wood by marine ascomycetes and their anamorphs. Fungal Divers 15: 1–14.

32. Burger, Jakob, Markus Siegert, Eckhard Ströfer and Hans Hasse. November 2010. Poly(oxymethylene) dimethyl ethers as components of tailored diesel fuel: Properties, synthesis and purification concepts. Fuel. 89(11): 3315–3319. https://doi.org/10.1016/j.fuel.2010.05.014.

33. Burke, Ultan, Kieran P. Somers, Peter O'Toole, Chis M. Zinner, Nicolas Marquet, Gilles Bourque et al. February 2015. An ignition delay and kinetic modeling study of methane, dimethyl ether, and their mixtures at high pressures. Combustion and Flame 162(2): 315–330. https://doi.org/10.1016/j.combustflame.2014.08.014.

34. Bušić, Arijana, Semjon Kundas, Galina Morzak, Halina Belskaya, Nenad Marđetko, Mirela Ivančić Šantek et al. June 2018. Recent Trends in Biodiesel and Biogas Production. Food Technol. Biotechnol. 56(2): 152–173. https://doi: 10.17113/ftb.56.02.18.5547.

35. Charles, A.S. and Hall John R. Benemann. October 2011. Oil from Algae? BioScience. 61(10): 741–742. https://doi.org/10.1525/bio.2011.61.10.2.

36. Chen, Wen, Cunwen Wang, Weiyong Ying, Weiguo Wang, Yuanxin Wu and Junfeng Zhang. 2009. Continuous Production of Biodiesel via Supercritical Methanol Transesterification in a Tubular Reactor. Part 1: Thermophysical and Transitive Properties of Supercritical Methanol. Energy & Fuels, 23 (1): 526-532. https://doi.org/10.1021/ef8005299.

37. Chen, Y.-H. Henry, John M. Reilly and Sergey Paltsev. September 2011. The prospects for coal-to-liquid conversion: A general equilibrium analysis. Energy Policy 39(9): 4713–4725. https://doi.org/10.1016/j.enpol.2011.06.056.

38. Cheng, C.H., T.B. Du, H.C. Pi, S.M. Jang, Y.H. Lin and H.T. Lee. 2011. Comparative study of lipid extraction from microalgae by organic solvent and supercritical CO2. Bioresour. Technol. 102(21): 10151–10153. https://doi: 10.1016/j.biortech.2011.08.064.

39. Chisti, Y. 2007. Biodiesel from microalgae. Bio. Technology Advances 25: 294–306.

40. Christensen, Earl, Aaron Williams, Stephen Paul, Steve Burton and Robert L. McCormick. 2011. Properties and performance of levulinate esters as diesel blend components. Energy Fuels 25(11): 5422–5428. https://doi.org/10.1021/ef201229j.

41. Christian P. Kubicek. 2013. Fungi and Lignocellulosic Biomass. John Wiley & Sons, Inc. Published Online: 13 JUL 2012.

42. Chun, Wan, Md. Asraful Alam, Xin-Qing Zhao, Xiao-Yue Zhang, Suo-Lian Guo, Shih-Hsin Ho et al. May 2015. Current progress and future prospect of microalgal biomass harvest using various flocculation technologies. Review. Bioresource Technology 184: 251–257. https://doi.org/10.1016/j.biortech.2014.11.081.

43. Coal liquefaction-From Wikipedia, the free encyclopedia. https://en.wikipedia.org/wiki/Coal_liquefaction.

44. Collotta, Massimo, Leonardo Busi, Pascale Champagne, Francesco Romagnoli, Giuseppe Tomasoni, Warren Mabee et al. 2017. Comparative LCA of Three Alternative Technologies for Lipid Extraction in Biodiesel from Microalgae Production. Energy Procedia. 113: 244–250.

45. Cotta, F., M. Matschke, J. Großmann, C. Griehl and S. Matthes. 2011. Verfahrenstechnische Aspekte eines flexiblen, tubulären Systems zur Algenproduktion. Process-related aspects of a flexible, tubular system for algae production. DECHEMA.

46. Coward, Thea, Jonathan G.M. Lee and Gary S. Caldwell. March 2013. Development of a foam flotation system for harvesting microalgae biomass. Algal Research. 2(2): 135–144. https://doi.org/10.1016/j.algal.2012.12.001.

47. Dadgar, Farbod, Rune Myrstad, Peter Pfeifer, Anders Holmen and Hilde J.Venvik. July 2016. Direct dimethyl ether synthesis from synthesis gas: The influence of methanol dehydration on methanol synthesis reaction. Catalysis Today 270: 76–84. https://doi.org/10.1016/j.cattod.2015.09.024.

48. Dahlia, M.El., Maghraby Eman and M.Fakhry. January–March 2015. Lipid content and fatty acid composition of Mediterranean macro-algae as dynamic factors for biodiesel production. Oceanologia. 57(1): 86–92. https://doi.org/10.1016/j.oceano.2014.08.001.

49. Das, Dr. Piyali, and Anubhuti Bhatnagar. August 2017. Different Feedstock & Processes for Production of Methanol & DME as Alternate Transport Fuels. *In*: Prospects of Alternative Transportation Fuels, Ist Edition, Chapter 7. Akhilendra Pratap Singh, Atul Dhar, Rashmi Avinash Agarwal, Mritunjay Kumar Shukla, Avinash Kumar Agarwal (Eds.). pp. 39. Springer. https://doi:10.1007/978-981-10-7518-6_8.

50. Davis, Timothy W., Dianna L. Berry, Gregory L. Boyer and Christopher J. Gobler. June 2009. The effects of temperature and nutrients on the growth and dynamics of toxic and non-toxic strains of Microcystis during cyanobacteria blooms. Harmful Algae. 8(5): 715–725. https://doi.org/10.1016/j.hal.2009.02.004.

51. De Falco, Marcello. Dimethyl Ether (DME) Production. http://www.oil-gasportal.com/dimethyl-ether-dme-production-2/.

52. Decker, Eva and Ralf Reski. 2008. Current achievements in the production of complex biopharmaceuticals with moss bioreactors. Bioprocess and Biosystems Engineering 31(1): 3–9. https://doi:10.1007/s00449-007-0151-y. PMID 17701058.

53. Demirbas, A. 2004. Combustion characteristics of different biomass fuels. Progress in Energy and Combustion Science 30(2): 219–230.

54. Demirbas, Ayhan. 2009. Political, economic and environmental impacts of biofuels: A review. Applied Energy 86: S108–S117. www.elsevier.com/locate/apenergy.

55. Demirbaş, Ayhan. July 2001. Biomass resource facilities and biomass conversion processing for fuels and chemicals. Energy Conversion and Management. 42(11): 1357–1378. https://doi.org/10.1016/S0196-8904(00)00137-0.

56. Dept. of Energy, USA, and Energy Efficiency Renewable Energy, Alternative fuel Data (http://www.afdc. energy.gov/fuels/emerging_dme.html)].

57. Dimethyl Ether (DME) Fact Sheet (http://www.biofuelstp.eu/factsheets/dme-fact-sheet.html).

58. Dimethyl ether From Wikipedia, the free encyclopedia; https://wiki2.org/en/Dimethyl_ether

59. Dimethyl Ether Technology and Markets. December 2008. 07/08-S3 Report, ChemSystems. [Techno-economic Analysis for the Thermochemical Conversion of Biomass to Liquid Fuels by Y. Zhu, S.A. Tjokro Rahardjo, C. Valkenburg, L.J. Snowden-Swan, S.B. Jones and M.A. Machinal. June 2011. Pacific Northwest National Laboratory Richland, Washington 99352].

60. Dong,Tao, Eric P. Knoshaug, Ryan Davis, Lieve M.L. Laurens, Stefanie Van Wychen, Philip T. Pienkos et al. November 2016. Combined algal processing: A novel integrated biorefinery process to produce algal biofuels and bioproducts. Research Article, Algal Research. 19: 316–323. [Open Access articles.ScienceDirect.com]. https://www.sciencedirect.com/science/journal/22119264/open-access.

61. Dutta, S., S. De, I. Alam, M.M. Abu-Omar and B. Saha. 2012. Direct conversion of cellulose and lignocellulosic biomass into chemicals and biofuel with metal chloride catalysts. J. Catal. 288: 8–15.

62. Ehara, K. and S. Saka. 2005. Decomposition behavior of cellulose in supercritical water, subcritical water, and their combined treatments. J. Wood Sci. 51: 148–153. doi:10.1007/s10086-004-0626-2.

63. Extraction of oil from fungi under supercritical conditions. https://www.sciencedirect.com/science/article/abs/pii/S0896844618307964; https://www.sciencedirect.com/science/article/abs/pii/089684469290037K; https://meridian.allenpress.com/jfp/article/68/4/790/171973/ Chemical-Composition-and-Antimicrobial-Activity-of; https://www.sciencedirect.com/science/article/abs/pii/S0308814605005960.

64. Fanick, Novel Renewable Additive for Diesel Engines. SAE 2014-01-1262. http://www.internationaljournalssrg. org/IJME/2017/Volume4-Issue4/IJME-V4I4P105.pdf; https://www3.epa.gov/otaq/fuels1/ffars/web-dies.html.

65. FAQ - International DME Association [https://www.aboutdme.org/index.asp?bid=234]

66. Fleisch, T.H., A. Basu and R.A. Sills. 2012. Introduction and advancement of a new clean global fuel: The status of DME developments in China and beyond. J. Natural Gas Science and Eng. 9: 94–107.

67. Fleisch,T.H., A. Basu, M.J. Gradassi and J.G.Masin. 1997. Dimethyl ether: A fuel for the 21st century. Studies in Surface Science and Catalysis. 107: 117–125. https://doi.org/10.1016/S0167-2991(97)80323-0.

68. Fundamental Aspects of Dimethyl Ether. January 2015. Department of Energy, USA. www.afdc.energy.gov/pdfs/3608.pdf.

69. Gerchman, Yoram, Barak Vasker, Mordechai Tavasi, Yael Mishael, Yael Kinel-Tahan and Yaron Yehoshua. 2017. Effective harvesting of microalgae: Comparison of different polymeric flocculants. Bioresource Technology 228: 141–146.

70. Gerken, Henri G., Bryon Donohoe and Eric P. Knoshaug. January 2013. Enzymatic cell wall degradation of Chlorella vulgaris and other microalgae for biofuels production. Planta. 237(1): 239–253.

71. Gikonyo, Barnabas (Ed.). 2014. Advances in Biofuel Production: Algae and Aquatic Plants. Johnson Matthey Technol. Rev. 58(3): 169. Apple Academic Press, Inc. New Jersey, USA, 2014, 398 pages, ISBN: 978-1-926895-95-6. https://doi:10.1595/147106714x682346.

72. Greenhouse Gas Mitigation Project at the International University of Bremen by International Research Consortium on Continental Margins. Retrieved on June 26th, 2007 from http://www.irccm.de/greenhouse/project. html © Mora Associates Ltd, 2007.

73. Gunstone, F.D., J.L. Harwood and A.J. Dijkstra. 2007. The Lipid Handbook with CD-ROM. 3rd edition. CRC Press, Boca Raton, FL.

74. Guoyin, Huang, Huan Xiang Zhou, Ziyi Wang, Haixia Liu, Yi Cao and Dairong Qiao. November 2016. Novel fungal lipids for the production of biodiesel resources by Mucor fragilis AFT, Environmental Progress and Sustainable Energy, AiChE, 35(6): 1784–1792.

75. Gupta, S.K., M. Kumar, A. Guldhe, F.A. Ansari, I. Rawat, K. Kanney et al. Feb. 2015. Development and effectiveness of low cost cationic polymer for harvesting micro algae. 5th Int. Conf. on plant and environmental pollution ICPEP-5, NBRI, Lucknow, India.

76. Gupta, S.K., M. Kumar, A. Guldhe, F.A. Ansari, I. Rawat, K. Kanney et al. September 2014. Design and development of polyamine polymer for harvesting microalgae for biofuels production. Energy Conversion and Management. 85: 537–544. https://doi.org/10.1016/j.enconman.2014.05.059.

77. Hahn-Hägerdal, B., M. Galbe, M.F. Gorwa-Grauslund and G. Lidén G. Zacchi. December 2006. Bio-ethanol – the fuel of tomorrow from the residues of today-Review. Trends in Bio. Technology 24(12): 549–556. https://doi.org/10.1016/j.tibtech.2006.10.004.

78. Halim, R., M.K. Danquah and P.A. Webley. Jan 2012. Extraction of oil from microalgae for biodiesel production: A review. Biotechnology Advances 30(3):709–732. https://doi: 10.1016/j.biotechadv.2012.01.001.

79. Han, Ting, Lu Haifeng, Ma Shanshan, Yuanhui Zhang, Liu Zhidan and Duan Na. January, 2017. Progress in microalgae cultivation photobioreactors and applications in wastewater treatment: A review. Int. J. Agric & Biol. Eng., Open Access at https://www.ijabe.org, Vol. 10 No.1.

80. Handbook of microalgal culture. June 2013. 2nd (ed.). Blackwell Science Ltd. 2013. ISBN 978-0-470-67389-8.

81. Hassannia, Jeff H. May 2009. Algae Biofuels Economic Viability: A Project-Based Perspective. https://www.greentechmedia.com/articles/read/algae-biofuels-economic-viability-a-project-based-perspective-4561.

82. Hattab, Mariam Al and Abdel Ghaly. 2015. Microalgae oil extraction pre-treatment methods: critical review and comparative analysis. Review article open access, J. of Fundamentals of Renewable Energy and Applications. 5(4): 26 pages. https://doi: 10.4172/2090-4541.1000172.

83. Hielscher Ultrasound technology. https://www.hielscher.com/algae_extraction_01.htm.

84. Höök, M. and K. Aleklett. 2010. A review on coal-to-liquid fuels and its coal consumption. Int. J. of Energy Research. 34(10): 848–864.

85. Hossain, Nazia, T.M. Indra Mahlia and Rahman Saidur. May 2019. Latest development in microalgae-biofuel production with nano-additives. Biotechnology for Biofuels 12(1). DOI: 10.1186/s13068-019-1465-0.

86. http://www.extension.org/algae-for-biofuel-production].

87. http://www.fao.org/uploads/media/0707_Wagner_Biodiesel_from_algae_oil.pdf.

88. http://www.japantransport.com/conferences/2006/03/dme_detailed_information.pdf, Conference on the Development and Promotion of Environmentally Friendly Heavy Duty Vehicles such as DME Trucks, Washington DC, March 17, 2006.

89. http://www.renewablesinfo.com/drawbacks_and_benefits/biofuels_from_algae_advantages_and_disadvantages.html.

90. https://auto.howstuffworks.com/fuel-efficiency/biofuels/10-disadvantages-of-biofuels5.html.

91. https://en.wikipedia.org/wiki/Algae_fuel and https://wikivisually.com/wiki/Algae_fuels.

92. https://en.wikipedia.org/wiki/Algae_fuel#Use_of_Byproducts.

93. https://en.wikipedia.org/wiki/Photobioreactor.

94. https://energsustainsoc.biomedcentral.com/articles/10.1186/s13705-019-0232-1.

95. https://iopscience.iop.org/article/10.1088/1742-6596/1276/1/012073/meta.

96. https://microbewiki.kenyon.edu/index.php/Biodiesel_from_Algae_Oil

97. https://www.futurebridge.com/industry/perspectives-energy/renewable-diesel-the-fuel-of-the-future/.

98. https://www.regi.com/products/transportation-fuels/renewable-diesel.

99. Huang, Suiliang, Min Wu, Changjuan Zang, Shenglan Du, Joseph Domagalski, Magdalena Gajewska et al. June 2016. Dynamics of algae growth and nutrients in experimental enclosures culturing bighead carp and common carp: Phosphorus dynamics. Int. J. of Sediment Research. 31(2): 173-180. https://doi.org/10.1016/j.ijsrc.2016.01.003.

100. IDA Fact Sheet DME/LPG Blends 2010 v1.

101. Jacquot, Jeremy. Oct. 2009. 5 Companies Making Fuel From Algae Now. http://www.popularmechanics.com/science/energy/a4677/4333722/ale.ht.

102. Jae-YonLee, ChanYoo, So-YoungJun, Chi-YongAhn and Hee-MockOh. January 2010. Comparison of several methods for effective lipid extraction from microalgae. Bioresource Technology 101(1) Supplement: S75–S77. https://doi.org/10.1016/j.biortech.2009.03.058.

103. Jämsä, Mikael, Fiona Lynch, Anita Santana-Sánchez, Petteri Laaksonen, Gennadi Zaitsev, Alexei Solovchenko. et al. September 2017. Nutrient removal and biodiesel feedstock potential of green alga UHCC00027 grown in municipal wastewater under Nordic conditions. Algal Research. 26: 65–73. https://doi.org/10.1016/j.algal.2017.06.019.

104. Jamshidi, L.C.L.A., C.M.B.M. Barbosa, L. Nascimento and J.R. Rodbari. 2013. Catalytic Dehydration of Methanol to Dimethyl Ether (DME) Using the Al62,2Cu25,3Fe12,5 Quasicrystalline Alloy. Chemical Engineering & Process Technology 4: 5. http://dx.doi.org/10.4172/2157 7048.1000164.

105. Janssen, A., M. Muether, A. Kolbeck, M. Lamping et al. Oct. 2010. The impact of different biofuel components in diesel blends on engine efficiency and emission performance," SAE Technical Paper 2010-01-2119. https://doi.org/10.4271/2010-01-2119.

106. Janssen, A., S. Pischinger and M. Muether. 2010. Potential of cellulose-derived biofuels for soot free diesel combustion. SAE Int. J. Fuels Lubr. 3(1): 70–84, https://doi.org/10.4271/2010-01-0335. Also In: Kinetically Controlled CI Combustion and Controls, 2010-SP-2280, SAE International Journal of Fuels and Lubricants-V119-4, SAE International Journal of Fuels and Lubricants-V119-4EJ.

107. Jeevan Kumar, S.P, Vijay Kumar Garlapatia, Archana Dasha, PeterScholzd and Rintu Banerjeea. 2017. Sustainable green solvents and techniques for lipid extraction from microalgae: A review. Algal Research 21: 138–147.

108. Jie, Liu; Liu Shenghua, Li Yi, Wei Yanju, Li Guangle and Zhu Zan. 2010. Regulated and Nonregulated Emissions from a Dimethyl Ether Powered Compression Ignition Engine. Energy Fuels 24: 2465–2469.

109. Joshi, Sunita and Durga Nath Dhar 1987. Unconventional Sources of Oils and fats, Strategies for Rural Development, J.K. Gehlawat and K. Kant (eds.). pp. 307–313. Arnold Publishers

110. Kajitani, S., C.L. Chen, M. Oguma, M. Alam and K.T. Rhee. 1998. Direct injection diesel engine operated with propane – DME blended fuel. SAE Paper 982536.

111. Kaushik, Nirmala, Soumitra Biswas and P.R. Basak, March 2010. New Generation Biofuels –Technology & Economic Perspectives. presented at the conference on 'Frontier Issues in Technology, Development and Environment' , held at Madras School of Economics, Chennai during March 19–21.

112. Kennedy, Jennifer. 2017. Three Types of Sea Weed (Marine Algae). Science, Tech, Math. Animals and Nature.

113. Khan, Mohd Moiz, Riyaj Uddin Khan, Fahad Zishan Khan, Moina Athar. 2013. Impacts of Biodiesel on the Environment. Int. J. of Environmental Engineering and Management ISSN 2231-1319, 4(4): 345–350 © Research India Publications, http://www.ripublication.com/ ijeem.htm.

114. Khemka, Ankita and Meenu Saraf. Feb. 2015. Induction of lipid production in Desertifilum tharense Msak01 and *Leptolyngbya* sp. under nitrogen and/or phosphorus limitation condition. 5th Int. Conf. on plant and environmental pollution ICPEP-5, NBRI, Lucknow, India.

115. Khot, Mahesh, Srijay Kamat, Smita Zinjarde, Aditi Pant, Balu Chopade and Ameeta RaviKumar. 2012. Single cell oil of oleaginous fungi from thetropical mangrove wetlands as a potential feedstock for biodiesel. Microbial Cell Factories. http://www.microbialcellfactories.com/content/11/1/71.

116. Kim, Hyung Jun, Su Han Park, Kwan Soo Lee and Chang Sik Lee. 2011. A study of spray strategies on improvement of engine performance and emissions reduction characteristics in a DME fuelled diesel engine. Int. J. Energy 36: 1802–1813.

117. Knothe, G., J.V. Gerpen and J. Krahl. 2005. The biodiesel Handbook. AOCS Press, Champaign, Illinois.

118. Knothe, Gerhard. 2010. Biodiesel and renewable diesel: A comparison. Progress in Energy and Combustion Science 36: 364–373.

119. Knothe, G. and K.R. Steidley. 2005. Lubricity of components of biodiesel and petrodiesel. The Origin of Biodiesel Lubricity. Energy & Fuels 19: 1192–1200.

120. Koul, Rashi, Naveen Kumar and R.C Singh. 2021. A review on the production and physicochemical properties of renewable diesel and its comparison with biodiesel. Energy Sources, Part A: Recovery, Utilization, and Environmental Effects 43(18): 2235–2255. https://doi.org/10.1080/15567036.2019.1646355.

121. Kumar, Dipesh, Singh, Bhaskar and Korstad, John 2017. Utilization of lignocellulosic biomass by oleaginous yeast and bacteria for production of biodiesel and renewable diesel. Renewable and Sustainable Energy Reviews, Elsevier 73(C): 654–671. Handle: RePEc:eee:rensus:v:73:y:2017:i:c:p:654-671. DOI: 10.1016/j.rser.2017.01.022.

122. Kumar, H., Sarma, A.K. andKumar, P. 2020. A comprehensive review on preparation, characterization, and combustion characteristics of microemulsion based hybrid biofuels. Renew. Sustain. Energy Rev. 117, 109498.

123. Kumar, Pankaj, M.R. Suseela and Kiran Toppo. 2011. Physico-chemical characterization of algal oil: a potential biofuel. Asian J. Exp. Biol. Sci. 2(3): 493–497. [Algae laboratory, National Botanical Research Institute, Lucknow, India].

124. Kumar, Pankaj, M.R. Suseela, Kiran Toppo, S.K. Mandotra, S.K. Mishra and Pushpa Joshi. Feb. 2015. Optimization of micro-algal cultivation for enhanced biomass production for sustainable biofuel. Vth Int. Conf. on plant and environmental pollution ICPEP-5. NBRI, Lucknow, India.

125. Larson, E.D., G. Liu and R.H. Williams. 2003. Synthetic fuel production by indirect coal liquefaction. Energy for Sustainable Development 7(4): 79–102, 165.

126. Li, Lan, Winnie Wong-Ng and Jeffrey Sharp. 2014. Polymer composites for energy harvesting, conversion, and storage. American Chemical Society. Division of Polymer Chemistry, ACS symposium series, 1161. Washington, DC : American Chemical Society.

127. Li, Y., F.G. Naghdi, S. Garg , T.C. Adarme, Vega, K.J. Thuracht, W.A. Ghafor et al. 2014. A comparative study: the impact of different lipid extraction methods on current microalgal lipid research. Microb. Cell Fact. http:// doi: 10.1186/1475-2859-13-14.

128. Li, Yecong, Wenguang Zhou, Bing Hu, Min Min, Paul Chen and Roger R. Ruan. September 2012. Effect of light intensity on algal biomass accumulation and biodiesel production for mixotrophic strains Chlorella kessleri and Chlorella protothecoidecultivated in highly concentrated municipal wastewater. Biotechnology and Bioengineering 109(9): 2222–2229. http://doi:10.1002/bit.24491.

129. Li, Yuping, Tiejun Wang, Xiuli Yin, Chuangzhi Wu, Longlong Ma, Haibin Li et al. March 2010. 100 t/a-Scale demonstration of direct dimethyl ether synthesis from corncob-derived syngas. Renewable Energy 35(3): 583–587. https://doi.org/10.1016/j.renene.2009.08.002.

130. Lorenzen, Jan, Nadine Igl, Marlene Tippelt, Andrea Stege, Farah Qoura, Ulrich Sohling et al. June 2017. Extraction of microalgae derived lipids with supercritical carbon dioxide in an industrial relevant pilot plant. Open Access, Bioprocess and Biosystems Engineering 40(6): 911–918.

131. Mandotra, S.K., M.R. Suseela, Pankaj Kumar and P.W. Ramteke. Feb. 2015. Effect of pH, light intensity and phosphate concentrations on the biomass and lipid production of green microalgal *Scenedesmus abundans*. 5th Int. Conf. on plant and environmental pollution ICPEP-5, NBRI, Lucknow, India.

132. Marsh, George. March/April 2009. Small wonders: biomass from algae. Renewable energy focus.

133. Meeuwse, Petra, P.M. Johan, Johannes Tramper Sanders and Arjen Rinzema. September/October 2013. Lipids from fungi: Tomorrow's source of biodiesel? Biofuels, Bioproducts and Biorefining 7(5): 475–626 © Society of Chemical Industry and John Wiley & Sons Ltd.

134. Meier, D. and O. Faix. April 1999. State of the art of applied fast pyrolysis of lignocellulosic materials—a review. Bioresource Technology 68(1): 71–77. https://doi.org/10.1016/S0960-8524(98)00086-8.

135. Meng, X., J. Yang, X. Xu, L. Zhang, Q. Nie and M. Xian. 2009. Biodiesel production from oleaginous microorganisms. Renew Energy 34: 1–5. https://doi:10.1016/j.renene.2008.04.014.

136. Mercer, Paula and Roberto E. Armenta. May 2011. Developments in oil extraction from microalgae: Review Article. European J. of Lipid Science and Technology 113(5): 539–547. https://doi: 10.1002/ejlt.201000455.

137. Migliori, M., A. Aloise, E. Catizzone and G. Giordano. 2014. Kinetic analysis of methanol to dimethyl ether reaction over H-MFI Catalyst. Ind. Eng. Chem. Res. 53: 14885–14891.

138. Mooij, Timde, Guusde Vries, Christos Latsos, René H.Wijffels and Marcel Janssen. April 2016. Impact of light color on photobioreactor productivity. Algal Research 15: 32–42. DOI: 10.1016/j.algal.2016.01.015. https://doi.org/10.1016/j.algal.2016.01.015.

139. Mosier, Nathan, Charles Wyman, Bruce Dale, Richard Elander, Y.Y. Lee, Mark Holtzapple et al. April 2005. Features of promising technologies for pre treatment of lignocellulosic biomass. Bioresource Technology 96(6): 673–686. https://doi.org/10.1016/j.biortech.2004.06.025.

140. MSU News - MSU-led team finds new type of fuel in Patagonia fungus, MSU's College of Engineering and researchers at Yale University. (www.montana.edu/news/.../msu-led-team-finds-new-type-of-fuel-in-patagonia-fungus).

141. Mubarak, M., A. Shaija and T.V. Suchithra. January 2015. A review on the extraction of lipid from microalgae for biodiesel production: Review article. Algal Research 7: 117–123. https://doi.org/10.1016/j.algal.2014.10.008.

142. Müller, Manfred and Ute Hübsch. 2005. Dimethyl Ether in Ullmann's Encyclopedia of Industrial Chemistry. Wiley-VCH, Weinheim. https://doi:10.1002/14356007.a08_541.

143. Müller, Manfred and Ute Hübsch. 2005. Dimethyl Ether in Ullmann's Encyclopedia of Industrial Chemistry. Wiley-VCH, Wein heim. https://doi.10.1002/14356007a08_541.

144. Muylaert, Koenraad, Dries Vandamme, Imogen Foubert and Patrick V. Brady. Harvesting of Microalgae by Means of Flocculation. Biomass and Biofuels from Microalgae. Chapter in Biofuel and Biorefinery Technologies book series (BBT, volume 2), pp. 251–273.

145. Myco-diesel from Fungus Gliocladium roseum research (http://www.westernfarmerstockman.com/story-montana-research-reveals-diesel-fuel-producing-fungus-9-20367).

146. Naghdi, Forough Ghasemi, Lina M. González González, William Chan and Peer M. Schenk. Nov. 2016. Progress on lipid extraction from wet algal biomass for biodiesel production. Microb. Biotechnol. 9(6): 718–726. Published online 2016 May 19. https://doi: 10.1111/1751-7915.12360.

147. Naik, S.N., Vaibhav V. Goud, Prasant K. Rout and Ajay K. Dalai. February 2010. Production of first and second generation biofuels: A comprehensive review. Renewable and Sustainable Energy Reviews 14(2): 578–597. https://doi.org/10.1016/j.rser.2009.10.003.

148. Nhien, L.C., N.V.D. Long and M. Lee. 2017. Novel heat–integrated and intensified biorefinery process for cellulosic ethanol production from lignocellulosic biomass. Energy Convers Manage, 141: 367–377. 10.1016/j.enconman.2016.09.077.

149. Nicolò, M.S., S.P.P. Guglielmino, V. Solinas and A. Salis. January 2017. 'Biodiesel from Microalgae' Living Reference Work Entry, Consequences of Microbial Interactions with Hydrocarbons, Oils, and Lipids: Production of Fuels and Chemicals. Part of the series Handbook of Hydrocarbon and Lipid Microbiology pp. 1–20.

150. Niyogi, Professor Kris. Aug. 2009. Selected viewer questions answered about algae biofuel. Department of Plant and Microbial Biology, University of California, Berkeley (Posted on internet on 28.8.2009).

151. Oberholster, Paul J., Anna-Maria Botha and Peter J. Ashton. 2009. The influence of a toxic cyanobacterial bloom and water hydrology on algal populations and macroinvertebrate abundance in the upper littoral zone

of Lake Krugersdrift, South Africa. [http://www.repository.up.ac.za/bitstream/handle/2263/9269/Oberholster_Influence(2009).pdf;sequence=1].

152. Olah, G.A., A. Geopart and G.K. Surya Prakash. 2006. Beyond Oil and Gas: The Methanol Economy. Weinheim, Wiley-VCH.

153. Oliveira, Fernando C. De and Suani T. Coelho. August 2017. History, evolution, and environmental impact of biodiesel in Brazil: A review. Renewable and Sustainable Energy Reviews 75: 168–179. https://doi.org/10.1016/j.rser.2016.10.060.

154. Origionoil - extraction of lipid from algae without killing it. July 2009. http://algaenews.blogspot.in/2009/07/more-algae-action-originoil-plans-to.html

155. Parivesh-A news letter from ENVIS Centre- Central Pollution Control Board, Alternative Transport of Fuels - An Overview. (http://cpcbenvis.nic.in/cpcb_newsletter/Alternative%20Transport%20Fuels%20An%20Overview.pdf).

156. Park, J.B.K., R.J. Craggs and A.N. Shilton. January 2011. Wastewater treatment high rate algal ponds for biofuel production. Bioresource Technology 102(1): 35–42.

157. Park, S.H. and C.S. Lee. 2014. Applicability of dimethyl ether (DME) in a compression ignition engine as an alternative fuel. Energy Conv. and Management 86: 848–863.

158. Park, Su Han, Hyung Jun Kim and Chang Sik Lee. 2011. Study on the dimethyl ether spray characteristics according to the diesel blending ratio and the variations in the ambient pressure, energizing duration, and fuel temperature. Energy Fuels 25: 1772–1780.

159. Patel, Akash Kumar, M.R. Suseela and Munna Singh. Feb. 2015. Screening of microalgae sp. from the effluent of carpet industry for production of biodiesel. 5th Int. Conf. on plant and environmental pollution ICPEP-5. NBRI, Lucknow, India.

160. Patel, Madhumita and AmitKumar. May 2016. Production of renewable diesel through the hydroprocessing of lignocellulosic biomass-derived bio-oil: A review. Renewable and Sustainable Energy Reviews 58: 1293–1307. https://doi.org/10.1016/j.rser.2015.12.146.

161. Patil, K.R. and Dr. S.S. Thipse. October 2012. The potential of DME-diesel blends as an alternative fuel for CI Engines. Int. J. of Emerging Technology and Advanced Engineering 2(10)35, ISSN 2250-2459. Website: www.ijetae.com.

162. Peral, E. and M. Martín. 2015. Optimal production of dimethyl ether from switchgrass-based syngas via direct synthesis. Ind. Eng. Chem. Res. 54: 7465–7475.

163. Phasey, J., D. Vandamme and H.J. Fallowfield. November 2017. Harvesting of algae in municipal wastewater treatment by calcium phosphate precipitation mediated by photosynthesis, sodium hydroxide and lime. Algal Research 27: 115–120.

164. Photobioreactor, From Wikipedia, the free encyclopedia https://en.wikipedia.org/wiki/Photobioreacto.

165. Potharaju, S. Sai Prasad, Jong Wook Bae, Suk-Hwan Kang, Yun-Jo Lee and Ki-Won Jun. December 2008. Single-step synthesis of DME from syngas on Cu–ZnO–Al $2O 3$/zeolite bifunctional catalysts: The superiority of ferrierite over other zeolites. Fuel Processing Technology 89(12): 1281–1286. https://doi:10.1016/j.fuproc.2008.07.014.

166. Pragya, Namita, Krishan K. Pandey and P.K. Sahoo. August 2013. A review on harvesting, oil extraction and biofuels production technologies from microalgae. Renewable and Sustainable Energy Reviews 24: 159–171. https://doi.org/10.1016/j.rser.2013.03.034.

167. Prakash, Saurabh, Ghanshyam Paswan and Kumar Nikhil. March 2014. Liquid Coal as a Green Energy: A Review. Int. J. of Engineering and Technical Research (IJETR) ISSN. 2(3): 2321–0869. https://www.erpublication.org/published_paper/IJETR021392.pdf.

168. Pulz, O. 2001. Photobioreactors: production systems for phototrophic microorganisms. Applied Microbiology and Biotechnology 57: 287–293. https://doi: 10.1007/s002530100702.

169. Pyrolysis oil- From Wikipedia, the free encyclopedia; https://en.wikipedia.org/wiki/Pyrolysis_oil.

170. Rai, Monika Prakash, Trishnamoni Gautom and Nikunj Sharma. 2015. Effect of Salinity, pH, Light Intensity on Growth and Lipid Production of Microalgae for Bioenergy Application. online J. of Biological Sciences 15(4): 260–267.

171. Ramanathan, Ranjith Kumar, Polur Hanumantha Rao and Muthu Arumugam. January 2015. Lipid extraction methods from microalgae: a comprehensive review. Frontiers in Energy Research. ISSN 2296-598X. Volume 2: 9 pages. Front. Energy Res. https://doi.org/10.3389/fenrg.2014.00061.

172. Ref. US Department of Agriculture's Agricultural Research Service, National Centre for Agricultural Utilization Research/Bioenergy. 2010. (https://www.ars.usda.gov/research/project/?accnNo=417041&fy=2010).

173. Reitan, Kjell Inge, Jose R. Rainuzzo and Yngvar Olsen. December 1994. Effect of Nutrient Limitation on Fatty Acid and Lipid Content of Marine Microalgae 30(6): 972–979. https://doi:10.1111/j.0022-3646.1994.00972.

174. Reports from Renewable energy focus March/April 2009 p.10.

175. Reports in Renewable Energy focus, Jan, Feb 2009.

176. Rhodes, Christopher J. May 2009. Oil from algae; salvation from peak oil? Science Progress 92(1): 39–90(52).

177. Robson, J., A. Alessi, C. Bochiwal, C. O'Malley and J.P.J. Chong. January 2017. Biomethane as an Energy Source, Living Reference Work Entry, Consequences of Microbial Interactions with Hydrocarbons, Oils, and Lipids: Production of Fuels and Chemicals, Part of the series Handbook of Hydrocarbon and, Lipid Microbiology pp. 1–12.

178. Rodolfi, Liliana, Graziella Chini Zittelli, Niccolò Bassi, Giulia Padovani, Natascia Biondi, Gimena Bonini et al. January 2009. Microalgae for oil: Strain selection, induction of lipid synthesis and outdoor mass cultivation in a low-cost photobioreactor. Biotechnology and Bioengineering 102(1): 100–112. DOI: 10.1002/bit.22033.

179. Rogers, Jonathan N., Julian N. Rosenberg, Bernardo J. Guzman, Victor H. Oh, Luz Elena Mimbela, Abbas Ghassemi et al. January 2013. A critical analysis of paddlewheel-driven raceway ponds for algal biofuel production at commercial scales. Algal Research 4(1)122. DOI: 10.1016/j.algal.2013.11.007.

180. Rong, F. and D.G. Victor. 2011. Coal liquefaction policy in China: Explaining the policy reversal since 2006. Energy Policy Elsevier.

181. Roy Chaudhary, A., K. Kundu and V.R. Dahake. Nov.2011. Algal Biodiesel: Future prospects and problems. Water and energy research digest, Water and Energy International 68(11).

182. Sahoo, Narendra Kumar, Sanjay Kumar Gupta, Ismail Rawat, Faiz Ahmad Ansari, Poonam Singh, Satya Narayan Naik et al. 2017. Sustainable Dewatering and drying of self-flocculating microalgae and study of cake properties. Journal of Cleaner Production. doi: 10.1016/j.jclepro.2017.05.015.

183. Sakaki, Keiji et al. September 1990. Supercritical fluid extraction of fungal oil using CO2, N2O, CHF3 and SF6. J. of the American Oil Chemists' Society 67(9): 553–557. [https://link.springer.com/article/10.1007/BF02540765].

184. Sánchez-Camargo, Andrea P., Hugo A. Martinez-Correa, Losiane C. Paviani and Fernando A. Cabral. March 2011. Supercritical CO2 extraction of lipids and astaxanthin from Brazilian redspotted shrimp waste (Farfantepenaeus paulensis). J. of Supercritical Fluids 56(2): 164–173. https://doi.org/10.1016/j.supflu.2010.12.009.

185. Santala, S., E. Efimova, V. Kivinen, A. Larjo, T. Aho, M. Karp et al. 2011. Improved triglycerol production in Acinetobacter baylyi ADP1 by metabolic engineering. Microbe Cell Fact. 10: 36. 10.1186/1475-2859-10-36.

186. Sayre, Richard T., Zoee Perrine and Sangeeta Negi. October 2012. Optimization of photosynthetic light energy utilization by microalgae. Algal Research 1: 134–142. https://doi:10.1016/j.algal.2012.07.002.

187. Scott, Stuart A., Matthew P. Davey, John S. Dennis, Irmtraud Horst, Christopher J. Howe, and David J. Lea-Smith et al. 2010. Biodiesel from algae: challenges and prospects. Current Opinion in Biotechnology 21: 277–286. www.sciencedirect.com.

188. Semelsberger, Troy A., Rodney L. Borup and Howard L. Greene. June 2006. Dimethyl Ether (DME) as an Alternative Fuel. J. of Power Sources 156(2): 497–511. DOI: 10.1016/j.jpowsour.2005.05.082.

189. Sethi, A. and Michael E. Scharf. February 2013. Biofuels: Fungal, Bacterial and Insect Degraders of Lignocellulose. Published online: DOI: 10.1002/9780470015902.a0020374.

190. Sezer, Ismet. 2011. Thermodynamic, performance and emission investigation of a diesel engine running on dimethyl ether and diethyl ether. Int. J. of Thermal Sciences 50: 1594–1603.

191. Shahid, Ayesha, Aqib Zafar Khan, Tianzhong Liu, Sana Malik, Ifrah Afzal and Muhammad A. Mehmood. 2017. Algae Based Polymers, Blends, and Composites-Chapter 7 – Production and Processing of Algal Biomass. Chemistry, Biotechnology and Materials Science, pp. 273–299. https://doi.org/10.1016/B978-0-12-812360-7.00007-0.

192. Sharma, Kalpesh K., Holger Schuhmann and Peer M. Schenk. 2012. High lipid induction in microalgae for biodiesel production. Energies 5(5): 1532–1553. doi: 10.3390/en5051532.

193. Sharpe, Ben. June 2013. Volvo Truck's plan to commercialize DME technology. The International Council on Clean Transportation, 10. https://theicct.org/blogs/staff/volvo-trucks-plan-commercialize-dme-technology.

194. Shriwastav, Amritanshu, Jeenu Thomas and Purnendu Bose. June 2017. A comprehensive mechanistic model for simulating algal growth dynamics in photobioreactors. Bioresource Technology 233: 7–14. https://doi.org/10.1016/j.biortech.2017.02.080.

195. Singh, D., D. Sharma, S.L. Soni, C.S. Inda, S. Sharma, P.K. Sharma and A. Jhalani. 2021. A comprehensive review of physicochemical properties, production process, performance and emissions characteristics of 2nd generation biodiesel feedstock: Jatropha curcas. Fuel, 285, 119110.

196. Singh, Devendra, K.A. Subramanian and M.O. Garg. January 2018. Comprehensive review of combustion, performance and emissions characteristics of a compression ignition engine fueled with hydroprocessed renewable diesel. Renewable and Sustainable Energy Reviews 81(2): 2947–2954. https://doi.org/10.1016/j.rser.2017.06.104.

197. Solana, M., C.S. Rizza and A. Bertucco. 2014. Exploiting microalgae as a source of essential fatty acids by supercritical fluid extraction of lipids: comparison between Scenedesmus obliquus, Chlorella protothecoides and Nannochloropsis salina. J. Supercrit Fluids. 92: 311–318. doi: 10.1016/j.supflu.2014.06.013.

198. Somashekar, D. et al. April 2001. Efficacy of extraction methods for lipid and fatty acid composition from fungal cultures. World J. of Microbiology and Biotechnology 17(3): 317–320. [link.springer.com/article/10.1023/A:1016792311744].

199. Song, J., Z. Huang, X. Qiao and W. Wang. 2004. Performance of controllable premixed combustion engine fueled with dimethyl ether. Energy Conversion and Management 45: 2223–2232.

200. Spatari, S., V. Larnaudie, I. Mannoh, M.C. Wheeler, N.A. Macken, C.A. Mullen and A.A. Boateng. December 2020. Environmental, exergetic and economic tradeoffs of catalytic- and fast pyrolysis-to-renewable diesel. Renewable Energy 162: 371–380. https://doi.org/10.1016/j.renene.2020.08.042.

201. Staples, Mark D., Robert Malina, Hakan Olcay, Matthew N. Pearlson, James I. Hileman, Adam Boies and Steven R.H. Barrett. 2014. Lifecycle greenhouse gas footprint and minimum selling price of renewable diesel and jet fuel from fermentation and advanced fermentation production technologies. Energy & Environmental Science. Issue 5.

202. Steinrücken, Pia, Svein Rune Erga, Svein Are Mjøs, Hans Kleivdal and Siv Kristin Prestegard. September 2017. Bioprospecting North Atlantic microalgae with fast growth and high polyunsaturated fatty acid (PUFA) content for microalgae-based technologies. Algal Research 26: 392–401, Elsevier B.V. doi: 10.1016/j.algal.2017.07.030.

203. Subhas, G. Venkat and S. Venkat Mohan. Jan. 2014. Lipid accumulation for biodiesel production by oleaginous fungus Aspergillus awamori: Influence of critical factors. Fuel. 116: 509–515. http://www.researchgate.net/journal/0016-2361_Fuel.

204. Subramaniam, R., S. Dufreche, M. Zappi and R. Bajpai. 2010. Microbial lipids from renewable resources: production and characterization. J. Ind. Microbiol. Biotechnol. 37: 1271–1287. 10.1007/s10295-010-0884-5.

205. Swaminathan, M.S. May 2011. Working paper on 'Bioenergy Resource Status in India'. Prepared for DFID by the PISCES RPC Consortium. Research Foundation, Chennai, India.

206. Teng, Ho, James C. McCandless and Jeffrey B. Schneyer. 2004. Thermodynamic Properties of Dimethyl Ether - An Alternative Fuel for Compression-Ignition Engines. SAE Technical Paper 2004-01-0093. 26 pages. https://doi.org/10.4271/2004-01-0093. (Also in: Compression Ignition Engine Performance for Use With Alternate Fuels-SP-1825, SAE 2004. Transactions Journal of Fuels and Lubricants-V113-4.

207. Thakur, Vijay Kumar and Manju Kumari Thakur. August 2014. Processing and characterization of natural cellulose fibers/thermoset polymer composites: Review. Carbohydrate Polymers 109: 102–117. https://doi.org/10.1016/j.carbpol.2014.03.039.

208. Tian, Tian and Taek Soon Lee. 2017. Advanced Biodiesel and Biojet Fuels from Lignocellulosic Biomass. Living Reference Work Entry, Consequences of Microbial Interactions with Hydrocarbons, Oils, and Lipids: Production of Fuels and Chemicals, Part of the series. Handbook of Hydrocarbon and Lipid Microbiology, pp. 1–25.

209. Tiron, Olga, Costel Bumbac, Elena Manea, Mihai Stefanescu and Mihai Nita Lazar. July 2017. Overcoming Microalgae Harvesting Barrier by Activated Algae Granules. Scientific Reports. 7, Article number: 4646. doi:10.1038/s41598-017-05027-3

210. Tran, Dang-Thuan, Kuei-LingYeh, Ching-LungChen and Jo-ShuChang. March 2012. Enzymatic transesterification of microalgal oil from Chlorella vulgaris ESP-31 for biodiesel synthesis using immobilized Burkholderia lipase. Bioresource Technology 108: 119–127. https://doi.org/10.1016/j.biortech.2011.12.145.

211. Ullah, Kifayat, Mushtaq Ahmad, Sofia Vinod KumarSharma, Pengmei Lu, Adam Harvey, Muhammad Zafar et al. August 2014. Algal biomass as a global source of transport fuels: Overview and development perspectives. Progress in Natural Science: Materials International 24(4): 329–339. open access, https://doi.org/10.1016/j.pnsc.2014.06.008.

212. Van der Drift, A. and H. Boerrigter. Jan. 2006. Synthesis Gas from Biomass for fuels and chemicals. ECN- Biomass, Coal and Environmental Research, ECN-C--06-001. https://www.researchgate.net/publication/259800555_Synthesis_gas_from_biomass_for_fuels_and_chemicals.

213. Van Straten, Gerrit and Sandor Herodek. June 1982. Estimation of algal growth parameters from vertical primary production profiles. Ecological Modelling. 15(4): 287–311. https://doi.org/10.1016/0304-3800(82)90086-2.

214. Vancov, Tony, Amy-SueAlston, Trevor Brown and Shane McIntosh. September 2012. Use of ionic liquids in converting lignocellulosic material to biofuels: Review. Renewable Energy 45: 1–6. https://doi.org/10.1016/j.renene.2012.02.033 b.

215. Vicente, G, L.F. Bautista, R. Rodriguez, F.J. Gutierrez, I. Sadaba, R.M. Ruiz-Vazquez et al. 2009. Biodiesel production from biomass of an oleaginous fungus. Biochem. Eng. J. 48: 22–27. 10.1016/j.bej.2009.07.014.

216. Viesturs, D. and Ligita Melece. 2014. Advantages and disadvantages of biofuels: Observations in Latvia. Engineering for Rural Development. Jelgava 29.-30.05. https://futureofworking.com/7-advantages-and-disadvantages-of-algae-biofuel.

217. Viviers, J.M.P. and J.H. Briers. Oct. 1982. Harvesting of algae grown on raw sewage. Water SA. 8(4): 178–186. National Institute for Water Research, CSIR, PO Box 395, Pretoria 0001.

218. Wackett, Lawrence P. December 2016. Biofuels (Butanol- Ethanol Production, Living Reference Work Entry, Consequences of Microbial Interactions with Hydrocarbons, Oils, and Lipids: Production of Fuels and Chemicals, Part of the series Handbook of Hydrocarbon and Lipid Microbiology pp. 1–6.

219. Wang, Tiejun, Yuping li, Longlong Ma and Chuangzhi Wu. September 2010. Biomass to dimethyl ether by gasification/synthesis technology—an alternative biofuel production route. Frontiers in Energy 5(3): 330–339. DOI: 10.1007/s11708-010-0121-y.

220. Ward, A.J., D.M. Lewis and F.B. Green. July 2014. Anaerobic digestion of algae biomass: A review. Algal Research 5: 204–214.

221. Williams, R.H. December 2003. A comparison of direct and indirect liquefaction technologies for making fluid fuels from coal. Energy for Sustainable Development 7(4). DOI: 10.1016/S0973-0826(08)60382-8.

222. Wiyarno, B., R.M. Yunus and M. Mel. 2011. Extraction of algae oil from *Nannocloropsis* sp.: A Study of Soxhlet and Ultrasonic-Assisted Extractions. J. of Applied Sciences 11 (21): 3607–3612.

223. Wu, Yin-Hu, Yin Yu and Hong-Ying Hu. February 2014. Effects of Initial phosphorus concentration and light intensity on biomass yield per phosphorus and lipid accumulation of *Scenedesmus* sp. LX1. BioEnergy Research 7(3). DOI: 10.1007/s12155-014-9411-2.

224. Wu, Yiping, Fubo Zhao, Shuguang Liu, Lijing Wang, Linjing Qiu, Georgii Alexandrov et al. 2018. Bioenergy production and environmental impacts', Geoscience Letters, Official J. of the Asia Oceania Geosciences Society (AOGS). 5: Article number 14.

225. Xin, Li, HuHong-ying, Gan Ke and Sun Ying-xue. July 2010. Effects of different nitrogen and phosphorus concentrations on the growth, nutrient uptake, and lipid accumulation of a freshwater microalga *Scenedesmus* sp. Bioresource Technology 101(14): 5494–5500. https://doi.org/10.1016/j.biortech.2010.02.016.

226. Xu, H., U. Lee and M. Wang. Dec. 2020. Life-cycle energy use and greenhouse gas emissions of palm fatty acid distillate derived renewable diesel. Renewable and Sustainable Energy Reviews 134: 110144.

227. Xu, Mingting, Jack H. Lunsford, D. WayneGoodman and Alak Bhattacharyya. February 1997. Synthesis of dimethyl ether (DME) from methanol over solid-acid catalysts. Applied Catalysis A: General. 149(2): 289–301. https://doi.org/10.1016/S0926-860X(96)00275-X.

228. Yaripour, F., F. Baghaei, I. Schmidt and J. Perregaard. February 2005. Catalytic dehydration of methanol to dimethyl ether (DME) over solid-acid catalysts. Catalysis Communications 6(2): 147–152. https://doi.org/10.1016/j.catcom.2004.11.012.

229. Yen H.W., S.C. Yang, C.H. Chen and J.S. Chang. 2015. Supercritical fluid extraction of valuable compounds from microalgal biomass. Bioresour. Technol. 184: 291–296. doi: 10.1016/j.biortech.2014.10.030.

230. Ying, Wang, Zhou Longbao and Wang Hewu. 2006. Diesel emission improvements by the use of oxygenated DME/diesel blend fuels. Int. J. Atmospheric Environment. 40: 2313–2320. DOI:10.1016/J.ATMOSENV.2005.12.016.

231. Ying, Wang, Zhou Longbao, Li Genbao and Zhu Wei. Dec. 2008. Study on the application of DME/diesel blends in a diesel engine. Int. J. Fuel Processing Technology 89(12): 1272–1280. DOI:10.1016/J.FUPROC.2008.05.023.

232. Yoon, Seung Hyun, June Pyo Cha and Chang Sik Lee. Nov. 2010. An investigation of the effects of spray angle and injection strategy on dimethyl ether (DME) combustion and exhaust emission characteristics in a common-rail diesel engine. Int. J. Fuel Processing Technology 91: 1364–1372.

233. Youn, In Mo, Su Han Park, Hyun Gu Roh and Chang Sik Lee. 2011. Investigation on the fuel spray and emission reduction characteristics for dimethyl ether (DME) fuelled multi-cylinder diesel engine with common-rail injection system. Int. J. Fuel Processing Technology 92(7): 1280–1287.

234. Young-Chul, Lee, Kyubock Lee and You-Kwan Oh. May 2015. Recent nanoparticle engineering advances in microalgal cultivation and harvesting processes of biodiesel production: A review. Bioresource Technology 184: 63–72. https://doi.org/10.1016/j.biortech.2014.10.145.

235. Zhang, Wennan. August 2010. Automotive fuels from biomass via gasification. Fuel Processing Technology 91(8): 866–876. https://doi.org/10.1016/j.fuproc.2009.07.010.

236. Zhang, J., X. Fang, X.-L. Zhu, Y. Li, H.-P. Xu, B.-F. Zhao et al. 2011. Microbial lipid production by the oleaginous yeast *Cryptococcus curvatus* O3 grown in fed-batch culture. Biomass Bioenergy 35: 1906–1911. 10.1016/j.biombioe.2011.01.024.

237. Zhang, Yan, Chuyang Y. Tang and Guibai Li. Jun 2012. The role of hydrodynamic conditions and pH on algal-rich water fouling of ultrafiltration. Water Research 46(15): 4783–4789. https://doi.org/10.1016/j.watres.2012.06.020.

238. Zhao, Xiaoming, Meifeng Ren and Zhigang Liu. 2005. Critical solubility of dimethyl ether (DME) + diesel fuel and dimethyl carbonate (DMC) + diesel fuel. Fuel 84(18): 2380–2383. DOI10.1016/j.fuel.2005.05.014.

239. Zhou, Chun-Hui, Xi Xia, Chun-Xiang Lin, Dong-Shen Tong and Jorge Beltramini. Aug. 2011. Catalytic conversion of lignocellulosic biomass to fine chemicals and fuels. Chemical Society Reviews. Issue 11. DOI: 10.1039/C1CS15124J.

240. Zhou, Dr. Chen, Dr. Nanyi Wang, Prof. Dr. Xiaoxing Liu, Prof. Dr. Jürgen Caro, Prof. Dr. Aisheng Huang. Oct. 2016. Efficient synthesis of dimethyl ether from methanol in a bifunctional zeolite membrane reactor. Angewandte German Chemical Society 55(41): 12678–12682. https://doi.org/10.1002/anie.201604753.

241. Zittelli, Graziella, Liliana Rodolfi, Niccolo Bassi, Natascia Biondi and Mario R. Tredici. 2012. Chapter 7 Photobioreactors for Microalgae Biofuel Production. In Michael A. Borowitzka, Navid R. Moheimani. Algae for Biofuels and Energy. Springer Science & Business Media. pp. 120–121. ISBN 9789400754799.

CHAPTER 6

Methane and Biogas

6.1 Sources of Methane [10,17,89,104,259,263]

It is well known that 'Methane (CH_4)' is a major 'greenhouse gas'. The 'global warming potential (gwp)' of CH_4 is about 21–25 times that of CO_2. However, methane is a very useful fuel. The main sources of *'non-renewable'* methane are mentioned below:

- Natural gas recovered from underground/sea/oceans by deep drilling (natural gas has about 94% methane).
- Gas from Gas hydrates in the oceans.
- Gas emissions from coal mining.
- Gas generated by gasification of nonrenewable fuels like coal, lignite, petroleum products, etc.
- Natural emissions from oceans when there is a rise in temperature.

Biogas, whose main useful content is methane, is emitted as detailed in the next paragraph. The biogas as well as the gas generated from gasification of organic matter are sources of renewable methane.

6.1.1 Sources and Uses of Biogas [23]

The main sources of renewable methane are mentioned below:

- Biogas* production by anaerobic decay. *(Biogas is a gas produced by anaerobic digestion (in the absence of oxygen) of organic material, largely comprising of methane and carbon dioxide. Biogas is often called 'marsh gas' or "swamp gas" because it is produced by the same anaerobic processes that occur during the underwater decomposition of organic material in wetlands. The typical composition of biogas is: CH_4: 50–70%; CO_2: 30–50 %; and small amounts of hydrogen (H_2), carbon monoxide (CO), compounds of sulphur, halogens, etc.)*
- Landfill gas, which is a kind of biogas, contains CH_4: 40–45%; CO_2: 35–50%; etc. *(The landfill gas is generated by the anaerobic decay of landfill (organic) materials, which is mostly municipal solid wastes.)*
- Gas generated by the thermal gasification of 'organic matter'. *(The composition of gas would depend on the composition of organic matter, the gasification method, the conditions under which the gasification is carried out and on the oxidant used. Since it is not generated by biological processes, it is not termed as 'biogas'; it is called 'renewable gas'.)*
- Gas generated in biological/chemical processes from renewable materials. (Gas composition would depend on the process, materials and other conditions.)

- Methane gas emission from wetlands and paddy cultivation by 'biomass decay' *(Recovery of this methane emission is not feasible as it is widely distributed over the field in small concentration)*.

- Methane gas emission by 'livestock'. *(Recovery of this methane is not feasible, because the emissions are small, non-continuous and distributed.)*

Pathways of Production or Recovery of Methane from Renewable Sources

➤ The production of biogas by anaerobic processes' from sewage, cattle dung, industrial effluents, etc., in an organised manner.

➤ The production of methane by the gasification of biomass/lignocellulosic waste material from plants by chemical, thermal, or biological processes.

➤ The recovery of landfill gas from landfill areas, which is also a kind of biogas.

The main use of methane is as a 'gaseous fuel' for boilers, gas engines, gas turbines and fuel cells. It is also used as a heating fuel in various applications. Biogas is made up of methane and carbon dioxide and other gases and impurities and is currently being produced mainly at landfills, wastewater treatment plants and biodigesters (anaerobic digesters), which are key to the process of making biogas from organic wastes, including agricultural wastes.

Biogas must be cleaned and upgraded to be used as a good fuel for power generation and transportation. (See *Para 6.11 on upgradation of biogas or landfill gas.*) With the appropriate equipment, it is fully compatible with conventional compressed natural gas (CNG) and liquefied natural gas (LNG). In addition, some biogas systems inject biogas directly into natural gas pipelines; biogas that meets pipeline natural gas quality specifications is called 'renewable natural gas' or 'biomethane'. Users of conventional natural gas, including CNG and LNG, can purchase renewable natural gas credits in much the same way that renewable electricity can be purchased.

6.1.2 *Environmental Benefits of Biogas Recovery and Use* [130,199]

As methane is a very powerful greenhouse gas compared to CO_2, capturing methane from any substrate (for example, waste), which would otherwise be emitted to the atmosphere and utilizing it for other purposes will obviously reduce the GHG impact. Biogas energy is considered to be carbon neutral since carbon emitted by its combustion comes from organic matter that fixes the carbon from atmospheric CO_2. Additionally, utilizing biogas will replace fossil fuels, a main contributor to greenhouse gases. The natural sink for 'methane' is a reaction with the 'OH' ion in 'troposphere' and soil.

It takes about 9 to 15 years' time to remove it from the atmosphere by conversion/degradation to 'carbon dioxide'. By burning methane, CO_2 is formed, which has smaller greenhouse potential than methane.

Biogas is a renewable fuel, which can be produced locally from wastes that are produced constantly (i.e., being renewed), as opposed to fossil fuels, which are a finite resource that often need to be transported over long distances. The use of biogas produces a significantly lower adverse environmental and climatic impact than fossil fuels.

Biogas generation additionally produces enriched organic manure, which can supplement or even replace chemical fertilizers.

Biogas generation is an efficient way of energy conversion (saves fuelwood) and it saves women and children from the drudgery of collecting and carrying firewood, exposure to smoke in the kitchen and time consumed in the cooking and cleaning of utensils. It leads to improvement in the environment, sanitation and hygiene. The household wastes and bio-wastes can be disposed of usefully and in a healthy manner. Anaerobic digestion inactivates pathogens and parasites and is quite effective in reducing the incidence of water-borne diseases. As the technology of biogas is cheap and simple, it is ideal for small-scale application and it can provide a source for decentralized power generation as well and it is a source of employment generation in the rural areas.

The use of biogas as fuel creates insignificant quantities of air pollutants compared to coal, lignite, or petroleum-based oils.

Note: on renewable information numbers (RINs)—*RINs are the credits that refineries use to meet the US Federal Renewable Fuel Standard (RFS), which mandates a certain percentage of the nation's transportation fuels come from renewable resources such as ethanol and biodiesel. The US EPA administers the current RFS program, called 'RFS2', in accordance with the Energy Independence and Security Act of 2007 (EISA) passed by the US Congress, which also created specific renewable fuel categories of cellulosic biofuel, biomass-based diesel and advanced biofuel. Each of these categories is required to reduce greenhouse gas emissions as compared to gasoline over the fuel's entire lifecycle (see Para 1.3 also). RINs are created when an advanced biofuel is sold as a vehicle fuel and the fuel provider can sell the RINs to petroleum refineries. (Based on the US. Environmental Protection Agency reports for the Renewable Fuel Standard Program (RFS2)).*

Biogas has one of the lowest lifecycle greenhouse gas emission profiles when used as a vehicle fuel and can create RIN credits under the "advanced biofuel" category.

6.1.3 Safety Precautions and Dangers in Biogas Generation and Handling [163,230]

The dangers of biogas are mostly similar to those of natural gas, but with an additional risk from the toxicity of its hydrogen sulphide fraction. Biogas can be explosive when mixed in the ratio of 1 part biogas to 8–20 parts of air. Special safety precautions have to be taken for entering an empty biogas digester for maintenance work. Naked flames should not be allowed at the project site when the plant is in running condition.

It is important that a biogas system should never have negative pressure as this could cause an explosion. Negative gas pressure may occur if too much gas is removed or leaked. Therefore, biogas should not be used at pressures below one inch (25.4 mm) column of water, measured by a pressure gauge.

Frequent smell checks must be performed on a biogas system. If biogas is smelled anywhere, windows and doors should be opened immediately. If there is a fire, the gas should be shut off at the gate valve of the biogas system.

6.2 Biogas Production by Biological Decomposition of Biomass [10,42,49,78,89,94,97,98,100,135,143,144,148,166,187,196,240,246,248, 259,263,266,287,309]

Biogas production by biological process is carried out by the action of various groups of anaerobic bacteria on organic materials. The anaerobic bacteria, which grow and act in the absence of oxygen, are slow growing. There is a greater degree of metabolic specialization in these anaerobic microorganisms. Most of the free energy present in the substrate goes over to the final product methane and less energy is available for the growth of microbial mass. The sludge disposal after digestion is not difficult.

In a biogas plant, the anaerobic digester is a closed unit where biodegradable organic substances and animal dung slurry are allowed to be acted upon by microorganisms. The liquid that comes out of the biodigester, having a small quantity of solids (forced out as more undigested material enters the tank), can be used as a convenient growth stimulant for nearby plants. The spent biomass/sludge after yielding the biogas is an excellent fertilizer. It may be mentioned here that the pathogens are eliminated during the biogas generation.

Note: *The first attempt in India to produce biogas by biological decomposition was made at the sewage treatment plant in Dadar, Mumbai, by scientists of Indian Agriculture Research Institute, New Delhi (IARI). The plant was commissioned in 1937. Detailed studies were made by the IARI scientists on the optimum production* *of methane from cow dung. Simple models of biogas plants were installed and commissioned by the Khadi and Village Industries Commission (KVIC), Mumbai and later in other parts of country from 1951 onwards— for example, at the ordinance factory in Kolkata and in several military dairy farms. The Indian Institute of Sciences, Bangalore and the National Dairy Research Institute, Karnal, further studied various factors involved in biogas production and since then, research in the area of biogas production was coordinated by Indian Council of Agriculture Research (ICAR) and involved several universities and institutions.*

The installation period totally depends on the size of the biogas plant to be installed and plants of 1000 kg per day and above may generally require an installation period of 45-60 days. Theoretically, there is no limit on the storage of biogas, but in practice, it is advisable to have a storage capacity that can keep around 12 hours of biogas generation.

6.2.1 *Rich Feedstocks for the Generation of Biogas* [2,12]

Theoretically, any organic material can be decomposed anaerobically to produce biogas, but some materials work better than others. In general, materials rich in energy and easily digestible are good for the purpose of biogas generation. Animal and human wastes are very suitable for it, as they contain lot of nutrients for microorganisms and decompose in a short time. Biodigesters can be fashioned from septic tanks, but the waste production is often not enough to produce enough biogas and cleaning agents (bleach, etc.) kill the anaerobic bacteria necessary for digestion. Plant waste matter is slow to digest and often has lower energy content. For feeding the biodigester, one cannot rely solely on plant wastes. Plant material can be used in combination with animal/human wastes. Acidic wastes should be avoided, as they would disturb the anaerobic processes.

Following are good feedstock sources for the generation of biogas:

a. Animal wastes like cattle dung, urine, goat and sheep droppings, piggery wastes, slaughterhouse wastes, fish wastes, left out animal food/wastes, etc.

b. Human wastes like faeces, urine, food and vegetable/fruit waste, left out organic materials, etc.

c. Biodegradable organic waste from industries/hotels/shops/eateries.

d. Biodegradable finely chopped wastes of marine/aquatic plants or weeds, agricultural waste and plant material used in combination with animal wastes.

The biogas generated from the waste of farm animals may contain about 60% methane. It can be made from most of the biodegradable wastes such as municipal solid waste, sewage, cattle and poultry waste, distillery spent wash, black liquor from paper mill, effluents from agro-industries, residues from agro-industries, etc.

The average amount excreta generated per day by full-grown animals for the purpose of estimation may be taken as: (i) cows: 8–10 kg, (ii) buffalos:10–15 kg, (iii) Horses: 5–7 kg, (iv) small animals like pigs/goats/sheep: 1.5–2 kg.

For the production of biogas, the average figure of requirement of cattle dung may be taken as 25–32 kg per day for a 1 cubic meter plant. The cattle dung takes about 30–50 days to ferment, depending on ambient temperature. For lower ambient temperatures, the retention period will be higher (for 10–15°C ambient, a retention period of 55 days may be required). The period of digestion can be reduced by mixing digested sludge with the waste. About 10 kg of dried manure/day can be obtained by feeding 25 kg fresh dung/day in a biogas plant.

For sludge digestion (mesophilic) and biogas generation, the working conditions are:

(i) Optimum range of pH value 7–7.2; general limits of pH 6.7–7.4

(ii) Optimum temperature range of 30–35°C

(iii) Solid content of slurry 8–10%

(iv) Carbon to nitrogen ratio (C/N) of about 30

(v) Total alkalinity concentration as $CaCO_3$ – 2000 – 3500 mg/liter

The slurry of cattle dung is made by mixing 1kg of wet dung with 1kg of water. For a good result (of gas generation), the carbon to nitrogen ratio may be maintained at about 30 by mixing different kinds of waste materials in required proportions in dung (see Table 6.1a). The cattle dung has about 1.7% nitrogen on dry basis. The carbon to nitrogen ratio C/N in dry dung may be taken as 25.

Apart from major components like methane and carbon dioxide, minor components are present as well. Hydrogen sulphide is formed during the microbial reduction of sulphur compound like sulphates, peptides, amino acids, etc. The halogens have their origin in waste materials. Ammonia may form during the bacterial degradation of proteins. Nitrogen and oxygen may come from the various compounds in waste and also from the air. There are also some other impurities like siloxanes and particles in biogas. Siloxanes are compounds containing silicon-oxygen bonds, which may be present in wastes. (Siloxanes are used in products such as deodorants and shampoos).

The composition of biogas obtained from different sources is slightly different, as indicated in Table 6.1b, but it largely depends on the composition of wastes. The equivalent quantities of various fuels for one of biogas are given in Table 6.1c.

Table 6.1a. Carbon to Nitrogen Ratio of Various Materials.

Se. No.	Material	Nitrogen Content (%)	Ratio of Carbon to Nitrogen
1.	Urine	15.18	8:1
2.	Cow dung	1.7	25:1
3.	Poultry manure	6.3	N.A.*
4.	Night soil	5.5–6.5	8:1
5.	Grass	4.0	12:1
6.	Sheep waste	3.75	N.A. *
7.	Mustard straw	1.5	20:1
8.	Potato tops	1.5	25:1
9.	Wheat straw	0.3	128:1

* N.A. Data Not Available

Table 6.1b. Composition of Biogas from Various Sources. [17]

Biogas components	Animal & Agricultural Wastes	Sewage	Landfills
Methane	55–75%	55–65%	40–45%
Carbon dioxide	25–45%	30–40%	35–50%
Nitrogen	0–10%	0–10%	0–20%
Hydrogen sulphide	0–1.5%	Up to 200 ppm	About 200 ppm depending on the composition of wastes
Water vapour	Saturated	Saturated	Saturated
Halogens	Trace amount	Up to 4ppm	Amount dependent on the composition of wastes
Higher hydrocarbons	Trace amount	Trace amount	Up to 200 ppm
Ammonia	Up to 100 ppm	Up to 100 ppm	Up to 5%

Impurities like siloxanes and various particulates may be present.

Table 6.1c. Equivalent Quantities.

Name of the fuel	Kerosene	Fire-wood	Cowdung cakes	Charcoal	Soft coke	Butane	Furnace Oil	Coal gas	Electricity
Equivalent quantities to 1 m^3 of Biogas	0.620 L	3.474 kg	12.296 kg	1.458 kg	1.605 kg	0.433 kg	0.4171	1.177	4.698 kWh

6.2.2 *Factors Affecting Biodigestion and Biogas Generation* [38,100,126,248]

(1) Temperature of the 'fermenting sludge'. *[Fermentation in the methane reactor occurs in mesophilic conditions for which the optimum temperature range is 30–35°C and also in thermophilic conditions for which the optimum temperature range is 50–55°C.]* Tropical climates generally have no problems with temperature because the anaerobic bacteria thrive in higher temperatures. But in subtropical areas (where the temperature in winter drops) and in low-temperate climates (where temperature may be generally low), there is a need for heating in colder months to achieve the required temperature, say 30–35°C.

(2) A 'dose' of fresh sludge, that is, the loading rate and ratio of fresh and septic sludge. *[The acid-forming bacteria proliferate fast and increase in numbers, but the methane- forming bacteria reproduce and multiply slowly and it is advantageous to increase the number of methane-forming bacteria by seeding it with digested sludge, which is rich in methane-forming bacteria].*

(3) The biogas potential of feedstock and the nature of substrate.

(4) The design of the digester.

(5) The uniformity of feeding and type of feedstocks.

(6) Nutrients. *[The major nutrients are carbon, hydrogen, oxygen, nitrogen, phosphorous and sulphur. Of these nutrients, nitrogen and phosphorous are generally in short supply. To maintain the proper balance of nutrients, an extra phosphorus-rich raw material like 'night soil' and nitrogen-rich material like chopped leguminous plants may be added along with cow dung to obtain a better production of biogas].*

(7) The total solid content of the feed material and type of feedstock.

(8) The 'intensity of mixing' of 'septic sludge' with 'fresh loads'.

(9) The carbon to nitrogen ratio. *(A C/N ratio of about 30 is desirable.)*

(10) *A ratio of N:P:S (typically about 15:5:3).*

(11) A pH value/hydrogen-ion concentration and total alkalinity and accumulation of acids inside the digester. *(The pH value should be kept close to 7. The addition of lime may be resorted to.)* For methanogesis, the oxidation reduction potential (ORP) should be low and no oxidizing agents should be added to the wastes.
.*[OxidationReduction Potential (ORP) is a measure of the tendency of a chemical species to acquire electrons and thereby be reduced. The reduction potential is measured in volts (V) or millivolts (mV). Each species has its own intrinsic reduction potential; the more positive the potential, greater is the species' affinity for electrons and tendency to be reduced. ORP is a common measurement of water quality].*

(12) The retention period in the reactor/digester.
.[The retention period is calculated by dividing the total capacity of the digester by the rate at which organic matter is fed into it. This figure should not be less than 2 to 4 days otherwise there will be a loss of methane-forming bacteria along with the slurry out flow. The retention period calculations are only valid after the initial fermenting periods, when the reactor/digester starts giving out gas properly. Different materials take different amounts of time to get well-fermented. For example, cow/buffalo dung may take 50 days, night soil may take 30 days, pig dung/poultry droppings may take 20 days, etc.].

(13) Toxicity due to the end product. *[If the digested slurry is allowed to remain in the digester beyond a certain period, it becomes toxic to the organisms and the fermentation rate and gas generation may fall. Pesticides and disinfectants from farms may kill bacteria. Synthetic materials are toxic to methanogenic bacteria].*

(14) The pressure in the reactor. *[The pressure on the surface of slurry affects the fermentation; lower pressure is better for fermentation.]*

(15) The presence of certain metals.

(16) The presence of ammonia in more than a certain concentration is toxic to methanogens and therefore, reduces methane formation.

Troubleshooting

In case biogas production drops in the biodigester, one should check the following:

- *Biogas Leakages:* If there is very little biogas, there may be a leak somewhere. If there are no problems with the water levels, one should check the gas fluid separation chamber and its water levels. Leakages should be checked in the biogas balloon and also in the biogas pipeline.

.Non-corrosive materials should be used in construction. Cement and plastic cause no harm to the mixture in the tank, but metals should be avoided for use in the construction of the tank or the piping system through which the biogas travels. Corrosion will lead to many problems and leakages.

- *Temperature Problems:* If temperatures reach below 20°C, there will be a drastic decrease in biogas production. A heating system and temperature control should be reviewed.

- *Problems with the pH of the Biodigester:* The pH value in the biodigester tank should be as close to neutral (7) as possible. Since the anaerobic processes in a biodigester produce acids, the most common pH problem is one of acidity. The biodigester's contents can be tested by litmus paper. A small amount of lime may be added to correct the pH value. Excessive amounts of lime will not be soluble in the mixture and may harm the bacteria and therefore, a lime concentration of 500 mg for every liter of mixture in the biodigester tank should never be exceeded.

- *Review in Light of the Abovementioned Factors and Basics:* There are a number of problems that can arise during the life of a biodigester. To investigate problems, it is best to think back to the basics of what makes a biodigester work (organic material, nutrients, septic sludge/fresh sludge, C/N ratio, strong seals, warmth/ temperature, pH, etc.). The factors affecting biodigestion and biogas generation already mentioned above should be seen through. Things that could possibly harm its functioning should be eliminated.

- *Avoiding the unnecessary introduction of chemicals:* The chemicals and antibiotics that are key to digestion and methanization may be harmful to bacteria. Therefore, unnecessary chemicals should not be introduced into the digester and wastes of livestock that have been given antibiotics recently should not be used.

- *Water sealing:* If Biogas doesn't fill the biogas balloon, the biogas outlet from the biodigester and the pipeline that connects the biogas balloon and the biodigester should be checked. If there is enough pressure to break the water column inside the gas fluid separation chamber, the gas can easily escape from the chamber and will not fill the biogas balloon.

Main Causes of the Failure of Cattle Dung–Based Biogas Plants in Some Villages of Developing Countries

- Animal-based farming has been almost abandoned in villages and farmers have discontinued rearing bullocks and buffalos for farming. Therefore, adequate amounts of cattle dung are not available for biogas plants.

- The installation of oversized plants without a proper assessment of cattle-dung availability.

- A plant may have been installed at a great distance from cattle sheds and hence, there is difficulty in feeding the dung to the plant.

- A longer gas pipeline will have higher pressure drops from the plant to the place of gas utilisation and not enough pressure available for gas flow.

6.2.3 *Effect of Presence of Metals on Biogas Production* [70,75,185,203,226,285,302]

The following effects of metals on biogas production have been reported by various researchers.

a. The effects of additions of the *cobalt (Co), nickel/molybdenum/boron (Ni/Mo/B) and selenium/ tungsten (Se/W)* on the biogas process and the associated microbial community was investigated by experiments on laboratory-scale reactors treating food industry waste. The highest methane production (predicted value: 860 ml per gram of volatile-solids) was linked to high selenium/ tungsten concentrations in combination with a low level of cobalt.

   Only a limited influence of the trace metal additions on the bacterial community composition was found, with two bacterial populations responding to the addition of a combination of *Ni/ Mo/B*, while the dominant archaeal populations were influenced by the addition of *Ni/Mo/B and/or Se/W*. The maintenance of methanogenic activity was largely independent of the archaeal community composition, suggesting a high degree of functional redundancy in the methanogens of the biogas reactors. Moderate, but significant effects of trace element additions were found on the methane production efficiency, (that is, a 7–15% increase as a result of the presence of a Se/W supplement.) [57]

b. An investigation was conducted on the effects of nanoparticles (NPs) of certain metal-oxides. It was found that the nanoparticles composed of *metal oxides ZnO, CuO, CoO, Mn$_2$O$_3$, CO$_3$O$_4$, Ni$_2$O$_3$ and Cr$_2$O$_3$* exerted significant growth-inhibitory effects by causing membrane damage or oxidative stress responses. The nanoparticles of *certain metals or their oxides* in very small concentration may promote microbes and activities of some enzymes. [75]

c. As per some studies, biogas production is influenced by some metals. The methanogenic population is increased by the addition of certain metals in very small quantities individually or in combination in bioreactors and thereby, causing an enhancement in biogas production. These metals *are iron, cobalt, calcium, magnesium, molybdenum, nickel.* [94]

d. It was observed that the addition of nickel at 2.5 ppm increased the biogas production from digesters fed with a water hyacinth and cattle-waste blend. This they attributed to the higher activity *of nickel-dependent metalenzymes* involved in biogas production. [187]

e. It was reported that 600 µg of cadmium per gram of dry matter and 400 µg of nickel per gram of dry matter increased the biogas production and methane content. [89]

f. It was found that the addition of *borax and diborane* at 0.2 g/L increased the gas production from digesters fed with water hyacinth as the substrate. [121,246]

6.3 Biogas from Industrial Wastewaters with a High Organic Load [58,79,167,241]

Wastewaters from sugar factories, paper and pulp factories, breweries and distilleries, etc., have a high organic load. These wastewaters, if pretreated by 'anaerobic method' (before the main treatment of wastewater), can generate biogas.

   The wastewater is first filtered through a 'screen', then sent to a 'buffer/pre-souring tank', from which the effluent is led to a bioreactor via a heat exchanger to bring it to a suitable temperature. The biogas generated in the reactor is sent to a gas holder from where the gas can be sent for utilization. The liquid effluent from the bioreactor is led to a settling tank via a 'degasifier', from which the anaerobically treated effluent is sent to the 'main wastewater treatment plant,' where it receives suitable other treatments, including aerobic treatment. The emissions from the buffer/pre-souring tank, degasifier and settling tank are released into the atmosphere via a 'biofilter' for suppression of odors. A rough estimate of biogas generation from industrial effluents is given in Table 6.1d.

   There is a possibility of biogas production from the 'cake from sugar cane juice filter' in the sugar factory. About 80 liters of biogas per kilogram of fresh cake can be obtained.

Table 6.1d. Possible Biogas Production from Some Industrial Wastes. [58,79,167,241]

Name of Industry	Capacity	Waste in m³/day	*COD mg/L	Organic load kg/day	Possible biogas production m³/day
Distillery	30 m³	450 spent wash	100,000	45,000	14,625
Sugar	2,500 MT	500	3,000	1,500	485
Paper and pulp	20 MT	300	30,000	9,000	2,925
Dairy	50 m³	150	4,000	600	195

COD — Chemical oxygen demand

For using anaerobic processes for industrial wastewater treatment, it is desirable to mix sewage in the wastewater from the industry.

The maximum use of the anaerobic sludge blanket process is being done in the following type of industries (apart from sewage treatment plants) for treatment of their wastewaters.

- Distilleries, breweries
- Cane sugar industries
- Beet sugar industries
- Pulp and paper industry
- Vegetable and fruits processing
- Fruit juices, soft drinks production
- Starch, glucose, fructose productions
- Yeast industry and bakeries
- Milk Dairies, milk processing and dairy products industries

The biogas generation from the *spent wash* of distilleries is described in Para 6.8.

The effluent created after biogas generation and primary treatment is good for irrigation as it contains a lot of nutrients for the plants.

* Note on BOD and COD

Biochemical oxygen demand (BOD) is the quantity of oxygen required by a definite volume of water for oxidizing the organic matter contained in it by microorganisms (seed) under specific conditions. For its determination, the dissolved oxygen content of the sample with or without dilution is measured before and after incubation at 20°C for five days (or at 27 for three days). It is expressed in mg of oxygen required per liter of sample.

Chemical oxygen demand (COD) is the amount of oxygen required by organic and inorganic matter in a sample of water for its oxidation by a strong chemical oxidizing agent such as $K_2Cr_2O_7$. It is expressed as ppm of oxygen taken from an oxidizing agent (such as $K_2Cr_2O_7$) in two hours.

There is a close relationship between organic matter and dissolved oxygen in water. Water pollution by organic matter is measured by biochemical oxygen demand (BOD).

If a BOD value of a sample of water is 100 mg/L, it indicates that the biodegradation of organic matter in one liter of the sample would consume 100 mg of oxygen.

The selection of the microorganism seed is very important. The results of the BOD test are not reproducible. The substance decomposed in the BOD test may be food used by microorganisms or certain chemicals that are readily attacked by oxygen, probably with the help of the enzymes released by the microorganisms. Sulphates, sulphites, sulphides, ferrous ions, etc., are easily oxidizable compounds.

Nitrates, nitrites and phosphates in domestic sewage or in industrial waste are the pollutants that primarily stimulate the growth of the microorganisms, which may cause eutrophication of the ponds/lakes, bloom of microorganisms, algae, plankton, etc. Bacterial decomposition of organic matter and its decay and degradation consumes dissolved oxygen and may create a condition when most fish cannot survive in water. The BOD values of sewage may be 200 mg/L to several hundred mg/liter. The BOD test is usually influenced by:

 (i)   *The type of microorganism seed*

 (ii) *Nitrification process*

 (iii) *Mineral matter content*

 (iv) *Toxic material presence*

 (v)   *pH value of water*

   Due to this, it is not good to completely rely on BOD values and therefore, the COD test should be readily performed, as the COD test is not very influenced by the above factors.
   It is not necessary that COD values correlate with BOD values. (The oxygen consumed for the oxidation of NH_3 for conversion to nitrates and nitrites is not included in the COD value.)
   Textile wastes, paper mill wastes and other wastes having higher levels of cellulose have a considerably higher COD value as compared to the BOD test value as the microorganisms may not fully consume the cellulose.
   The BOD test is used to measure the degree of wastewater strength as its value approximates the amount of oxidizable organic matter. BOD values are useful in:

 (i)   *Process design*

 (ii) *Loading calculations*

 (iii) *Measurement of treatment efficiency*

 (iv) *Self-purifying capacity of streams of water and stream pollution control*

The COD test is used in:

 (i)   *Calculation of the efficiency of treatment plants*

 (ii) *Proposing standards for discharging various effluents*

 (iii) *Design and management of treatment plants*

6.4 Kinetics of Anaerobic Fermentation [21,29,34,35,36,57,66,67,72,74, 91,108,129,154,162,172,176,244,245,247,284]

a. Some kinetic models have been developed for the anaerobic fermentation process. The 'Monod model' shows a hyperbolic relationship between the exponential microbial growth rate and substrate concentration.

For low substrate concentration, the model has two kinetic parameters, that is, the microorganism growth rate and the half velocity constant, which are deterministic in nature and predict the conditions of the timing of maximum biological activity and its cessation. The rate of substrate utilization (r_S) is given by the following equation in this model:

$$r_S = (q_{max} S_x/K) + S,$$

where S is the limiting substrate concentration, K is the half velocity constant, x is the concentration of bacterial cells and q_{max} is the maximum substrate utilization rate.
For a high substrate concentration, the equation is as follows :

$$r_S = q_{max} . x$$

The Monod model suffers from the drawback that one set of kinetic parameters are not sufficient to describe the biological process for both short and long retention times and that kinetic parameters cannot be obtained for some complex substrates.

To remove this drawback, Hashimoto developed an alternative equation describing the kinetics of methane fermentation involving more parameters as follows:

For a given loading rate (S_0/HRT), the daily volume of methane per volume of digester depended on the biodegradability of the material (B_0) and kinetic parameters μm and K.

$$r_V = \{B_0 \cdot (S_0/\text{HRT})\} \cdot [1- \{K/(\text{HRT}. \mu m - 1 + K)\}]$$

Where, r_V *is the volumetric methane production rate per volume of digester per day (volume of* CH_4 *in liters per unit volume of digester in liters/day.)*
S_0 *is the influent total volatile solids (VS) concentration, in gram per liter (g* L^{-1}*),*
B_0 *is ultimate methane yield in liters per gram of volatile solids added (liters of* CH_4 g^{-1} *VS added),*
HRT is the hydraulic retention time in days,
μm is the maximum specific growth of microorganism per day,
K is the kinetic parameter (dimensionless).
For the generation of biogas from manure, *μm and K may* vary as:

$$\mu m = 0.013\ t - 0.129$$
$$K = 0.6 + 0.0206\ \exp(0.051\ So)$$

where, t is the temperature in °C. [168]

   b. Various other equations have been suggested for predicting biogas production kinetics. A few examples of such kinetic equations are given below: [91,142,188]
   1. The linear equation of the biogas production rate in the ascending and descending limb can be expressed by the equation given below. It is assumed that biogas production rate will increase linearly with an increase in time and after reaching a maximum. After this, it would decrease linearly to zero with an increase in time. [91]

$$y = a + bT \tag{1}$$

Where y = biogas production rate in ml/g/day; T = time in days for digestion; a (ml/g/day) and b (ml/g/day²) are the constants obtained from the intercept and slope of the graph of y vs. T. For the ascending limb, b is positive and negative for the descending limb.

   2. The exponential plot for the ascending and descending limbs can be presented by equation (2). Here it is assumed that biogas production rate will increase exponentially with increase in time and after reaching the maximum value, it will decrease to zero exponentially with increase in time. [142]

$$y = a+ b\ \exp(cT) \tag{2}$$

where, y = biogas production rate (ml/g/day); T = time needed for digestion (days); a, b = constants (ml/g/day); c = constant (L/day). For the ascending limb, c is positive and it is negative for the descending limb.

   3. The Gaussian equation shown in Equation (3) can be applied to simulate biogas production rates, including both ascending and descending limbs, assuming that biogas production rates will follow the normal distribution over the hydraulic retention time.

$$y = a \exp [-0.5 \{(T-T_0)/b\}^2] \tag{3}$$

here,
y = biogas production rate (ml/g of volatile solids /day);
T = time needed for digestion (days), that is, hydraulic retention time; a ((ml/g of volatile solids/day); b (days) are constants, T_0 = is the time at which maximum biogas production occurs.

   4. Logistic growth equation, the **Gompertz equation** and exponential rise to maximum are used to simulate the cumulative biogas production. Equation (4) represents the linear model, Equation (5) represents the **Gompertz model** and Equation (6) represents the exponential rise to the maximum model: [29]

$$y = a / \{1 + b \exp(-kT)\} \tag{4}$$

Where, y = cumulative biogas production (mL/g of volatile solids); k = kinetic rate constant (1/day); T = HRT (days); a,b are the constants.

$$y = a \exp(-be^{-ct}) \tag{5}$$

Where y is cumulative biogas production (mL/g of volatile solids), T is hydraulic retention time in days, a and b are positive numbers, a is the biogas production potential in mL/g, b is the minimum time required to produce biogas (days) and c sets the growth rate (y-scaling), which are constants.

$$Y = a \{1 - \exp(-kT)\} \tag{6}$$

Where **y** is the cumulative biogas production (mL/g of volatile solids), **a** is the biogas production potential (ml/g), k is the kinetic rate constant and **T** is the hydraulic retention time in days. [115]

6.5 Chemistry of Biogas Production from Organic Wastes [22,153,155,175,194,259,266,276,288]

Main steps in biogas production from organic waste are:

(1) Hydrolysis

(2) Acidogenesis

(3) Acetogenesis

(4) Methanogenesis

The processes involved in these steps are described below.

[It may be mentioned here that <u>mesophilic bacteria</u> act in the temperature range of 20 to 40°C (for which the optimum temperature range is 30–35°C); the <u>thermophilic bacteria</u> act best in the temperature range of 50 to 55°C and the <u>psychrophilic bacteria</u> (cold-loving) best grow between 4 to 10°C.]

1. Hydrolysis of Organic Waste

The hydrolysis converts the organic matter to soluble compounds like glucose, fructose, galactose, amino acids and fatty acids by action of the enzymes excreted by the bacteria (like lipolytic bacteria, proteolytic bacteria, celluloytic bacteria, etc.).

(a) *Carbohydrates* of the polysaccharide class present in wastes are converted to simple sugars of monosaccharides types by the action of bacteria/enzymes/acids during hydrolysis.

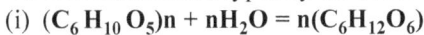

(i) $(C_6 H_{10} O_5)n + nH_2O = n(C_6H_{12}O_6)$

.*Polysaccharides on hydrolysis are converted to glucose, fructose, maltose, etc.*
.*starch, cellulose*

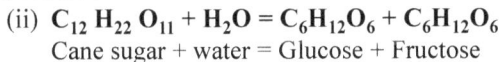

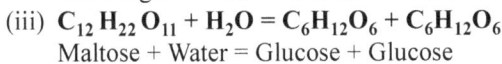

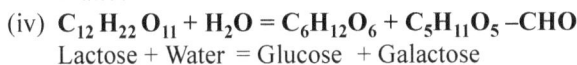

(ii) $C_{12} H_{22} O_{11} + H_2O = C_6H_{12}O_6 + C_6H_{12}O_6$
Cane sugar + water = Glucose + Fructose

(iii) $C_{12} H_{22} O_{11} + H_2O = C_6H_{12}O_6 + C_6H_{12}O_6$
Maltose + Water = Glucose + Glucose

(iv) $C_{12} H_{22} O_{11} + H_2O = C_6H_{12}O_6 + C_5H_{11}O_5 - CHO$
Lactose + Water = Glucose + Galactose

(b) *Proteins and Lipid in Waste or Wastewater:* Lipo-proteins contain lipids and amino acids. Lipids are fatty substances. It may be noted that lipids, if present in waste, can be simple lipids or complex lipids. Simple lipids are esters of glycerol with higher fatty acids. Complex lipids contain nitrogen and phosphorous. Lipids on hydrolysis will give fatty acids like formic acid, acetic acid, propionic acid, higher fatty acids, etc. All proteins, in general, on hydrolysis by bacteria/enzyme/acid will be broken up into amino acids.

2. Fermentation or Acidogenesis

The monosaccharide sugars formed as a result of hydrolysis of 'carbohydrates' and amino acids and other acids as a result of hydrolysis of 'protein and lipids' on *fermentation* (i.e., acidogenesis) by *fermentation bacteria* will form products (*of acidogenesis*) like volatile fatty acids, lactic acid, alcohols, hydrogen, carbon dioxide, ammonia, hydrogen sulphide, etc.

3. Acetogenesis

The products of the acidogenesis process are further fermented by *hydrogen-producing bacteria* and are converted to acetates, acetic acid, hydrogen and carbon dioxide.

4. Methanogenesis

After the fermentation processes of acidogenesis and acetogenesis, the products obtained (such as acetates, acetic acid, formic acid, methyl alcohol, ethyl alcohol, butyl alcohol, formaldehyde, etc.) are acted upon by methanogenic bacteria, producing mainly methane (CH_4) and carbon dioxide (CO_2).

Biochemical Reactions in Methanogenesis:

The following types of reactions will take place during methanogenesis:

i. $2CH_3COONa + H_2O = \quad Na_2CO_3 \quad + CO_2 + \; 2CH_4$

.(Sodium acetate)   (Sodium Carbonate)   (Methane)

ii. $CH_3COOH = CH_4 + CO_2$

.(acetic acid)

iii. $(HCOO)_2Ca + H_2O = \quad CaCO_3 \quad + CO_2 + 2H_2$

.(Calcium formate)   (calcium carbonate)

iv. $CO_2 + 4H_2 = CH_4 + 2H_2O$

v. $HCOOH \quad + 3H_2 = CH_4 + 2H_2O$

.(Formic acid)

vi. $2C_2H_5OH \quad + CO_2 = CH_4 + 2CH_3COOH$

.(Ethyl Alcohol) . (acetic acid)

vii. $CH_3OH \quad + H_2 = CH_4 + H_2O$

.(Methyl alcohol)

viii. $2C_4H_9OH \quad + CO_2 = CH_4 + 2C_3H_7COOH$

.(Butyl Alcohol) . (Butyric acid)

ix. $HCHO \quad + 2H_2 = CH_4 + H_2O$

.(Formaldehyde)

The famous general equations of 'Buswell and Boruff' [107] for conversion of organic matter to methane by anaerobic digestion is as follows:

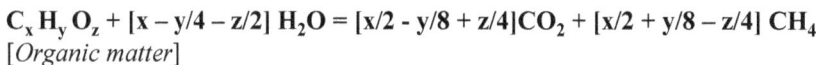

$C_x H_y O_z + [x - y/4 - z/2] H_2O = [x/2 - y/8 + z/4]CO_2 + [x/2 + y/8 - z/4] CH_4$
[*Organic matter*]

For cellulose, this reaction can be written as:

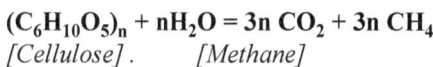

$(C_6H_{10}O_5)_n + nH_2O = 3n\, CO_2 + 3n\, CH_4$
[*Cellulose*] .   [*Methane*]

Note: However, this reaction is very slow.

The chemistry of the anaerobic treatment of wastewater and biomethanization is shown in Figure 6.1.

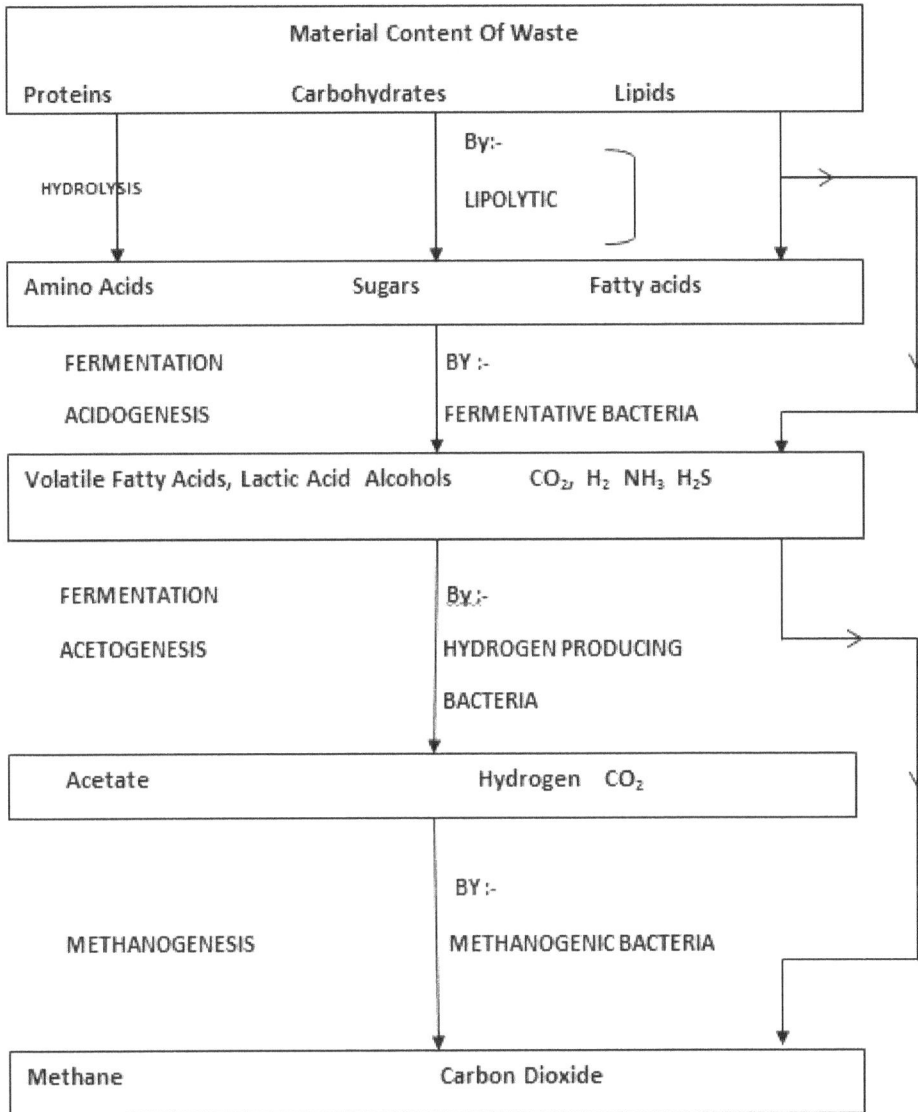

Fig. 6.1. Chemistry of Anaerobic Treatment for Waste Water–Biomethanization.

Biogas generation by anaerobic treatment of industrial wastewater is shown in Figure 6.2.

6.6 Methanization Bacteria [99,148,193,211,212,242,283,291] .

The methane forming bacteria species include the following:

Methanobacterium sohngeniin. sp.; *Methanobacterium smelianskii n.* sp.; *Methanobacterium escherichia formica; Methanobacterium formicum n.* sp.; *Methanobacterium suboxydans n.* sp.; *Methanobacterium propionicum n.* sp.; *Methanococcus mazei n.* sp.; Methanosarcina sp., *Methanosaeta* sp., *Methanobacterium thermoautotrophicum*, etc.

Each species of microorganism acts on different substances present in the organic matter to form methane. Table 6.2 shows the various methanization bacteria and the substances on which they act to produce methane and also the substances by which they cannot produce methane.

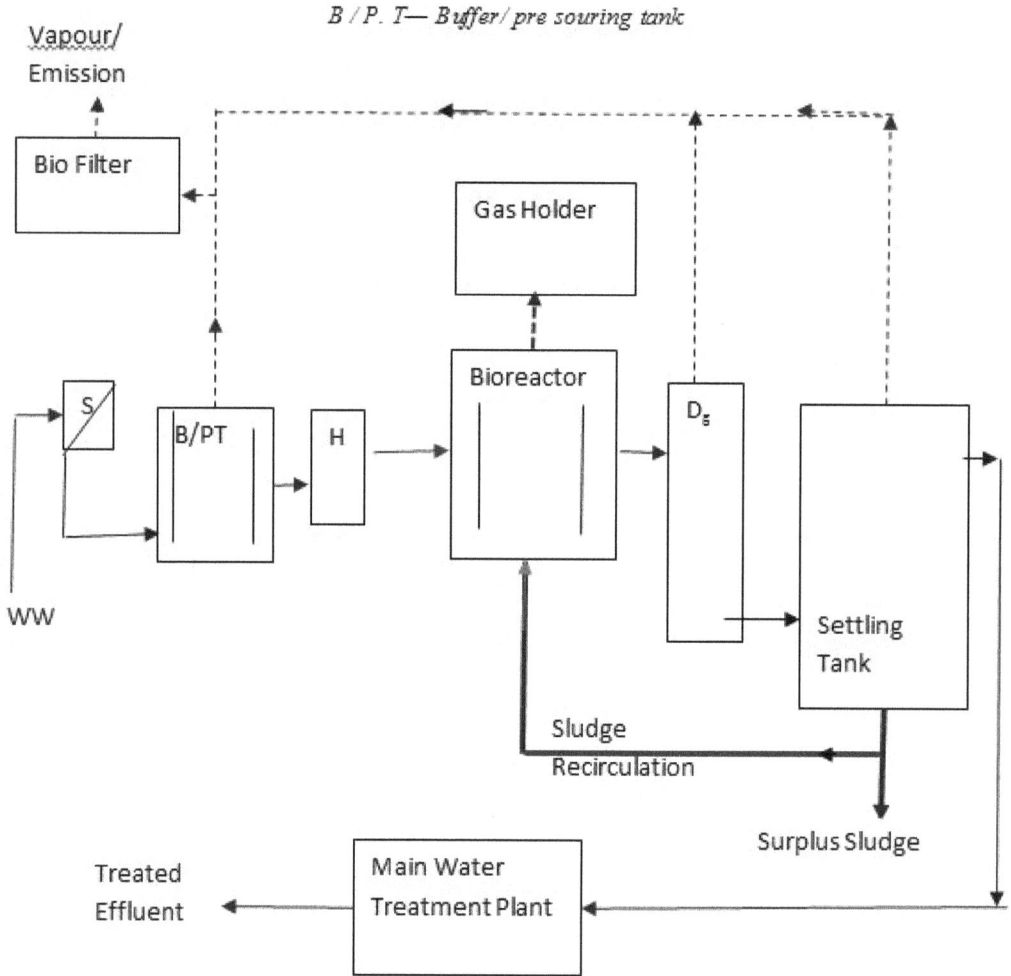

Fig. 6.2. Biogas Generation by Anaerobic Pre-treatment of Industrial Waste Water Before Final Treatment.

6.7 Types of Biogas Plants [8,82,95,225,233,267]

6.7.1 Classification of Biogas Plants According to Process [107,109]

According to process, the biogas plants may be mainly classified as continuous or batch-type process plants. The continuous plants may be of a single-stage process or double-stage process. The continuous process will produce gas continuously, it requires a smaller digestion chamber and requires a lesser digestion period. However, the initial fermenting period is allowed before any output can be obtained in any process.

(a) *Single-Stage Continuous Process:* In this process, the digestion of organic materials and conversion into biogas is completed in a single chamber. The raw materials are fed regularly and spent residue also has to exit out. The single-stage process suits the animal dung–type feed, but with agricultural residues problems arise.

(b) *Double-Stage Continuous Process:* In this process, the acidogenic stage and methanogenic stages are physically separated in two chambers. The first stage is completely the mixed type, but the second stage, the, methanogenic stage is stratified. The dilute acidic effluent goes to

Table 6.2. Methane-Forming Microorganisms. [99,148,211,283,291]

Species	Substances from which methane is produced	Tested substances which do not give methane
1. *Methanobacterium sohngenii* n. sp. This species includes: (a) Thermophilic methane bacterium [61,62] (b) methane bacterium [63,64]	(i) Salts of the following acids: formic, acetic, butyric, caprylic (octanoic), caproic (hexanoic), capric (decanoic) acids (ii) Mixture of carbon dioxide and hydrogen (iii) Salts of the acids like formic, acetic, propionic (slowly) butyric, (very slowly) isobutyric, oxalic, lactic, gluconic acids (iv) Salts of the acetic, butyric acids; ketones; acetone	(i) Salts of the following acids: propionic, valeric, pelargonic (acetylenic carboxylic acid) (ii) Ethyl alcohol
2. *Methanobacterium omelianskii* n. sp. This species includes the second methane bacterium [63,64]	(i) Alcohols: ethyl, propyl, isopropyl, butyl, isobutyl, amyl (ii) Mixture of carbon dioxide and hydrogen (iii) Salts of formic acid; propyl alcohol	(i) Mixtures of the following acids: formic, acetic, propionic, butyric, valeric, malonic, succinic (ii) Alcohols and carbohydrates: methyl glycerol, D-Mannitol*, glucose (iii) Mixture of carbon dioxide and hydrogen (iv) Yeast autolysate
3. *Methanobacterium formicum (Escherichia formica)*[63]	In hydrogen atmosphere formate, formaldehyde, methyl alcohol, hexa-methylene-tetra amine, carbon dioxide, or carbon monoxide.	Compounds containing more than one carbon atom.
4. *Methanobacterium formicum* n. sp. [63]	(i) Salts of formic acid (ii) Mixture of carbon dioxide with hydrogen or water	Salts of other fatty acids
5. *Methanobacterium suboxydans* n. sp.	Salts of the following acids: butyric, valeric, caproic	Salts of acetic and propionic acids
6. *Methanobacterium propionicum* n. sp.	Salts of propionic acid	
7. *Methanococccus mazei* n. sp. This species includes: a. coccus (Groenwege) b. coccus [63]	(i) Salts of acetic and butyric acids (ii) Butyrate and acetone	Ethyl and butyl alcohols
8. *Methanococcus vannielii* n. sp.	Formats, mixture of carbon dioxide and hydrogen	(i) Salts of the following acids: acetic, propionic, butyric, succinic, (ii) Methyl & ethyl alcohols: (iii) Carbohydrates: glucose
9. *Methanosarcina methanica* This species includes sarcina [63]	(i) Salts of acetic and (possibly) butyric acids (ii) Methyl alcohol (iii) Salts of propionic acid	Ethyl alcohol
10. *Methanosarcina barkeriin* sp.	Salts of acetic acid (difficult), methyl alcohol, mixture of carbon dioxide with hydrogen or water	Salts of other organic acids and other alcohol

Notes—*D-Mannitol (alcohol): $C_6H_{14}O_2$; Capric acid (decanoic acid): $C_{10}H_{20}O_2$; valeric acid: $C_5H_{10}O_2$; caproic acid (hexanoic acid): $C_6H_{12}O_2$; pelargonic acid (acetylenic carboxylic acid): **CHCCOOH**

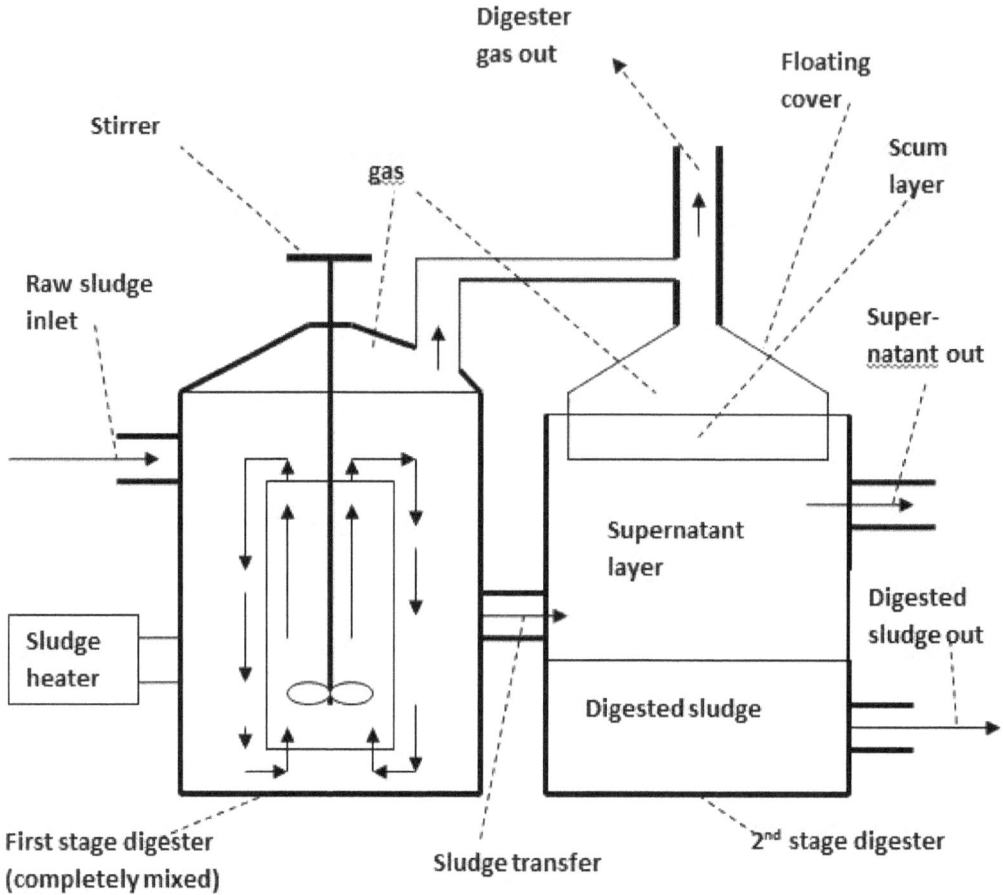

Fig. 6.3. Two Stage Anaerobic Digester.

the second chamber where methanogenesis takes place. The double-stage process can better tolerate a fibrous material feed along with cattle dung. A schematic diagram of a two-stage anaerobic digester is shown in Figure 6.3.

(c) **Batch-Type Process Plants:** In this process, the gas production is intermittent and feed is also provided at intervals. The digesters are charged and sealed for 40 to 60 days and during the process, no further material is added or removed from it. While unloading the spent material, about 10% material is left in the digester. A battery of digesters is charged with feed materials along with some seed material (fermented slurry) and some lime and urea. The digesters are charged and emptied one by one in a synchronous manner to maintain the regular supply of gas through a common gas holder. This design is expensive and only large-scale plants will be economically viable. The long fibrous material (like plant materials, fruits & vegetable wastes, aquatic weeds, industrial solid organic waste, etc.) along with dung/night soil can also be well-digested in batch process.

6.7.2 *Typical Methanization Reactor for Sewage and Industrial Sludge* [12,63,175,177,200,201,240,270,281,306]

The optimum conditions for anaerobic decomposition of organic matter of sewage waste are provided in a methane reactor. There are numerous designs of methane reactors.

In one typical design of the methane reactor, heating is provided by the hot water or steam. Hot water is circulated in the tubes while 'steam' is normally delivered straight into the sludge for heating. The contents are stirred by mechanical agitators or hydraulic pumps. Pumps are usually used to move the bottom layers into the upper layers which loosen the fermentation biomasses by the liberation of ample gases. Sludge pumps are used for the delivery and discharge of sludge. In hot countries, heating may not be required when the ambient temperature is high.

The main condition for the operation of methane reactors is the presence of 'septic sludge', which is heavily populated with 'microorganisms' adapted to the given pollutants. Septic sludge is prepared during the starting period of the sewage treatment plant. Ripened sludge (septic sludge) can be taken from another (working) methane reactor or from another source, for example, from a 'sewage well'. *[Since fresh sludge is fermented very slowly (over several months) and therefore, it is mixed with septic sludge to cut down the time period.]*

If 'septic sludge' and 'fresh sludge' are taken in the proportion of 2:1, the microorganisms quickly adapt to the given pollutant and the starting period of the reaction is thus considerably shortened.

The starting period is accompanied by acid fermentation in which volatile fatty acids are accumulated in the sludge, thus decreasing the pH value by neutralization of some of the alkalinity.

The whole mass has a foul odor because of the liberation of '*Indole* **(Benzopyrrol)**', hydrogen sulphide gas (H_2S) and mercaptans, etc. Fermentation in the methane reactor occurs in mesophilic conditions (optimum temperature 30 to 35°C) and 'thermophilic' conditions (optimum temperature 50 to 55°C). The optimum conditions for the process are selected with respect to the sanitary requirements and the methods of subsequent treatment and utilization.

The decomposing part of the sludge consists mainly of carbohydrates, fats and proteins. Being treated under the same conditions, these components are mineralized at different rates and attain various degrees of decomposition during the same time. The causative agents of methane formation in methane tanks are the same groups of microbes, which are involved in mineralization in septic tanks and settling tanks except that the process is more intense here because of the favorable conditions provided. The process can be accelerated by the addition of concentrated enzyme produced by the bacterial decomposition of organic matter. But these concentrated enzymes will be quickly spent; hence, only activated sludge containing specially prepared bacterial culture concentrates may be added.

Yield of Biogas and Manure from Bagasse and Sugar Industry Trash

National Sugar Institute, Kanpur, UP, India, has evolved a viable technology for producing biogas from wastes of sugar industry comprising of 'bagasse and trash'. In a trial of biogas plant in the institute itself, 10–12 tons of bagasse and trash from the sugar industry produced 2000–2400 cubic meter of biogas and 20–25 tons of moist biomanure (One cubic meter biogas is roughly equivalent to half a liter of petrol in heat value.) The biogas so produced can be used in gas engines to produce mechanical power or be burnt to produce heat.

6.7.3 *Various Constructional Models of Biogas Plants* [279]

There are a number of designs of biogas plants such as:

(a) Fixed dome type biogas plant

(b) Floating gas holder plant

(c) Dry fermentation based fixed dome biogas plant

(d) Upflow anaerobic sludge blanket (UASB) type biogas plant, etc.

These designs also have many variations.

Fixed Dome Type Plant

In the fixed dome type plant digester, the gas holder and digester are combined. This type is suited to the batch process, especially when daily feeding in small quantities is adopted. It has a masonry structure. Dung slurry feed is allowed to ferment in the digester. The gas formed by fermentation rises and collects in the dome, which develops a slight pressure and causes diffusion of the gas. The pressure range may be 0–90 cm of water column. When gas is withdrawn, the pressure will fall. The pressure in the dome is not found to create any serious problem in small plants. Such type of plants are normally constructed underground and thus, are not affected by low atmospheric temperatures. There are no moving parts in this design of plant.

A sketch of a typical circular fixed drum anaerobic digester is shown in Figure 6.4A, while a sketch of a variation developed in India, called 'Deen Bandhu Biogas Plant Design' is shown in Figure 6.4B.

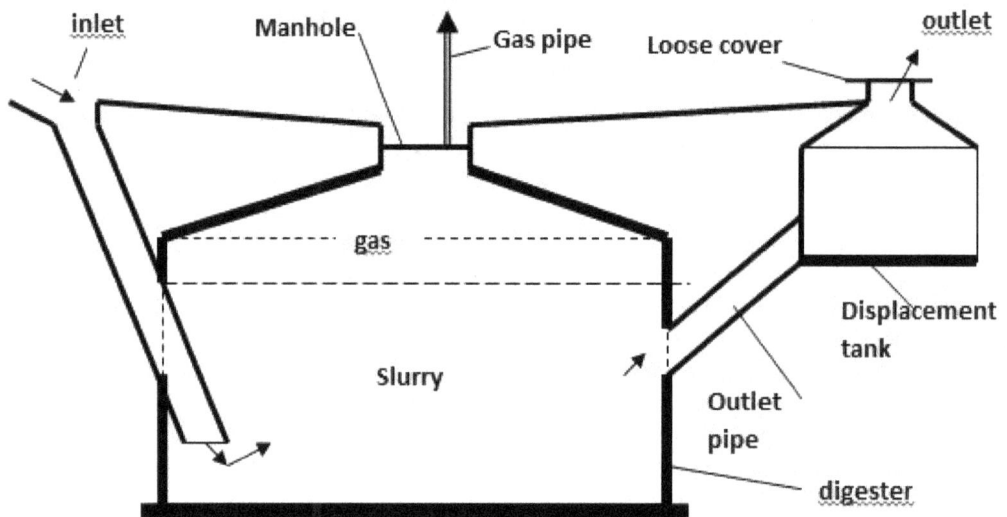

Fig. 6.4 A.   Typical Fixed Drum Type Anaerobic Digester.

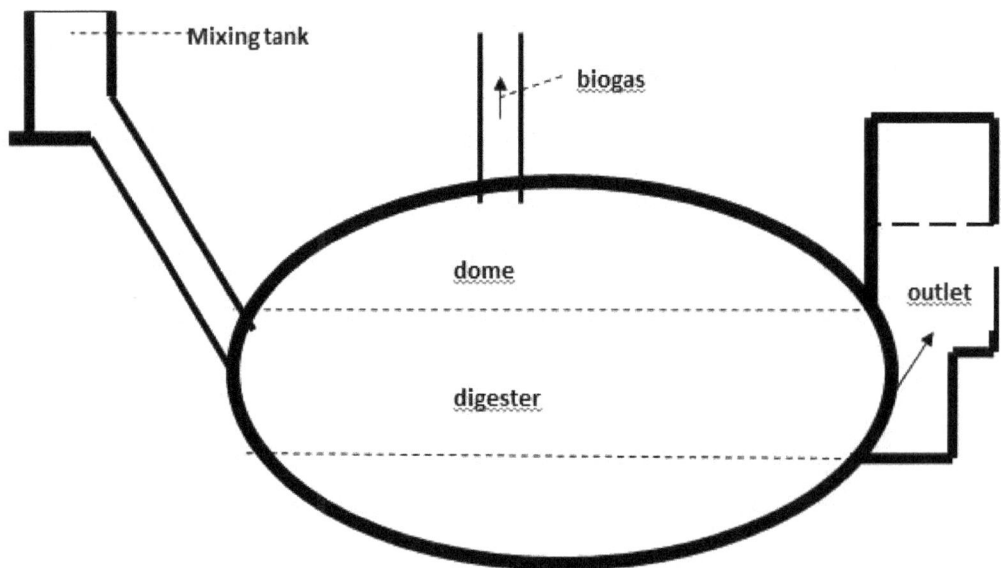

Fig. 6.4 B.   Biogas Plant Design with a Typical Circular Fixed Drum Anaerobic Digester (Deen Bandhu Biogas Plant Design).

Floating Gas Holder Type Plant

Floating gas holder/drum-type models are suitable for non-rocky areas with a low water table. The digester is like 'a well provided with inlet and outlet pipe'. The digester has a masonry construction, while the gas holder is made of mild steel. A sketch of a circular digester with a floating gas holder without a water seal, developed in India by Khadi Village Industries Commission (KVIC) is shown in Figure 6.4C.

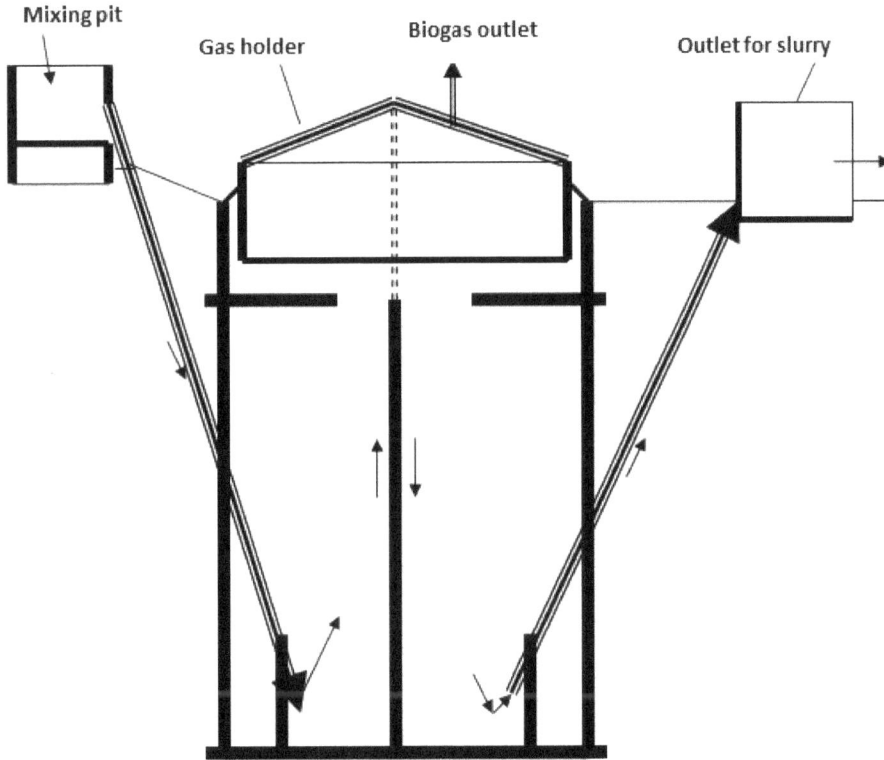

Fig. 6.4C. Circular Digester with a Floating Gas Holder (KVIC-design).

6.8 Biogas from Sewage/Wastewaters [10,78,125,167,220,240,241,280]

Biogas can be generated from wastewater/sewage through an anaerobic treatment. There are several types of anaerobic treatments for sewage/wastewater such as: lagoon, completely mixed digestion process, anaerobic contact process, anaerobic filter and up-flow anaerobic sludge blanket process.

Out of the above processes, the anaerobic sludge blanket process is quite suitable for sewage treatment and simultaneously generating and recovering biogas. In this process, wastewaters with a wide range of COD* (*low to very high but above 500 mg/L and even up to 50,000 mg/L*) can be used to ensure that solid content is not higher than 5000 mg/L. Such a reactor consists of a tank having digestion compartment at the bottom and a gas-liquid-solid separator at the top. The bottom compartment has a sludge layer, which is called the 'Sludge blanket'. The wastewater enters the tank at the bottom under the sludge blanket/layer and passes upward through the layer. The deflectors/baffles are provided to separate solids/gases. The liquid flow is from the bottom to the top and the biogas and effluent go out from the top. The reactor has multiple gas hoods for the separation of biogas. The biogas is tapped, compressed and led to the gas holder.

Cigra Process of Biogas

Biogas generation from sewage water by the 'anaerobic up-flow sludge blanket process' is shown in the Figure 6.5. About 1m³ of biogas may be produced from 3500 US gallons of wastewater.

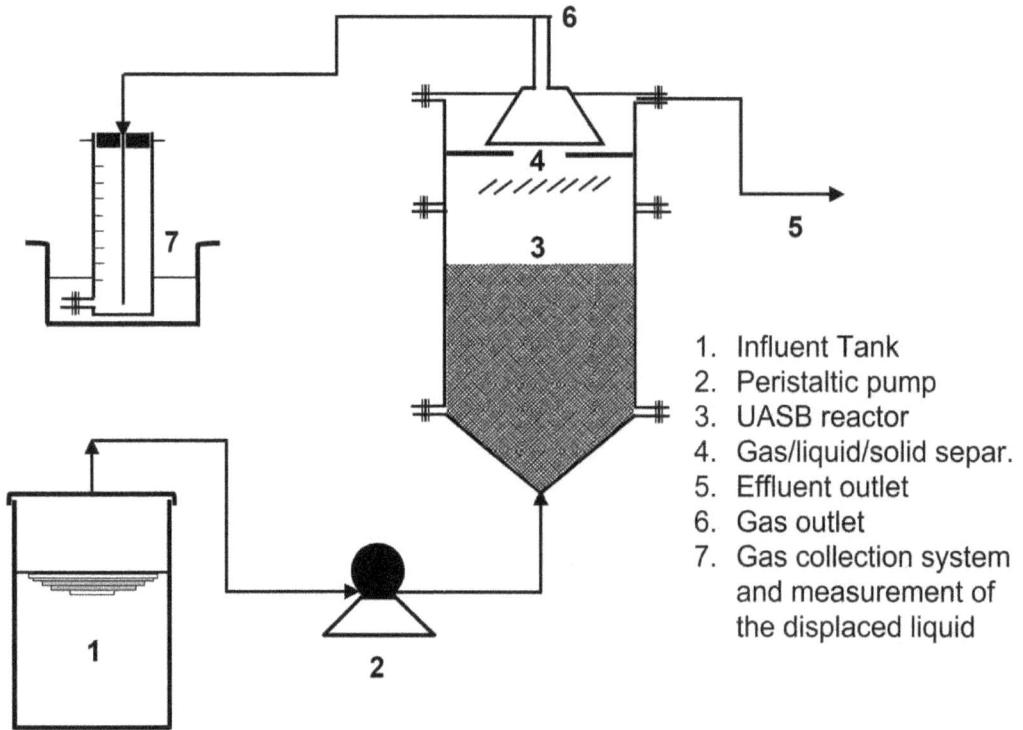

Fig. 6.5. Schematic of Upflow Anaerobic Sludge Blanket Process.
Source: *https://www.iwapublishing.com/news/flow-anaerobic-sludge-blanket-reactor-uasb*

1. Influent Tank
2. Peristaltic pump
3. UASB reactor
4. Gas/liquid/solid separ.
5. Effluent outlet
6. Gas outlet
7. Gas collection system and measurement of the displaced liquid

Production from Sewage

The Cigra process has an in-ground lagoon covered by a plastic sheet with an inflow arrangement for sewage liquid and an outflow arrangement for anaerobically degraded effluent. The biogas generated by anaerobic degradation is recovered through a biogas outlet and is compressed and led to the gas holder. The anaerobic reaction takes place in the lagoon itself. The excess sludge is removed, which can then be used as fertilizer. The effluent from the lagoon can be sent for further aerobic treatment.

6.9　Biogas from 'Distillery Spent Wash' by 'Degremont's Anapulse Technology' [1,3,221,222]

For distillery spent wash, the suspended growth pulsated sludge blanket reactor (*named 'Anapulse treatment process'*) is quite attractive for both biogas generation and reduction of the organic load of the effluent. The process claims to achieve maximum biogas generation and 90% degradation of organic biodegradable pollutants.

Specific features of the Anapulse technology provided by Degremont India Ltd. (SCO-4 Sect 14, Gurgaon, Haryana, India) are:

1. It provides an acid reactor for the hydrolysis-acidification phase, an Anapulse digester for the acetogenesis-methano-genesis phase and a biogas handling system.

2. A pulsating sludge blanket, by cyclic feeding of the raw effluent at the digester bottom, provides effective mixing and contact of the concentrated biomass with incoming effluent while eliminating any need for any moving part inside the digester.

3. There is internal recirculation of the effluent, while mixing at the top end is achieved by a 'gas lifts mechanism', using biogas generated in the digester in a close-loop.

4. An integral clarification zone with special 'lamella modules (tube settlers)' increase the liquid separation effectiveness and ensure separated sludge retention in the digester, thus minimizing the sludge carry over.

5. Lime is used for initial pH correction at the start-up, but later on, the pH correction is achieved by recycling a part of the 'Anapulse digester overflow', which is alkaline (because of alkalinity generated during methano-genesis) in the acid reactor.

6. The effluent from the acid reactor is pumped into a digester at the bottom as it is essentially an up-flow anaerobic sludge blanket process.

7. This technology has been used in many distilleries and claims to achieve biogas generation of 0.45 to 0.50 m³/kg of COD* (*chemical oxygen demand*) destroyed. The technology is supposed to achieve a reduction of 90% BOD* (*biological oxygen demand*) and a reduction of over 65% COD in the effluents.

The flow cycle of the plant would be typically as follows:

The raw effluent goes through a screen, the screened effluent goes to a chamber, from this chamber, effluent is sent to a heat exchanger by feed pumps and the cooled effluent from the heat exchanger goes to acid reactor. Lime is dosed in the effluent from the acid reactor. Part of the effluent from the acid reactor is recirculated in this reactor itself and the rest of it is fed to the Anapulse digester through digester feed pumps. Here, the gas that is evolved is sent to the gas holder and the separated sludge can be sold as manure. The treated effluent is sent for further treatment. A portion of this treated effluent is recycled in the acid reactor.

6.10 Landfill Gas [10,68,149,150,206,207,261,263,271,272,273,274,275]

Landfilling by garbage/municipal solid wastes is a very significant source of biogas, termed as 'landfill gas,' of which about half is methane. The balance of this gas is CO_2 with high contamination by CO, compounds of sulphur, nitrous oxide, halogens, water vapour, etc. The collection of landfill gas is very challenging, but it is a big source of biogas/methane and therefore, should be tapped for utilization, otherwise it will escape into the atmosphere and cause global warming. It may be mentioned here that methane is a much more powerful greenhouse gas than CO_2. In USA, landfilling accounts for about 25% of all methane sources of emission.

Landfills are designed and operated to control potential disease vectors, protect surface water and groundwater, control litter and protect air quality so that there is no open burning as toxins may be released to the air from trash. A safe and efficient landfill has a subbase liner (cell liner); intermediate and final cover (over the trash); final cover (approximately 4 feet deep or an equivalent); leachate (liquid) collection and treatment; litter control; management; inspection programs; safety programs; and methane gas collection/utilization. A leachate drainage system should be created around the dumps.

Fine-grained soils with a high clay content having low permeability (to not allow water to "leak" into groundwater aquifers) are very suitable for landfills.

Hot ashes should not be dropped in landfills as they may cause fires. Fire safety is very important in and around landfills and waste dumps as there are lots of materials and even equipment that can catch fire. The emissions from the landfill may also catch fire.

The wastes should be sorted out to remove plastics, glass, metals, e-wastes, medical wastes, etc., before dumping.

In order to keep watch over the percolation of leachate in groundwater near the landfills or waste dumps, the groundwater should be sampled and tested at regular intervals (say four to six months).

6.10.1 Mechanism of Generation of Biogas in Landfills [14,23,235]

The biogas generation in landfills is predominantly caused by the biodegradation of organic matter in the municipal solid waste. There are four phases of this biodegradation:

• *Phase I:* Aerobic digestion starts shortly after the landfill is filled with waste. This will continue till all the oxygen in the voids in the garbage is depleted. During this phase, the heterotrophic aerobic bacteria/microorganisms produce gas that may have CO_2 (about 30%), water vapor, nitrates and other oxygenated compounds and very small quantity of methane. During this process, when the oxygen level drops below 10–15% (v/v), anaerobic microorganisms get activated. The anaerobic decomposition may continue and last for 6–18 months on the waste at the bottom of the landfill.

• *Phase II:* The aerobic digestion changes to anaerobic digestion. The heterotrophs take oxygen from the nitrates (NO_3^-) and sulphates (SO_4^{2-}) and reduce them to nitrogen (N_2) and hydrogen sulphide (H_2S) in the presence of moisture. Carbon dioxide is also produced in this phase, which may form carbonic acid (H_2CO_3) with moisture. Some organic acids may also be formed. The pH value of leachate decreases.

• *Phase III:* The activity of anaerobic microorganism accelerates in this phase (acidogenesis) and the anaerobes hydrolyse the complex organic compounds like cellulose, fats and proteins into simpler compounds. Organic acids like acetic acid (CH_3COOH), butyric acid (C_3H_7COOH), lactic acid [$CH_3CH(OH)COOH$] and a small quantity of fulvic acid along with some other complex organic acids are produced. Some amount of hydrogen is also produced. Due to the presence of carbonic acid and organic acids, the pH value drops to about 5. The BOD, COD and leachate conductivity increases. Due to the acidity, the metals and other inorganic constituents dissolve away in this phase and consequently the free acid content drops.

• *Phase IV:* As the acidity drops, the methanogenic bacteria multiply and a methanogenic substrate including acetates, methanol and formate forms. The methanogenic bacteria take over and degrade the acetic acid and other volatile organic acids and generate methane and carbon dioxide. With the rise of the pH value, the metals that had dissolved away previously precipitate and the BOD and COD fall. Some formation of organic acids, however, continues simultaneously with the formation of methane.

• *Maturation Phase:* Ultimately, the nutrients for bacteria as well biodegradable organic wastes decline and the gas formation falls off. Some complex stable acids like humic acids and fulvic acids remain in the leachate and the gas formation stops.

Table 6.2a gives an idea about how the composition of the gas generated in landfill varies after the closure of the pit for the first 48 months. A large landfill may continue to produce adequate biogas for over 15–20 years. Slow decomposition and gas production may even take place for some years after that.

6.10.2 Factors Affecting Gas Generation in Landfills [4,151,271,295]

In general, the methane generation rate will be affected by the moisture content in the landfill waste, temperature and pH value prevailing in the landfill, availability of nutrients for methanogens, atmospheric conditions like barometric pressure, temperature and precipitation and the age of landfill at the time of reckoning. The following factors primarily affect the landfill gas generation:

1. *Moisture content:* A moisture content in the waste of about 50–60% is ideal for maximum methane generation, but generally, the municipal solid waste contains an average moisture content of about 25% w/w. If the moisture content in the landfill increases on any account beyond 70%, the gas generation and decay process are significantly slowed down.

Table 6.2a. Typical Landfill Gas Composition after the Closing of the Pit. [149,150,151,210]

Time reckoned from the closing of the pit in months	% Nitrogen	% Carbon dioxide	% Methane
0–3	5.2	88	5
3–6	3.8	76	21
6–12	0.4	65	29
12–18	1.1	52	40
18–24	0.4	53	47
24–30	0.2	52	48
30–36	1.3	46	51
36–42	0.9	50	47
42–47	0.4	51	48

. In arid areas, the moisture content of the waste may be only up to 15%. The decay would be negligible with very little methane emission and the waste may get fossilized. In the lower layer of the landfill, the leachate may increase the moisture content and so some decay may take place. The moisture content is by far the most important factor in the generation of biogas.

2. *Temperature:* The temperature prevailing in landfill may determine the type of bacterial prominence. This, in turn, may affect the gas production. The optimum temperature range for aerobic decay is 54–71°C, while the same for anaerobic decomposition is 30–41°C. As the temperature rises to the range of 35–65°C, there is a rise in the thermophilic bacterial population and an increase in methanogenic activity.

. It may be mentioned that various compounds present in the waste may react chemically and generate heat, heat is also generated by biochemical reactions taking place in the landfill. The heat generated accelerates the rate of chemical reactions.

3. *pH Value Prevailing in the Landfill:* Depending on the biological activity taking place in the landfill, the pH value varies in the range of 5–9. At a lower pH (5–7) value, acidophilic bacteria predominate and there is a reduction of methanogenesis, while in the pH range of 7–8, the methanogenic bacteria predominate. The hydrogen-producing bacteria like *C.thermocellum* and hydrogen-consuming bacteria like *M. thermoautitrophicum*, grow only above the pH value of 6. *At a low pH value, the metal present may dissolve in the leachate and become toxic for bacteria (which are important for biogas generation).*

4. *Nutrients:* Microorganisms in the landfills require nutrients for their growth. The main nutrients are carbon, hydrogen, oxygen, nitrogen and phosphorous and the micro-nutrients are elements like sodium, potassium, sulphur, calcium, magnesium, etc. Most of these elements are present in the garbage and moisture in the landfill. *The addition of sludge from the wastewater treatment plant, sewage treatment process and activated sludge or manure can boost the nutrient supply and enhance the population of bacteria for biodegradation.*

5. *Atmospheric conditions:* Barometric pressure, temperature and precipitation affect landfills. The layer near the surface of the landfill is directly affected. The surface concentration of gas components is reduced and advection (horizontal movement) is created near the surface. The biochemical reaction rate is reduced by a fall in temperature or in moisture content. Precipitation may affect the generation of biogas by supplying moisture/water and dissolved oxygen to the waste (as per factor # 1 mentioned above).

6. *Age of waste:* As already described earlier, there are four phases in the biodegradation of landfills. The peak production of landfill gas will take place 5–7 years after closing the waste in the pit. It may be mentioned that some biogas generation may be there even before the pit is closed, but the gas collection may be difficult before closing the pit.

6.10.3 *Migration of Landfill Gas* [41,68,134,189,206,207,234]

The gas from landfill not only diffuses to the surface, but may also move laterally (called gas migration). The migration direction depends on the permeability of the soil and the garbage filled in the pit. The migration is very likely to occur if the pit has not been provided with a lining. There is more lateral migration of gas in coarse and porous soil than in fine grained soil. When the pit is closed on the top with dense soil or concrete and there is no escape to the surface (no gas collection and lining on the sides of the pit is damaged), migration can easily take place. The higher moisture content in the top layers inhibits the upward movement; similarly, a frozen surface will restrict the upward escape of gas. These conditions will induce the lateral migration of gas.

Apart from soil permeability and moisture content, the following factors also affect the movement of landfill gas:

  i. Depth of groundwater and depth of the pit
 ii. Condition of waste, layering, compaction, porosity, voids and cracks and fissures due to differential settlement, etc.
iii. Manmade feature of the pit like the presence of sewers, culverts, etc.
 iv. Daily addition of garbage to the landfill
  v. Soil cover and the cap of the landfill
 vi. Landfill gas recovery arrangements

Mechanisms of Gas Movements

There are three gas movement mechanisms, namely 'molecular effusion', 'molecular diffusion' and 'convection'.

  1. **Molecular effusion** can take place where the pit has been compacted, but not covered, at the interface between the atmospheric air and landfill. The rate of release according to Raoult's law would be based on the vapor pressure of the compounds present. The principal release mechanism for dry solids is the direct exposure of the waste vapor to the atmosphere.

  2. **Molecular diffusion** is the flow of the biogas from a high concentration area to low concentration area. It is dependent on the difference in the concentrations of biogas and therefore, the biogas generated in the landfill will naturally move to an area where there is low gas concentration or no gas. The movement can be upward or lateral, but will ultimately release into the atmosphere if the same is not sucked/pumped out for utilization. The different constituents of biogas have separate diffusion rates.

  3. **Convection** takes place from a higher pressure area to a lower pressure area (there is a *pressure gradient*) and therefore, the biogas will also flow from the landfill (which is a higher pressure area because of the generation of biogas) to a lower pressure area. The movement by convection is faster than diffusion, although its mostly in the same direction.

6.10.4 *Process of Landfill Gas Recovery* [26,44]

A large pit or crater at the outskirts of city is used for filling garbage/organic wastes. The pits should be lined before filling with garbage to restrict the lateral migration of gas. After the pit is full, it should be covered with soil and consolidated. An impermeable cap of concrete or another material should be provided.

Some vents with non-return valves may be provided at a place where gas accumulates to prevent the migration of gas, but no air should be allowed to enter the system.

After 2–3 months, depending on the climate, borings are done at many places in the entire area over the crater and the perforated pipes are lowered. The biogas is pumped out by suction through these pipes. In the meantime, another crater or pit is used for filling the garbage/organic wastes. When the gas from the particular pit is extracted (the period would depend on the size of the landfill and various other factors), the area can be used for other purposes.

The landfill area after recovery of the gas can also be used for tree plantation *[see Para 2.1.8 Chapter 2]. An attractive option is to recover the compost formed in the landfill for use as manure in agriculture.*

If no mechanical pumping system is provided, the gas recovery system is called the *'passive gas collection system (PGCS)'*. Here, the gas is supposed to move to the collection point on its own.

In the *'active gas collection system'*, the gas is sucked out mechanically through pumps. It is obvious that the active system is much better than passive system. This system has good features like better control over lateral migration, more collection of gas, more reliability, less odor, etc.

6.10.5 *Models for Pre-estimation of Landfill Gas Production* [128,169,243,264]

Various models are being used for estimating the gas production rate from a landfill site:

1. Theoretical stoichiometric model. [261]

2. USEPA Landfill Gas Emission Model (LandGEM) Version3.02 May 2005. [272]

3. The Mexico Landfill Gas Model, Version 2.0 March, 2009. [275]

4. The Ecuador Landfill Gas Model, Version1.0, Feb 2009. [273]

5. The Central America Landfill Model, Version2.0, March 2007. [274]

[Here, only theoretical stoichiometric models have been provided. If the reader is interested in other models, he may consult the references given against each.]

Theoretical Stoichiometric Model

Generally, the anaerobic decay of the biodegradable organic fraction of municipal solid waste is represented by the following type of transformation:
Different types of microorganisms

Organic matter + water + nutrients→ → →CH_4 + NH_3 + H_2S + CO_2 + CO + Other hydrocarbons + new cells + resistant organic matter + heat

The model determines the theoretical methane generation capacity based on stoichiometry using the empirical formula representing the decomposition of waste **($C_aH_bO_cN_d$)**. For practical purposes, the overall conversion of the organic waste in landfills may be assumed to form methane, carbon dioxide, water vapor/water and ammonia, neglecting other factions. This is represented by the following equation:

$$C_aH_bO_cN_d = n\ C_wH_xO_yN_z + m\ CH_4 + s\ CO_2 + r\ H_2O + (d\text{-}nx)\ NH_3$$

Where: (i) Composition of the organic matter before the process is represented by $C_aH_bO_cN_d$, while the composition after the process is represented by $C_wH_xO_yN_z$ on a molar basis and (ii) s = (a – nw – m), and r = (c – ny – 2s)

If it is assumed that the organic wastes are stabilized completely (i.e., all of it has been converted) and that half of the carbon converts to methane and the rest of the carbon is converted to other components, while all the nitrogen is converted to ammonia and all the oxygen in water and organic component is fully utilized (without taking any oxygen from air), the following equation would be representative (*also assuming that conditions are stoichiometric*):

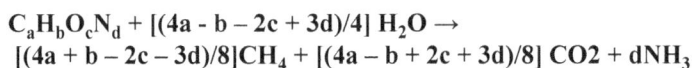

$$C_aH_bO_cN_d + [(4a\text{ - }b – 2c + 3d)/4]\ H_2O \rightarrow$$
$$[(4a + b – 2c – 3d)/8]CH_4 + [(4a – b + 2c + 3d)/8]\ CO2 + dNH_3$$

6.11 Conversion of Carbon Dioxide into Methane [301,297,236,254]

6.11.1 Conversion of CO_2 to CH_4 by Microorganisms [119,186,190,252,301]

A group of microorganisms called 'archaea,' converts carbon dioxide (CO_2) into methane (CH_4) on supply of a minute current, without requiring any organic matter, in the following manner.

- The cathode and anode are put in separate chambers of a two-chamber cell.
- The cathode is covered with a biofilm of archaea and is immersed in water + inorganic nutrients.
- The anode is immersed in water.
- On application of voltage, a flow of electric current results in the evolution of methane.

Electricity should preferably be supplied from a renewable source (solar- or wind-power source). Methane can be generated and stored for use.

The process is in an initial stage of research and requires further research and development for the actual field utilisation of this reverse process to create fuel from carbon dioxide. (*This in fact is a reverse of the microbial fuel cell*). [166]

6.11.2 Photocatalytic Conversion of CO_2 into Methane [20,118,120,299,303]

The photocatalytic conversion of CO_2 using solar energy, (*artificial photosynthesis*), is the most attractive route for the transformation of CO_2 to fuels and chemicals. Many studies have been devoted to the semiconductor-based photocatalytic reduction of CO_2 with H_2O. Several kinds of semiconductors like TiO_2, Ga_2O_3, $ZnGe_2O_4$, $ZnGa_2O_4$, etc., have been reported for their roles in the photocatalytic conversion of CO_2, although their activity is very low.

Co-catalysts are known to play a crucial role in semiconductor-based photocatalysis. The enhancing effects of noble or coinage metal co-catalysts such as Pd, Pt, Au, Ag, or Cu have been studied for the photocatalytic conversion of CO_2 with H_2O to hydrocarbons (in comparison to semiconductors such as TiO_2). The noble or coinage metal co-catalyst may facilitate the separation of photo-generated electrons and holes by trapping electrons, enhancing the photocatalytic activity.

Binary co-catalysts of platinum (Pt) and cupreous oxide (Cu_2O) with a core–shell structure significantly enhance the photocatalytic reduction of CO_2 with H_2O into CH_4 and CO. The Cu_2O shell provides sites for the preferential activation and conversion of CO_2, whereas the Platinum (Pt) core extracts the photo-generated electrons from TiO_2. The deposition of the Cu_2O shell on Pt nanoparticles markedly suppresses the reduction of H_2O to H_2 (see Figure 6.6). [303]

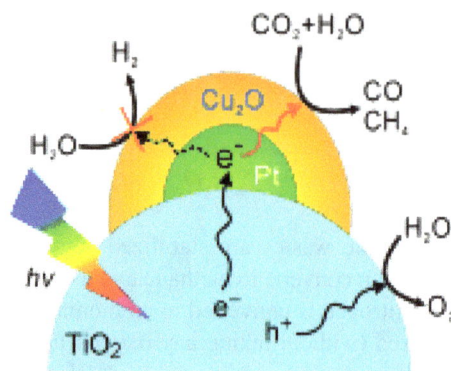

Fig. 6.6. Deposition of Cu_2O Shell on Pt Nanoparticles.

6.12 Upgradation of Biogas/Landfill Gas [18,31,170,231,237,255]

As already mentioned in *Para 6.2.1*, there are a number of noncombustible components, impurities and dilutants in biogas/landfill gas. These components, if removed, would render the gas as a more useful fuel.

a. ***Removal of Water Vapour:*** The generated biogas is saturated with water vapor. The water vapor, if not removed, may condense in gas pipelines and cause corrosion. Water vapor can be removed by cooling, compression, absorption, or adsorption. Increasing the pressure or decreasing the temperature of biogas will cause the condensation of vapor, which can be removed as a condensate.

b. ***Removal of Hydrogen Sulphide (H_2S):*** The concentration of hydrogen sulphide in the biogas can be decreased by precipitation either by the addition of ferric or ferrous chloride or ferrous sulphate. Iron sulphide forms precipitates as it is almost insoluble. The method is primarily used as a first measure in digesters with a high sulphur concentration.

H_2S can be adsorbed on activated carbon or on iron hydroxide/ferric oxide–coated wooden chips. It may be removed by washing with a caustic soda solution.

Hydrogen sulphide can be removed by a biological treatment as it can be oxidized by microorganisms of the species *Thiobacillus* and *Sulfolobus*.

H_2S can also be removed while removing CO_2.

c. ***Removal of Ammonia:*** A separate cleaning step for ammonia is not necessary as it is soluble in water and gets removed during the removal of water vapor or during the removal of other components during upgrading.

d. ***Removal of Siloxanes:*** When siloxanes are burned, silicon oxide, a white powder, is formed, which can create a problem in gas engines. Siloxanes can be removed by cooling the gas, by adsorption on activated carbon, or on even spent activated carbon. They can also be adsorbed on activated aluminium or silica gel. Siloxanes are also removed during the removal of hydrogen sulphide. They can be absorbed by liquid hydrocarbons.

e. ***Removal of Particulates:*** The particulate matter in gas can cause mechanical wear in engines. Particulates can be removed by mechanical filters.

f. ***Reducing the CO_2 content in Gas:*** The main upgrading action for improving the calorific value of the gas is the removal of or reducing the carbon dioxide content in it. The CO_2 can be removed by cold-water scrubbing, alkaline washing/scrubbing, or scrubbing by organic solvents like polyethylene glycol. The cold-water scrubbing is the cheapest method and a common way of upgrading the gas as it also removes particulates, ammonia and other soluble components, besides CO_2. There are various other methods for the removal of carbon dioxide such as absorption by amines, the cryogenic separation of gases, pressure swing adsorption (PSA), etc. For separation of gases, also see *Para 7.15, of Chapter 7.*

6.13 Pyrolysis of Wastes [9,173,182,239,261,278]

6.13.1 *Process of Pyrolysis* [290]

The waste containing organic materials like tyres, plastics, paper, agricultural wastes, crop residues, refuse, food-processing wastes, sewage sludge, municipal solid waste, etc., can be converted to bio-oil and gas by pyrolysis. This process breaks down the long chain of organic polymers into simple compounds. [*Pyrolysis is a process in which the material containing carbon is heated in the absence of oxygen.*]

Low-temperature pyrolysis is carried out in the temperature range of 470–555°C. It converts the solid feedstock into a mixture of solid, liquid and gaseous products. In low-temperature pyrolysis, the formation of aliphatic compounds predominates. The solid waste is shredded and fed into a pyrolysis kiln via a refuse/waste-feeding facility. The gas, so released, is mostly methane with some propane, butane etc. These gases are cleaned by passing through a dust precipitator and then cooled to condense the vapors. The condensed liquid consists of oil and tar. The solid residue is coke, which comprises carbon and some incombustible (inert) material.

The liquids generated predominate over the gases in low-temperature pyrolysis in the abovementioned range of temperatures (470–555°C), but in high-temperature pyrolysis performed at about 1000°C temperature, the generation of gases predominates over the liquids. The presence of certain catalytic additives accelerate the pyrolysis reaction at lower temperatures. Faster depolymerisation of plastics waste occurs by pyrolysis to produce fuel from it in their presence. The liquid fuel produced can be subjected to fractional distillation to obtain various fractions like gasoline, kerosene, diesel, etc. [249] *[Also see Para 5.9.3 of Chapter 5, for 'Liquid fuels from biomass by fast pyrolysis'.]*

6.13.2 *Wood Pyrolysis* [293]

If the pyrolysis is done without combustion, it will require reducing the biomass/wood to chip size and then drying and heating them in iron 'retorts' in the absence of air by external heat to about 400°C. The resulting products are:

(a) Wood gas which mainly consists of methane, ethane, hydrogen, carbon monoxide, carbon dioxide and nitrogen. [*The wood gas is used as a fuel for heating the retort for pyrolysis.*]

(b) Pyroligneous acid (cooled vapors) consisting of methyl alcohol, acetone, acetic acid, water and other impurities.

(*For pyrolysis of wood, the typical percentage composition of this product (Pyroligneous acid) may be: methyl alcohol (3%); acetone (5%); acetic acid (10%); and the balance comprises water and impurities.*)

(c) Wood tar, a thick black liquid that on further distillation yields a mixture of creosols/ creosote (a timber preservative).

(d) Solid residue, that is, charcoal and pitch (mainly, carbon and ash).

Separation of Products

To separate the products, the pyroligneous acid is treated with milk of lime $[Ca(OH)_2]$, which reacts with acetic acid (CH_3COOH) present in the acid to form calcium acetate $[(CH_3COO)_2Ca]$ and water.

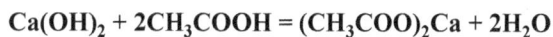

$$Ca(OH)_2 + 2CH_3COOH = (CH_3COO)_2Ca + 2H_2O$$

The resulting liquid on fractional distillation separates acetone at 56°C and methyl alcohol (CH_3OH) at 65°C, leaving behind calcium acetate and water, etc.

Calcium acetate, when heated with concentrated sulphuric acid (H_2SO_4), regenerates the acetic acid (CH_3COOH).

$$(CH_3COO)_2Ca + H_2SO_4 = 2CH_3COOH + CaSO_4 \text{ (Calcium Sulphate)}$$

Further purification of acetic acid can be done with suitable chemical methods/physical methods.

It may be noted that the end products depend on the temperature of pyrolysis. At about 400–500°C, *'organic liquids' predominate, but at about 1000°C 'gases predominate'*. A part of the gases and char are normally used for heating. A schematic diagram showing the pyrolysis of wood is given in Figure 6.7.

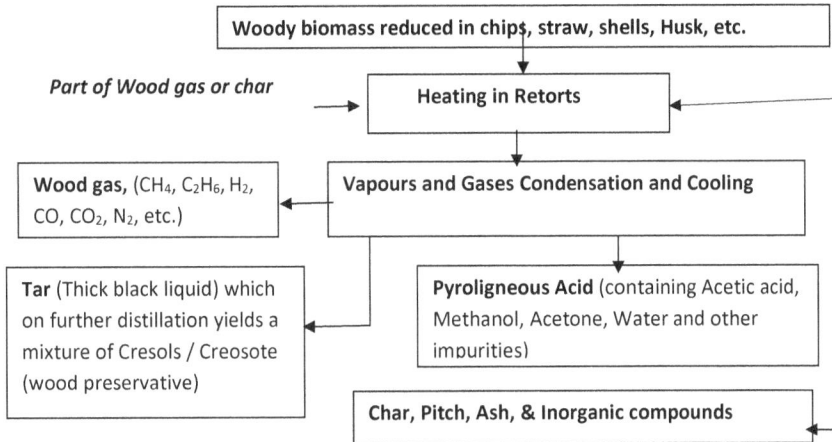

Fig. 6.7. Schematic Diagram of Pyrolysis.

6.14 Thermal Gasification Technology for Lignocellulosic Agricultural or Forest Wastes [6,43,127,160,184,202,258,261,265, 287,307]

6.14.1 *Chemical Reaction in Gasification* [87]

The conversion of waste wood, forest wastes and agricultural wastes can be done to gaseous form by a thermal process for convenience of use. The gases so obtained mainly comprise of combustible components like carbon monoxide, hydrogen, a small amount of methane, non-combustible components like nitrogen (mainly because of the use of air in gasification) and carbon dioxide. The mixture of ($CO + H_2$) produced is termed 'syngas'. The general equation of waste material gasification would be as follows:

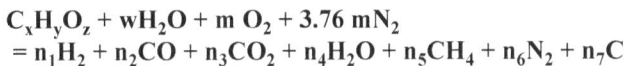

$$C_xH_yO_z + wH_2O + m\ O_2 + 3.76\ mN_2$$
$$= n_1H_2 + n_2CO + n_3CO_2 + n_4H_2O + n_5CH_4 + n_6N_2 + n_7C$$

where, w = amount of water per kilo mole of waste material,

    m = amount of oxygen per kilo mole of waste,

    n_1, n_2, n_3, n_4, n_5, n_6 and n_7, are coefficients of gaseous products and soot (stoichiometric coefficients in kilo moles).

The gasification is done by partial combustion of carbon in the biomass by oxygen and the reaction of steam.

Reactions

The reactions taking place in the gasification process are complex and may be of the following nature:

  (i) $C + O_2 = CO_2 + 393,800$ kJ/ kg mol. (oxidation)   *[Exothermic]*

  (ii) $C + CO_2 = 2CO - 172,600$ kJ/ kg mol. (Boudouard reaction)   *[Endothermic]*

  (iii) $2C + O_2 = 2CO + 221,200$ kJ/ kg mol. (partial oxidation)   *[Exothermic]*

  (iv) $C + H_2O = CO + H_2 - 131,400$ kJ/ kg mol. (Water gas)   *[Endothermic]*

  (v) $CO + H_2O = CO_2 + H_2 + 41,200$ kJ/ kg mol. (Water shift reaction)   *[Exothermic]*

  (vi) $C + 2H_2O = CO_2 + 2\ H_2 - 78,700$ kJ/ kg mol.   *[Endothermic]*

.(The above technology can be used to convert biomass to hydrogen.)

(vii) $CO_2 + H_2 = CO + H_2O$ - 41,200 kJ/ kg mol. *[Endothermic; an undesirable reaction]*

(viii) *Methane formation by the combination of carbon and hydrogen*:

.$C + 2H_2 = CH_4$ + 75,000 kJ/kg mol. (in char at 500–600°C)   *[Exothermic]*

Pyrolysis takes place in the temperatures between 200–600°C (after drying of biomass) before oxidation. The oxidation zone temperature is 900–1200°C. The reduction zone (after oxidation) has a temperature of 900–600°C.

If improvement in the calorific value of gaseous fuel is desired, it can be done by using 'oxygen', instead of 'air'. The separation of nitrogen from air is cheaper than the separation of nitrogen from the resultant gaseous fuel. However, the air separation needs a lot of energy. Separation of air into its components is done only when it is really necessary; otherwise, air is used as such.

Pyrolysis, gasification and combustion are distinguished from one another mainly by the temperatures achieved during each process and by the amount of oxygen available relative to that required for full oxidation of the carbonaceous fuel.

Difference between Gasification and Incineration

There are key differences between gasification and incineration that make gasification a much cleaner and efficient process. The main points are:

- Incineration is designed to maximize the conversion of waste to CO_2 and water vapor. On the other hand, gasification is designed to maximize the conversion of waste to pure CO and H_2. The outputs of gasification are cleaner compared to the outputs of incineration. Lime can be added to gasifier to absorb sulphur dioxide.

- Incineration uses large quantities of excess air and has a highly oxidizing environment. Gasification uses limited quantities of oxygen and has a reducing environment.

- The temperature ranges of incineration and gasification are different.

- In incineration, the flue gas clean-up is done at atmospheric pressure. In gasification, the syngas clean-up is done at a high pressure.

- In incineration, the treated flue gas is discharged into the atmosphere. In gasification, the treated syngas is used for production of other compounds/fuels or used as fuel. This makes gasification a cleaner and efficient process.

- In incineration, fuel sulphur is converted to SOx and discharged with the flue gas. In gasification, the sulphur compounds can be absorbed within the kiln itself or can be recovered. These would include hydrogen sulphide and some amount of sulphur dioxide. These gases can be converted to elemental sulphur or sulphuric acid.

- In incineration, the bottom ash and fly ash is collected, treated and disposed of as hazardous wastes. In gasification, the noncombustible matter may be converted to slag, which is non-leachable, non-hazardous and suitable for use in construction activities. A part of the particulate matter is recycled in the gasifier; the rest of it may be disposed or processed for the reclamation of any useful component.

6.14.2 *Types of Gasifiers* [39,52,105,106,238,268]

Several types of gasifiers have been developed for gasification of solid fuels and biomass. The main types of gasifiers used are briefly described below:

- Entrained Flow Gasifier
- Bubbling Fluidized Bed Gasifier
- Circulating Fluidized Bed Gasifier
- Dual Fluidized Bed Gasifier

- Fixed Bed Gasifier
- Plasma Gasifier

(1) Entrained Flow Gasifier

In this gasifier, feed material must be dried and powdered/pulverized or made into slurry form and atomised. Powdered biomass can be fed into a gasifier with pressurised oxygen and/or steam. The gasifier operates under pressure and requires oxygen instead of air for syngas production. It can be run as a cyclonic reactor. This gasifier is run at a high temperature to produce tar-free syngas, while the ash is removed as slag. Lime may be added as a fluxing agent to control the viscosity of the slag. The conversion efficiency to syngas is high. The conditions in the gasifier favor the production of hydrogen instead of methane and thus, the syngas has a low content of methane. The turbulent flame at the top of the gasifier burns some of the biomass and provides large amounts of heat to attain a high temperature (1200–1500°C) for fast conversion of biomass into very high-quality syngas.

These types of gasifiers have been used mainly for coal gasification. These gasifiers would need a pilot flame for start-up. They can accept a mixture of feedstocks (coal and biomass), provided the particle size is less than 1 mm, the feed moisture content is less than 15% and the composition remains fairly steady over time.

These types of gasifiers are not generally used for biomass gasification as fuel preparation conditions may be difficult to achieve for biomass. However, some companies have developed them for biomass. *(See Chapter 5, Para 5.9.6)*.

(2) Bubbling Fluidized Bed Gasifier (BFB)

A fluidized bed operates when a quantity of fine solid particulate material like sand is forced to behave as a fluid by the introduction of pressurized air, oxygen, or steam through the particulate material, just to keep the material agitated.

In this gasifier, the velocity of gas is relatively low and a bed of fine inert material sits at the gasifier bottom, with air, oxygen or steam being blown upwards through the bed just fast enough to agitate the material. The velocity in this gasifier may typically be 1 to 3 m/s.

Biomass is fed from the side, which mixes and combusts or forms syngas that leaves upwards. Good mixing will lead to faster pyrolysis. The feed material should be pre-processed to have a small particle size. The gasifier can accommodate variation in particle size and moisture content of feed material up to the following limits:

Particle size from 50 to 150 mm and moisture content from 10 to 55%.

In the bubbling type gasifier, the heat and mass transfer and specific heat capacity is good. It has an easy start-up and good turndown. It operates at a moderate temperature (say below 900°C) with a good temperature control. The gases generated have a higher proportion of methane, but tars are present. Operation at temperatures below 900°C avoids ash melting and sticking and also reduce NOx formation. It can also be run as a pressurised reactor.

If oxygen is used in place of air as an oxidant, a higher generation of hydrogen can be realized. If the biomasses with low ash fusion and ash deformation temperatures are used as feed, there is still a risk of bed agglomeration.

(3) Circulating Fluidized Bed Gasifier (CFBG)

In CFBG, the air/steam/oxygen is blown through the bed of fine inert material (sand) upwards fast enough to suspend material throughout the gasifier. The velocity is in the range of 5 to 10 m/s in the CFBG leads to good mixing and faster pyrolysis. Due to higher velocity, there is a higher loading of particulates and therefore the gasifier should be designed to prevent erosion. The biomass is fed inside from the side and gets suspended. It combusts or reacts to form syngas and the required heat is also provided by combustion.

The fine material consisting of sand and feed material that escapes from the furnace is collected in a cyclone ahead and is returned to the bed, while the gasified feed and other gases go through further cleaning and treatment. These gasifiers operate in the moderate temperature range of 850–900°C, avoid ash agglomeration and also reduce the formation of NOx. The feed material size should be kept below 20 mm, but the gasifier can accept various types of feed materials. In the case of the CFB gasifier, the carbon burn-out is better than in the case of the BFB type. These gasifiers can be designed for pressurized operation if the syngas is required to be pressurized for downstream use. Foster Wheeler have been providing CFB technology and others are also entering the field. CFB gasifiers can be designed for larger capacities.

(4) Dual Fluidized Bed Indirect Gasifier (DFBIG)

The *DFBIG* has two parts: a gasification chamber and a combustion chamber. Biomass is first put in the gasification chamber, where it is converted to syngas (using steam) and char. Here the syngas is free from nitrogen.

The char moves over to the combustion chamber of the CFB or BFB type and is burnt there in the air. The bed material is heated and is then recycled in the gasification chamber to provide indirect heat to the gasification chamber.

The DFBIG runs at temperatures less than 900°C to avoid ash slagging. The gasifier produces higher quantities of methane and is therefore suitable for methane production by downstream methanization synthesis. Because of lower temperature operations, the formation of NOx is low. These gasifiers can be designed for pressurised operations.

(5) Fixed Bed Gasifier (FBG)

The fixed bed type of gasifier requires a lumpy feed and not feed in crushed or pulverized form because the gas flow through the bed is quite slow. Here, air or oxygen can be used as an oxidant, but generally air is used for biomass. There are four thermal zones in the FBG, namely:

- Drying zone: evaporation of feed moisture

- Pyrolysis zone: heating of feed material to 300–400°C without addition of oxygen/air and converting it to pyrolysis gases, which would be laden with liquid hydrocarbons, tar and char

- Reduction zone where the operating temperature is above 800°C: most of the char is also converted to syngas

- Combustion zone where temperature is above 1000°C: the balance tar and char are burnt and provide the heat required for reaction in other zones.

There can be two versions of this type of gasifier, that is, the up-shaft gasifier and the down-shaft gasifier.

<u>Up-Shaft Gasifier:</u> In this type, the feed material is injected from the top, while the oxidant is injected at the bottom and syngas goes out from the top. The biomass and syngas move in the opposite direction. Some of the char would burn off as it falls and provides process heat. The methane and tar-rich gas leaves at the top of the gasifier and the ash falls from the grate for collection in the bottom hopper.

<u>Down-Shaft Gasifier:</u> In this type, the feed material is added at the top and the oxidant can be injected either from the side or from top, while the syngas flows out from the bottom. Some of the biomass feed will burn off as it falls and then forms a bed of hot char (reaction zone), through which the gases would have to pass through. The ash is collected below in the hopper under the grate.

The gases from the 'down-shaft gasifier' pass downward through the combustion zone, so the tar gets burnt out and a very insignificant quantity of tar remains in the outgoing gases (say about 1 g Nm^{-3}), while the outgoing gases from the '*up-shaft design*', which flow out from top, contain a much higher quantity of tar (say, 100 g Nm^{-3}). The biomass is fed at the top of the gasifier and the air, oxygen, or steam intake is also at the top or from the sides; hence, the biomass and gases move in the same direction. The '*down shaft gasifier*' outgoing gases contain a higher quantity of particulate matter and have a higher temperature compared to the '*up-shaft gasifier's*' outgoing gases. Due to this the efficiency of the down-shaft design gasifier is lower than that of the up-shaft design gasifier. However, the almost tar-free output gases from the down-shaft design gasifier is thought to be a great advantage, outweighing the advantage of higher efficiency and lower particulate matter provided by the up-shaft design gasifier.

6.14.3 *Effects of Gasifier Feedstock Characteristics* [159,289]

The following characteristics of the feed material (biomass) are important for gasification in gasifiers:

1. *Bulk density of biomass:* It affects the energy density, material handling, storage and transport.

2. *Shape and size of biomass:* It affects the operation of the gasifier and the pressure drop across the bed.

3. *Moisture content:* It affects the quality of the gas produced.

4. *Volatile matter content:* It affects the level of tar produced.

5. *Content of elements like halogens, sulphur, arsenic, lead, mercury, etc., which generate pollutants:* These affect the quality/pollutant level of gas produced as well as that of the ash discharged from the environmental angle.

6. *Content of potassium, sodium, calcium, iron, magnesium, silica, phosphorus:* These affect the gasifier-operating conditions, ash softening temperatures, ash fusion temperatures and other ash characteristics.

7. *Ash content and its fusion characteristics:* Both affect the operation of the gasifier.

6.14.4 *Generation of Dioxins/PAHs and other Pollutants during Gasification and Cleaning* [101,214]

It may be mentioned that during the burning of biomass, pollutants like oxides of sulphur, oxides of nitrogen, chlorine compounds, tars, particulates, etc., are generated. There may also be some formation of polycyclic aromatic hydrocarbons (PAHs) and dioxins like polychlorinated dibenzo P-dioxins called 'PCDDs' and polychlorinated dibenzo-furans called 'PCDFs'. 'Dioxin' is commonly used to refer to: <u>Dioxins and dioxin-like compounds,</u> *which are a diverse range of chemical compounds known to exhibit "dioxin-like" toxicity. Chemically a dioxin is a* <u>heterocyclic 6-membered ring,</u> *where two carbon atoms have been substituted by oxygen atoms:* <u>1,2-Dioxin, 1,4-Dioxin.</u>

The gasifier output gases may have all the abovementioned impurities. *If lime is added in the bed material of the gasifier, oxides of sulphur, hydrochloric acid gas, chlorine and other halogens, etc., will be absorbed. However, the addition of lime may make the ash fluid and create difficulties in fluidized bed gasifiers (both the bubbling type and circulating type). Gas-cleaning equipment like the bag filter house and wet scrubbers are mostly used to clean the gases.*

The injection of activated carbon ahead of bag filters can remove *many pollutants including hydrogen sulphide, mercury and dioxins.*

The water scrubbing can remove particulates and soluble gaseous pollutants and a large content of dioxins too. After passing through the wet scrubber, the gases can be heated again by the 'gas-to- gas heat exchanger'.

Dioxins can also be removed by passing the gases through a catalyst tower after the removal of particulates, sulphur dioxide, hydrogen chloride, etc. (The catalysts are poisoned by these impurities). But the catalysts give good performance only in a temperature range of 230–270°C. This can be possible only by introducing gas to gas heat exchangers in the circuit.

The concentrations of hydrogen sulphide in the gas can be decreased by precipitation by the addition of ferric or ferrous chloride or ferrous sulphate. Iron sulphide forms precipitates as it is almost insoluble. H_2S can be adsorbed on activated carbon or on iron hydroxide/ferric oxide– coated wooden chips. It may be removed by washing with caustic soda or a sodium carbonate solution.

Mostly, the hydrogen sulphide (H_2S) and CO_2 occur simultaneously in the gases. Both of these gases are soluble in the sodium carbonate solution. The selectivity of H_2S for absorption compared to CO_2 depends on the ratio of the H_2S to CO_2 content of the gases being treated. At a higher pH value, that is, if the pH is higher than 12, the absorption of CO_2 is higher.

The removal of the balance hydrogen sulphide, if required, can be done by absorption using amines, but only after the removal of sulphur dioxide. The amines used commercially for the absorption of H_2S are: the primary amine (for example) Mono-ethanol-amine (MEA), the secondary amine (for example, Di-ethanol-amine (DEA) or Di-propanol-amine (DIPA)) and the tertiary amine (for example, methyl-di-ethanol-amine (MDEA)).

Municipal solid wastes (MSW) can also be gasified like any other organic waste. MSW can also be used directly as a fuel in a refuse-fired boiler to raise steam for the process or for power production. The boiler gases would have to be cleaned up by the gas-cleaning equipment. If a lot of PVC is present in the wastes being gasified, a vast amount of hydrochloric acid gas will be generated, which can then be recovered for commercial use from the gases or will have to be absorbed or removed by gas-cleaning equipment. It is known that dioxins decompose at temperatures above 700°C. If the waste is gasified at high temperatures and the exhaust gases are cooled quickly and filtered, the dioxins are avoided from the gases. Some municipalities adopt this method for the incineration of wastes, but it is costly.

Some of the methods discussed for upgrading biogas in Para 6.11 can be used for cleaning and upgrading gases from the gasifier.

6.14.5 *Requirements of Syngas Quality* [51,223,257]

It may be pointed out that the cleaning of the synthesis gas (syngas) obtained from the gasifier will be essential for most of its uses. The extent of cleaning requirements for the outgoing gases from gasifiers will depend on their downstream use.

If the gases are to be used for steam raising in a boiler, the syngas can be used even without any cleaning, that is, in raw form (for example, conventional dust removal), but condensation up to burner should be avoided by maintaining the gases at a high temperature.

For use in a gas engine or gas turbine, the syngas should be cleaned to remove dust, tar, oxides of sulphur, alkali metals, halogens and their compounds, etc., to a reasonable extent.

If the syngas (which is supposed to be mainly a mixture of carbon monoxide and hydrogen) is to be used in *the Fischer Tropsch (FT) process*, the cleaning limits are stringent. The pollutants like tar, particulate matter, sulphur oxides, halogen and their compounds, alkali metals, heavy/toxic metals (particularly mercury and arsenic), ammonia and hydrogen cyanide should be almost completely removed (See Table 6.3). Even the diluents like carbon dioxide, nitrogen and methane need to be removed to attain higher process efficiencies.

For methanol synthesis or other catalytic processes, the purity limits of syngas may be a little more stringent than even the FT process's typical limits mentioned in the Table 6.3. The typical limits of pollutants and diluents of syngas for various downstream purposes are indicated in Table 6.3.

6.14.6 *Research and Developments in Biomass Gasification* [11,32,59,61,93,133,140,141,170,208,218]

A. Gas and Tar Cracking [33,62,92,115]

Catalytic Tar cracking

Catalytic hydrocracking has been practiced in oil refineries for several decades. Basically, the catalytic hydrocracking process cracks the high-boiling, high molecular weight oxygenates, phenolics, ethers, polyaromatic hydrocarbons (PAH) and large PAHs into lower-boiling, lower molecular weight olefinic and aromatic hydrocarbons and then, hydrogenates. The high-boiling,

Table 6.3. Typical Syngas Quality Requirements for Different Downstream Uses. [51,223,257]

Parameters	Typical Raw Syngas Quality	Limits for Engine	Limits for FT Process	Limits for Methanol Synthesis
Tars	1000 to 10,000 mg/Nm³	15 mg/Nm³	1 mg/Nm³ (15°C below dew point)	1 mg/Nm³ (15°C below dew point)
Particulates		15 mg/Nm³	0.1 mg/Nm³	0.1 mg/Nm³
Sulphur **	750 mg/Nm³	50 mg/Nm³	0.1 mg/Nm³	0.1 mg/Nm³
Halides	2500 mg/Nm³	15 mg/Nm³	0.01 mg/Nm³	0.001 mg/Nm³
Alkali metals		------	0.01 mg/Nm³	0.1 mg/Nm³
Mercury	0.3 mg/Nm³	Guidelines for waste incineration	Poison	Poison
Arsenic	1 mg/Nm³			
Ammonia	-------	NH_3 + HCN 10 mg/Nm³	10 mg/Nm³	10 mg/Nm³
Hydrogen cyanide	-------		0.01 mg/Nm³	0.01 mg/Nm³
Nitrogen	Up to 50%	50% max.	Inert gases reduce efficiency	Inert gases reduce efficiency
Carbon dioxide	Up to 15%	--------	Reduces efficiency. As low as possible	4 to 8% for max activity
Ratio of H_2:CO	0.5–1.5 :1	-------	1.5–2 : 1	$(H_2\text{-}CO_2)$: $(CO+CO_2)$ as 2:1
Methane	0 to 5%	-------	Reduces efficiency. As low as possible	Reduces efficiency. As low as possible
Temperature	500°C gasifier outlet	25°C	200 to 350°C	150 to 270°C
Pressure	Atmospheric or pressurized up to 40 bars	Atmospheric	10 to 50 bars	50 to 100 bars

high molecular weight paraffinic hydrocarbons can also be cracked into lower-boiling, lower molecular weight paraffinic hydrocarbons.

The tar levels of gas obtained by the gasification of biomass can be brought down significantly by the use of catalytic hydrocracking. The cracked tar is partly converted into additional fuels for use in engines, since the decomposition processes of tar are similar to those of crude oil refining. Any sulphur, nitrogen and oxygen present in the hydrocracking feedstock are also hydrogenated to a large extent and form gaseous hydrogen sulphide (H_2S), ammonia (NH_3) and water vapor, which are subsequently removed. The final hydrocracking products are essentially free of sulphur and nitrogen impurities and consist mostly of paraffinic and naphthenic hydrocarbons.

Generally, the hydrocracking conversion process is exothermic. The hydrogen, which is consumed by hydrocracking, is produced by the carbon-steam reaction and water-gas shift reaction during the gasification process. Therefore, there is no need for additional hydrogen.

The overall tar conversion increases with catalyst temperature and therefore, higher operating temperature within the limits for a particular catalyst is better. Lowering the flow rate and increasing the residence time in the reactor enhances the tar conversion.

Stainless steel turnings of 5 to 10 mm size electrochemically plated with palladium (Pd) can be used as a catalyst. (The stainless steel turning may be taken from the waste of the machining industry.) [254]

Gasification/reforming with the simultaneous production of 'carbon nanotubes' and hydrogen has also been developed and tested as per the following investigations:

(1) The production of hydrogen and carbon nanotubes can be done by the gasification of the plastic waste material. Special catalysts are used in the process. The incorporation of Al into the SBA-15 matrix by a two-step "pH-adjusting" hydrothermal method produces strongly acidic sites.

The yield of carbon nanotubes (CNT) strongly depends on the surface acidity. The catalyst 10Ni/Al–SBA-15(10)–P exhibits the highest activity for CNT production accompanied by a relatively high concentration. Moreover, the polyol catalysts exhibited higher dispersions of active metal sites and showed higher activity for CNT growth compared to the impregnated catalysts. The 10Ni/Al–SBA-15(10)–P catalyst has a smooth structure with few defects and a homogeneous outer diameter. The Ni loading on the support also influences the CNT yield as well as its quality. Overall, the incorporation of Al into the SBA-15 framework increases the surface acidity, promoting the activity of CNT and production, simultaneously, during waste plastic gasification. [255]

(2) While producing hydrogen from methane, nano carbon materials with an attractive texture and structure can be produced instead of carbon dioxide. Suitable catalysts for the purpose are based on the nanometer-scale nickel particles prepared from a hydrotalcite (like the anionic clay precursor), which have been tested. The conversion of methane to hydrogen increases with reaction temperature but beyond 923K, the nickel catalyst cannot work. The modification of the catalyst with doping of copper increases the activation temperature and leads to the production of nano carbon with an attractive structure. [259, 224]

B. Advanced Gasification by Plasma Torches [15,71,161]

The advanced gasification of wastes uses 'plasma torches' which raise the temperature well above those reached using conventional technologies. Through this process, the waste is not burnt, but actually gets broken down into some of its basic component gases. The syngas produced is mainly a mixture of carbon monoxide and hydrogen along with small quantities of other gases.

In this method, the waste is shredded and is reacted with oxygen or air using plasma torches in a reactor. The resulting syngas is cooled and cleaned up of dust. The syngas can be used as a fuel directly or may be further converted to other forms of fuels. The carbon monoxide in the syngas when reacted with steam can produce 'hydrogen'. [see water shift reaction *Para. 6.14.1*]. It is claimed that the technology is more efficient and has lower associated carbon emissions. A schematic flow diagram of the process is shown in Figure 6.8.

Plasma Gasifier

The plasma gasifier on the above principle operates by gasifying the feed material in plasma generated, for example, by a high energy electric arc. An inert gas (for example argon) is passed through the electric arc, which heats it up to a very high temperature (for example, 1500–5000°C).

Biomass (even untreated) added to the gasifier comes in contact with the plasma gas/arc and its organic material together with moisture is converted to gaseous form, while inert materials are converted to vitreous slag. The plasma gasifiers can use the oxygen element of the moisture in the biomass feed and may require very little additional oxidants. The biomass material is reduced to its elemental components in gaseous form. Any type of waste material can be gasified which, otherwise, is difficult. However, there are some disadvantages of this method, such as: high energy consumption/power requirement for plasma torches, high level of chlorine in the gas (especially with some types of biomass), objectionable presence of dioxins and heavy metals, etc. [*Plasma gasification uses plasma torches, but it is also possible to use plasma arcs in a subsequent process step for syngas clean-up.*]

In an oxygen-starved environment, the thermal disintegration of carbonaceous materials into fragments of compounds generates 'plasma'* which when it cools, recombines into a gas, consisting mainly of CO and H_2. With 'plasma gasification', it is possible to transform all types of wastes into new fuel. The gases from a gasifier may be made to generate steam, distilled water, hydrochloric acid (HCL), sulphur compounds, etc. in addition to new fuels. Electricity can also be produced from the steam generated.

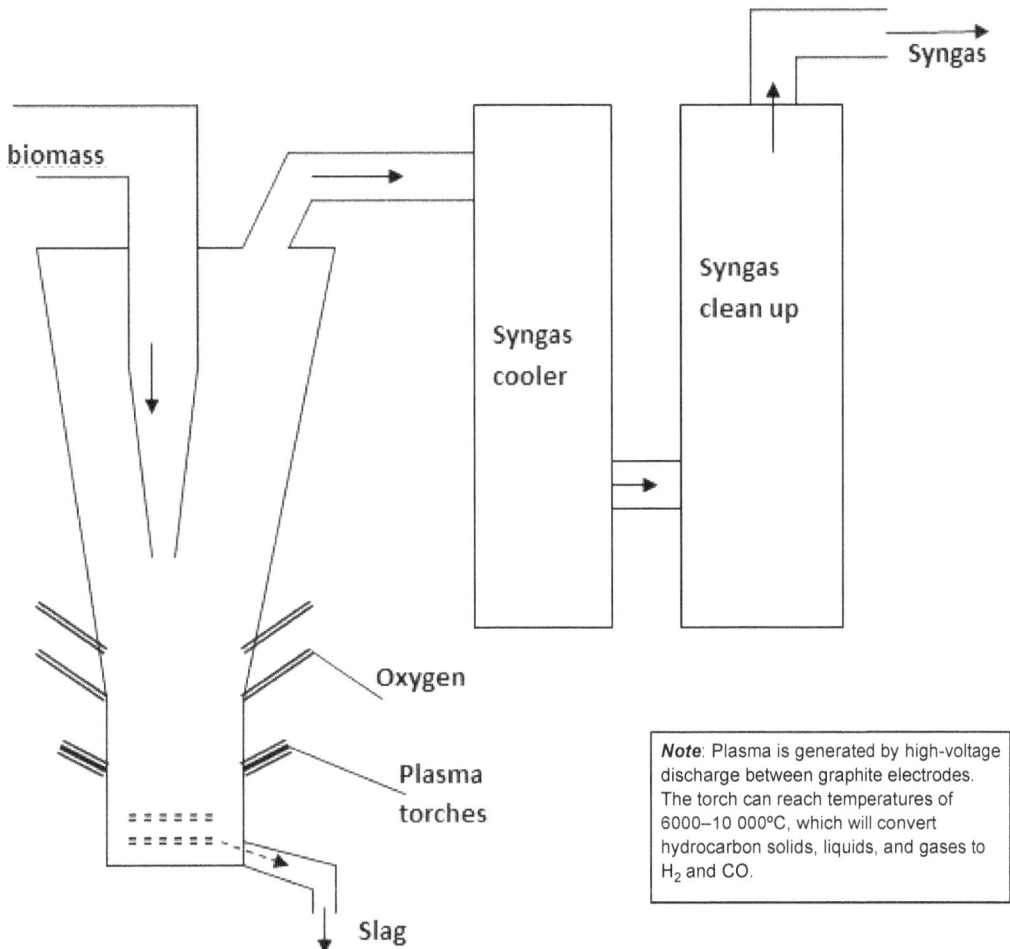

Fig. 6.8. Advance Gasification by Plasma Torches.

Plasma gasification significantly reduces air emissions, including carbon emissions and greenhouse gas emissions, as well as emissions of nitrous oxides, sulphur dioxide, mercury and particulate matter.

*Note: *Plasma*

Plasma is the fourth state of nature, by far the most common form of matter. Plasma in the stars and in the tenuous space between them makes up over 99% of the visible universe and perhaps most of that which is not visible.

In physics and chemistry, plasma is a state of matter similar to gas, in which a certain portion of the particles are ionized. On heating, a gas dissociates its molecular bonds, rendering it into its constituent atoms. Further heating leads to ionization (a loss of electrons), turning it into a plasma that contain charged particles, positive ions and negative electrons.

Plasma consists of a collection of free-moving electrons and ions (atoms that have lost electrons). Energy is needed to strip the electrons from atoms to make plasma. The energy can be of various origins: thermal, electrical, or light (ultraviolet light or intense visible light from a laser). With insufficient sustaining power, plasmas recombine into neutral gas. Ordinary solids, liquids and gases are both electrically neutral and too cool or dense to be in a plasma state.

C. *Flash Pyrolysis* [9,28,54,60,123,137,174,178,205,262]

A pilot plant of 20 kg/h capacity for flash pyrolysis of industrial and municipal waste and agro-waste based on a gas-heated fluidized bed was developed and tested successfully and the characteristics and yields of the 'liquid products/bio-oils' obtained from the various types of wastes were determined. The wastes such as rice husk saw dust, spent tyres, jatropha cake and bagasse can be used. (The trial runs with jatropha cake as well as spent tyres were successful.)

This is similar to fast pyrolysis described in *Para.5.9.3 of Chapter 5.*

A research and development project named "Development of pilot scale pyrolysis test unit for production of bio-fuels and value added by-products from Jatropha seed shells" was taken up by the Tata Energy Research Institute (TERI), New Delhi, India. Gas (useable as fuel) was also evolved along with liquid products in flash pyrolysis.

6.14.7 *Suitability of Producer Gas for Gas Engines* [215,219]

A long-term evaluation of 6B series 25 kW gas engine showed that they could be run on 'producer gas' and such engines have been installed at various places (a turbo-charged version of this engine was also developed). Biomass can be gasified and directly utilized for running of the gas engine, as shown schematically in Figure 6.9. The typical approximate composition of the producer gas is: 29.1% CO, 10.2% CO_2, 1% H_2, 59.6% N_2 and the balance comprises other gases.

Dual-fired Engines

Megawatt-size power plants driven by internal combustion dual-fired engines (1.2 MW) are already in existence for several diesel-replacement projects (e.g., in Tirunelveli, Tamil Nadu). The dual-fired engines can switch over from liquid fuel to gaseous fuel and vice versa for keeping emissions low.

6.15 Conversion of 'Biomass Gasifier Exit Gases' to Methanol, Methane, or other Chemicals [141,179,213,222]

The gases obtained by the gasification of biomass can be re-formed into methanol, methane, or other chemicals (if required); but in order to do this, it is better to use oxygen instead of air in the gasifier to keep the gasifier gases mostly free from nitrogen.

1. *Methanol:* The hydrogen (H_2) and carbon monoxide (CO) components obtained from the gasification of biomass can be converted to methanol by a catalytic reaction at a high pressure and high temperature, as mentioned in Para 5.9.2 of Chapter 5. The syngas purity for methanol synthesis should be as per *Table 6.3 in Para 6.14.5.*

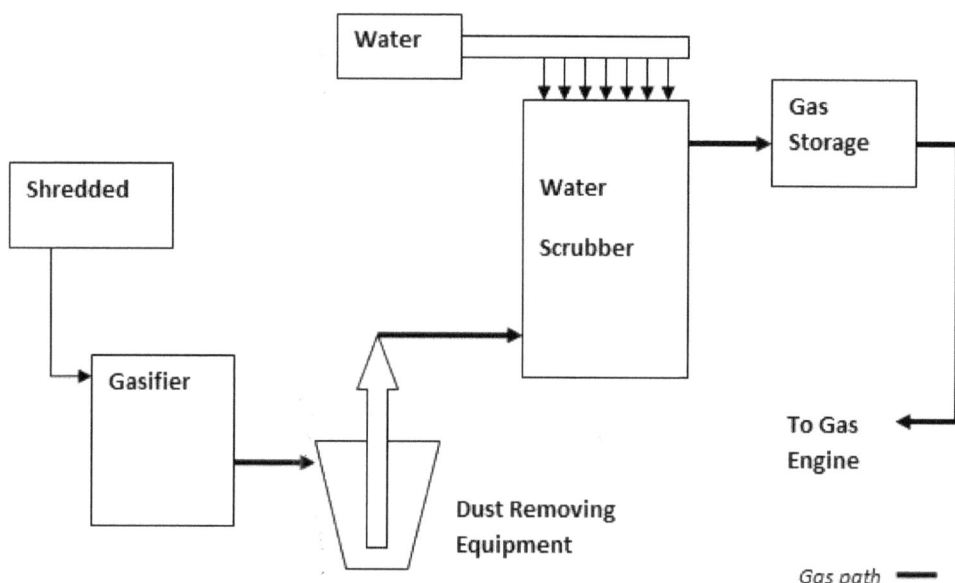

Fig. 6.9. Direct Utilization of Biomass Gasifier Gas for Running Engine.

. $CO + 2H_2 = CH_3OH$ (Methanol)

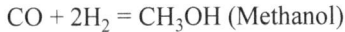

2. **Methane:** Methane formation by components of gasifier exit gases can be done by the following methods:

a. Methane can be synthesized from gasifier gases by passing a mixture of CO and H_2 (largely free from nitrogen) over nickel at about 300°C.

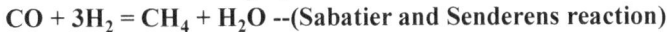

$CO + 3H_2 = CH_4 + H_2O$ --(Sabatier and Senderens reaction)

b. The Methane can also be formed by passing a mixture of CO_2 and H_2 over nickel (Ni) deposited on magnesia (MgO) at about 328°C.

$CO_2 + 4H_2 = CH_4 + 2H_2O$

c. The synthesis of methane can be done by striking an electric arc between carbon electrodes (at about 1200°C) exclusively in an atmosphere of hydrogen or it can be synthesised by heating carbon at 475°C in an atmosphere of hydrogen, in the presence of nickel.

$C + 2H_2 = CH_4$

Hydrocarbon gases, mostly methane and ethane, can also be directly produced from woody matter by treatment at high temperature (600 °C) and pressure (50 atmosphere) in an atmosphere of hydrogen gas.

$C + 2H_2 = CH_4 + 75,000$ kJ/kg mol.

Hydrogen, for the above use, can be generated in a reactor with CO and steam.

$CO + H_2O = CO_2 + H_2 + 41,200$ kJ/ kg mol.

3. **Heavy Oils:** Heavy oils can be produced in a reactor by the methods described in *Para 5.8, (Chapter 5)*. The heavy oils so produced can be refined by the usual methods.

6.16 Methane Recovery from Gas Hydrates [80,103,116,122,136,145,156,157,158,165,197,216,224,256]

6.16.1 *Gas Hydrates* [24,45,46,84,180]

Methane gas can form a hydrate called 'methane clathrate – $CH_4.nH_2O$' in cold and high-pressure conditions. Typically, one volume of gas hydrates can discharge about 160 volumes of methane and 0.8 volume of freshwater at standard temperature and pressure (STP). Typically, favorable conditions for the formation of methane hydrates exist in cold climates at ocean floors and also on land, deep below the ground, in aquifers. These gas hydrates are found in certain parts of the world, like some places on the ocean floor and permafrost areas of the Arctic.

'Methane hydrates' are found below oceans and sometimes on land, in underground aquifers. Recovery of methane is possible from oceans like 'Arctic Ocean' where it exists as hydrate at the bottom of the sea. Methane in the ocean is produced by microbes on the sea floor. These microbes break down the organic matter that sinks down from the sunlit zone near the surface. Organic matter is composed, for example, of the remains of dead algae and animals, as well as their excrements. In the deepest areas of the ocean, below approximately 2000 to 3000 metres, only a very small amount of organic remains reach the bottom, because most of it is broken down by other organisms on its way down through the water column. As a rough estimate, only around 1 percent of the organic material produced at the surface may end up in the deep sea and therefore, there cannot be any significant accumulation of methane in deep oceans. *Methane hydrates, therefore, primarily occur on the continental slopes, those areas where the continental plates meet the deep-sea regions as there is sufficient organic matter accumulating at the bottom and the combination of temperature and pressure is favorable. In very cold regions like the Arctic, methane hydrates even occur on the shallow continental shelf (less than 200 meters of water depth) or on the land in permafrost, the deep-frozen Arctic soil that does not thaw even in the summer.*

It is estimated that there could be more potential fossil fuel contained in the methane hydrates than in the world's reserves of coal, oil and natural gas. Depending on the mathematical model employed, present calculations of their abundance range between 100 and 530,000 Gigatons of carbon in methane hydrates. The most likely figures for the amount of trapped carbon in methane hydrate are between 1000 and 5000 Gigatons, that is, around 100 to 500 times as much carbon as is released into the atmosphere annually by the burning of coal, oil and gas.

Although many deposits seem to be inaccessible at present and the present day technology of extraction seems very expensive, various countries have been trying to develop suitable mining/extraction techniques in order to be able to use methane hydrates as a source of energy in future. [46]

As already pointed out in *Chapter 1 (Para 1.6),* the recovery of methane from 'non-renewable sources' like trapped methane from coal mines and methane gas hydrates from oceans/seas should be done with a sense of urgency because of these resources being potential dangers to environment. If the methane from these sources is released into the atmosphere inadvertently due to any reason (or due to global warming itself), it would cause a greater greenhouse effect and accelerated global warming than its planned utilization and conversion to carbon dioxide. *[The aim of tapping these sources of energy is, therefore, in conformity with the utilization of renewable sources of energy. It is for this reason that the subject has been briefly mentioned in this book].*

6.16.2 *Discoveries and Estimates of Methane Hydrates*
[24,46,50,55,110,116,122,132,146,147,180,183,227,294]

In the past 25 years, numerous estimates of the amount of methane contained in global gas hydrate deposits and the amount that can be economically extracted have been predicted. Published estimates of total methane in marine gas hydrates range from ~ 3100 Tm^3 to ~ 7,650,000 Tm^3, while the range for the smaller permafrost reservoirs is 14 Tm^3 to 34,000 Tm^3. [227] Global estimates for methane contained in both permafrost and oceanic reservoirs have now converged on ~ 15,000 Tm^3 (10 terra tons), but may range from 0.5 to 24 terra tons. [50] For the US. Exclusive Economic Zone offshore and North Slope permafrost region combined, Collett (1995) estimated ~9000 Tm^3 of methane in gas hydrates, [50] an estimate now lowered to ~ 5700 Tm^3 based on observations during drilling. Even this latter figure, which represents more than one-third of the nominal global estimate given by Kvenvolden and Lorenson (2001), is ~ 150 times the 95% confidence-level estimate of the US conventional natural gas reserve (Collett 2002) and ~ 900 times the current annual natural gas consumption in the US. [46, 50, 227]

In 2007, a US project found clathrate reserves in Alaska with 80% of the ice's pore space packed with methane. Tim Collett, a clathrate specialist at the US Geological Survey, who was part of the team, said there might be reserves all along the Alaska North Slope, including beneath existing oil installations at Prudhoe Bay and, alarmingly for environmentalists, the Arctic National Wildlife Refuge. Collett estimated that there was between 0.7 and 4.4 trillion m^3 of methane hydrate in Alaska alone.

New research has identified tremendous stores of 'gas hydrates' throughout the world, including in the US and Japan. The potential of extraction of natural gas from gas hydrates in Alaska's North Slope region is estimated at 2.42 trillion m^3. Gas hydrates have also been found by the US off the eastern coast and in the Gulf of Mexico.* Gas hydrate research has picked up speed in some countries, particularly in Japan and USA.

In 2004, a German and Chinese team found methane venting from the seabed off the coast of Taiwan in the South China Sea and in 2006, Indian researchers found a layer of methane clathrates 130 meters thick off its east coast in an area known as the Krishna Godavari (KG) Basin. Collett calls these "one of the world's richest marine gas hydrate accumulations".

In India, gas hydrates were found in the Mahanadi Basin and Andaman Basin as well. The occurrence of gas hydrates varies with reference to the different settings with a maximum thickness of 120 m in fractured clays in the KG basin at a water depth of 1000 m and 40 m below the seafloor. [146]

According to a 'blog, *UNEP Global Outlook on Methane Gas Hydrates*', the total amount of methane contained in the world's gas hydrates is equivalent to 200 to 2,200 times the current annual global energy consumption from all sources (18th March 2015). [110]

The shape of the methane hydrate lump is shown in Figure 6.10A, [294] when brought up from the seafloor as retrieved during an expedition to the Hydrate Ridge, off the coast of Oregon in the US. Figure 6.10B shows the methane hydrates on the seafloor in the Gulf of Mexico. [237] Figure 6.10C shows the findings of methane hydrates occurring in the world. [294]

6.16.3 Reports of Methane Extraction from Hydrates [25,69,80]

a. *A press release from Department of energy (US) on May 2nd 2012 states:*
"In early 2012, a joint project between the United States and Japan produced a steady flow of methane by *injecting carbon dioxide into the methane hydrate accumulation*. The carbon dioxide replaced the methane in the hydrate structure and liberated methane to flow to the surface. This test was significant because it allowed the production of methane without the instabilities associated with a melting gas hydrate." [180]

b. *UNEP Year Book 2014: Emerging Issues in Our Global Environment provides an update on methane from hydrates:*
In March 2013, the world's first offshore methane hydrate production test was conducted off the coast of Honshu Island, Japan. The test site was chosen based on seismic and drill-well data indicating methane hydrate-rich sedimentary layers in this area. About 120,000 m^3 of methane gas was produced from the hydrate-bearing sediments.

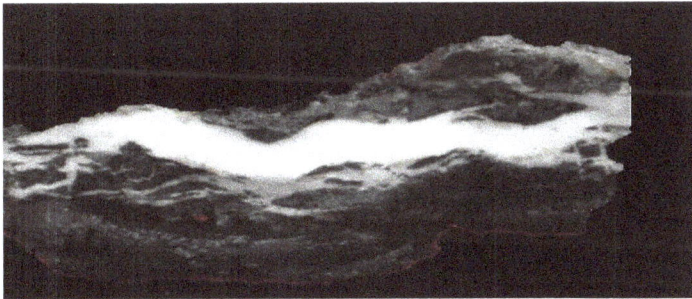

Fig. 6.10A. Shape of Methane Hydrate Lump Brought up from the Sea Floor.
[**Source**: *worldoceanreview.com/.../climate-change-and-methane-hydrates*]

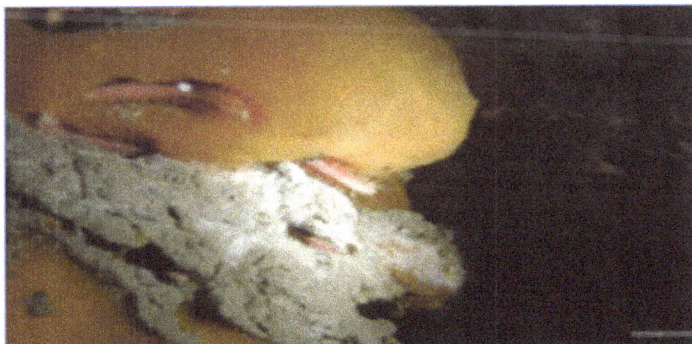

Fig. 6.10B. Methane Hydrate on Sea Floor in Gulf of Mexico. [294]
[***Source:*** *Adobe Export PDF: Hawaii Natural Institute Honolulu - HT 96822*]

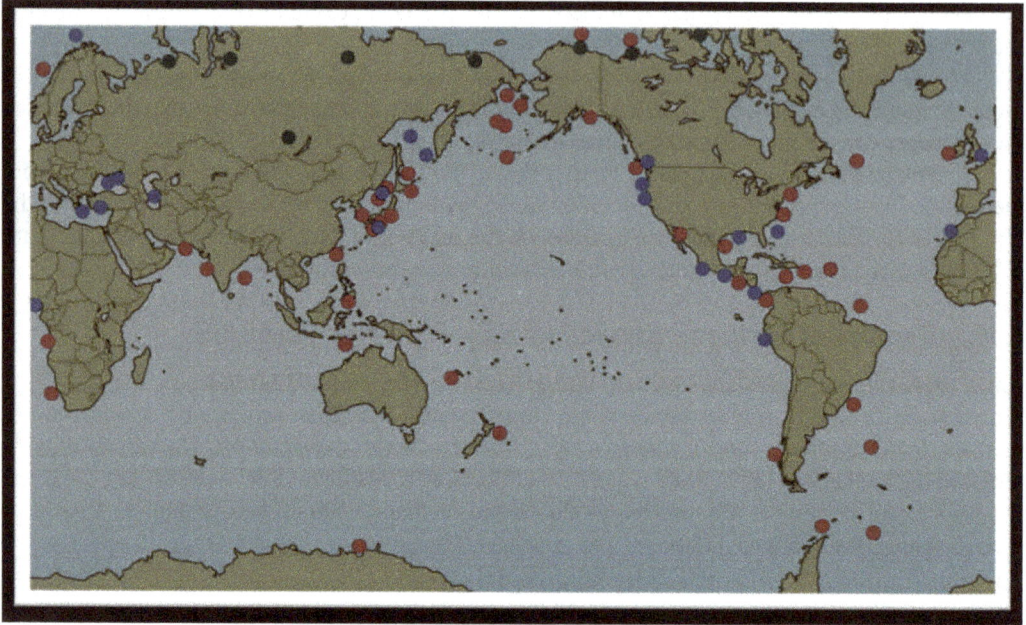

Fig. 6.10C. Occurrence of Methane Hydrates in the World. [294]
(Methane hydrate occurs in all the oceans as well as on land. The green dots show occurrences in the northern permafrost regions. Occurrences identified by geophysical methods are indicated by red dots.
The occurrences shown by blue dots were verified by direct sampling.)
[**Source**: *worldoceanreview.com/.../climate-change-and-methane-hydrates*]

Scientists and engineers are analysing the data collected. At the same time, an international team of researchers has been studying sediment samples containing gas hydrates obtained from layers beneath the deep seafloor in the 'Nakao Trough' off Japan. Highly sophisticated techniques were required to retrieve these samples and keep them in their natural, stable conditions.

Japan and other countries are assessing the extent of available methane hydrate deposits while simultaneously looking at technologies for commercially viable natural gas production. A long-term production test (e.g., one lasting over 18 months) may be required to be carried out to prove that the sustained methane production from gas hydrates is viable. This is the critical research and development step on the path to eventual commercialization. Japan has announced plans to make its extraction technology commercially viable by the end of this decade.

In the early 2000s, a team of Canadian and Japanese scientists succeeded in extracting methane from the Mallik gas hydrate site by heating the reservoir. *Still better results were obtained in 2008 by lowering the reservoir's pressure without resorting to heating.* After the experiment ended, technicians on-site expressed confidence that production could have continued even longer. The success of this second land-based experiment indicated *that decompression techniques may be a more viable route to commercialization of methane hydrates.*

With the necessary technology and favourable market conditions, natural gas production based on extraction of methane from methane hydrates may become economically viable in some regions from some reservoirs.

However, the complex questions remain to be answered—not only about how to achieve production of natural gas from methane hydrates, but also about the future environmental impacts of continuing to use natural gas as a fuel. Renewable energy like solar and wind energy must be developed instead of fossil fuels based energy systems.

International cooperative efforts are necessary to address environmental issues, including the links between fossil fuel combustion and climate change. Such cooperative efforts are carried out by the '*Global Methane Initiative*' and the '*Global Carbon Project*', which produces regularly updated calculations of the global methane budget and trends. [296]

c. *Reports from Japan:*

Success of Japan in producing methane from methane hydrate is listed below:

Ministry of Economy, Trade and Industry, Japan (METI)

The test was carried out in the Mackenzie Delta, Canada via a collaborative research agreement between Japan, Canada, Germany, USA and India. Hydrate depressurization also was tested.

FY 2001–2002: Two-dimensional and three-dimensional seismic surveys in the eastern Nankai Trough (Japan) were conducted by Ministry of Economy, Trade and Industry, Japan (METI).

2004: METI exploratory test wells "Tokai-oki to Kumano-nada Japan)" were drilled.

2007: On 5 March 2007, METI announced that the estimated volume of methane hydrate in-place gas resources in the eastern Nankai Trough was approximately 40 Tcf (about 1.1 trillion m^3).

METI exploratory test wells "Tokai-oki to Kumano-nada" drilled in 2004 revealed that the high saturation methane hydrate in sand reservoirs found in 2000 comprised alternating turbidite sand and mud layers. These zones of concentrated methane hydrate in the eastern Nankai Trough area were considered to have a high potential for development.

Development of novel sampling and analytical tools, such as the Pressure-Temperature Core Sampler (PTCS) to recover hydrate cores while maintaining in situ pressure and temperature was a very successful method. The PTCS device developed earlier improved and recorded an 80% core recovery success rate, maintaining sample in situ pressure 90% of the time during "Tokai-oki to Kumano-nada" test well drilling in 2004.

Japan's own well production simulator, MH21-HYDRES, was created to predict the behavior of methane hydrate-bearing layers.

2006–2007: The second onshore methane hydrate production test (the first winter) was carried out in Canada with the goal of testing the depressurization method at a field scale and refining a gas production simulator code. Over approximately 12.5 h of depressurization, 830 m^3 of gas was produced; however, the production test was interrupted by a significant amount of sand that flowed into the well earlier than expected, causing malfunction of the pump. Completion of this test consequently was postponed until 2008.

2008: The second onshore methane hydrate production test (the second winter) was completed in Canada. About 1.3×10^4 m^3 of gas was produced continuously for six days employing the depressurization method. To prevent produced sand flow into the well, a sand screen was employed. The depressurization method was confirmed to be a valid gas production technique through the two onshore production tests conducted in Canada.

In FY 2007, METI announced that, based on data from their offshore surveys and the analysis of these data and model results, the volume of methane hydrate in-place resources in the eastern Nankai Trough area was approximately 1.1 trillion m^3. This amount is equivalent to about a 10-year supply of Japan's annual natural gas consumption in FY 2012.

FY 2012–FY 2015: The first offshore production test was conducted. On 19 March 2013, JOGMEC made a preliminary announcement reporting that approximately 120,000 m^3 of gas was produced from a methane hydrate layer utilizing the depressurization method; production continued for about six days. It subsequently was officially disclosed that "a cumulative 119,500 m^3 of gas at atmospheric conditions was produced" and detailed technical reports on this offshore production test were prepared and made available in English.

On 6th November 2014, JOGMEC Japan (a committee to promote and to oversee methane hydrate R&D organized by Japan Oil, Gas and Metals National Corporation) and the National Energy

Technology Laboratory of the USA Department of Energy signed a Memorandum of Understanding for a joint long-term onshore production test in Alaska, USA.

On 29 June 2017, METI announced the completion of this test, reporting preliminary values of produced gas of approximately 35,000 m³ for the first production well over 12 days and 200,000 m³ for the second production well over 24 days. [227] A long-term onshore methane hydrate production test on the North Slope of Alaska was planned after FY 2017. [13]

A Japanese study has estimated that at least 40 trillion cubic feet (1.1 trillion cubic meters) of methane hydrates lie in the eastern Nankai Trough off the country's Pacific coast, equal to about 11 years of Japanese gas consumption. Japan's trade ministry reported success on May 8, 2017 in producing gas by extracting methane gas from methane hydrate deposits offshore Japan's central coast. A drilling crew in Japan reported a similar successful operation on 4th May, 2017 offshore the Shima Peninsula for extracting methane from gas hydrate/combustible ice. [37]

Note: In 2012, it was reported in the press that Japan successfully extracted methane hydrate from its seabed by deep drilling. It was a white solid substance, which burnt with a pale flame [90].

Report from China: China had announced on 24th May, 2017 that their floating gas extraction platform in the South China Sea had borne highly promising results and not just for experimental purposes, but for potential commercialization. Engineers drilled to the bottom of the sea and depressurized the hydrates right there, bringing the gas to the surface. According to the reports in Chinese media, they managed to get as much as 35,000 cubic meters of gas per day. [304]

Report from India: Following a collaborative exploration in 2014, with the United States Geological Survey (USGS), the Japan Drilling Company (JDC) and the Japan Agency for Marine-Earth Science and Technology (JAMSTEC), the Indian petroleum ministry issued a statement that the expected reserves of gas hydrates in India is 1894 trillion cubic meters (1 cubic metre of hydrates can contain 160 cu.m. of gas). Second expedition in 2015 confirmed the presence of large, highly saturated gas hydrate accumulations throughout the Krishna-Godavari (K-G) Basin with a potential which could be significantly more than the largest known natural gas field.

Owing to the lack of technical expertise, some experts believe that only a fraction of hydrate reserves will be commercially exploitable as most of the reserves are located in areas where the extraction process is difficult. Further, extraction may cause other problems like oceanic landslides which may result in the leakage of vast amount of harmful methane gas into the atmosphere. If the R&D efforts are viable, it can provide a big push for making India self-sufficient in energy. India has also entered into an agreement with Canada to develop hydrate extraction technology. Japan, which hopes to start commercial production from its offshore hydrates by 2027, is actively collaborating with the US and India since 2018. (Source: https://www.idsa.in/askanexpert/the-prospects-of-methane-hydrates).

6.16.4 *Method of Extraction of Methane from Methane Hydrates* [81,103,165,181,216]

Research is going on to find the most economical technology to extract natural gas from gas hydrates. One method of extracting methane from its hydrate involves the depressurization of deposits. Theoretically, the use of the usual gas-drilling method seems possible, but many problems and details are yet to be resolved.

As CO_2 also forms the CO_2 hydrate ($CO_2. nH_2O$) under similar conditions (and is stable), it seems very attractive to exchange the methane molecule with CO_2 and release methane from gas hydrates ($CH_4. nH_2O$) for use. Injection of CO_2 in the hole drilled for methane hydrates can release methane from the hydrates as the affinity of CO_2 in water is more than that for methane. The natural exchange of CO_2 with CH_4 hydrate is exothermic. To lower the cost of the separation of CO_2, a mixture of CO_2 and N_2 may be injected. This will reduce CO_2 in the atmosphere and the methane so produced may be reserved for power generation, from where the carbon dioxide produced is recovered, compressed and sent down in exchange of methane. This process is environmentally friendly and will help in the efforts to check global warming.

Phase changes are the key requirement to understand reservoir behavior when applying the gas hydrate extraction technique. Until recently, there were two methods of extracting methane from hydrates from offshore locations that were considered feasible. One is to drill a hole into the hydrate deposit to release the pressure, allowing the methane to separate from the clathrate and flow up the wellhead. The second is to warm the hydrate by pumping steam or hot water, again releasing the methane from its icy matrix.

By using the thermal stimulation technique,* it is possible to confine the energy delivery into the gas-hydrate-bearing reservoir to dissociate the reservoir for methane production, but a lot of heat energy is required for the process, making it uneconomical. The two processes of heating and depressurization may be combined to improve the economy—by mostly using selective depressurization and heating only.

In 2002, Canadian, American, Japanese, Indian and German researchers tested both techniques in the field, at a drill site called Mallik on the outer extremity of the Mackenzie river delta in the Canadian Arctic. Both were successful, but the energy costs of the heating method nearly outweighed the energy gained from the methane released, making depressurization the more attractive option.

Note: *If microwave heating instead of conventional heating can be used, methane hydrate can be decomposed and dissociated soon with the help of microwaves. The rate of hydrate dissociation will increase with increasing microwave power. For the microwave radiation, there is a linear relationship between temperature and time. The more the power, the greater is the decomposition rate. Microwaves will also increase the permeability and porosity of hydrates. The energy ratio of microwave heating is more than that of water bath heating, but it is lower than the theoretical value of the thermal stimulation production. Selective heating is possible with microwaves.* [103,165,216].

6.17 Coal Bed Methane [47,48,111,112,114,139,253]

Coal bed methane (CBM) is the gas lying trapped in virgin coal/lignite while the coal mine methane (CMM) is methane gas trapped in seams that have been worked on. *It is released into the atmosphere during coal mining, threatening the life of miners, besides leading to the emission of greenhouse methane gas.* There is no difference in the composition of the CBM and CMM gas as such and only the geo-mining conditions are little different. CMM also uses gas from abandoned mines called AMM (abandoned mines methane) and methane from the ventilation exhaust of mine fans, which is referred to as VAM (ventilated air methane).

Coal mine methane has always been considered as a danger faced during underground coal mining, as it can create a serious threat to mining safety and productivity due to its explosion risk. Methane in a mine can create a localized zone of high concentration in an area of low air velocities and quantities. The concentration of methane in these zones may be in the range of 5% to 15%, known as the explosive range. In this range, methane can be ignited easily with the presence of an ignition source to create a violent methane explosion that may propagate in the presence of combustible coal dust. Lack of oxygen in the area can also kill the miners even if the explosion does not take pace. Ventilation in coal mines has to keep methane levels well below the explosive limit by diluting methane emissions that occur during mining. A ventilated air methane extraction system extracts the diluted methane. The system may be designed to either destroy the methane (by converting it to CO_2) or recover the energy from the methane.

Since methane emissions from VAM represent most of the methane emissions from coal mines, much attention has been given to explore ways to capture and utilize low concentrations of methane under variable flow conditions.

Methane Capture from Coal Beds

Methane adsorbed into a solid coal matrix (***coal macerals***) can be released if the coal seam is depressurised. Methane may be extracted by drilling wells/boreholes into the coal seam. Water pressure in the bed has to be reduced by pumping water from the well. The decrease in pressure allows methane to desorb from the coal and flow as a gas up the well to the surface. Methane can then be compressed and piped to storage or can be directly used. If required, the extracted methane can be upgraded.

Methane should not be put in the water pipeline, but it should flow up the back of the well (casing) to the compressor station. The dewatering should not make the water level too low, otherwise the methane may travel up the tubing into the waterline and make the well 'gassy'. Although methane may be recovered in a water-gas separator at the surface, it may cause the scouring, wearing and breakdown of pumps.

Coal bed methane (CBM) is deemed a 'natural gas'. Many countries in the world have been recovering AMM and CMM over the years.

Enhanced CBM (Coal Bed Methane) Recovery

Enhanced CBM recovery can be done (similar to enhanced oil recovery applied to oil fields) by injecting CO_2 into a bituminous coal bed. The CO_2 will occupy pore space and also adsorb onto the carbon in the coal by displacing methane. The carbon has more affinity for CO_2 than for methane (approximately twice), thus potentially allowing for enhanced methane recovery. This method would simultaneously sequester CO_2 underground and would add to the efforts to halt global warming.

Purpose of Recovery

Vigorous efforts are required to recover methane from the abovementioned sources because:

 i. During coal mining, its release threatens the life of miners besides the emissions creating a greenhouse effect.

 ii. The recovery of CBM and CMM opens a good *alternative energy source and at the same time making mining comparatively more environmentally friendly and safe.*

 iii. The methane in ventilated air is very lean but the recovery of 'VAM' in gas-filled mines is essential as it can improve *the environment and make the mines safer for miners.*

Notes:

1. In-situ-Estimation of Methane in Coal [111]

The quantity of gas is determined by the Meisner and Kim formula by using the moisture content, volatile content, volume of methane adsorbed on wet coal, fixed carbon, thickness of coal and temperature.

Meinser (1984) observed that the amount of methane gas is related to volatile matter.

$$V_{CH4} = -325.6 \times log\ (V.M/37.8)$$

Estimation of the in situ gas content of the coal can be evaluated by using Kim's equation (Kim 1977):

$$V = (100 - M - A)/100 \times [\ Vw\ /Vd]\ [K(P)^N - (b \times T)]$$

Where,

V	=	*Volume of methane gas adsorbed (ml/g)*
M	=	*Moisture content (%)*
A	=	*Ash content (%)*
Vw/Vd	=	*1/(0.25 × M + 1)*
Vw	=	*Volume of gas adsorbed on wet coal (ml/g)*
Vd	=	*Volume of gas adsorbed on dry coal (ml/g)*

The values of K and N depend on the rank of the coal and can be expressed in terms of ratio of fixed carbon (FC) to volatile matter (VM)

K	*=*	*0.8 (F.C/V.M) + 5.6 Where*
F.C	*=*	*Fixed carbon (%)*
VM	*=*	*Volatile matter (%)*
N	*=*	*Composition of coal (for most bituminous coals, N = (0.39 – 0.013 × K))*
b	*=*	*Adsorption constant due to temperature change (ml/g/°C)*
T	*=*	*Geothermal gradient × (h/ 100) + To*
T	*=*	*Temperature at given depth*
To	*=*	*Ground temperature*
h	*=*	*Depth (m)*

2. The CMM may be found in the world wherever coal mines exist. In India, the possibility of recovery of CBM is in coal mines of the eastern region and the Godavari basin as well as in the lignite mines of western India. Seven such areas have been identified. The abandoned mines methane (AMM) can be recovered in India, particularly, from mines around Jharia and Raniganj in the eastern coal mine region.

6.18 Gas to Liquid Fuel (GTL) [76,85,86,88,113]

Any gaseous fuel, whether from renewable sources (like biomass) or nonrenewable sources (like natural gas, gas hydrates, coal mine methane, or coal bed methane), can be converted to liquid fuels. The sulphur content of liquid fuels so obtained would generally be very low, but would depend upon the source of the gaseous fuel. It may be mentioned here that syngas obtained after reforming biomass can be converted to liquid fuel by similar processes. The pyrolysis, gasification and FT process involved in the conversion of biomass to liquid fuel has already been dealt with in Para 5.9 of Chapter 5.

6.18.1 Advantage of Converting Upgraded Biogas to Liquid Fuel for Use in Vehicles [5,19]

If natural gas, methane from gas hydrates from the sea, or methane from renewable sources like biogas, landfill gas, etc., is converted to liquid fuel, it becomes very convenient for transportation, onboard storage in vehicles and use in internal combustion engines in vehicles or industries. The gas to liquids (GTL) process converts methane to a diesel like fuel. The fuel produced has energy density comparable to conventional diesel. The process is scalable and can be applied to both large and small sources of gas or deposits. Even the flare gas can be used as feed for the process. However, the conversion of gas to liquid fuel increases the net carbon dioxide emission by about 9%. In spite of the slight increase in the net greenhouse gas emissions by using GTL fuel in vehicles, it may still be preferable because of simplicity and economics of storage and use. *(Refer to the *Note and Table 6.4 below.)*

The conventional petroleum diesel fuel is becoming increasingly unacceptable because of pollution by diesel exhaust gases that contain particulate matter, polycyclic aromatic hydrocarbons (PAH), sulphur compounds, heavy metals, sulphur dioxide (SO_2), sulphur trioxide (SO_3), etc. The liquid fuel produced by the GTL process (Fischer-Tropsch (FT) diesel), when burnt in engines, produces lesser particulate matter and a negligible amount of SO_2 as compared to petroleum diesel oil. The low aromatic and almost negligible sulphur content of the fuel results in reduced toxicity as well as corrosion problems. (For details of the FT process, see Para 5.9 of chapter 5.)

*Note: Impact on GHG Emission by Change of Petroleum Liquid Fuel to Alternative fuel

The impact on greenhouse gas (GHG) emission by change in fuel in the transport sector has been computed by the EPA (of USA) in EP A420-F-07-035, April 2007. Table 6.4 indicates the change in greenhouse gas emissions if petroleum fuels like petrol/diesel are replaced by the alternative fuels indicated in column 2 of Table 6.4.

Table 6.4. Impact on GHG Emissions by Change of Petroleum Liquid Fuel to Alternative fuel.

Se. No.	Alternative Fuel used	Change in GHG emission	Remarks
1	Biodiesel	(–) 68%	[a]Biomass growing captures CO_2 from the atmosphere, thus, reducing carbon emission. This is in addition to CO_2 capture technology used in the process.
2	Sugar ethanol	(–) 56%	
3	Cellulosic ethanol	(–) 91%	
4	Gaseous hydrogen	(–) 41%	
5	Compressed natural gas (CNG)	(–) 29%	[b] Different results stem, in part, from varied assumptions of the type of CTL (coal to liquid fuel) process employed and the source of electricity used for the carbon capture process.
6	Liquefied natural gas (LNG)	(–) 23%	
7	Corn ethanol	(–) 22%	
8	Liquefied petroleum gas (LPG)	(–) 20%	
' 9	Methanol	(–) 09%	
10	Liquid hydrogen	(+) 07%	* Natural gas converted to liquid fuel (GTL), is a very clean and convenient fuel, but it increases the net carbon dioxide emission by about 9%.
11*	**GTL diesel**	**(+) 09%**	
12	Electricity	(–) 47%	
13	Coal+ biomass[a] to liquid fuel with CO_2 capture [include 38 % energy from switch grass]	(–) 100%	
14	Coal + biomass[a] to liquid fuel with CO_2 Capture [include 28% energy from switchgrass]	(–) 76%	
15	Coal+ biomass[a] to liquid fuel with CO_2 capture [include 10% energy from switch grass]	(–) 22 %	
16	Coal to liquid fuel with Carbon capture[b]	(–) 5% to (+) 4%	
17	Coal to liquid fuel without CO_2 capture[b]	(+) 80% to 113%	

It may be noted from the table that the biomass conversion to liquid fuel, biodiesel, cellulosic ethanol, etc., have very positive impacts on the reduction of greenhouse gases.

However, the conversion of coal to liquid fuel without carbon capture actually increases the net carbon dioxide emission very much and has a very negative impact. GTL, that is, natural gas converted to liquid fuel, is a very clean and convenient fuel, however, it also increases the net carbon dioxide emission by about 9%.

6.18.2 Gas to Liquid (GTL) fuel conversion Process Technology [56,65,76,77,83,85,102,117,217]

The GTL fuel conversion process comprises mainly of three steps:

- The natural gas/methane from a renewable source is reformed to produce a synthesis gas (syngas) by partial oxidation or steam reforming or a combination of the two processes having 'hydrogen/carbon monoxide' in the ratio of 2:1 (approximately). The production of syngas is central to the process. The reforming reaction converts a feed material like biomass, coal, natural gas, etc., to a mixture of carbon monoxide and hydrogen ($CO + H_2$), which is termed as 'synthesis gas' (syngas). The process is usually an endothermic process and the energy is provided externally. The water gas shift reaction monitors the ratio of the H_2 to CO in the syngas.

- The resulting syngas is fed to the Fischer-Tropsch reactor and is converted to mostly straight-chain olefins in the presence of a catalyst. The catalyst is either iron or cobalt based and the reaction is highly exothermic. The temperature and pressure and the catalyst used determine whether a light or heavy synthetic fuel is produced. *(For example, at 330°C, with an iron catalyst, mostly gasoline and olefins are produced, whereas at 180–250°C, with a cobalt catalyst, mostly diesel and waxes are produced.)*

- The high molecular weight liquid products can be hydrocracked*' in a simple low-pressure process to produce naphtha, kerosene and diesel that are virtually free of sulphur and aromatics. (*See *Note on hydrocracking*)

*Note: The Hydrocracking Process [27,53]

The 'hydro-cracking' process is also used in modern day refineries. It is a catalytic cracking process assisted by the presence of hydrogen gas at an elevated partial pressure. The function of hydrogen is the purification of the hydrocarbon feed from 'sulphur' and 'nitrogen'. In 'hydro-cracking' a bi-functional catalyst is normally used. The catalysts, in the presence of hydrogen, are capable of rearranging and breaking chains of hydrocarbons and adding hydrogen to aromatic compounds and olefins for conversion to naphthalenes and alkanes.

The products of this process are saturated hydrocarbons ranging from ethane and LPG, to heavier hydrocarbons consisting mostly of isoparaffins with a low sulphur content. However, the main products produced from hydrocracking are 'jet fuel' and 'diesel', along with some LPG, kerosene and high-octane gasoline fractions. The reaction conditions, that is, temperature, pressure and catalyst activity would be different for obtaining different products. This process is used in Europe and Asia (including India) because of a high demand for diesel and kerosene. In the US, fluid catalytic cracking is more common because of the higher demand for gasoline.

Upgraded Biogas (or Upgraded Gas from Biomass Gasifier) from or Natural Gas Reforming Process [5]

There are several reforming processes through which syngas can be produced from upgraded biogas (or upgraded gas from a biomass gasifier) or natural gas with methane as the major component. But the steam reforming, partial oxidation and autothermal reforming (ATR) are widely used in practice. Some reforming methods also have their merits like the reduction of greenhouse gas emission, energy consumption and process yield. Various methods give a different ratio of hydrogen to carbon monoxide in the syngas.

i. Steam Reforming [171,198,249,250,251]

In this process, the gas to be reformed (i.e., upgraded biogas or natural gas) reacts with steam at 800–900°C and pressure of around 30 bars in presence of nickel, cobalt, or copper as the catalyst to produce synthesis gas, a mixture of CO and H_2. The reaction is endothermic.

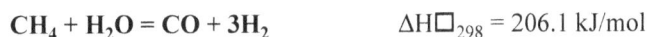

$$CH_4 + H_2O = CO + 3H_2 \qquad \Delta H\square_{298} = 206.1 \text{ kJ/mol}$$

Syngas reacts further to give more hydrogen and carbon dioxide via the water gas shift (WGS) reaction, which is a side reaction in steam reforming. In the forward direction, the reaction is exothermic reaction.

$$CO + 3H_2 + H_2O = CO_2 + 4H_2 \qquad \Delta H\square_{298} = -41.2 \text{ kJ/mol}$$

Steam reforming of 'gas to be reformed' produces syngas with a H_2:CO molar ratio close to 3. Generally, the steam reforming unit consists of two sections, namely the radiant section and the convective section. In the radiant section, the feed is reformed by superheated steam to syngas. Heat is recovered from the hot product gases in the convective section for preheating the feed and reactant for the generation of superheated steam. Due to the high temperature requirement, the process is expensive, although it has the advantage of yielding a high ratio of H_2/C in the syngas.

ii. Reforming by Partial Oxidation [40,152,164,191,192]

Partial oxidation occurs when a sub-stoichiometric fuel-oxygen mixture is partially combusted at a high temperature in a reformer. The reaction is exothermic and therefore, requires a lesser amount of heat, but the reforming process is considered expensive because of the requirement of oxygen that would have to be produced by a separate air separation unit. The molar ratio of H_2/CO in the syngas from this process is around 2. Due care is required in this process of reforming as the methane and oxygen mixture is explosive:

$$2CH_4 + O_2 = 2CO + 4H_2$$

There are two zones in the partial oxidation reactor. In the first zone, the hydrocarbons, oxygen and possibly a small amount of steam (formed during the reaction and also formed by moisture in the feed) react together. The second zone is a heat exchanger that recovers the heat from the products after the reaction. The non-catalytic type reformer requires very high temperatures in the range of 1200–1500°C. Such non-catalytic reformers have been developed by some firms (like Texaco and Shell).

The partial oxidation reformer with the use of a catalyst works in the temperature range of 800–900°C. The catalyst can get poisoned by the presence of sulphur and therefore, catalytic partial oxidation can be used only if the sulphur content of gas to be reformed (natural gas/ upgraded biogas) is below 50 ppm.

iii. Carbon Dioxide Reforming [40,152,164,195,198,305,308]

Carbon dioxide reforming of 'gas to be reformed' involves the production of synthesis gas (CO and H_2) with a 'H_2 to CO ratio of 1' by reacting methane with CO_2 at high temperatures over a suitable catalyst (such as Ni, Cu and other noble metals). The synthesis gas so produced has its use in the manufacture of oxygenates and other liquid hydrocarbons, such as acetic acid.

The main reaction (which is endothermic) is depicted by:

$$CO_2 + CH_4 = 2CO + 2H_2$$

The energy for this endothermic reaction can be provided by using the natural gas from the feed stream. As both CO_2 and CH_4 are greenhouse gases and utilized in this process in generating syngas, the advantage of this process is obvious. The carbon dioxide reforming has, therefore, found a superlative use for gases. However, the problem in this process is of some undesirable reactions that may occur along with the main reaction. This includes the cracking of methane, which leads to the deposition of carbon as coke.

iv. Autothermal Reforming (ATR) [16,124]

Owing to its simplicity, lesser use, dependence on catalysts and lower capital investment, autothermal reforming is the most widely used and dominant natural gas reforming technique.

The reforming reactions take place in a reactor, which primarily consists of two distinctive zones, namely the '*combustion zone*' and '*catalytic thermal zone*'.

- **a.** *Combustion Zone:* Here the feedstock (natural gas/upgraded biogas) is mixed with steam and the feed gas is *partially oxidized with oxygen* (reaction is exothermic) to provide energy for the subsequent endothermic reactions.

- **b.** *Catalytic and Thermal zone:* The unreacted methane and the steam from the feed then pass over a catalyst layer and react to form synthesis gas. This process generally gives a 'H_2 to CO ratio of 3'. Additionally, a water gas shift reaction takes place in this zone to boost the production of hydrogen and the lowering of CO.

A schematic diagram showing production of liquid fuel from upgraded biogas, or natural gas, etc. **(GTL)** is given in Figure 6.11.

v. Chemical- looping reforming [3,64,138,228,260]

If the oxygen is provided indirectly by a solid chemical compound for partial oxidation of the hydrocarbon feed, the process is called 'innovated catalytic partial oxidation reforming' or 'chemical-looping reforming'. Some solid metal oxides like NiO, Fe_2O_3, CuO, Mn_3O_4, etc., can be used as an oxygen carrier. In this type of reforming, two interconnected fluidized beds with particles of NiO and $MgAl_2O_4$ may be used as the bed material, oxygen carrier and reformer catalyst (reactor temperature of 820–930°C). In the fuel reactor, the oxygen carrier metal oxide is reduced

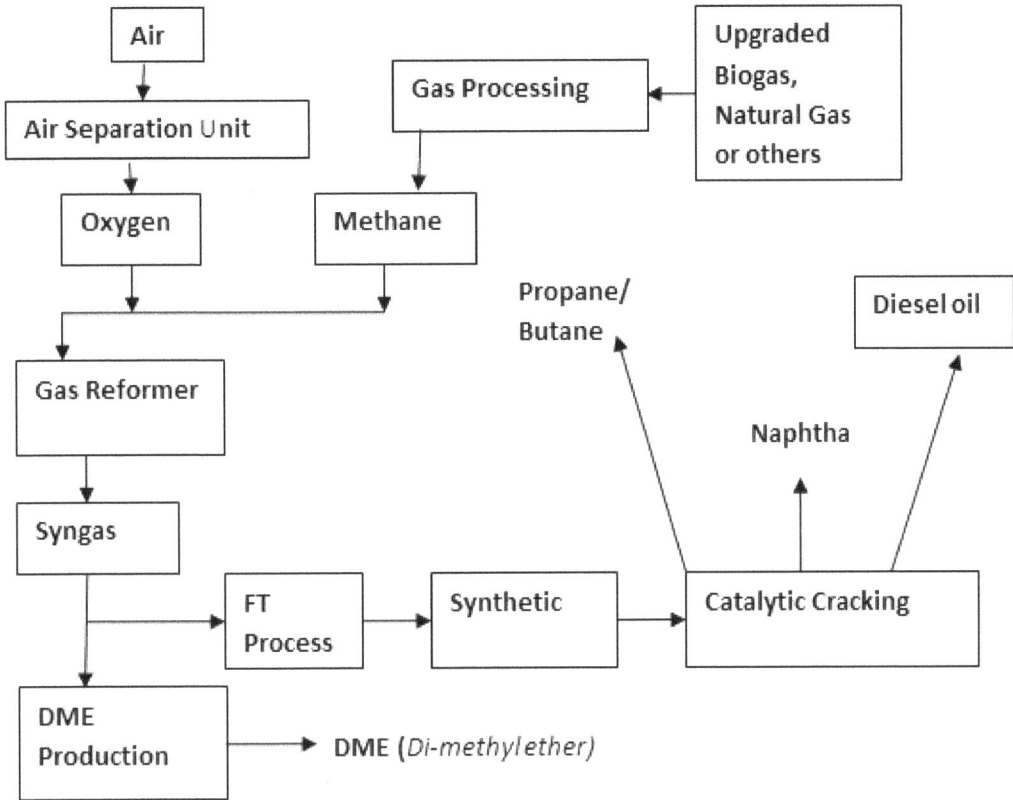

Fig. 6.11. Synthetic Liquid Fuel Production from Upgraded Biogas etc. (GTL).

by the fuel, which is partially oxidized to hydrogen and carbon monoxide. Some of the feed may be fully oxidized to form carbon dioxide and water/steam. In the air reactor, the oxygen carrier is regenerated (i.e., reoxidized to form metal oxide) with air. The formation of carbon soot may be avoided by the addition of steam to feed gas.

Partial Oxidation of Fuel: $CH_4 + NiO = CO + 2H_2 + Ni$

Regeneration of metal oxide: $2Ni + O_2 = 2NiO$

Chemical reforming has a merit of avoiding the 'expensive air separation unit'.

Necessity of Syngas Cleaning for the FT Process [229]

Cleaning of syngas is very important as the pollutants like oxides of sulphur, ammonia, hydrogen sulphide gases, particulate matter and tars cause poisoning of the catalyst. It may be mentioned that the lower temperature in the fuel gasification causes tar formation, while the presence of sulphur and nitrogenous matter in the fuel being gasified is responsible for the formation of other gaseous pollutants. Even the alkaline compounds, if present in syngas, interfere with the catalyst activities in the FT process and therefore these compounds require removal. Some feedstock for gasification (like straw) contains a high level of alkaline compounds. *The alkaline compounds normally may condense on particulate matter during cooling of syngas below 600ºC and may be removed along with the particulate matter. (Refer to Table 6.3.)*

6.18.3 *Proprietary GTL Processes* [73,76,96,269,286]

Each company producing GTL has its own proprietary FT technology. Most processes use a slurry-phase reactor** with a cobalt-based catalyst. Slurry is made up of molten FT wax with a catalyst suspended in it. This allows for good contact between catalyst and syngas, which is bubbled through

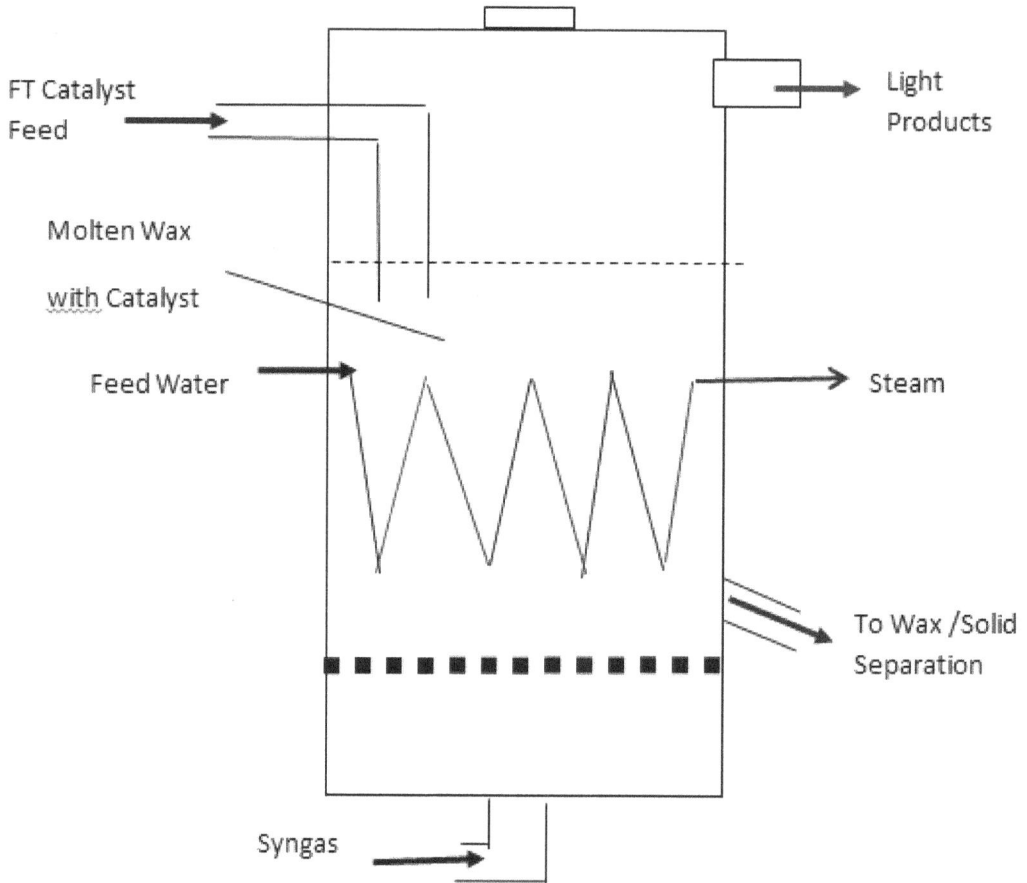

Fig. 6.12. Slurry Bubble Column Reactor of F T Process for GTL Conversion.

it. The exceptions are Shell and BP, whose processes use a fixed-bed reactor, while Rentech Inc. uses an iron-based catalyst. Most companies use autothermal reforming (ATR), as described in Para (iv) above, rather than steam reforming because it is less expensive for larger plants. The steam reforming capital cost is almost linear with the capacity because the amount of tubing required is proportional to the plant capacity. *(**See **Note** on the slurry phase reactor).*

A schematic diagram of a typical slurry bubble column reactor for the FT process for gas to liquid fuel conversion (GTL) is shown in Figure 6.12.

Characteristics of a few Proprietary Processes

a. Sasol's Process

Sasol is a supplier of 'Synfuel technology'. They provide technology to manufacture petroleum products from coal in South Africa. The firm has built a series of Fischer-Tropsch coal-to-oil plants, but now it is providing natural-gas-to-oil (GTL) technology also.

Sasol has commercialized four reactor types with the slurry phase distillate process being the most recent. Its products are more olefinic than those from the fixed bed reactors and are hydrogenated to straight-chain paraffins. Its 'slurry phase distillate process' converts natural gas into liquid fuels, primarily to a superior-quality diesel-like fuel.

The other technology uses the 'Sasol advanced synthol (SAS) reactor' to produce mainly light olefins and gasoline fractions. Sasol has developed high-performance cobalt-based and iron-based catalysts for these processes.

b. Shell Process

Shell has carried out R&D since the late 1940s on the conversion of natural gas, leading to the development of the 'Shell middle distillate synthesis (SMDS)' route, a modified F-T process. But unlike other F-T synthesis routes aimed at gasoline as the principal product, SMDS focuses on maximizing yields of middle distillates, notably kerosene and gas oil.

The process consists of three steps: the production of syngas with a H_2:CO ratio of 2:1; syngas conversion to high molecular weight hydrocarbons via F-T using a high-performance catalyst; and hydrocracking* and hydroisomerization to maximize the middle distillate yield. The products are highly paraffinic and free of nitrogen and sulphur. *(*See Note on hydrocracking.)*

c. Exxon Process

Exxon has developed a commercial F-T system from natural gas feedstock. Exxon claims its slurry design reactor and proprietary catalyst systems result in high productivity and selectivity along with significant economy of scale benefits.

They employ a three-step process: fluid bed synthesis gas generation by catalytic partial oxidation; slurry phase F-T synthesis; and fixed bed product upgrade by 'hydro-isomerization'. The process can be adjusted to produce a range of products. More recently, Exxon has also developed a new chemical method based on the Fischer-Tropsch process to synthesize diesel fuel from natural gas. Exxon claims that better catalysts and improved oxygen-extraction technologies reduce the capital cost of their process.

d. Syntroleum Process

The Syntroleum Corporation of USA has a natural-gas-to-diesel technology based on the F-T process. It is claimed to be competitive as it has a lower capital cost due to the redesign of the reactor; using an air-based autothermal reforming process instead of oxygen for the synthesis gas preparation (to eliminate the significant capital expense of an air separation plant); and due to use of their special high-yield catalyst.

The firm has now licensed its proprietary process for converting natural gas into other synthetic crude oils and transportation fuels.

e. Rentech Process

Rentech (Denver, CO, USA) has developed an F-T process using a molten wax slurry reactor and precipitated iron catalyst to convert gases and solid carbon-bearing material into straight-chain hydrocarbon liquids. In their process, long straight-chain hydrocarbons are drawn off as a liquid heavy wax while the shorter-chain hydrocarbons are withdrawn as overhead vapors and condensed to soft wax, diesel fuel and naphtha.

*Note: on Slurry Phase Reactor [73,286]

The slurry reactor is the advanced reactor technology for gas-to-liquid fuel processes. Its advantages include simple construction, excellent heat transfer performance, online catalyst addition and withdrawal and reasonable interphase mass transfer rates with low-energy input, which make it very suitable for gas-to-liquid processes. However, the multiphase flows are very complex. Under industrial conditions, the high pressure, temperature and solid concentration have notable and complex influences on the gas bubble behaviours, gas holdup, liquid velocity and mass and heat transfer. There are some challenging engineering problems like 'Gas liquid mass transfer in gas -liquid-solid slurry system especially when solid concentration is high' and separation of fine catalyst particles from the viscous wax in FT synthesis.

6.19 Summary

This chapter is devoted to methane/biogas. Methane is a powerful greenhouse gas, but is a very good fuel and can be obtained both from renewable and nonrenewable sources. The main renewable sources of methane that can be tapped are: (i) biogas production by anaerobic decay/biological or chemical processes from renewable materials, (ii) thermal gasification of organic materials and (iii) landfills, generally from municipal solid wastes.

Good feedstocks for biogas generation mainly by biodegradation are:

(a) Animal wastes like cattle dung, urine, goat and sheep droppings, piggery wastes, slaughterhouse wastes, fish wastes, left out animal food/wastes, etc.

(b) Human wastes like faeces, urine, waste food/vegetable/fruit/other organic wastes, etc.

(c) Biodegradable organic waste from industries/ hotels/ shops/ eateries.

(d) Biogas from distillery spent wash, sewage and wastewaters during treatment.

(e) Municipal solid wastes (MSW) landfills. (The gas recovered is called landfill gas.)

Theory of methanization of organic wastes, kinetics of fermentation, factors affecting bio-digestion and biogas production and various methods of production of biogas from various resources have been discussed in detail. The method of tapping landfill gas from municipal solid wastes (MSW), mechanism of generation of landfill gas, composition of gas, phenomena of gas migration, etc., have also been discussed in detail.

The discussion on production of gaseous fuel through thermal gasification of lignocellulosic waste materials/biomass covers various aspects such as the chemistry of gasification, types of gasifiers and their feedstock characteristics, generation of dioxins and PAHs during gasification, requirement of syngas quality for various purposes and also R&D in biomass gasification in the area of gas and tar cracking, advance gasification by plasma torches, flash pyrolysis, advance/plasma gasification. Conversion of 'biomass gasifier output gases' to methanol, methane, or other chemicals has also been discussed.

The nonrenewable sources which are of particular interest for safeguarding the environment are: methane recovery from gas hydrates from oceans and methane recovery from coal mining. The recovery of methane from these sources is essential and should be carried out with a sense of urgency because they are potential dangers to the environment. If the methane from these sources is released into the atmosphere inadvertently due to any reason (or due to global warming itself), it would cause a greater greenhouse effect and accelerated global warming than its planned utilization and conversion to carbon dioxide. The aim of tapping these sources of energy is, therefore, in conformity with the broader objective of arresting global warming by controlling the emission of greenhouse gases. It for this reason that the subject has been briefly mentioned in this book.

Several reports about the discoveries of methane hydrates and efforts for tapping of methane from this resource have been mentioned. This recovery of methane from methane hydrates is still in its infancy and developing stage, while the recovery of methane from coal mining is sufficiently developed.

From the point of view of ease of use, it is very advantageous to convert methane obtained from any source (renewable or nonrenewable) to liquid fuel. The technology is dubbed as GTL (gas to liquid) technology and has also been discussed in this chapter. Various reforming processes (like steam reforming, reforming by partial oxidation, carbon dioxide reforming, autothermal refining, chemical looping reforming, etc.), proprietary processes (like Sasol's process, Shell process, Syntroleum process and Rentech process) and slurry phase reactor have been described.

QUESTIONS

1. *What are the resources for methane from (a) nonrenewable sources and (b) renewable sources?*

2. *What is biogas?*

3. *What is its typical composition? What air pollutants are present in it?*

4. *What is a biodigester? What materials can be fed into it to get a good output? What are the rich feedstocks that decompose in a short time and contain nutrients?*

5. *Can you use plant wastes to generate biogas in a digester?*

6. *What are BOD and COD? What is their significance and use?*

7. What are the best temperature ranges for the growth of mesophilic bacteria, thermophilic bacteria and psychrophilic bacteria? Give examples of methanogenic microorganisms.

8. Give examples of industries, from which the effluents and wastes can generate biogas?

9. What are the factors affecting biogas generation in a biodigester?

10. What are the effects of presence of metals like Co, Se, W, B and Ni and metal oxides like ZnO, CuO, CoO, Mn_2O_3, Co_3O_4, Ni_2O_3 and Cr_2O_3 on biogas generation?

11. If the biogas output of a biodigester drops, what troubleshooting measure should be taken?

12. What are the possible dangers in biogas generation and handling? What precautions should be taken to ward off these dangers?

13. Describe the biotechnology for production of biogas from renewable sources giving chemistry behind it.

14. Give some examples of different types of biogas plants.

15. Give an example of biogas generation from sewage through 'Anaerobic Up-flow Sludge Blanket Process'.

16. What is a suspended growth pulsated sludge blanket reactor? Can you use this for biogas generation from distillery wastes?

17. Describe a method of biogas production from industrial wastewaters/organic waste and enumerate its environmental advantage.

18. How would you produce biogas from organic solid wastes including human/animal wastes?

19. Write a short note on the utilization of sewage for the production of biogas.

20. Are there any kinetic models for the anaerobic fermentation process of biogas generation?

21. What is landfill gas and its typical composition? How would you harness this resource for fuel and manure?

22. What are the different phases in the biodegradation of municipal solid waste landfill? How does the composition of landfill gas differ with time after closing the pit?

23. What are the factors that affect the generation of landfill gas?

24. Describe the phenomena of migration of landfill gas and the mechanisms of gas movements.

25. What type of soil is best suited to a landfill?

26. Why is fire safety very important in landfills?

27. What precautions are required to protect groundwater from leachate produced by landfills?

28. Describe the process of landfill gas recovery from municipal solid wastes.

29. Give the theoretical stoichiometric model for pre-estimation of landfill gas production.

30. Describe the gasification technology for methane production from biomass.

31. What is pyrolysis? What is the difference between combustion, gasification and pyrolysis?

32. What is the difference between incineration and gasification?

33. What are the main types of gasifiers?

34. What are the main differences among the bubbling fluidized bed gasifier (BFBG), circulating fluidized bed gasifier (CFBG) and dual fluidized bed indirect gasifier (DFBIG)?

35. Describe a circulating fluidized bed gasifier (CFBG) for biomass.

36. What are the main features of fixed bed biomass gasifier of down-shaft and up-shaft types?

37. What are the effects of characteristics of feedstocks on gasifier performance?

38. What air pollutants are generated in gasifier units?

39. What is the typical requirement of syngas quality for different purposes?

40. *What are the different types of pyrolysis?*

41. *Write brief note on advanced gasification by breaking down the wastes using plasma torches.*

42. *What is plasma?*

43. *How does the plasma gasification help the environment?*

44. *How would you synthesize methanol and methane from gasifier gases?*

45. *What are methane hydrates and what are the environmental advantages of recovering them?*

46. *Write a short note about the discovery of gas hydrate across the world.*

47. *How can they be recovered and usefully exploited? What are the difficulties in exploiting this source of energy?*

48. *Which methods of recovering the methane from gas hydrates have been used so far?*

49. *Define: Coal Mine Methane, Abandoned Mines Methane, Ventilated Air Methane.*

50. *Describe method of converting methane gas to liquid fuel (GTL).*

51. *What are advantages of GTL?*

52. *What do you understand by reforming? Which types of reforming are mostly used in the case of conversion of natural gas to liquid fuels. Describe steam reforming.*

53. *Write short notes on: (a) Carbon dioxide reforming. (b) Partial oxidation reforming. (c) Chemical looping Reforming.*

54. *What are the merits of autothermal reforming (ATR) as against other reforming processes?*

55. *What are the main differences in the following processes for converting gas to liquid fuels: (i) Shell Process, (ii) Exxon Process, (iii) Syntroleum Process, (iv) Rentech Process?*

56. *Describe the principle behind the slurry phase reactor for gas to liquid fuel conversion.*

57. *How does the gas from renewable sources like biogas/land fill gas/gasification of organic matter help environmental degradation and reduce the greenhouse effect?*

58. *Describe how would you upgrade the biogas/landfill gas and make it a suitable transportation fuel?*

59. *What are Renewable Information Numbers (RIN)?*

60. *How can biogas play significant role in fulfilling energy needs in rural areas?*

61. *Give the principle of conversion of carbon dioxide into methane?*

References

1. Acharya, Bhavik K., Sarayu Mohana and Datta Madamwar. July 2008. Anaerobic treatment of distillery spent wash—a study on upflow anaerobic fixed film bioreactor. Bioresour Technol. 99(11): 4621–6. doi: 10.1016/j.biortech.2007.06.060. Epub 2007 Aug 31. PMID: 17765535.

2. Achinas, Spyridon, Vasileios Achinas and Gerrit Jan Willem Euverink. June 2017. A technological overview of biogas production from biowaste. Green Chemical Engineering.-Review. Engineering 3(3): 299–307. https://doi.org/10.1016/J.ENG.2017.03.002.

3. Adanez, Juan, AlbertoAbad, FranciscoGarcia-Labiano, PilarGayan, Luis F.de Diego. April 2012. Progress in Chemical-Looping Combustion and Reforming technologies- Review. Progress in Energy and Combustion Science 38(2): 215–282. https://doi.org/10.1016/j.pecs.2011.09.001.

4. Adhikari, Bikash, Khet Raj Dahal and Sanjay Nath Khanal. September 2014. A review of factors affecting the composition of municipal solid waste landfill leachate. ISSN: 2319-5967 ISO 9001:2008. Certified Int. J. of Engineering Science and Innovative Technology (IJESIT). 3(5): 273.

5. Adnan, Amir Izzuddin, Mei Yin Ong, Saifuddin Nomanbhay, Kit Wayne Chew and Pau Loke Show. Oct. 2019. Technologies for Biogas Upgrading to Biomethane: A Review. Bioengineering. file:///C:/Users/Rutag/AppData/Local/Temp/bioengineering-06-00092-v2.pdf.

6. Alauddin, Zainal A.B.Z., Pooya Lahijani, Maedeh Mohammadi and Abdul Rahman Mohamed. December 2010. Gasification of lignocellulosic biomass in fluidized beds for renewable energy development: A review. Renewable and Sustainable Energy Reviews 14(9): 2852–2862. https://doi.org/10.1016/j.rser.2010.07.026.

7. Alitalo, A., M. Niskanen and E. Aura. 2015. Biocatalytic methanation of hydrogen and carbon dioxide in a fixed bed bioreactor. Bioresour. Technol. 196: 600–605. DOI: 10.1016/j.biortech.2015.08.021.

8. Amaratunga, M. 1986. Structural behaviour and stress conditions of fixed dome type of biogas units. In: Elhalwagi, M.M. (Ed.). Biogas Technology, Transfer and Diffusion, London & New York. pp. 295-301.0001182; ISBN: 1-85166-000-3.

9. Andrade, L.A., F.R.X. Batista, T.S. Lira, M.A.S. Barrozo and L.G.M. Vieira. 2018. Characterization and product formation during the catalytic and non-catalytic pyrolysis of the green microalgae *Chlamydomonas reinhardtii*. Renew. Energ., 119: 731–740.

10. Angelidaki, I., M. Alves, D. Bolzonella, L. Borzacconi, J.L. Campos and A.J. Guwy. March 2009. Defining the biomethane potential (BMP) of solid organic wastes and energy crops: a proposed protocol for batch assays. Research Article 59(5): 927–934. https://doi.org/10.2166/wst.2009.040.

11. Arena, Umberto. April 2012. Process and technological aspects of municipal solid waste gasification: A review. Waste Management 32(4): 625–639. https://doi.org/10.1016/j.wasman.2011.09.025.

12. Armaha, Edward Kwaku, Emmanuel Kweinor Tetteha and Bright Boafo Boamahb. December 2017. Overview of biogas production from different feedstocks. Int. J. of Scientific and Research Publications 7(12): 158 ISSN 2250-3153. www.ijsrp.org.

13. Arora, Amit, Chandrajit Balomajumder and Swaranjit Singh. March 2015. Techniques for Exploitation of Gas Hydrate (Clathrates) an Untapped Resource of Methane Gas. [https://www.omicsonline.org/open-access/techniques-for-exploitation-of-gas-hydrate-clathrates-an-untapped-resource-of-methane-gas-1948-5948-1000190.php?aid=55011].

14. Asgari, Mohammad Javed, Kamran Safavi and Forogh Mortazaeinezahad. 2011. Landfill Biogas production process. Int. Conf. on Food Engineering and Biotechnology IPCBEE. Volume 9, IACSIT Press, Singapore. http://www.ipcbee.com/vol9/40-B10035.pdf.

15. Astashynski, Valiantsin M., Andrey V. Gusarov and Nikolai N. Cherenda (eds.). 2009. Thermal Plasma Gasification of Biomass for Fuel Gas Production. High Temperature Material Processes: An International Quarterly of High-Technology Plasma Processes. 13(3-4). ISSN Print: 1093-3611, ISSN Online: 1940-4360.

16. Azarhoosh, M.J., H. Ale Ebrahim and S.H. Pourtarah. 2016. Simulating and optimizing auto-thermal reforming of methane to synthesis gas using a non-dominated sorting genetic algorithm II method. Chemical Engineering Communications 203(1): 53–63. https://doi.org/10.1080/00986445.2014.942732.

17. Bharathiraja, B., T. Sudharsana, J. Jayamuthunagai, R. Praveenkumar, S. Chozhavendhan and J. Iyyappan. July 2018. Biogas production–A review on composition, fuel properties, feed stock and principles of anaerobic digestion. Renewable and Sustainable Energy Reviews 90: 570–582. https://doi.org/10.1016/j.rser.2018.03.093.

18. Bala, Renu, Vaishali Gautam and Monoj Kumar Mondal. January 2019. Improved biogas yield from organic fraction of municipal solid waste as preliminary step for fuel cell technology and hydrogen generation. Int. J. of Hydrogen Energy 44(1): 164–173. https://doi.org/10.1016/j.ijhydene.2018.02.072.

19. Balkenhoff, Birgit and Diarmid Jamieson. Upgraded biogas as renewable energy. SLR Consulting. https://www.iswa.org/uploads/tx_iswaknowledgebase/5-325paper_long.pdf.

20. Bi, Feng, Muhammad Fahad Ehsan, Wei Liu and Tao He. September 2014. Visible-Light Photocatalytic Conversion of Carbon Dioxide into Methane Using Cu 2 O/TiO 2 Hollow Nanospheres. Chinese Journal of Chemistry, DOI: 10.1002/cjoc.201400476.

21. Billington, R.S. February 1988. A review of the kinetics of the methanogenic fermentation of lignocellulosic wastes. Review paper. J. of Agricultural Engineering Research. 39(2): 71-84. https://doi.org/10.1016/0021-8634(88)90131-X.

22. Bingemer, H.G. and P.J. Crutzen. February 1987. The production of methane from solid wastes. J. of Geophysical Research 92(D2): 2181–2187.

23. Biogas From Wikipedia, the free encyclopaedia; https://en.wikipedia.org/wiki/Biogas

24. Blog- Frozen Heat UNEP global outlook on methane gas hydrates. 2014. Volume One. William Waite, Ray Boswell, Scott Dallimore (eds.). (http://www.methanegashydrates.org/blog).

25. Boswell, Ray, Koji Yamamoto, Sung-Rock Lee, Timothy Collett, Pushpendra Kumar and Scott Dallimore. 2014. Chapter 8 - Methane Hydrates. Future Energy (2nd Edition), Improved, Sustainable and Clean Options for our Planet, pp 159-178. https://doi.org/10.1016/B978-0-08-099424-6.00008-9.

26. Boyle, W.C. October 1976-1977. ENERGY RECOVERY FROM SANITARY LANDFILLS - A REVIEW. Microbial Energy Conversion, Proc. of a Seminar Sponsored by the UN Institute for Training and Research (UNITAR) and the Ministry for Research and Technology of the Federal Republic of Germany Held in Göttingen. pp 119-138. https://doi.org/10.1016/B978-0-08-021791-8.50019-6.

27. Bricker M., V. Thakkar, J. Petri. 2014. Hydrocracking in Petroleum Processing. In: S. Treese, D. Jones, P. Pujado (eds). Handbook of Petroleum Processing. Springer, Cham. https://doi.org/10.1007/978-3-319-05545-9_3-1.

28. Bridgwater, A.V. and S.A. Bridge. 1991. A review of biomass pyrolysis and pyrolysis technologies. In: Biomass Pyrolysis Liquids Upgrading and Utilisation, Elsevier Science Publishing Co., New York.

29. Budiyono, Iqbal Syaichurrozi and Siswo Sumardiono. 2014. Research Article Kinetic Model of Biogas Yield Production from Vinasse at Various Initial pH: Comparison between Modified Gompertz Model and First Order Kinetic Model. Research J. of Applied Sciences, Engineering and Technology. 7(13): 2798-2805. DOI:10.19026/rjaset.7.602 (URL:http://creativecommons.org/licenses/by/4.0/). 2798.

30. Cadena S., F. Cervantes, L. Falcón and J. García-Maldonado. 2019. The Role of Microorganisms in the Methane Cycle. Front. Young Minds. 7:133. doi:10.3389/frym.2019.00133.

31. Cavenati, Simone , Carlos A. Grande and Alírio E. Rodrigues. August 2005. Upgrade of Methane from Landfill Gas by Pressure Swing Adsorption. Energy Fuels. 19(6):2545-2555. https://doi.org/10.1021/ef050072h.

32. Chan, Fan Liang and Akshat Tanksale. October 2014. Review of recent developments in Ni-based catalysts for biomass gasification. Renewable and Sustainable Energy Reviews. 38:428-438. https://doi.org/10.1016/j.rser.2014.06.011.

33. Chen, Guan-Yi, Cong Liu, Wen-Chao Ma, Bei-Bei Yan and Na Ji. November 2015. Catalytic Cracking of Tar from Biomass Gasification over a HZSM-5-Supported Ni–MgO Catalyst. Energy Fuels 29(12): 7969-7974. https://doi.org/10.1021/acs.energyfuels.5b0083.

34. Chen, Wen-Hsing, Shen-Yi Chen, Samir Kumar Khanal and Shihwu Sung. December 2006. Kinetic study of biological hydrogen production by anaerobic fermentation. Int. J. of Hydrogen Energy. 31(15): 2170-2178; https://doi.org/10.1016/j.ijhydene.2006.02.020.

35. Chen, Yih Ren and Andrew G. Hashimoto. 1978. Kinetics of methane fermentation. https://www.semanticscholar.org/paper/Kinetics-of-methane-fermentation-ChenHashimoto/d860ba393749b8ff5677c36ffb70e0ec745de04d.

36. Chen, Yud-Ren, Vincent H.Varel and Andrew G. Hashimoto. December 1980. Methane Production from Agricultural Residues. A Short Review. Ind. Eng. Chem. Prod. Res. Dev.19(4): 471-477. https://doi.org/10.1021/i360076a001.

37. China just extracted gas from 'Flammable ice'. https://www.sciencealert.com/china-has-just-tapped-into-natural-gas-found-in-flammable-ice.

38. Cho, Sunja, Seonghwan Park, Jiyun Seon, Jaechul Yu and Taeho Lee. September 2013. Evaluation of thermal, ultrasonic and alkali pre-treatments on mixed-microalgal biomass to enhance anaerobic methane production. Bioresource Technology. 143:330-336. https://doi.org/10.1016/j.biortech.2013.06.017.

39. Chopra, Sangeeta and Anil Kumar Jain. May 2007. A review of fixed bed gasification systems for biomass. Agricultural Engineering International : The CIGR e-journal 9(5). https://www.researchgate.net/publication/228668672_A_review_of_fixed_bed_gasification_systems_for_biomass.

40. Choudhary, Vasant R. and Kartick C.Mondal. September 2006. CO_2 reforming of methane combined with steam reforming or partial oxidation of methane to syngas over $NdCoO_3$ perovskite-type mixed metal-oxide catalyst. Applied Energy. 83(9):1024-1032. https://doi.org/10.1016/j.apenergy.2005.09.008.

41. Christophersen, Mette and Peter Kjeldsen. Dec. 2001. Lateral gas transport in soil adjacent to an old landfill: factors governing gas migration. Int. Solid Waste Association. https://doi.org/10.1177/0734242X0101900615.

42. Christy, P.Merlin, L.R., Gopinath and D.Divya. June 2014. A review on anaerobic decomposition and enhancement of biogas production through enzymes and microorganisms. Renewable and Sustainable Energy Reviews. 34:167-173. https://doi.org/10.1016/j.rser.2014.03.010.

43. Chung, J. N. January 2014. A theoretical study of two novel concept systems for maximum thermal-chemical conversion of biomass to hydrogen. Front. Energy Res. https://doi.org/10.3389/fenrg.2013.00012.

44. Ciuła, Józef, Violetta Kozik, Agnieszka Generowicz, Krzysztof Gaska, Andrzej Bak, Marlena Pa ́zdzior. et al. 2020. Emission and Neutralization of Methane from a Municipal Landfill-Parametric Analysis. Energies, 13:6254. doi:10.3390/en13236254.

45. Clathrate hydrate From Wikipedia, the free encyclopedia. https://en.wikipedia.org/wiki/Clathrate_hydrate.

46. Climate change and methane hydrates. World Ocean Review. http://worldoceanreview.com/en/wor-1/ocean-chemistry/climate-change-and-methane-hydrates/.

47. Coal Bed Methane. https://www.studentenergy.org/topics/coal-bed-methane.

48. Coalbed methane From Wikipedia, the free encyclopedia https://en.wikipedia.org/wiki/Coalbed_methane.

49. Cohen, Tico. Sept. 2004. Waste to Energy, a waste solution success in Thailand. Refocus 5(5):26–28. DOI: 10.1016/S1471-0846(04)00220-3.

50. Collett T.S.1971-1992. Energy resource potential of natural gas hydrate. American Association of Petroleum Geologists Bulletin 86.

51. Coq, Laurence Le and Ashenafi Duga. January 2012. Syngas Treatment Unit for Small Scale Gasification-Application to IC Engine Gas Quality Requirement. J. of Applied Fluid Mechanics. 5(1):95-103.

52. Corella, José , José M. Toledo and Gregorio Molina. September 2007. A Review on Dual Fluidized-Bed Biomass Gasifiers. Ind. Eng. Chem. Res., 46(21): 6831–6839. https://doi.org/10.1021/ie0705507.

53. Cracking (chemistry). From Wikipedia, the free encyclopedia; https://en.wikipedia.org/wiki/Cracking_(chemistry)

54. Curtis, L.J. , and D.J. Miller. 1988. Transport model with radiative heat transfer for rapid cellulose pyrolysis. Ind. Eng. Chem. Res. 27:1775.

55. Dallimore, S. R. and T. S. Collett. 1995. Intrapermafrost gas hydrates from a deep core hole in the Mackenzie Delta, Northwest Territories, Canada. Geology. 23(6):527-530. https://doi.org/10.1130/0091-7613(1995)023<0527:IGHFAD>2.3.CO;2.

56. Dancuart, L.P. and A.P.Steynberg. 2007. Fischer-Tropsch Based GTL Technology: a New Process. Studies in Surface Science and Catalysis. 163:379-399. https://doi.org/10.1016/S0167-2991(07)80490-3.

57. De Gioannis G., A. Muntoni, G. Cappai and S. Milia. 2009. Landfill gas generation after mechanical biological treatment of municipal solid waste. Estimation of gas generation rate constants. Waste Manage. 29:1026–1034.

58. Demirer, Göksel N. , Metin Duran, Engin Güven, Örgen Ugurlu, Ulas Tezel and Tuba H. Ergüder. November 2000. Anaerobic treatability and biogas production potential studies of different agro-industrial wastewaters in Turkey. Biodegradation. 11(6): 401–405.

59. Devi, Lopamudra, Krzysztof J. Ptasinski and Frans J.J.G. Janssen. February 2003. A review of the primary measures for tar elimination in biomass gasification processes. Biomass and Bioenergy, 24(2):125-140. https://doi.org/10.1016/S0961-9534(02)00102-2.

60. Diebold, J. , and A.V. Bridgwater. 1996. Overview of fast pyrolysis of biomass for the production of liquid fuels. In A.V. Bridgwater and D.G.B. Boocock (eds.). Developments in Thermochemical Biomass Conversion, Blackie Academic, London.

61. Dong, Leilei and Hao Liu Saffa Riffat. August 2009. Development of small-scale and micro-scale biomass-fuelled CHP systems – A literature review. Applied Thermal Engineering. 29(11–12): 2119-2126. https://doi.org/10.1016/j.applthermaleng.2008.12.004.

62. Dou, Binlin, Jinsheng Gao, Xingzhong Sha and Seung Wook Baek. December 2003. Catalytic cracking of tar component from high-temperature fuel gas. Applied Thermal Engineering. 23(17): 2229-2239. https://doi.org/10.1016/S1359-4311(03)00185-6.

63. Duan, Nina, Bin Dong, Bing Wu and Xiaohu Dai. January 2012. High-solid anaerobic digestion of sewage sludge under mesophilic conditions: Feasibility study. Bioresource Technology, 104: 150-156. https://doi.org/10.1016/j.biortech.2011.10.090.

64. Dueso, Cristina, María Ortiz, Alberto Abad, Francisco García-Labiano, Luis F.de Diego, Pilar Gayán and JuanAdánez. April 2012. Reduction and oxidation kinetics of nickel-based oxygen-carriers for chemical-looping combustion and chemical-looping reforming. Chemical Engineering Journal. 188: –154. https://doi.org/10.1016/j.cej.2012.01.124.

65. Dyer, Paul N., Robin E. Richards, Steven L. Russek and Dale M Taylor. October 2000. Ion transport membrane technology for oxygen separation and syngas production', Solid State Ionics.) 134(1-2): 21–33. https://doi.org/10.1016/S0167-2738(00)00710-4.

66. Echiegu, E.A. 2015. Kinetic Models for Anaerobic Fermentation Processes-A Review. American Journal of Biochemistry and Biotechnology, 11(3), 132-148. https://doi.org/10.3844/ajbbsp.2015.132.148.

67. Echiegu, Emmanuel Amagu. 2015. Review Kinetic Models for Anaerobic Fermentation Processes-A Review. American J. of Biochemistry and Biotechnology. 11(3):132.148. DOI: 10.3844/ajbbsp.2015.132.148.

68. El-Fadel, M. and R. Khoury. June 2010. Modeling Settlement in MSW Landfills: a Critical Review. J. Critical Reviews in Environmental Science and Technology. 30-2000(3):327-361. Published online: https://doi.org/10.1080/10643380091184200

69. Englezos, Peter. September 2019. Extraction of methane hydrate energy by carbon dioxide injection-key challenges and a paradigm shift Chinese Journal of Chemical Engineering. 27(9): 2044-2048. https://doi.org/10.1016/j.cjche.2019.02.031.

70. Espinosa, A., L. Rosas, K.Ilangovan and A.Noyola. 1995. Effect of trace metals on the anaerobic degradation of volatile fatty acids in molasses stillage. Water Science and Technology. 32(12): 121-129. https://doi.org/10.1016/0273-1223(96)00146.

71. Fabry, Frédéric, Christophe Rehmet, Vandad Rohani and Laurent Fulcheri. September 2013. Waste Gasification by Thermal Plasma: A Review. Waste and Biomass Valorization 4(3): 421–439.

72. Fedailaine, M., K. Moussi, M. Khitous, S. Abada, M. Saber and N. Tirichine. 2015. Modeling of the Anaerobic Digestion of Organic Waste for Biogas Production. Procedia Computer Science. 52:730-737. https://doi.org/10.1016/j.procs.2015.05.086.

73. Fedou, Stèphane, Eric Caprani, Damien Douziech and Sebastien Boucher Axens. 2008. Conversion of syngas to diesel, An overview of Fischer-Tropsch technologies for the production of diesel from syngas using a variety of feedstocks. [https://www.axens.net/document/19/conversion-of-syngas-to-diesel.../english.html].

74. Feng, Lei, Yuan Gao, Wei Kou, Xianming Lang, Yiwei Liu, Rundong Li et al. 2017. Application of the Initial Rate Method in Anaerobic Digestion of Kitchen Waste. Biomed. Res Int., 3808521. doi: 10.1155/2017/3808521.

75. Feng, Xin Mei, Anna Karlsson, Bo H. Svensson and Stefan Bertilsson. September 2010. Impact of trace element addition on biogas production from food industrial waste – linking process to microbial communities. Alfons Stams (ed.). (http://onlinelibrary.wiley.com/doi/10.1111/j.1574-6941.2010.00932.x/full).

76. Fischer–Tropsch process From Wikipedia, the free encyclopedia; https://en.wikipedia.org/wiki/Fischer%E2%80%93Tropsch_process.

77. Forghani, A.A., H. Elekaei and M.R. Rahimpour. May 2009. Enhancement of gasoline production in a novel hydrogen-permselective membrane reactor in Fischer–Tropsch synthesis of GTL technology. Int. J. of Hydrogen Energy. 34(9): 3965-3976. https://doi.org/10.1016/j.ijhydene.2009.02.038.

78. Fountoulakis, M.S., I.Petousi and T.Manios. October 2010. Co-digestion of sewage sludge with glycerol to boost biogas production. Waste Management. 30(10):1849-1853. https://doi.org/10.1016/j.wasman.2010.04.011.

79. Fuchs, W., H. Binder, G. Mavrias and R. Braun. February 2003. Anaerobic treatment of wastewater with high organic content using a stirred tank reactor coupled with a membrane filtration unit. Water Research, 37(4): 902-908. .https://doi.org/10.1016/S0043-1354(02)00246.

80. Fujioka, Tomoo, Kazuya Jyosui, Hiroyuki Nishimura and Kazuyoku Tei. September 2003. Extraction of Methane from Methane Hydrate Using Lasers. Japanese Journal of Applied Physics 42(9A):5648-5651. DOI: 10.1143/JJAP.42.5648.

81. Fujioka, Tomoo, Kazuya Jyosui, Hiroyuki Nishimura and Kazuyoku Tei. September 2003. Extraction of Methane from Methane Hydrate Using Lasers. Japanese Journal of Applied Physics, 42(9R): 5648.

82. Fulford, D. 1985. Fixed Concrete Dome Design. Biogas - Challenges and Experience from Nepal. Vol I. United Mission to Nepal, pp.3.1-3.10.

83. Gabriel, Kerron J., Mohamed Noureldin, Mahmoud MEl-Halwagi, Patrick Linke, Arturo Jiménez-Gutiérrez and Diana Yered Martínez. August 2014. Gas-to-liquid (GTL) technology: Targets for process design and water-energy nexus. Current Opinion in Chemical Engineering. 5: 49-54. https://doi.org/10.1016/j.coche.2014.05.001.

84. Gas Hydrate https://www.sciencedirect.com/topics/earth-and-planetary-sciences/gas-hydrate.

85. Gas to liquids From Wikipedia, the free encyclopedia; https://en.wikipedia.org/wiki/Gas_to_liquids

86. Gas to Liquids – GTL; http://www.natgas.info/gas-information/what-is-natural-gas/gtl.

87. Gasification-From Wikipedia, the free encyclopedia https://en.wikipedia.org/wiki/Gasification.

88. Gas-to-Liquids (GTL) allows the converson of Natural Gas into Liquid hydrocarbons and Oxygenates through Chemical Reactions; from: Advanced Natural Gas Engineering , 2009.

89. Geetha, G. S., K.S. Jagadeesh and T.K.R. Reddy. 1990. Biomass. 21: 157-161. 50. [Biogas production technology: An Indian perspective – jstor].

90. Geuss, Megan. May 2017. Japan, China have extracted methane hydrate from the seafloor- Gas hydrates difficult to extract but estimated abundance makes mining attractive. https://arstechnica.com/science/2017/05/energy-dense-methane-hydrate-extracted-by-japanese-chinese-researchers/. https://arstechnica.com/science/2017/05/energy-dense-methane-hydrate-extracted-by-japanese-chinese-researchers/].

91. Ghatak, Manjula Das and P. Mahanta. Jul-2014. Comparison of Kinetic Models For Biogas Production Rate From Saw Dust. Ijret: Int. J. of Research In Engineering and Technology. 03(07). Eissn: 2319-1163 | Pissn: 2321-7308@ http://www.Ijret.Org.

92. Ghoneim, Salwa A., Radwa A. El-Salamony and Seham A. El-Temtamy. 2016. Review on Innovative Catalytic Reforming of Natural Gas to Syngas. Vol.04(01) Article ID:63774,24 pages. World Journal of Engineering and Technology. 10.4236/wjet.2016.41011. http://creativecommons.org/licenses/by/4.0/.

93. Golding S. D., I. T. Uysal, C. Boreham, K.A. Baublys and J.S. Esterle. 2009. Implications of natural analogue studies for CO2 storage in coal measures with enhanced coal bed methane. In: Proc. of 2009 Asia Pacific Coalbed Methane Symposium and 2009 China Coalbed Methane Symposium, Xuzhou, Jaingsu, China, Sep 2009. 2:712-725.

94. Gooch, Curt, Jennifer Pronto, Brent Gloy, Norm Scott, Steve McGlynn, Christopher Bentley. June 2010. Feasibility Study of Anaerobic Digestion and Biogas Utilization Options for the Proposed Lewis County Community Digester. Cornell University Ithaca, New York.

95. GTZ: Biogas Digest Volume II: Biogas - Application and Product Development. https://energypedia.info/wiki/File:Biogas_gate_volume_2.pdf.

96. Gullapalli, S. and A. Falender. GTL innovation produces clean base oils from natural gas. Shell Global Solutions, Houston, Texas. http://gasprocessingnews.com/features/201410/gtl-innovation-produces-clean-base-oils-from-natural-gas.aspx.

97. Gunaseelan, V. N. 1988. Anaerobic Digestion of Gliricidia Leaves for Biogas and Organic Manure Biomass, Biomass. 17: 1–11.

98. Gunaseelan, V. Nallathambi. 1997. Anaerobic digestion of biomass for methane production: A review. Biomass and Bioenergy. 13(1–2):83-114. https://doi.org/10.1016/S0961-9534(97)00020-

99. Guneratnam, A.J., E. Ahern, J. A. FitzGerald, S.A. Jackson, A. Xia, A.D.W. Dobson. et al. 2017. Study of the performance of a thermophilic biological methanation system. Bioresour. Technol. 225: 308–315.

100. Hagos, Kiros, Jianpeng Zong, Dongxue Li, Chang Liu and Xiaohua Lu. September 2017. Anaerobic co-digestion process for biogas production: Progress, challenges and perspectives. Renewable and Sustainable Energy Reviews. 76:1485-1496. https://doi.org/10.1016/j.rser.2016.11.184.

101. Hale, Sarah E., Johannes Lehmann, David Rutherford, Andrew R. Zimmerman, Robert T. Bachmann, Victor Shitumbanuma et al. February 2012. Quantifying the Total and Bioavailable Polycyclic Aromatic Hydrocarbons and Dioxins in Biochars. Environ. Sci. Technol.46(5), 2830–2838. https://doi.org/10.1021/es203984k.

102. Hall, Kenneth R. October 2005. A new gas to liquids (GTL) or gas to ethylene (GTE) technology. Catalysis Today. 106(1–4): 243-246. https://doi.org/10.1016/j.cattod.2005.07.176.

103. He, S., D.-Q. Liang, D.-L. Li and L.-L. Ma. Jun 2013. The Formation of Natural Gas Hydrate From SDS-Solutions and Decomposition By Microwave Heating in a Static Reactor. Petroleum Science and Technology. 31(16). 1655-1664. https://doi.org/10.1080/10916466.2010.551238.

104. Heilig, Gerhard K. November 1994. The greenhouse gas methane (CH_4): Sources and sinks, the impact of population growth, possible interventions. Population and Environment . 16:109–137.

105. Hejazi, Bijan, John R. Grace, Xiaotao Bi and Andrés Mahecha-Botero. 2017. Kinetic Model of Steam Gasification of Biomass in a Bubbling Fluidized Bed Reactor. Energy Fuels. 31(2): 1702–1711. https://doi.org/10.1021/acs.energyfuels.6b03161.

106. Hernández, Juan J., Guadalupe Aranda-Almansa, Antonio Bula. June 2010. Gasification of biomass wastes in an entrained flow gasifier: Effect of the particle size and the residence time. Fuel Processing Technology. 91(6):681-692. https://doi.org/10.1016/j.fuproc.2010.01.018.

107. Heyer R. , D. Benndorf, F. Kohrs, J. De Vrieze, N. Boon, M. Hoffmann et al. 2016. Proteotyping of biogas plant microbiomes separates biogas plants according to process temperature and reactor type. Biotechnology for Biofuels. 9:155. doi: 10.1186/s13068-016-0572-4.

108. Hill, D.T. January 1983. Simplified monod kinetics of methane fermentation of animal wastes. Agricultural Wastes. 5(1): 1-16. https://doi.org/10.1016/0141-4607(83)90009-4.

109. Holm-Nielsen, J .B., T. Al Seadi and P. Oleskowicz-Popiel. November 2009. The future of anaerobic digestion and biogas utilization. Bioresource Technology. 100(22):5478-5484. https://doi.org/10.1016/j.biortech.2008.12.046.

110. http://www.reuters.com/article/japan-methane-hydrate-idUSL4N1IA35A   and   http://www.independent. co.uk/news/science/japan-china-combustible-ice-frozen-fossil-fuel-extract-seafloor-energy-methane-hydrate-a7744456.html0.

111. https://en.wikipedia.org/w/index.php?title=Coalbed_methane_extraction&oldid=709537640.

112. https://www.clarke-energy.com/coal-gas/

113.. https://www.sciencedirect.com/topics/engineering/gas-to-liquids.

114. https://www.worldcoal.org/coal/coal-seam-methane.

115. Huang, Jiu, Klaus Gerhard Schmidt and Zhengfu Bian. 2011. Removal and Conversion of Tar in Syngas from Woody Biomass Gasification for Power Utilization Using Catalytic Hydrocracking. Energies. 4:1163-1177. doi:10.3390/en4081163. [www.mdpi.com/1996-1073/4/8/1163/pdf].

116. Hydrates contain vast store of world gas resources https://www.ogj.com/home/article/17225397/hydrates-contain-vast-store-of-world-ga resources.

117. Iandoli, Carmine L. and Signe Kjelstrup. June 2007. Exergy Analysis of a GTL Process Based on Low-Temperature Slurry F−T Reactor Technology with a Cobalt Catalyst. Energy Fuels. 21(4):2317–2324. https://doi.org/10.1021/ef060646y.

118. In, Dr. Su-Il, Dimitri D. Vaughn II, Prof. Raymond E. Schaak. April 2012. Hybrid CuO-$TiO_{2-x}N_x$ Hollow Nanocubes for Photocatalytic Conversion of CO_2 into Methane under Solar Irradiation. Angewandte International edition Chemie. 51(16): 3915-3918. https://doi.org/10.1002/anie.201108936.

119. Irfan, Umair. August 2016. Engineered Bacterium Turns Carbon Dioxide into Methane Fuel -If scaled up, batches of bacteria could convert CO_2 emissions into fuel, in a single step. ClimateWire. Scientific American.

120. Izumi, Yasuo. January 2013. Recent advances in photocatalytic conversion of carbon dioxide into fuels with water and/or hydrogen using solar energy and beyond. Article in Coordination Chemistry Reviews. 257:171-186. DOI: 10.1016/j.ccr.2012.04.018.

121.   Jain, S. K., G.S. Gujral, N.K. Jha and P. Vasudevan. 1992. Biores. Technol. 41: 273–277.

122. Janicki, Georg, Stefan Schlüter, Torsten Hennig, Hildegard Lyko, and Görge Deerberg. 2011. Simulation of Methane Recovery from Gas Hydrates Combined with Storing Carbon Dioxide as Hydrates. Journal of Geological Research. |Article ID 462156 | In special issue on Gas Hydrate on Continental Margins, Michela Giustiniani (ed.). https://doi.org/10.1155/2011/462156.

123. Janse, A.M.C., R.W.J.Westerhout and W.Prins. May 2000. Modelling of flash pyrolysis of a single wood particle. Chemical Engineering and Processing: Process Intensification. 39(3): 239-252. https://doi. org/10.1016/S0255-2701(99)00092-6.

124. Jaouen,Nicolas, LucVervisch, Pascale Domingo. 2017. Auto-thermal reforming (ATR) of natural gas: An automated derivation of optimised reduced chemical schemes. Proceedings of the Combustion Institute. 36(3):3321-3330.

125. Jenicek, P., J. Bartacek, J. Kutil, J. Zabranska and M. Dohanyos. 2012. Potentials and limits of anaerobic digestion of sewage sludge: Energy self-sufficient municipal wastewater treatment plant? Water Sci Technol. 66(6):1277-1281. https://doi.org/10.2166/wst.2012.317.

126. JoãoNeiva de Figueiredo and Sérgio Fernando Mayerle. November 2014. A systemic approach for dimensioning and designing anaerobic bio-digestion/energy generation biomass supply networks. Renewable Energy. 71:690-694. https://doi.org/10.1016/j.renene.2014.06.031.

127. Johnson, J.M.-F., M.D. Coleman, R. Gesch, Abdulla Jaradat, Rob Mitchell, Don Reicosky et al. 2007. Biomass-bioenergy crops in the United States: a changing paradigm. Am. J. Plant Sci. Biotechnol 1(1): 1 –28.

128. Kamalan, Hamidreza. July 2016. A New Empirical Model to Estimate Landfill Gas Pollution. J. of Health Science Surveillance Sys. 4(3). https://www.researchgate.net/publication/305956550_A New Empirical Model to Estimate Landfill_Gas_Pollution.

129. Karadagli, Fatih and Bruce E. Rittmann. 2007. A mathematical model for the kinetics of Methanobacterium bryantii M.o.H. considering hydrogen thresholds. Biodegradation. 18:453–464.

130. Karapidakis E.S., A.A. Tsave, P.M. Soupios and Y.A. Katsigiannis. 2010. Energy efficiency and environmental impact of biogas utilization in landfills. Int. J. Environ. Sci. Tech., 7 (3): 5 99-6 08.

131. Khandekar, Yogesh. S. and Dr. Narendra. P. Shinkar. Jan 2019. Methods of Distillary Spent Wash: A Review. Int. J. of Engineering Science Invention (IJESI) ISSN (Online): 8(1): 2319 – 6734.

132. Khlystov, Oleg, Marc DeBatist, Hitoshi Shoji, Akihiro Hachikubo, Shinya Nishio, Lieven Naudts. et al. January 2013. Gas hydrate of Lake Baikal: Discovery and varieties. J. of Asian Earth Sciences. 62(30):162-166. https://doi.org/10.1016/j.jseaes.2012.03.009.

133. Kirkels, Arjan F. and Geert P.J.Verbong. January 2011. Biomass gasification: Still promising? A 30-year global overview. Renewable and Sustainable Energy Reviews. 15(1):471-481. https://doi.org/10.1016/j.rser.2010.09.046.

134. Kjeldsen, Peter and Erling V.Fischer. 1995. Landfill gas migration—Field investigations at Skellingsted landfill, Denmark. Waste Management & Research. 13(5):467-484. https://doi.org/10.1016/S0734-242X(05)80025-4.

135. Kleerebezem, Robbert , Bart Joosse, Rene Rozendal and Mark C. M. Van Loosdrecht. December 2015. Anaerobic digestion without biogas? Reviews in Environmental Science and Bio/Technology. 14(4):787–801.

136. Koh, Dong-Yeun, Hyery Kang , Dae-Ok Kim, Minjun Cha, Prof. Huen Lee. Aug. 2012. Recovery of Methane from Gas Hydrates Intercalated within Natural Sediments Using CO_2 and a CO_2/N_2 Gas Mixture. Chemistry Europe, European Chemical Societies Publishing. 5(8): 1443-1448. https://onlinelibrary.wiley.com/doi/abs/10.1002/cssc.201100644.

137. Koufopanos, C.A. , N. Papayannakos, G. Maschio, and A. Lucchesi. 1991. Modelling of the pyrolysis of biomass particles. Studies on kinetics, thermal and heat transfer effects. Can. J. Chem. Eng., 69:907.

138. Krenzke, Peter T., Jesse R.Fosheim, Jane H.Davidson. November 2017. Solar fuels via chemical-looping reforming. Solar Energy. 156: 48-72. https://doi.org/10.1016/j.solener.2017.05.095.

139. Krüger M, S. Beckmann, B. Engelen, T. Thielemann, B. Cramer, A. Schippers, et al. 2008. Microbial methane formation from hard coal and timber in an abandoned coal mine. Geomicrobiology Journal. 25:315-321.

140. Kruse, Andrea. January 2009. Hydrothermal biomass gasification. The Journal of Supercritical Fluids, 47(3):391-399. https://doi.org/10.1016/j.supflu.2008.10.009.

141. Kumar, Ajay, David D. Jones and Milford A. Hanna. July 2009. Thermochemical Biomass Gasification: A Review of the Current Status of the Technology. Energies 2(3):556-581. nhttps://doi.org/10.3390/en20300556.

142. Kumar, S., A.N. Mondal, S.A. Gaikward, S. Devotta and R.N. Singh. 2004. Qualitative assessment of methane emission inventory from municipal solid waste disposal sites: a case study. Atmos. Environ. 38: 4921–4929.

143. Kumar, Anil, Nitin Kumar, Prashant Baredar and Ashish Shukla. May 2015. A review on biomass energy resources, potential, conversion and policy in India. Renewable and Sustainable Energy Reviews 45: 530–539. https://doi.org/10.1016/j.rser.2015.02.007.

144. Kushare, B.E., A.M. Jain and V. Kushare. Oct. 2004. Power generation through municipal waste by bio methanization – case study of Nasik Municipal Corp. Int. Conf. on Renewable Energy. CBIP, New Delhi.

145. Kvenvolden, Keith A. November–December 1995. A review of the geochemistry of methane in natural gas hydrate. Review paper, Organic Geochemistry. 23(11–12):997-1008. https://doi.org/10.1016/0146-6380(96)00002-2.

146. Kvenvolden, K.A. and T.D. Lorenson. 2001. The global occurrence of natural gas hydrates. In: C.K.Paull and W.P. Dillon (eds.). Natural Gas Hydrates: Occurrence, Distribution and Detection. Monograph 124, American Geophysical Union, Washington, DC.

147. Kvenvolden, Keith A. and Thomas D. Lorenson. 2001. Global Occurrences of Gas Hydrate. document id ISOPE-I-01-069, Int. Society of Offshore and Polar Engineers. The Eleventh Int. Offshore and Polar Engineering Conf. 17-22 June, Stavanger, Norway.

148. Labatut, Rodrigo A. January 2012. Anaerobic Biodegradability of Complex Substrates: Performance and Stability at Mesophilic and Thermophilic Conditions. A Ph.D. Dissertation of Cornell University.

149. Landfill Gas Energy Technologies, https://www.google.co.in/search?q=landfill+gas&oq=LANDFILL+GAS&aqs=chrome.0.0l6.5759j0j4&sourceid=chrome&ie=UTF-8.

150. Landfill gas From Wikipedia, the free encyclopedia; https://en.wikipedia.org/wiki/Landfill_gas.

151. Landfill Gas Primer - An Overview for Environmental Health Professionals https://www.atsdr.cdc.gov/HAC/landfill/html/ch2.html.

152. Larentis, Ariane Leites, Neuman Solangede Resende, Vera Maria MartinsSalim, José CarlosPinto. July 2001, Modeling and optimization of the combined carbon dioxide reforming and partial oxidation of natural gas. Applied Catalysis A: General. 215(1–2): 211-224. https://doi.org/10.1016/S0926-860X(01)00533-6.

153. Lastella, G., C. Testa, G. Cornacchia, M. Notornicola, F. Voltasio and Vinod Kumar Sharma. January 2002. Anaerobic digestion of semi-solid organic waste: biogas production and its purification. Energy Conversion and Management. 43(1):63-75. https://doi.org/10.1016/S0196-8904(01)00011-5.

154. Lawrence, Alonzo W. and Perry L. McCarty. Feb., 1969. Kinetics of Methane Fermentation in Anaerobic Treatment. Journal of Water Pollution Control Federation. 41(2). Research Supplement to: 41(2): Part II pp. R1-R17, Wiley. https://www.jstor.org/stable/25036255.

155. Lay, J.J., Y.Y.Li, T.Noike, J.Endo and S.Ishimoto. 1997. Analysis of environmental factors affecting methane production from high-solids organic waste. Water Science and Technology. 36(6–7): 493-500; https://doi.org/10.1016/S0273-1223(97)00560-X.

156. Lee, Huen Prof., Yongwon Seo Dr., Yu-Taek Seo Dr., Igor L. Moudrakovski Dr. and John A. Ripmeester Dr. October 2003. Recovering Methane from Solid Methane Hydrate with Carbon Dioxide. Angewandte International Edition Chemie, 42(41): 5048-5051. https://doi.org/10.1002/anie.200351489.

157. Lee, Huen, Do-youn Kim and Young-June Park (Inventor). Method for recovering methane gas from natural gas hydrate. https://patents.google.com/patent/US7988750B2/en.

158. Lee, Sang-Yong and Gerald D.Holder. June 2001. Methane hydrates potential as a future energy source. Fuel Processing Technology, 71(1–3): 181-186. https://doi.org/10.1016/S0378-3820(01)00145-X.

159. Lee, See Hoon, Sang Jun Yoon, Ho Won Ra, Young Il Son, Jai Chang Hong and Jae Goo Lee. August 2010. Gasification characteristics of coke and mixture with coal in an entrained-flow gasifier. Energy. 35(8): Pages 3239-3244. https://doi.org/10.1016/j.energy.2010.04.007.

160. Lee, Sze Ying Lee, Revathy Sankaran, Kit Wayne Chew, Chung Hong Tan, Rambabu Krishnamoorthy, Dinh-Toi Chu et al. 2019. Waste to bioenergy: a review on the recent conversion technologies. BMC Energy 1: Article number 4.

161. Lemmens, Bert, Helmut Elslander, Ive Vanderreydt, Kurt Peys, Ludo Diels, Michel Oosterlinck et al. 2007. Assessment of plasma gasification of high caloric waste streams. Waste Management, 27(11):1562-1569. https://doi.org/10.1016/j.wasman.2006.07.027.

162. Lequerica J. L., S. Vallés and A. Flors. January 1984. Kinetics of rice straw methane fermentation. Applied Microbiology and Biotechnology. 19(1):70–74. https://link.springer.com/article/10.1007/BF00252820.

163. Lessons For Safe Design and Operation of Anaerobic Digesters. 2012. SYMPOSIUM SERIES NO. 158 Hazards XXIII IChemE.https://www.icheme.org/media/9063/xxiii-paper-67.pdf

164. Li, Baitao, Xiujuan Xu, Shuyi Zhang. January 2013. Synthesis gas production in the combined CO_2 reforming with partial oxidation of methane over Ce-promoted Ni/SiO_2 catalysts. International Journal of Hydrogen Energy. 38(2): 890-900. https://doi.org/10.1016/j.ijhydene.2012.10.103.

165. Liang, Deqing, Song He, DongLiang Li. March 2009. Effect of microwave on formation/decomposition of natural gas hydrate. Chinese Science Bulletin 54(6):965-971. DOI: 10.1007/s11434-009-0116-4.

166. Linville, J. L., Y. Shen, P.A. Ignacio-de Leon, R.P. Schoene and M. Urgun-Demirtas. 2017. In-Situ Biogas Upgrading during Anaerobic Digestion of Food Waste Amended with Walnut Shell Biochar at Bench Scale. Waste Manage. Res. 35: 669– 679. DOI: 10.1177/0734242X17704716.

167. Luostarinen, S., S. Luste and M. Sillanpää. January 2009. Increased biogas production at wastewater treatment plants through co-digestion of sewage sludge with grease trap sludge from a meat processing plant. Bioresource Technology, 100(1): 79-85. https://doi.org/10.1016/j.biortech.2008.06.029.

168. Mahadevaswamy, M. and L.V. Venkataraman. 1990. Biol. Wastes. 32(4): 243–251.

169. Mahar, Rasool Bux , Abdul Razaque Sahito, Dongbei Yue and Kamranullah Khan. February 2016. Modeling and simulation of landfill gas production from pretreated MSW landfill simulator. Frontiers of Environmental Science & Engineering, 10(1):159–167.

170. Marín, D.; Posadas, E.; Cano, P.; Pérez, V.; Blanco, S.; Lebrero, R.; Muñoz, R. Seasonal Variation of Biogas Upgrading Coupled with Digestate Treatment in an Outdoors Pilot Scale Algal-Bacterial Photobioreactor. Bioresour. Technol. 2018, 263, 58– 66, DOI: 10.1016/j.biortech.2018.04.117.

171. Marsch, H D and H.J. Herbort. 1982. Produce synthesis gas by steam reforming natural gas. Hydrocarbon Process. 61:6 United States https://www.osti.gov/biblio/7016992-produce-synthesis-gas-steam-reforming-natural-gas.

172. Martin, Antonio, Rafael Borja and Charles J. Banks. May 1994. Kinetic model for substrate utilization and methane production during the anaerobic digestion of olive mill wastewater and condensation water waste. J. of Chemical Technology and Biotechnology. 60(1):fmi,1-116. https://doi.org/10.1002/jctb.280600103.

173. Martínez, Juan Daniel, Neus Puy, Ramón Murillo, Tomás García, María Victoria Navarro and Ana Maria Mastral. July 2013. Waste tyre pyrolysis – A review. Renewable and Sustainable Energy Reviews. 23:179-213. https://doi.org/10.1016/j.rser.2013.02.038.

174. Maschio, G. , A. Lucchesi, and C.A. Koufopanos. 1994. A study of kinetic and transfer phenomena in the pyrolysis of biomass particles. In A.V. Bridgwater (Ed.). Advances in Thermochemical Biomass Conversion. p. 746. Blackie Academic, New York.

175. Mata-Alvarez, J., S. Macé and P Llabrés. August 2000. Anaerobic digestion of organic solid wastes. An overview of research achievements and perspectives. Bioresource Technology. 74(1):3-16. https://doi.org/10.1016/S0960-8524(00)00023-7.

176. McCarty, P. L. and F. E. Mosey. 1991. Modelling of Anaerobic Digestion Processes (A Discussion of Concepts). Water Sci Technol 24 (8): 17-33. https://doi.org/10.2166/wst.1991.0216.

177. Meabe, E., S. Déléris, S. Soroa and L.Sancho. November 2013. Performance of anaerobic membrane bioreactor for sewage sludge treatment: Mesophilic and thermophilic processes. J. of Membrane Science, 446:26-33. https://doi.org/10.1016/j.memsci.2013.06.018.

178. Melaaen, M.C. , and M.G. Grønli. 1996. Modelling and simulation of moist wood drying and pyrolysis. In A.V. Bridgwater, D.G.B. Boocock (eds.). Developments in Thermochemical Biomass Conversion. p. 132. Blackie Academic, London.

179. Melnichuk, L. J., K.V. Kelly and R.S. Davis. System and method for converting biomass to ethanol via syngas. - US Patent 8,088,832, 2012 – Google Patents.

180. Methane Hydrate: The World's Largest Natural Gas Resource geology.com. Oil and Gas.

181. Method and system for producing methane gas from methane hydrate formations. Inventor: J. Kenneth Wittle and Christy W. Bell. https://patents.google.com/patent/US7322409B2/en.

182. Microwave pyrolysis of wastes. https://patents.google.com/patent/US3843457A/en3.

183. Milkov, A.V. June 2000. Worldwide distribution of submarine mud volcanoes and associated gas hydrates. Marine Geology. 167(1–2):29-42. https://doi.org/10.1016/S0025-3227(00)00022-0.

184. Molino, Antonio, Vincenzo Larocca, Simeone Chianese and Dino Musmarra. 2018. Review Biofuels Production by Biomass Gasification: A Review, Energies 11(4):811. doi:10.3390/en11040811 www.mdpi.com/journal/energies.

185. Mudhoo, A. and S. Kumar. November 2013. Effects of heavy metals as stress factors on anaerobic digestion processes and biogas production from biomass. Int. J. of Environmental Science and Technology. 10(6):1383–1398.

186. Mueller, Julia Ruth. 2012. Microbial Catalysis of Methane from Carbon Dioxide - The Future of Renewable Energy is Inside You. BS Thesis, The Ohio State University.

187. Nagamani, B. and K. Ramasamy. July 1999. Biogas production technology: An Indian perspective. Current Science. 77(1): 44-55. Current Science Association. https://www.jstor.org/stable/24102913.

188. Nagmani, B. and K. Ramaswamy. July 1999. Biogas Production Technology: An Indian perspective. Current Science 77(1):44-55. Current ScienceAssociation. https://www.jstor.org/stable/24102913.

189. Nastev, Miroslav, René Therrien, René Lefebvre and Pierre Gélinas. November 2001. Gas production and migration in landfills and geological materials. J. of Contaminant Hydrology, 52(1–4):187-211. https://doi.org/10.1016/S0169-7722(01)00158-9.

190. Nelabhotla, Anirudh Bhanu Teja and Carlos Dinamarca. 2019. Bioelectrochemical CO2 Reduction to Methane: MES Integration in Biogas Production Processes. Open Access Journal Appl. Sci. 9: 1056. doi:10.3390/app9061056.

191. Nematollahi, Behzad, Mehran Rezaei and Majid Khajenoori. February 2011. Combined dry reforming and partial oxidation of methane to synthesis gas on noble metal catalysts. Int. J. of Hydrogen Energy. 36(4): 2969-2978. https://doi.org/10.1016/j.ijhydene.2010.12.007.

192. Nematollahi, Behzad, MehranRezaei and Majid Khajenoori. February 2011. Combined dry reforming and partial oxidation of methane to synthesis gas on noble metal catalysts. Int. J. of Hydrogen Energy. 36(4): 2969-2978. https://doi.org/10.1016/j.ijhydene.2010.12.007.

193. Neves L., R. Oliveira and M.M.Alves. October 2004. Influence of inoculum activity on the bio-methanization of a kitchen waste under different waste/inoculum ratios. Process Biochemistry, Volume 39(12): Pages 2019-2024, https://doi.org/10.1016/j.procbio.2003.10.002.

194. Nielfa, A., R.Cano and M.Fdz-Polanco. March 2015. Theoretical methane production generated by the co-digestion of organic fraction municipal solid waste and biological sludge. Biotechnology Reports. 5: 14-21. https://doi.org/10.1016/j.btre.2014.10.005.

195. Nikoo, M. Khoshtinat and N.A.S.Amin. March 2011. Thermodynamic analysis of carbon dioxide reforming of methane in view of solid carbon formation. Fuel Processing Technology. 92(3): 678-691. https://doi.org/10.1016/j.fuproc.2010.11.027.

196. Nwoye, C.I., U. Ofoegbu and C.I. Okoli. August 2011. Biotreatment of Organic Waste Materials: An Effective Recycling Technology for Biogas Production. Advances in Science and Technology 5(1S):45 – 53.

197. Ota, Masaki; Yuki Abe, Masaru Watanabe, Richard L.Smith Jr. and Hiroshi Inomata. February 2005. Methane recovery from methane hydrate using pressurized CO_2. Fluid Phase Equilibria, 228–229: 553-559. https://doi.org/10.1016/j.fluid.2004.10.002.

198. Özkara-Aydınoğlu, Şeyma. December 2010. Thermodynamic equilibrium analysis of combined carbon dioxide reforming with steam reforming of methane to synthesis gas. International Journal of Hydrogen Energy. 35(23):12821-12828. https://doi.org/10.1016/j.ijhydene.2010.08.134.

199. Paolini, Valerio, Francesco Petracchini, Marco Segreto, Laura Tomassetti, Nour Naja & Angelo Cecinato. Apr 2018. Environmental impact of biogas: A short review of current knowledge. Journal of Environmental Science and Health, Part A- Toxic/Hazardous Substances and Environmental Engineering. 53(10): 899-906. https://doi.org/10.1080/10934529.2018.1459076

200. Parawira, W., I.Kudita, M.G.Nyandoroh and R.Zvauya. February 2005. A study of industrial anaerobic treatment of opaque beer brewery wastewater in a tropical climate using a full-scale UASB reactor seeded with activated sludge. Process Biochemistry, 40(2):593-599. https://doi.org/10.1016/j.procbio.2004.01.036.

201. Passeggi, Mauricio, Iván López and Liliana Borzacconi. May 2012. Modified UASB reactor for dairy industry wastewater: performance indicators and comparison with the traditional approach. J. of Cleaner Production. 26:90-94. https://doi.org/10.1016/j.jclepro.2011.12.022.

202. Patel, Madhumita, Xiaolei Zhang and Amit Kumar. January 2016. Techno-economic and life cycle assessment on lignocellulosic biomass thermochemical conversion technologies: A review. Renewable and Sustainable Energy Reviews. 53: 1486-1499. https://doi.org/10.1016/j.rser.2015.09.070.

203. Patel, Vikram B., Anami R. Patel, Manisha C. Patel and Datta B. Madamwar. October 1993. Effect of metals on anaerobic digestion of water hyacinth-cattle dung. Applied Biochemistry and Biotechnology. 43(1):45–50.

204. Pavlostathis, S. G. and E. Giraldo-Gomez. 1991. Kinetics of Anaerobic Treatment. Water Sci Technol. 24(8):35-59. https://doi.org/10.2166/wst.1991.0217.

205. Peacocke, G.V.C. , and A.V. Bridgewater. 1994. Design of a novel ablative pyrolysis reactor. In A.V. Bridgewater (Ed.). Advances in Thermochemical Biomass Conversion p. 1134.

206. Perera, L. A. K., G. Achari, and J. P. A. Hettiaratchi. 2002. Determination of Source Strength of Landfill Gas: A Numerical Modeling Approach. J.Env.Engg. 128 (5).

207. Perera, M D.N, J P.A Hettiaratchi and G Achari. 2002. A mathematical modeling approach to improve the point estimation of landfill gas surface emissions using the flux chamber technique. J.of Environmental Engineering and Science. 1(6): 451 -463. https://doi.org/10.1139/s02-033.

208. Perna, Alessandra, Mariagiovanna Minutillo and Elio Jannelli. January 2016. Hydrogen from intermittent renewable energy sources as gasification medium in integrated waste gasification combined cycle power plants: A performance comparison. Energy. 94(1):457-46. https://doi.org/10.1016/j.energy.2015.10.143.

209. Pfeffer, John T. June 1974. Temperature effects on anaerobic fermentation of domestic refuse. Biotechnology and Bioengineering, 16(6): 771-787. https://doi.org/10.1002/bit.260160607.

210. Pichtel, J. 2007. Waste Management Practices. CRC Press, Boca Raton, FL, USA.

211. Platen, H. and B. Schink. December 1987. Methanogenic degradation of acetone by an enrichment culture. Archives of Microbiology. 149(2):136–141. https://link.springer.com/article/10.1007/BF00425079.

212. Pommier S., D. Chenu, M. Quintard and X. Lefebvre. June 2007. A logistic model for the prediction of the influence of water on the solid waste methanization in landfills. Biotechnology and Bioengineering. 97(3):473-482. https://doi.org/10.1002/bit.21241.

213. Pozzo, Matteo and Andrea Lanzini Massimo Santarelli. April 2015. Enhanced biomass-to-liquid (BTL) conversion process through high temperature co-electrolysis in a solid oxide electrolysis cell (SOEC). Fuel. 145(1):39-49. https://doi.org/10.1016/j.fuel.2014.12.066.

214. Prabhansu, Malay Kr. Karmakar, Prakash Chandra and Pradip Kr.Chatterjee. June 2015. A review on the fuel gas cleaning technologies in gasification process. J. of Environmental Chemical Engineering. 3(2):689-702. https://doi.org/10.1016/j.jece.2015.02.011.

215. Producer gas from Wikipedia, te free encyclopedia; https://en.wikipedia.org/wiki/Producer_gas.

216. Rahim, Ismail, Shinfuku Nomura, Shinobu Mukasa, Hiromichi Toyota. August 2014. A Comparison of Methane Hydrate Decomposition Using Radio Frequency Plasma and Microwave Plasma Methods. 15th International Heat Transfer Conference, IHTC-15 at Kyoto, Japan. DOI: 10.1615/IHTC15.pls.009897.

217. Rahimpour, M.R. and H.Elekaei. June 2009. A comparative study of combination of Fischer–Tropsch synthesis reactors with hydrogen-permselective membrane in GTL technology. Fuel Processing Technology, 90(6):747-761, https://doi.org/10.1016/j.fuproc.2009.02.011.

218. Randhava, Sarabjit S., Richard L. Kao, Todd Harvey, Ajaib S. Randhava and Surjit S. Randhava (Inventor). February 2010. Method for converting biomass into synthesis gas using a pressurized multi-stage progressively expanding fluidized bed gasifier followed by an oxyblown autothermal reformer to reduce methane and tars. Patents by Inventor Sarabjit S. Randhava. https://patents.google.com/patent/US20100040510A1/en.

219. Rao, Sridhar Gururaja. 2003. Producer Gas based Spark-Ignited Reciprocating Engines. A Thesis Submitted for the Degree of Doctor of Philosophy In the Faculty of Engineering, IISc. Bangalore. [Also Rao, Sridhar Gururaja, H.V. Sridhar, S. Dasappa and P.J. Paul. March 2005. Development of producer gas engines. Proceedings of the Institution of Mechanical Engineers Part D Journal of Automobile Engineering 219(3):423-438. DOI: 10.1243/095440705X6596].

220. Rasi, Saija, Jenni Lehtinen and Jukka Rintala. December 2010. Determination of organic silicon compounds in biogas from wastewater treatments plants, landfills and co-digestion plants. Renewable Energy. 35(12):2666-2673. https://doi.org/10.1016/j.renene.2010.04.012.

221. Rathore, Dheeraj. 2012. Utilization of distillery spent wash as a source of biogas and fertilizer for sugarcane (Saccharum officinarum). Biodiversity & Sustainable Energy Development. DOI: 10.4172/2157-7625.S1.002.

222. Ravi Kumar, M. and Ermias Girma Aklilu. December 2016. Production of Biogas Fuel from Alcohol Distillery Plant. Int. J. of Scientific and Research Publications. 6(12):202 ISSN 2250-3153 www.ijsrp.org.

223. Review of technology for the gasification of biomass and wastes. June 2009. Final report of NNFCC project 09/008, E4tech. [A project funded by DECC, project managed by NNFCC and conducted by E4Tech]. website; gasificationnnfc090609.

224. Review of the Methane Hydrate Program in Japan MDPI- http://www.google.co.in./search?q=review+a+review+of+the+methane+hydrate+program+in+japanai+oyama+*+id+and+stephen+m.+masutani+hawaii+natural+energy+institute,+university+of+hawaii,+honolulu,+hi+96822,+usa%3B+stepenm%40hawaii.edu+*correspondence%3A+aioyama%40hawaii.edu%3B+tel.3A%2B1-808-956-5711&ie=&oe=

225. Ringkamp, M.–FH Hildesheim Holzminden. 1989. Types of Biogas Digesters and Plants. Regional Biogas Extension Programme GCR - Final Report on Statistical and Structural Examination of Caribbean Biogas Plants. 60 P. https://energypedia.info/wiki/Types_of_Biogas_Digesters_and_Plants.

226. Risco, MarioLuna-del, Kaja Orupõld and Henri-Charles Dubourguier. May 2011. Particle-size effect of CuO and ZnO on biogas and methane production during anaerobic digestion. Journal of Hazardous Materials, 189(1–2):603-608. https://doi.org/10.1016/j.jhazmat.2011.02.085.

227. Ruppel, Carolyn. 2007. Tapping methane hydrates for unconventional natural gas Elements. Publisher Mineralogical Society of America. https://doi.org/10.2113/gselements.3.3.193.

228. Rydén, Magnus , Anders Lyngfelt, Tobias Mattisson. September 2006. Synthesis gas generation by chemical-looping reforming in a continuously operating laboratory reactor. Fuel. 85(12–13):1631-1641. https://doi.org/10.1016/j.fuel.2006.02.004.

229. S.D.Sharma, M.Dolan, D.Park, L.Morpeth, A.Ilyushechkin, K.McLennan, D.J.Harris and K.V.Thambimuthu. January 2008. A critical review of syngas cleaning technologies — fundamental limitations and practical problems. Powder Technology. 180(1–2):115-121. https://doi.org/10.1016/j.powtec.2007.03.023.

230. Safety Practices for On-Farm Anaerobic Digestion Systems. Dec. 2011. https://www.epa.gov/sites/default/files/2014-12/documents/safety_practices.pdf

231. Sahota, Shivali, Goldy Shah, Pooja Ghosh, Rimika Kapoor, Subhanjan Sengupta, Priyanka Singh et al. March 2018. Review of trends in biogas upgradation technologies and future perspectives. Bioresource Technology Reports. 1: 79-88. https://doi.org/10.1016/j.biteb.2018.01.002.

232. Salimi, Mohammad, Farid Safari, Ahmad Tavasoli and Alireza Shakeri. September 2016. Hydrothermal gasification of different agricultural wastes in supercritical water media for hydrogen production: a comparative study. Int. J. of Industrial Chemistry. 7(3): 277–285.

233. Sasse, L. - GATE, Bremer. 1988. Arbeitsgemeinschaft für Überseeforschung Entwicklung (BORDA): Biogas Plants - Design and Details of Simple Biogas Plants. 2nd edition, 85 P., ISBN: 3-528-02004-0.

234. Scheutz, Charlotte, Peter Kjeldsen, Jean E. Bogner, Alex De Visscher, Julia Gebert and Helene A. Hilger. July 2009. Microbial methane oxidation processes and technologies for mitigation of landfill gas emissions. SAGE Journals, Waste Management Research. 27(5) https://doi.org/10.1177/0734242X09339325.

235. Schirmer, W. N., J. F. T. Jucá, A. R. P. Schuler, S. Holanda and L. L. Jesus. Apr./June 2014. Methane production in anaerobic digestion of organic waste from Recife (Brazil) landfill: evaluation in refuse of diferent ages. Braz. J. Chem. Eng. 31(2). São Paulo. http://dx.doi.org/10.1590/0104-6632.20140312s00002468.

236. Scientists Discover Method to Turn into Methane - A new catalyst can turn carbon dioxide or carbon monoxide into methane. https://www.popularmechanics.com/science/green-tech/news/a27412/catalyst-turn-co2-into-methane/

237. Sekoai, Patrick T., Nicolaas Engelbrecht, Stephanus P. du Preez, and Dmitri Bessarabov. July 2020. Thermophilic Biogas Upgrading via ex Situ Addition of H_2 and CO_2 Using Codigested Feedstocks of Cow Manure and the Organic Fraction of Solid Municipal Waste. ACS Omega 2020, 5(28):17367–17376. https://doi.org/10.1021/acsomega.0c01725

238. Sharma, Vikrant and Vijay K. Agarwal. Sep 2019. Numerical Simulation of Coal Gasification - A Circulating Fluidized Bed Gasifier. Process Systems Engineering. Braz. J. Chem. Eng. 36 (3). https://doi.org/10.1590/0104-6632.20190363s20180423.

239. Sharuddin, Shafferina Dayana Anuar, Faisal Abnisa, Wan Mohd Ashri Wan Daud and Mohamed Kheireddine Aroua May 2016. A review on pyrolysis of plastic wastes. Energy Conversion and Management, 115: 308-326. https://doi.org/10.1016/j.enconman.2016.02.037.

240. Shen, Y.; J.L. Linville, M. Urgun-Demirtas, R.P. Schoene and S. W. Snyder. 2015. Producing Pipeline-Quality Biomethane via Anaerobic Digestion of Sludge Amended with Corn Stover Biochar with in-Situ CO_2 Removal. Appl. Energy. 158: 300– 309. DOI: 10.1016/j.apenergy..08.016.

241. Shen, Yanwen, Jessica L. Linville, MeltemUrgun-Demirtas, Marianne M.Mintz and Seth W.Snyder. October 2015. An overview of biogas production and utilization at full-scale wastewater treatment plants (WWTPs) in the United States: Challenges and opportunities towards energy-neutral WWTPs. Renewable and Sustainable Energy Reviews. 50:346-362. https://doi.org/10.1016/j.rser.2015.04.129.

242. Shin, Hang-Sik , Sae-EunOh and Chae-Young Lee. 1997. Influence of sulfur compounds and heavy metals on the methanization of tannery wastewater. Water Science and Technology, 35(8): 239-245. https://doi.org/10.1016/S0273-1223(97)00173-X.

243. Sil, Avick, Sunil Kumar and Rakesh Kumar. January 2014. Formulating LandGem model for estimation of landfill gas under Indian scenario. Int. J. of Environmental Technology and Management. 17(2/3/4):293 – 299. DOI: 10.1504/IJETM.2014.061800.

244. Simeonov, I. 2010. Modelling and control of the anaerobic digestion of organic wastes in continuously stirred bioreactors. Chapter 2. In S. Tzonkov (ed.). Contemporary approaches to modeling, optimization and control of biotechnological processes, 41-76. Prof. Marin Drinov Acad. Publ. House, Sofia.

245. Simeonov, Ivan and Dimitar Karakashev. Mathematical Modelling of the Anaerobic Digestion Including the Syntrophic Acetate Oxidation. Department of Environmental Engineering, DTU Environment, Technical University of Denmark, 2800 Lyngby, Denmark.

246. Singh, S. K., Singh, A. and Pandey, G. N. 1993. Renew. Energy. 3: 45–47.

247. Song, Hyohak, Seh Hee Jang, Jong Myoung Park and Sang Yup Lee. May 2008. Modeling of batch fermentation kinetics for succinic acid production by Mannheimia succiniciproducens. Biochemical Engineering Journal, 40(1):107-115. https://doi.org/10.1016/j.bej.2007.11.021

248. Sorathia, Harilal S., Pravin P. Rathod and Arvind S. Sorathiya. July-Sept, 2012. Biogas Generation and Factors Affecting the Biogas Generation – A Review Study. Int. J. of Advanced Engineering Technology E-ISSN 0976-3945 IJAET. 3(3):72-78.

249. Steam Methane Reforming - Hydrogen Production-Producing hydrogen through SMR. https://www.engineering-airliquide.com/steam- methane-reforming-hydrogen-production

250. Steam reforming; From Wikipedia, the free encyclopedia; https://en.wikipedia.org/wiki/Steam_reforming

251. Steam reforming; From Wikipedia, the free encyclopedia; https://en.wikipedia.org/wiki/Steam_reforming

252. Stevens, Scott H., Dr. James G. Ferry and Dr. Martin Schoell. May 2012. Methanogenic Conversion of CO2 Into CH4 - A Potential Remediation Technology for Geologic CO2 Storage Sites. US Department of Energy. Advanced Resources International, Inc. https://www.osti.gov/servlets/purl/1041046.

253. Su, X.; Zhao, W.; Xia, D. The Diversity of Hydrogen-Producing Bacteria and Methanogens within an in Situ Coal Seam. Biotechnol. Biofuels 2018, 11, 1– 18, DOI: 10.1186/s13068-018-1237-2.

254. Su-Il, Professor. August 2018. Converting carbon dioxide into methane or ethane selectively. Science News from research organizations. https://www.sciencedaily.com/releases/2018/08/180813100246.htm.

255. Sun, Qie, Hailong Li, JinyingYan, Longcheng Liu, Zhixin Yu and Xinhai Yu. November 2015. Selection of appropriate biogas upgrading technology-a review of biogas cleaning, upgrading and utilisation. Renewable and Sustainable Energy Reviews. 51: 521-532. https://doi.org/10.1016/j.rser.2015.06.029.

256. Sun, Yonghong, Xiaoshu Lü and Wei Guo. March 2014. A review on simulation models for exploration and exploitation of natural gas hydrate. Arabian Journal of Geosciences. 7: 2199–2214.

257. Syngas; From Wikipedia, the free encyclopedia; https://en.wikipedia.org/wiki/Syngas.

258. Taherzadeh, M.J. and K. Karimi. 2008. Pretreatment of lignocellulosic wastes to improve ethanol and biogas production: a review. Int. J. Mol. Sci. 9:1621–51.

259. Tampio, E. A., Blasco, L., Vainio, M. M., Kahala, M. M., Rasi, S. E. Volatile Fatty Acids (VFAs) and Methane from Food Waste and Cow Slurry: Comparison of Biogas and VFA Fermentation Processes. GCB Bioenergy 2019, 11, 72– 84, DOI: 10.1111/gcbb.12556.

260. Tang, Mingchen, Long Xu, Maohong Fan. August 2015. Progress in oxygen carrier development of methane-based chemical-looping reforming: A review. Applied Energy. 151:143-156. https://doi.org/10.1016/j.apenergy.2015.04.017.

261. Tchobanoglous G, H. These and S. Vigil. 1993. ʻIntegrated solid waste management: Engineering principles and management issues, McGraw-Hill, Inc., New York. NY, USA.

262. Thurner, F. , and U. Mann. 1981. Kinetic investigation of wood pyrolysis. Ind. Eng. Chem. Process Des. Dev. 20: 482.

263. Tigabu, Aschalew D., Frans Berkhout and Pieter van Beukering. January 2015. Technology innovation systems and technology diffusion: Adoption of bio-digestion in an emerging innovation system in Rwanda. Technological Forecasting and Social Change, 90(Part A): 318-330. https://doi.org/10.1016/j.techfore.2013.10.011.

264. Tikhe, Kshitija S. , Dr. B. S. Balapgol and Dr. S. T. Mali. 2015. Review of Different Landfill Gas Estimation Models. IOSR Journal of Mechanical and Civil Engineering (IOSR-JMCE) e-ISSN : 2278-1684, p-ISSN : 2320–334X PP 71-79 www.iosrjournals.org Innovation in engineering science and technology (NCIEST-2015).

265. Tilman D., J. Hill and C. Lehman. 2006. Carbon-negative biofuels from low-input high-diversity grassland biomass. Science 314:1598–600.

266. Tong, Xinggang, Laurence H. Smith and Perry L. McCarty. 1990 Methane fermentation of selected lignocellulosic materials. Biomass. 21(4):239-255. https://doi.org/10.1016/0144-4565(90)90075-UG .

267. Types of Biogas Digesters and Plants;https://energypedia.info/wiki/Types_of_Biogas_Digesters_and_Plants.

268. Types of Gasifiers https://www.google.co.in/search?q=Types+of+Gasifiers&oq=Types+of+Gasifiers&aqs=chrome..69i57.1870j0j4&sourceid=chrome&ie=UTF-8.

269. Uberman, R., A. F. Modrzewski, and S. Žiković. 2019. Evaluate GTL processes compared with conventional refining. http://www.gasprocessingnews.com/features/201606/evaluate-gtl-processes-compared-with-conventional-refining.aspx.

270. Uemura, Shigeki and Hideki Harada. May 2000. Treatment of sewage by a UASB reactor under moderate to low temperature conditions. Bioresource Technology, 72(3):275-282. https://doi.org/10.1016/S0960-8524(99)00118-2.

271. US Environmental Protection Agency (USEPA). 1994. Design, operation and closure of municipal solid waste landfills. EPA/625/6-85/006, office of research and development, USEPA, Washington, DC, USA.

272. US Environmental Protection Agency (USEPA). 2005. Landfill-Gas Emissions Model- Version 3.02. User's Guide, EPA-600/R-05/047, Office of Research and Development Washington DC, USA.

273. US Environmental Protection Agency (USEPA). Feb 2009b. User's Manual Ecuador Landfill Model Version 1.0.

274. US Environmental Protection Agency (USEPA). March 2007. User's Manual Central America Landfill Gas Model Version 1.0.

275. US Environmental Protection Agency (USEPA). March 2009a. User's Manual Mexican Landfill Model Version 2.0.

276. Valo, Alexandre, Hélène Carrère, Jean Philippe Delgenès. September 2004. Thermal, chemical and thermo-chemical pre-treatment of waste activated sludge for anaerobic digestion. J. of Chemical Technology and Bio Technology. https://doi.org/10.1002/jctb.1106.

277. Van Buren, A. and M. Crook. 1985. A Chinese Biogas Manual – Popularizing Technology in the Countryside. Intermediate Technology Publications Ltd. London (UK), 1979, sixth impression 1985, 135 P. ISBN: 0903031655. Types of Biogas Digesters and Plants - energypedia.info. https://energypedia.info › wiki › Types_of_Biogas_Dig.

278. Van Zwieten, L., S. Kimber, S. Morris, K. Y. Chan, A. Downie, J. Rust, et al. February 2010. Effects of biochar from slow pyrolysis of papermill waste on agronomic performance and soil fertility. Plant and Soil, 327(1–2):235–246.

279. Various Constructional Models of Biogas Plants https://www.google.co.in/search?q=Various+Constructional+Models+of+Biogas+Plants&oq=Various+Constructional+Models+of+Biogas+Plants&aqs=chrome..69i57j33.5611j0j4&sourceid=chrome&ie=UTF-8.

280. Venkatesh, G. and Rashid Abdi Elmi. September 2013. Economic–environmental analysis of handling biogas from sewage sludge digesters in WWTPs (wastewater treatment plants) for energy recovery: Case study of Bekkelaget WWTP in Oslo (Norway). Energy. 58(1):220-235; https://doi.org/10.1016/j.energy.2013.05.025.

281. Vyrides, I. and D.C. Stuckey. March 2009. Saline sewage treatment using a submerged anaerobic membrane reactor (SAMBR): Effects of activated carbon addition and biogas-sparging time.Wate Research. 43(4): 933-942. https://doi.org/10.1016/j.watres.2008.11.054.

282. Wagenaar, B.M., W. Prins and W.P.M. van Swaaij. 1994. Pyrolysis of biomass in the rotating cone reactor. Chem. Eng. Sci. 49:5109.

283. Waksman, Selman A. 1940. The microbiology of cellulose decomposition and some economic problems involved. The Botanical Review. 6:637–665. [Coolhaas, C., Zur Kenntnis der Dissimilation Fettsaurer. Centr. Bakt. II.75: 161–170, 344–360;76: 38–44. 1928].

284. Wang, Jianlong and Wei Wan. May 2009. Kinetic models for fermentative hydrogen production: A review. Int. J. of Hydrogen Energy, 34(8): 3313-3323. https://doi.org/10.1016/j.ijhydene.2009.02.031.

285. Wang, Tao, Dong Zhang, Lingling Dai, Yinguang Chen, and Xiaohu Dai. May 2016. Effects of Metal Nanoparticles on Methane Production from Waste-Activated Sludge and Microorganism Community Shift in Anaerobic Granular Sludge. Sci Rep. 6: 25857. doi: 10.1038/srep25857 (https://www.ncbi.nlm.nih.gov/pmc/articles/PMC4863170/).

286. Wang, Tiefeng, Jinfu Wang and Yong Jin. August 2007. Slurry Reactors for Gas-to-Liquid Processes: A Review. Ind. Eng. Chem. Res. 46(18): 5824–5847. https://doi.org/10.1021/ie070330t.

287. Wannapokin, Anongnart, Rameshprabu Ramaraj, Yuwalee Unpaprom. February 2017. Research Article An investigation of biogas production potential from fallen teak leaves (Tectona grandis). School of Renewable Energy, Maejo University, Thailand. DOI: http://dx.doi.org/10.7324/ELSR.2017.31110.

288. Ward, A.J., D.M.Lewis and F.B.Green. July 2014. Anaerobic digestion of algae biomass: A review. Algal Research. 5: 204-214. https://doi.org/10.1016/j.algal.2014.02.001

289. Weerachanchai, Piyarat, Masayuki Horio and Chaiyot Tangsathitkulchai. February 2009. Effects of gasifying conditions and bed materials on fluidized bed steam gasification of wood biomass. Bioresource Technology, 100(3): 1419–1427. https://doi.org/10.1016/j.biortech.2008.08.002.

290. What is Pyrolysis? https://www.azocleantech.com/article.aspx?ArticleID=336.

291. Wikén, T. 1940. Untersuchungen über Methangärung und die dabei wirksamen Bakterien. Arch Mikrobiol 11: 312–317.

292. Wilkomirsky, I., E. Moreno and A. Berg. February 2014. Bio-Oil Production from Biomass by Flash Pyrolysis in a Three-Stage Fluidized Bed Reactors System. J. of Materials Science and Chemical Engineering. 2: 6–10. http://dx.doi.org/10.4236/msce.2014.22002.

293. Wood Pyrolysis; https://www.sciencedirect.com/topics/engineering/wood-pyrolysis.

294. Worldoceanreview.com/.../climate-change-and-methane-hydrates.

295. Wreford, Katherine. 1996. An analysis of the factors affecting landfill gas composition and production and leachate characteristics. https://open.library.ubc.ca/cIRcle/collections/ubctheses/831/items/1.0086990.

296. www.unep.org/yearbook/2014/PDF/chapt5.pdf.

297. Yamada, Kensei, Shuhei Ogo, Ryota Yamano, Takuma Higo and Yasushi Sekine. FEB-2020. Low-temperature conversion of carbon dioxide into methane in an electric field. Chemistry Letters. DOI: 10.1246/cl.190930. http://www.waseda.jp/top/en.

298. Yang, Ren-Xuan, Kui-Hao Chuang and Ming-Yen Wey. 2016. Carbon nanotube and hydrogen production from waste plastic gasification over Ni/Al–SBA-15 catalysts: effect of aluminum content. DOI: 10.1039/C6RA04546D (Paper) RSC Advances.6(47):40731-40740. [https://doi.org/10.1039/C6RA04546D].

299. Yuan, Lan and Yi-JunXu. July 2015. Photocatalytic conversion of CO_2 into value-added and renewable fuels. Applied Surface Science, Volume 342(1):154-167. https://doi.org/10.1016/j.apsusc.2015.03.

300. Yuan, Qing, Chang YuSun, Bei Liu, Xue Wang, Zheng-WeiMa, Qing-LanMa, et al. March 2013. Methane recovery from natural gas hydrate in porous sediment using pressurized liquid CO_2 Energy Conversion and Management. 67: 257–264. https://doi.org/10.1016/j.enconman.2012.11.018.

301. Zabranska, Jana and Dana Pokorna. 2017. Bioconversion of carbon dioxide to methane using hydrogen and hydrogenotrophic methanogens: Research review paper. J. Biotechnology Advances. https://doi.org/10.1016/j.biotechadv.2017.12.003.

302. Zayed, G. and J. Winter. June 2000. Inhibition of methane production from whey by heavy metals – protective effect of sulfide. Applied Microbiology and Biotechnology. 53(6):726–731.

303. Zhai, Qingge, Shunji Xie, Wenqing Fan, Prof. Dr. Qinghong Zhang, Yu Wang, Dr. Weiping Deng. et al. 2013. Photo-catalytic Conversion of Carbon Dioxide with Water into Methane: Platinum and Copper Oxide Co-catalysts with a Core–Shell Structure. Angew. Chem. Int. Ed. 52: 5776–5779. doi: 10.1002/anie.201301473. [State Key Laboratory of Physical Chemistry of Solid Surfaces, Innovation Center of Chemistry for Energy Materials, National Engineering Laboratory for Green Chemical Productions of Alcohols, Ethers and Esters, College of Chemistry and Chemical Engineering, Xiamen University, Xiamen 361005 China]. http://onlinelibrary.wiley.com/doi/10.1002/anie.201301473/abstract.

304. Zhang, Dagang, Yongjun Chen & Tianyu Zhang. April 2014. Floating production platforms and their applications in the development of oil and gas fields in the South China Sea. J. of Marine Science and Application. 13: 67–75. https://link.springer.com/article/10.1007/s11804-014-1233-2.

305. Zhang, Jianguo , HuiWang and Ajay K.Dalai. July 2007. Development of stable bimetallic catalysts for carbon dioxide reforming of methane. Journal of Catalysis. 249(2): 300-310. https://doi.org/10.1016/j.jcat.2007.05.004.

306. Zhang, Lei, Tim L.G.Hendrickx, Christel Kampman, Hardy Temmink and Grietje Zeeman. November 2013. Co-digestion to support low temperature anaerobic pretreatment of municipal sewage in a UASB–digester. Bioresource Technology. 148: 560-566. https://doi.org/10.1016/j.biortech.2013.09.013.

307. Zhang, Wennan. August 2010. Automotive fuels from biomass via gasification. Fuel Processing Technology, 91(8):866-876. https://doi.org/10.1016/j.fuproc.2009.07.010.

308. Zhang, Z.L., X.E.Verykios. December 1994. Methane activation-Carbon dioxide reforming of methane to synthesis gas over supported Ni catalysts. Catalysis Today. 21(2–3):589-595. https://doi.org/10.1016/0920-5861(94)80183-5.

309. Zheng, Yi., Jia Zhao, Fuqing Xu and Yebo Li. June 2014. Pretreatment of lignocellulosic biomass for enhanced biogas production- Review. Progress in Energy and Combustion Science, 42:35-53, https://doi.org/10.1016/j.pecs.2014.01.001.

CHAPTER 7

Hydrogen as a Renewable Fuel

7.1 Properties of Hydrogen [59,109]

7.1.1 Hydrogen Fires

It is well known that 'hydrogen' on combustion yields only water or steam and no greenhouse gas is evolved, so it is absolutely clean and a pollution-free fuel.

$$2H_2 + O = 2H_2O$$

However, hydrogen is not available in nature as a gas and it is produced by releasing it from water or other compounds of hydrogen. It requires energy, which should preferably be provided from 'green sources' (such as a solar or wind power source) to not add more greenhouse gases emissions to atmosphere.

The characteristics and properties of hydrogen are quite favorable for its use as a green fuel, but in practice there are some problems in adopting it. Hydrogen has the widest explosive/ignition mix range with air of all the gases except acetylene. That means that whenever air and hydrogen mix, say on a hydrogen leak, a spark occurring there would lead to an explosion, not merely a flame. This makes the use of hydrogen particularly dangerous in enclosed areas such as tunnels or underground spaces. Some differences with common fuels include the fact that pure hydrogen-oxygen flames burn in the ultraviolet (pale blue) colour range and are nearly invisible to the naked eye, thus requiring a flame detector to detect them if a hydrogen leak is burning. Hydrogen is colorless and odorless and therefore, leaks cannot be detected by color and smell.

Specifically, the amount of energy liberated during the reaction of hydrogen, on a mass basis, is about 2.5 times the heat of combustion of common hydrocarbon fuels like gasoline, diesel, methane, propane, etc. Therefore, for a given duty, the mass of hydrogen required is only about one third of the mass of hydrocarbon fuel needed.

The high-energy content of hydrogen also implies that the energy of a hydrogen gas explosion is about 2.5 times that of common hydrocarbon fuels on an equal mass basis. Therefore, the hydrogen gas explosions are more destructive and carry further. However, the duration of a conflagration tends to be inversely proportional to the combustive energy, so that hydrogen fires subside much more quickly than hydrocarbon fires.

a. Burning Speed of Hydrogen

The flame speed is the sum of the burning speed and displacement velocity of the unburned gas mixture. The burning speed varies with gas concentration and drops off at both ends of the flammability range. The burning speed/velocity is the speed at which a flame travels through a combustible gas mixture. (Burning speed is different from flame speed.) The burning speed indicates the severity of an explosion since high-burning velocities have a greater tendency to support the transition from deflagration to detonation in long tunnels or pipes. Below the lower flammability limit (LFL)* and above the higher flammability limit (UFL),** the burning speed is zero.

The burning speed of hydrogen is 2.65–3.25 m/s, which is higher than that of methane or gasoline at stoichiometric conditions. Thus, hydrogen fires burn quickly and cause severe explosion, but the fires tend to be relatively short-lived.

Note: *The flammability range of a gas is defined in terms of its lower flammability limit (LFL) and its upper flammability limit (UFL). The LFL of a gas is the lowest gas concentration that will support a self-propagating flame when mixed with air and ignited. Below the LFL, there is not enough fuel present to support combustion and the fuel/air mixture is too lean.*

** The UFL of a gas is the highest gas concentration that will support a self-propagating flame when mixed with air and ignited. There is not enough oxygen present to support combustion above UFL and the fuel/air mixture is too rich.

Between the LFL and UFL, is the flammable range in which the gas and oxygen are in the right proportions to burn when ignited.

b. *Quenching Gap for Hydrogen*

The quenching gap (or quenching distance) actually describes the flame extinguishing properties of a fuel when used in an internal combustion engine. Specifically, the quenching gap relates to the distance from the cylinder wall that the flame extinguishes due to heat losses. However, the quenching gap has no specific relevance for use with fuel cells.

The quenching gap of hydrogen at 0.635 mm (0.025 in) is approximately one third that of other fuels, such as gasoline. Thus, hydrogen flames travel closer to the cylinder wall before they are extinguished, making them more difficult to quench than gasoline flames. This smaller quenching distance can also increase the tendency for backfiring since the flame from a hydrogen-air mixture can more readily get past a nearly closed intake valve than the flame from a hydrocarbon-air mixture.

7.1.2 *Generation of Electrostatic Charge in Hydrogen*

Hydrogen has low electroconductivity; therefore, the flow or agitation of hydrogen gas or liquefied hydrogen may generate electrostatic charges that result in sparks. Flammable mixtures of hydrogen and air can get easily ignited on the generation of sparks.

The electrostatic charge should have an easy path to the ground in order to avoid sparks. All hydrogen-conveying equipment must, therefore, be thoroughly grounded.

7.1.3 *Asphyxiation* due to Hydrogen Leak*

In an enclosed area, leaks of hydrogen pose a danger of asphyxiation since the hydrogen diffuses quickly to fill the volume. The potential for asphyxiation in unconfined areas however, would be almost negligible due to the high buoyancy and diffusivity of hydrogen.

Note: *Asphyxiation: When the hydrogen leaks, it quickly diffuses and displaces air and the oxygen level falls. Oxygen levels below 19.5% are biologically inactive for humans. Effects of oxygen deficiency may include rapid breathing, diminished mental alertness, impaired muscular coordination, faulty judgment, depression of all sensations, emotional instability, and fatigue. As asphyxiation progresses, dizziness, nausea, vomiting, prostration, and loss of consciousness may result, eventually leading to convulsions, coma, and death. At concentrations below 12%, immediate unconsciousness may occur with no prior warning symptoms.*

7.1.4 *Comparative Properties of Hydrogen and Methane* [141]

Table 7.1a shows comparative properties of hydrogen and methane as fuel.

Table 7.1a. Comparative Properties of Hydrogen and Methane. [141]

Se. No.	Particulars	Hydrogen	Methane
1.	Chemical formula	H_2	CH_4
2.	Molecular weight	2.016	16.04
3.	Gas density in kg/m³ @ 20°C and 1atm.	0.0808	0.643
4.	Diffusivity (m²/s) × 10⁵	6.11	1.60
5.	Stoichiometric fuel volume fraction %	29.5%	9.48%
6.	Lower heating value per m³(MJ/m³)	9.9	32.6
7.	Lower heating value per kg (MJ/kg)	119.93	50..02
8.	Adiabatic flame temperature (K)	2380	2226
9.	Flammability lean limit (LFL)—by % volume Flammability rich limit (UFL)—by % volume Flame temperature in air °C	4% 75% 2045	5.3% 15% 1875
10.	Max. flame velocity (m/s)	3.06	0.39
11.	Min. ignition temperature (K)	845	905
12.	Min. ignition energy (mJ)	0.02	0.29
13.	Autoignition temperature in °C (of stoichiometric mixture)	585	540
14.	Octane number	130 + lean burn	125
15.	Electrostatic charge generation due to flow of fluid or agitation	yes	no
16.	Specific heat at constant pressure J/g K)	14.89	2.22
17.	Gross calorific value of hydrogen	141.8 MJ/ kg; (12.74 MJ/m³)	
18.	Net calorific value of hydrogen	118.8 MJ/kg; (9.9 MJ/m³)	
19.	Flame emissivity of hydrogen is lower than that of both methane and gasoline, while the emissivity of gasoline is higher than that of both methane and hydrogen		
20.	Hydrogen is explosive in concentration of	15–59 % of air	
21.	Thermal conductivity at 25°C	19 (W/m K)	
22.	Viscosity at 25°C	0.00892 centipoise	
23.	Specific heat at constant pressure at 25°C	14.3 k J/kg K	
24.	Specific volume of hydrogen @ 20°C and 1atm. is	12.1 m³/kg	
25.	Critical temperature of hydrogen is	(-)241°C	
26.	Boiling point of hydrogen	(-) 253°C	
27.	Solidifying temperature of hydrogen/melting point of solid hydrogen	(-) 259°C	
28.	Operating temperature for liquid hydrogen at a pressure of 2 bar	(-) 250°C	
29.	Density of liquid hydrogen at a pressure of 2 bar (-) 250°C	71 g/L	
30.	Energy density of liquid hydrogen at pressure of 2 bar (-) 250°C	2.36 kWh/L or 33.3 kWh/ kg	

7.1.5 *Safety Characteristics of Hydrogen, Methane, and Gasoline* [8,121,122]

A comparison of safety characteristics of gasoline, methane, and hydrogen has been made in the Table 7.1b, while the properties of hydrogen which affect its safety aspects are given in Table 7.1c.

From the properties mentioned above, it is clear that hydrogen can be a good fuel but it requires a lot of care in its handling, mixing it with oxygen, and a rise in its temperature because of the abovementioned properties. The Canadian Hydrogen Safety Program had concluded that hydrogen fueling can be as safe as CNG if safety measures are properly provided and observed.

Table 7.1b. Comparative Safety Characteristics of Gasoline, Methane and Hydrogen.
1—Safest, 2—Less safe, 3—Least safe. [8,121,122]

No.	Characteristics	Gasoline	Methane	Hydrogen
1.	Toxicity	3	2	1
2.	Toxicity of combustion	3	2	2
3.	Density	3	2	1
4.	Specific heat	3	2	1
5.	Ignition limits	1	2	3
6.	Ignition energy	2	1	3
7.	Flame temp.	3	1	2
8.	Flame emissivity	3	2	1
	Total	21	14	14

Table 7.1c. Properties of Hydrogen Affecting the Safety Aspects of Hydrogen as a Fuel.

Safe	Unsafe
Lower density – safe	Wider ignition limits – less safe*
Higher specific heat – safer	Lower ignition energy – less safe
Lower flame emissivity – safer	Lower ignition temp. – less safe
	High flame temp. – less safe
	Generation of electrostatic charge on agitation and flow – less safe
	Higher speed of burning – less safe
	Very low quenching gap – less safe
	Higher diffusion – less safe

The European Commission emphasizes the critical importance of *'education' in lifting technical safety barriers for the development of the hydrogen economy*. It is expected that with proper education about the safety measures, the general public may be able to use hydrogen technologies in everyday life with at least the same level of safety and comfort as fossil fuels.

In short, Hydrogen is being promoted as the environmentally friendly fuel of the future, because:

• Hydrogen is the earth's tenth-most abundant element and is the most abundant element in the universe.

• Can be generated from water and returns to water when it is burnt.

• Available in vast quantities from the water in the world's oceans.

• Can be used in fuel cells to generate electricity.

• Can be used as the fuel in internal combustion engines to replace petrol or diesel.

• Contains more than three times the energy of most hydrocarbon fuels on a weight basis.

• Would be available even when fossil fuels are exhausted.

• It is odorless and nontoxic.

• Hydrogen is a by-product of the refinery's industrial process and many other industries (like chlorine/caustic soda plants, semiconductor plants, etc.), which is an additional advantage.

7.1.6 *Safety Management of Hydrogen* [8,121,122]

As already stated, hydrogen requires very careful handling with due precautions as it is a highly inflammable gas. Some special equipment and controls are required if they are used. It can be carried in high-pressure cylinders, but larger quantities of gas is transported by pipelines. Hydrogen escapes from a leak of a fixed size three times faster than natural gas and therefore, it would fill any closed space very fast and would reach the lower flammability limit (which is 4% for it) very soon and thereby causing a danger of an explosion. However, the energy contained in the closed space at the lower flammability limit for hydrogen is one-fourth of that for natural gas. If compared on an equal mass basis, the high-energy content of hydrogen implies that the energy of a hydrogen gas explosion is about 2.5 times that of common hydrocarbon fuels and is more destructive and carries forward fast, as already mentioned.

The main danger from hydrogen is largely because of its very low ignition energy (less than one-tenth that for natural gas); therefore, even a tiny spark would ignite it and may cause an explosion.

The majority of fires in hydrogen systems in various industries are caused by leakages from valves, flanges, seals, and fittings. Hydrogen can penetrate seals that are considered normally airtight because hydrogen has very low viscosity and the lowest molecular weight on which the leakage from the seal would depend. The density of hydrogen being very low in comparison to air, it rises up rapidly and because of its high diffusion coefficient, hydrogen diffuses very fast in the atmosphere.

a. High Risk of Fires at Hydrogen Production Plants in Various Industries

Electrolysis process is used by several industries and power stations to generate hydrogen for their internal use, while in some industries hydrogen is a by-product. Wherever there is hydrogen, there are high potential risks of fires. As an essential precaution, hydrogen and oxygen or other oxidants like chlorine must be isolated, collected separately, and handled separately. Special attention is required for fire prevention, fire detection, and firefighting for the electrolysis rooms or where the hydrogen is produced. Still more attention should be paid to the places where hydrogen is at high temperature such as compressors, heat exchangers, furnaces, reformers, etc., for hydrogen.

b. Risks in Transport and Handling of Liquid Hydrogen

Liquid hydrogen has to be maintained at a very low temperature as its critical temperature, which, as already mentioned, is (−) 241°C. The air in the vicinity of liquid hydrogen may get liquefied, but the oxygen is more readily liquefied than nitrogen and therefore, the oxygen-rich liquid air in the vicinity of hydrogen poses a potential flammability danger. The situation for this may be caused by:

  i. when the liquid hydrogen tank or pipeline have not been properly insulated and a small leak occurs;

  ii. when the liquid hydrogen is being filled in an empty tank without purging out the air with an inert gas.

c. Measures to Prevent Hydrogen Accidents

For fault elimination inside a system having hydrogen as a fuel and an oxidant, the following measures are required:

  1. Elimination of the source of ignition.
  2. Effective separation of hydrogen and oxidants and prevention of the oxidant like air, oxygen, chlorine, etc., to enter the system from outside.
  3. Use of an inert gas like nitrogen/helium as a purging agent, and use of an inert gas cover.
  4. Pressure of oxidant should be kept lower than hydrogen in the system.
  5. Provision of very sensitive and reliable hydrogen leak detection system and its proper maintenance.
  6. Adequate measures for fire prevention and provision of fire-detection and firefighting systems.
  7. Efficient and proper maintenance of firefighting systems along with regular mock drills. *[Fire can be suppressed/extinguished by means of CO_2, powder-type extinguishers.]*
  8. Use of flame traps, flame suppressors, rapid closing devices.

9. System should be designed to withstand force of explosion and pressure surge.
10. Provision of explosion suppression system at the complex.
11. Provision of a dedicated explosion suppression system for pipeline.
12. Proper training of operating and maintenance personnel and their regular drill.

7.2 Use of Hydrogen in Transport Vehicles [8,10,13,52,79,98,99,109,110, 112,113,129,163,174,193,204,216,218,223,224]

Hydrogen can be used as a fuel for transport vehicles. Blending hydrogen (up to 30%) in compressed natural gas (CNG) to have an on-hand experience has been taken up as a 'project' in select vehicles (buses, cars, three-wheelers) by Ministry of Transport, India through the Society of Indian Automobile Manufacturers (SIAM) and Indian Oil Co. Ltd.

Banaras Hindu University (BHU), Varanasi, has developed motorcycles for demonstration, which are fuelled by hydrogen stored in metal 'Hydride tanks'. These motorcycles engines were also modified at the university to suit the 'change of fuel'. Five departments of the university (Automotive Engineering, Mechanical Engineering, Chemical Engineering, Physics, and Botany) are involved and participating in the project 'Hydrogen energy research, development and demonstration project'.

Tata Motors and ISRO have already developed and tested a hydrogen bus.

The distribution of hydrogen for the purpose of transportation is being taken up in many countries such as Portugal, Iceland, Norway, Denmark, Germany, US (California), Japan, and Canada. (*The hydrogen, as a by-product of the refinery's industrial process, offers an additional advantage.*)

Toyota launched its first production fuel-cell vehicle, the Toyota Mirai, in Japan, at the end of 2014 and began sales in California, mainly the Los Angeles area, in 2015. (The car has a range of 312 miles (502 km) and takes about five minutes to refill its hydrogen tank.) Hydrogen- fuel-cell-electric hybrid scooters are being made, such as the Suzuki Burgman Fuel cell scooter and the FHybrid.

In March 2015, China South Rail Corporation (CSR) demonstrated the world's first hydrogen-fuel-cell-powered tramcar at an assembly facility in Qingdao. A total of 83 miles of tracks for the new vehicle have been built in seven Chinese cities. As early as 2007, Pearl Hydrogen Power Sources of Shanghai, China, unveiled a hydrogen bicycle at the 9th China International Exhibition on Gas Technology, Equipment and Applications.

In Brno, Czech Republic, a vehicle 'Mercedes-Benz O530 Citaro' powered by hydrogen fuel cells is being used. The hydrogen-based-fuel-cell vehicles are being developed by some automobile manufactures who have tied up with manufacturers of fuel cells.

The UK started a fuel cell pilot program in January 2004; the program ran two fuel cell buses, while Western Australia's Department of Planning and Infrastructure has been operating fuel cell buses on trial in Perth as part of its sustainable transport.

In fuel-cell-powered vehicles, the internal combustion engines are replaced by 'Fuel cell stacks' as a prime source of power. The fuel-cell-based vehicles may take time to get commercialized because of the cost factor. The fuel cell used in cars has an added advantage of a superior power to weight ratio apart from being much more efficient and less polluting than internal combustion engines. This will cut down pollution in the transport sector. Table 7.2 gives the names of companies, names of vehicles, and state/type of hydrogen fuel used.

Hydrogen Application in Aircrafts/Rockets/Space shuttle

In February 2008, Boeing tested a manned flight of a small aircraft powered by a hydrogen fuel cell. Unmanned hydrogen planes have also been tested. For large passenger airplanes however, *The Times* reported that "Boeing said that hydrogen fuel cells were unlikely to power the engines of large passenger jet airplanes, but could be used as backup or auxiliary power units onboard."

Table 7.2. Hydrogen-Fuel-Cell-Powered vehicles.

Name of company	Name of vehicle	Type of hydrogen fuel
Toyota	FCHV (Kluger V)	Compressed hydrogen
Honda	FCX	Compressed hydrogen
General Motors	Hydro Gen 3	Compressed hydrogen
Daimler- Chrysler	NECR 4	Liquid hydrogen
Daimler- Chrysler	NECR 4a	Compressed hydrogen
Daimler- Chrysler	NECR 4 and Commander Jeep	Menthol (which may be obtained by converting hydrogen to menthol)
General Motors	Hydro Gen 3	Compressed hydrogen
Ford	Focus FCV (Hybrid)	Compressed hydrogen
Daimler-Chrysler	Chrysler Natrium	Hydrogen stored in metal hydrides/sodium borohydrides

Many large rockets use liquid hydrogen as fuel, with liquid oxygen as an oxidizer. An advantage of hydrogen rocket fuel is its highly effective exhaust velocity compared to other fuels.

Liquid hydrogen and oxygen were also used in the space shuttle to run the fuel cells that power the electrical systems. The by-product of the fuel cell is water, which is used for drinking and other applications in space requiring water.

7.3 Production of Hydrogen [2,3,4,7,14,16,19,20,23,25,29,30,31,34,35,37, 38,41,42,45,50,51,54,55,56,68,73,74,77,78,79,80,82,84,85,86,87,102,104, 109, 114,115,116,117,118,119,130,131,133,134,136,137,138,139,142,143, 144,145, 147,148,150,153,154,157,158,160,166,168,174,175,179,180,181, 183,184,186, 187,188,189,197,199,201,202,203,206,207,208,209,216,218, 225,226,228,229,230,231,232,234,238,239]

There are many methods by which hydrogen can be produced. These are given below and the details are discussed in the following paragraphs:

  i. Water splitting or electrolysis
 ii. Solar photolysis for hydrogen production
iii. Thermolysis of water—thermochemical production of hydrogen
 iv. Biomass conversion to hydrogen
  v. Hydrogen production from biomass through the gasification route
 vi. Hydrogen production from glycerol
vii. Hydrogen production from biomass gasification by disposal of emitted CO_2
viii. Initial Conversion of lignocellulosic materials to alcohols and then to hydrogen
 ix. Production of hydrogen from biomass through biological fermentation
  x. Hydrogen production from fossil fuels

Pathways for Production of Hydrogen from Renewables

It may be mentioned that there are several pathways for the production of hydrogen from renewable energy as shown in Figure 7.1.

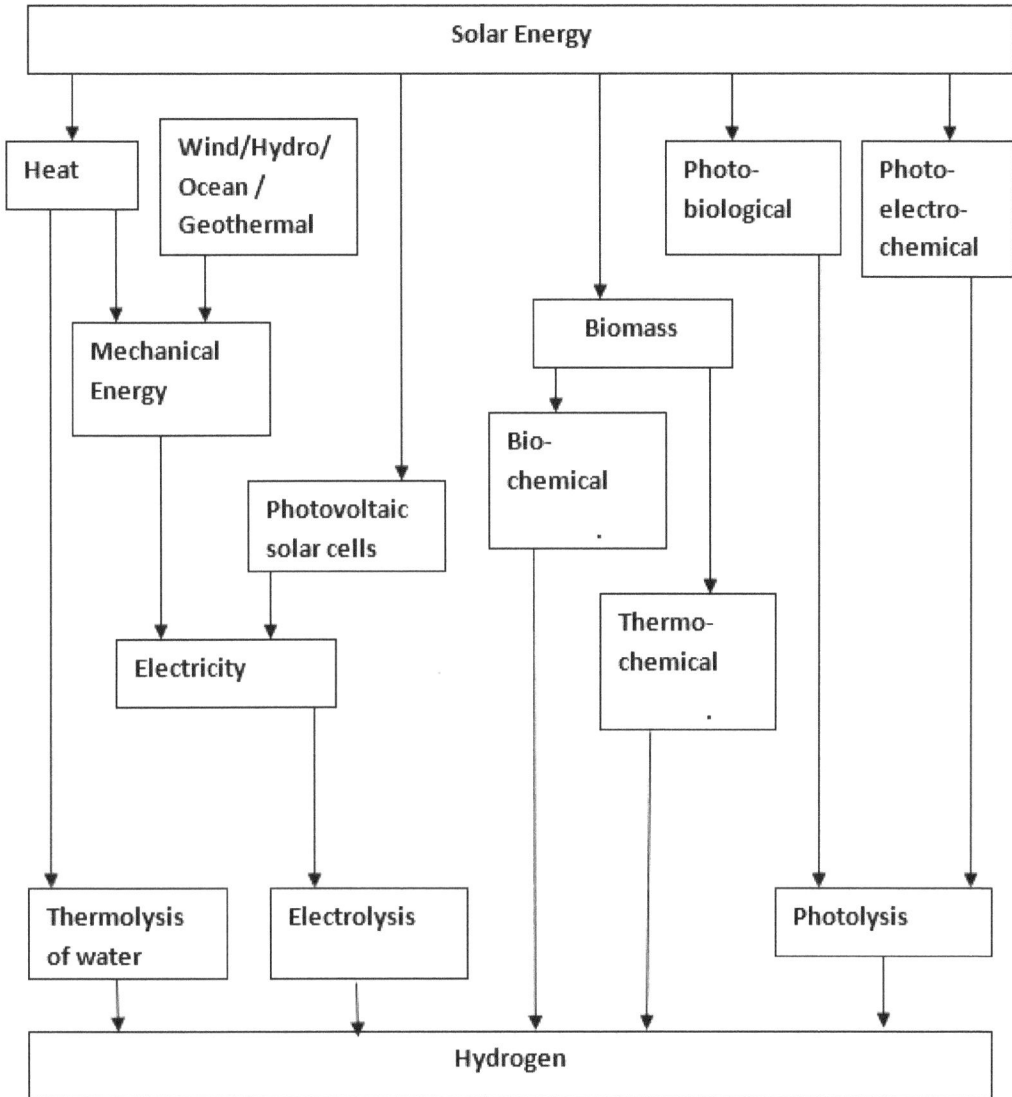

Fig. 7.1. Pathways to Hydrogen Production from Renewables.

7.4 Hydrogen from Water by Electrolysis

7.4.1 Standard Electrolysis

In this method, water containing a small quantity of sulphuric acid (H_2SO_4), or an alkali like potassium hydroxide (KOH of 10 to 25% strength) is electrolysed in a cell using an iron sheet cathode (negative electrode) and a nickel-plated iron sheet as an anode (positive electrode). The anode and cathode are separated from each other by a porous diaphragm (for example, one made of asbestos to avoid the spontaneous back recombination of hydrogen and oxygen). On passing direct current through the electrolytic cell, oxygen is liberated at the anode and the hydrogen is liberated at the cathode in the volumetric ratio of oxygen to hydrogen as 1:2. The required hydrogen is collected at the cathode as the main product, while the oxygen is collected as a by-product at the anode. The rate at which hydrogen is produced is directly proportional to the current passing between the electrodes (*according to Faraday's Law*).

The reactions which take place within the cell are given below:

(a) *With the use of* H_2SO_4,

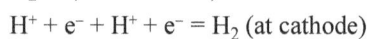

$H_2SO_4 = 2H^+ + SO_4{}^{2-}$

$H^+ + e^- + H^+ + e^- = H_2$ (at cathode)

$SO_4{}^{2-} — 2e^- = SO_4$; $2SO_4 + 2 H_2O = 2H_2SO_4 + O_2$, (at anode)

(b) *With the use of sodium hydroxide (NaOH)*

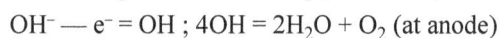

$NaOH = Na^+ + OH^-$; $Na^+ + e^- = Na$

$2 Na + 2H_2O = 2 NaOH + H_2$ (at cathode)

$OH^- — e^- = OH$; $4OH = 2H_2O + O_2$ (at anode)

(c) *With use of potassium hydroxide (KOH)*

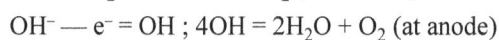

$KOH = K^+ + OH^-$; $K^+ + e^- = K$

$2 K + 2H_2O = 2 KOH + H_2$ (at cathode)

$OH^- — e^- = OH$; $4OH = 2H_2O + O_2$ (at anode)

The alkaline electrolyzer are commonly used for the electrolysis of water for producing hydrogen for local use.

There is a small gap between the electrode and the separator and the gases evolved are released from the gap. As the current density increases, the gas bubbles tend to form a continuous and highly resistive film of gas on the surface of both the electrodes, resulting in the slowing of the splitting reaction. The current densities therefore have to be limited to a suitable value to get the best results. In more efficient cells, the gap between the separator and the electrodes is reduced to almost zero and both the electrodes are made porous to enable the gases to evolve from the back of the electrodes. The interpolar distance is lower in such cells and higher current densities can be used.

Hydrogen from Brine Electrolysis

Large quantities of hydrogen (as well as chlorine) is obtained as a by-product during the production of caustic soda (NaOH) by the electrolysis of brine (sodium chloride 'NaCl' solution). The reactions that take place are mentioned below:

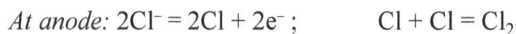

Ionization of 'NaCl' -------------------------------- $NaCl = Na^+ + Cl^-$

At anode: $2Cl^- = 2Cl + 2e^-$; $Cl + Cl = Cl_2$

At cathode: $2Na^+ + 2e^- = 2Na$; $2Na + 2H_2O = 2 NaOH + H_2$

Electrical Energy Consumed in the Production of Hydrogen

The production of one kilogram of hydrogen by electrolysis consumes approximately 50 kWh of electrical energy (for example, 4.23 kWh per Nm^3 of hydrogen, normal temperature being taken as 25°C), while the calorific value of the hydrogen produced is 39 kWh/kg. Obviously there are losses in the electrolysis process (the energy consumption being much higher than the calorific value of the hydrogen evolved), but the purity of hydrogen produced by electrolysis may be about 99.9% or more. The conversion efficiency of the electrolysers used to create hydrogen is between 60% and 80%, depending on the current and the materials used for the electrolytes and the electrodes.

The electrolyte for the electrolysis cell can be water + sodium chloride (NaCl) or water + potassium hydroxide (KOH) of various strengths from 5–15%. The energy consumption in electrolysis falls with the increase in the surface area of the electrodes and a pair of electrodes.

Note: *In an experiment with two pairs of stainless steel electrodes each of 500 cm², the lowest energy consumptions were:*

For 15% NaCl solution: 8.3 to 9.3 kWh/N m³ of H₂

For 10% KOH solution: 7.8 to 9.3 kWh/N m³ of H₂

The consumption for 10% KOH solution further decreased by increasing pairs of electrodes: (a) 7.6 kWh/ Nm³ of H₂ for three pairs of electrodes; (b) 3.2 kWh/Nm³ of H₂ for five pairs of electrodes.

In spite of the above values, the usual actual average consumption of power in practice with best efficiency achievable is of the order of 48–50 kWh/kg of hydrogen.

It is always better environmentally that the electrical energy generated and utilised for the electrolysis of water for the production of hydrogen should be from renewable sources. There is no environmental advantage in utilising electricity from hydro sources, or other continuous and reliable sources for producing hydrogen.

7.4.2 *High-pressure Electrolysis* [84]

Hydrogen can be generated by high-pressure electrolysis. The difference in high-pressure electrolysis and the standard electrolysis is that of the compressed hydrogen output around 120–200 bar in the former case. By pressurizing the hydrogen in the Electrolyzer, the need for an external hydrogen compressor is eliminated; the average energy consumption for the internal compression is less (say about 3%).

It may be mentioned that at a high pressure, the porous diaphragm presents a problem of mixing of oxygen and hydrogen and the safety issues need to be tackled. The hydrogen purity also decreases on increasing the pressure. However, some companies are commercially producing the pressurized units that operate at 1.2 MPa, with an output of 65 Nm³/h of hydrogen.

7.4.3 *Proton Exchange Membrane Water Electrolysis* [34]

In a proton exchange membrane water electrolysis cell, the two electrodes (cathode and anode) are pressed against a proton-conducting polymer electrolyte, forming a membrane electrode assembly. This assembly is immersed in de-ionized/pure water. When the direct current is passed, the splitting of water takes place, the hydrogen ions move across the polymer proton exchange membrane and appear at the cathode, while the oxygen evolves from the anode.

Various polymer materials for membranes have been developed. The most popular proton exchange membrane material 'Nafion' has been developed by DuPont de Nemours Co. This material is a co-polymer of tetra-fluoro-ethylene containing grafted sulfonic acid functional groups and possesses high chemical stability. In spite of protons remaining inside the membrane, the acidity of the material is high and the noble metal catalysts are required at both the anode and cathode. The catalyst is deposited on the surface of the membrane with the ion-exchange electrolyte. The catalyst at the anode may be iridium (Ir), while for the cathode, it may be platinum (Pt). As there is no liquid electrolyte, the electrodes should be tightly held to the membrane and a high surface contact between the catalyst (electronic conductor) and the electrolyte should be provided. During the operation, the formation of a small amount of hydrogen peroxide by reduction of oxygen can lead to the gradual degradation of the membrane. The swelling of membrane can also occur in contact with water. These problems may cause a decline in cell efficiency.

7.4.4 *High-temperature Electrolysis* [82,142]

Hydrogen can be generated through high-temperature electrolysis from energy supplied in the form of heat as well as electricity. As some of the energy in high-temperature electrolysis is supplied in the form of heat, smaller amount of energy is required to be converted twice (from heat to electricity, and then to chemical form), and so potentially in this process of hydrogen production, the 'energy requirement per kilogram of hydrogen produced is less'.

A type of high-temperature electrolyzer is a 'Solid-oxide water electrolysis cell', where the electrolyte is zirconia (ZnO_2), stabilized with yttrium and scandium oxides. The solid oxide electrolysis cell normally works in the temperature range of 800–1000°C. The bipolar plates are usually of stainless steel and the solid electrolyte is manganite-coated stabilized zirconia. The oxide ions diffuse across the zirconia electrolyte.

(*The resistivity of the solid electrolyte is higher than the alkaline solutions and ion-exchange membranes and therefore, thin ceramic membranes are required for the electrolysis process to reduce resistance.*)

7.4.5 *Hydrogen Production by Biocatalyzed Electrolysis* [199]

Besides regular electrolysis, the electrolysis using microbes is another possibility for the production of hydrogen. Biocatalyzed electrolysis is a novel biological hydrogen production process with the potential to efficiently convert a wide range of dissolved organic materials in wastewaters. Biocatalyzed electrolysis may become an attractive technology in the future for hydrogen production from a wide variety of wastewaters. Even substrates, formerly regarded to be unsuitable for hydrogen production due to the endothermic nature of the involved conversion reactions, can be converted with this technology. Biocatalyzed electrolysis achieves this by utilizing electrochemically active microorganisms that are capable of generating an electrical current from the oxidation of organic matter. When this biological anode is coupled to a *proton- reducing cathode by means of a power supply, hydrogen is generated.*

In theory, the biocatalyzed electrolysis of 'acetate' requires applied voltages that can be quite low (range 0.14–0.6V), while hydrogen production by the means of conventional water electrolysis, in practice, requires applied voltages well above 1.6 V.

Biocatalyzed electrolysis is a microbial fuel cell-based technology for the generation of hydrogen gas and other reduced products out of electron donors. Some examples of electron donors are acetate/acetic acid and wastewater. An external power supply can support the process and therefore circumvent the 'thermo-dynamical constraints that could have rendered the generation of hydrogen unlikely'.

At the anode, the electro-chemically active bacteria in the substrate containing acetic acid consumes it as carbon source and releases CO_2, protons (H^+), and electrons (e^-). (*The fermentation of organic matter produces acetic acid.*) At the cathode, the protons ($2H^+$) and electrons ($2e^-$) combine to form hydrogen (H_2). The applied voltage may be 0.6 Volts or so. The anodic oxidation of acetate/acetic acid may not be hampered by ammonium concentrations up to a certain concentration of ammonium ions.

With biocatalyzed electrolysis, hydrogen may be generated after running through the microbial fuel cell and a variety of aquatic plants and other plants including reeds, sweetgrass, cordgrass, rice straw, tomatoes, lupines, algae, etc., can be hydrolyzed and fermented to generate sugars and acetic acid.

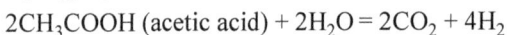

$$C_6H_{12}O_6 + 2H_2O = 2CH_3COOH \text{ (acetic acid)} + 2CO_2 + 4H_2$$
$$2CH_3COOH \text{ (acetic acid)} + 2H_2O = 2CO_2 + 4H_2$$

(*If required, the hydrogen produced at the cathode could be converted into methane.*)

7.4.6 *Chemical-assisted Water Electrolysis* [242]

Hydrogen (H_2) is widely regarded as an attractive alternative to carbon-based fuels for the future energy system due to its high-energy density and zero pollution. The global hydrogen demand is continuously increasing each year and will play an important role in energy consumption. Conventionally, most of the hydrogen is produced by steam reforming natural gas or other fossil fuels, which requires high temperature and pressures. In the process, it is itself contaminated by CO, which affects its subsequent applications.

The splitting of water or electrolysis can provide a promising alternative for producing high-purity H_2. Water electrolysis technology includes alkaline water electrolysis, proton exchange membrane water electrolysis, and high-temperature water electrolysis. This process is energy-intensive due to the sluggish kinetics of both the hydrogen evolution reaction (HER) and oxygen evolution reaction (OER) at the cathode and anode, respectively. This requires high overpotential to generate high current densities. Besides, the generation of H_2/O_2 gas mixture is a drawback of the conventional operating modes.

An emerging alternative for conventional water electrolysis is to integrate thermodynamically and kinetically more favorable anodic oxidation reactions to replace the challenging but less valuable OER by introducing reductive chemicals to the system. This system can lower the cell voltage and avoid the formation of the H_2/O_2 gas mixture. Additionally, value-added products can be produced in the anodic vicinity.

This process of chemical-assisted water electrolysis technology needs further investigation to realize its practical applications. Identification of suitable reductive chemicals and developing efficient electrocatalysts is the priority and it is important to understand the electrochemical anodic reaction mechanism. Also, the design of an efficient HER electrocatalysts, robust and stable membrane, and an efficient and scalable electrolyzer are required to use this promising technology.

7.5 Solar Photolysis for Hydrogen Production [55,78,232]

Solar photolysis involves the use of 'Solar photons' to produce hydrogen directly via: 'Photo electro–chemical systems' or 'Photo biological systems'.

7.5.1 *Photoelectrochemical (PEC) Production of Hydrogen* [25,41,56,138,180,188]

Using electricity produced by photovoltaic systems offers the cleanest way to produce hydrogen. Water is broken into hydrogen and oxygen by electrolysis aided by light in a photo- electrochemical cell (PEC).

a. Photo electrochemical photolysis involves the dissociation of water into hydrogen and oxygen directly at the surface of the semiconductor through irradiation by a solar photon:

  i. This may be taken as electrolysis without the 'electrolyser'.

  ii. The semiconductor material acts as a 'catalyst' to produce hydrogen directly at the 'semiconductor and water surface interface'.

b. A large number of photo-semiconductors were found to act as photocatalysts for the production of hydrogen, but most of these photocatalysts were found active only under UV light irradiation; the activity under visible light irradiation is very small. The UV radiation is only a tiny portion (about 4%) of solar radiation. The very low efficiency of hydrogen yield under visible light is a bottleneck and better semiconductor materials with higher efficiency under visible light have to be found.

c. The photocatalyst semiconductor material that can split water under UV radiation include titanium oxide and titanates, tantalates and niobates, oxides of indium, gallium, germanium, tin, of special configuration, cadmium sulphide.

d. Some photo-semiconductor materials that can absorb a wide range of solar spectrum radiation (including visible light) would be better for water splitting. Such materials include titanium disilicide ($TiSi_2$), graphitic carbon nitride (g-C_3N_4), and mixed lithium-iron-phosphate $\{(Li_9Fe_3(P_2O_7)_3(PO_4)_2\}$.

  i. Production of hydrogen via conventional PV cells is at present very costly. Efforts are now being made to develop hydrogen generators using 'Hydrogen solar tandem cell technology'. The tandem solar cell produces hydrogen directly from water and sunlight at about 7.5% efficiency without any carbon dioxide emission (it uses photocatalysis by nanocrystalline thin films. The nanocrystalline semiconductor materials using the tandem cell concept has a good potential.

ii.   The use of two sandwiched cells allows the transparent front cell to capture the blue or violet range of light and allows the green and red range to pass through to the second cell behind it.

iii. The second cell uses a surface of dye-sensitized nanocrystalline titanium dioxide.

iv.   This approach using tandem may be more economical.

e.   Research has been going on to find a most appropriate semiconductor that has/is
   (i)   Suitable 'Photo-electro–chemical properties'
   (ii)   Robustness to withstand a severe physical and chemical environment
   (iii) Economical
   (iv)   Research aimed toward developing a higher efficiency 'multi-junction cell technology' (underway in the photovoltaic industry)

f.   Gallium arsenide (GaAs), gallium indium phosphide (GaInP$_2$), amorphous silicon, amorphous silicon carbide, etc., have been found to be promising.

g.   GaAs and GaInP$_2$ are very costly although their efficiency is quite high.

7.5.2 *Photobiological Method of Hydrogen Production* [14,29,154,179,181]

Photobiological methods use photosynthetic organisms like some 'cyanobacteria' species referred to as 'blue-green algae'* and green algae to photo-produce hydrogen. Hydrogen production in green algae occurs within the chloroplast and is a light-dependent process. It may be noted that 'hydrogenase enzymes' catalyze hydrogen production in this prototroph (green algae). The algal hydrogenase has an iron-iron cluster (Fe-hydrogenase)** at its catalytic site, while the cyanobacteria enzyme contains a nickel-iron cluster. The algae for its growth does not need any carbon-based material as it can take CO_2 and nitrogen from air. Other needs for growth are: water, minerals (available in water Itself), and light energy.

The cyanobacterial strain *Nostoc muscurum* TISTR8871 produces higher biomass and has higher starch accumulation (for example, up to 32.9%) during the photosynthetic growth of green algae.

Starch in microalgae is hydrolysed and converted to organic acids by 'anaerobic fermentation' by bacteria such as lactic acid bacteria (i.e., *Bacillus acidi lactiti*) and the organic acid is then converted to hydrogen by biological catalysis in light.

The scientists at the University of Berkeley, USA, succeeded in tricking '*Chlamydomonas reinhardtii*' (common green algae) into producing hydrogen instead of CO_2. When deprived of sulphur (nutrient), the algae responded by stripping hydrogen from water, but 60% of the photons captured were wasted as heat due to too much of chlorophyll present in the cells. If the optical properties of algae are modified, the efficiency of hydrogen production can be increased.

Some photosynthetic bacteria like purple sulphur, purple non-sulphur, and green-sulphur bacteria evolve molecular hydrogen under nitrogen-deficient conditions using light energy and reduce organic compounds.

Notes:

(1) Some cyanobacteria species referred to as 'blue green algae' can be single-celled or colonial, depending upon the species and environmental conditions. These may form sheets, filaments, or even balls. Normally, they are vegetative cells forming photosynthetic cells under favorable conditions. Photosynthesis occurs in membranes called 'thylakoids' with chlorophyll being employed to absorb the sun's rays. Photosynthesis in cyanobacteria uses water as the electron donor and produces oxygen as a by-product. When climate becomes harsh, the cyanobacteria may form spores. Another type of cell in cyanobacteria is a thick-walled heterocyst containing the enzyme nitrogenase, which is vital for nitrogen fixation.*

The second group of enzymes present in cyanobacteria is hydrogenases. These enzymes are present in two different forms: (i) Uptake hydrogenases and (ii) Bidirectional, that is, reversible hydrogenases. The uptake hydogenase in cyanobacteria is responsible for the oxidation of hydrogen while the reversible hydrogenase synthesizes a good amount of hydrogen.

(2)**Fe-hydrogenase—The families of Fe-hydrogenase include: (i) Cytoplasmic, soluble, monomeric Fe-hydrogenase found in anaerobes like Clostridium pasteurianum and Megasphaera elsdenii, which are extremely sensitive to inactivation by oxygen and catalyse both hydrogen evolution and uptake (ii) Periplasmic hetrodimeric. Fe-hydrogenase from the Desulphovibrio spp., which can be aerobically purified and can catalyse hydrogen oxidation. Cytochrome C_3 & C_6 substances present in living cell act as electron donors or acceptors for Fe-hydrogenase.

The photons absorbed split water molecules and produce protons, electrons, and oxygen molecule. The electrons are carried away by electron carriers (present), which transfer the electron to the Fe-hydrogenase. The Fe-hydrogenase accepts the electron and with the available protons, synthesizes molecular hydrogen.

(3) Photo-fermentation with Rhodobacter sphaeroides SH2C can be employed to convert fatty acids with smaller molecules into hydrogen.

(4) Blue green algae differ from green algae in several aspects. In addition to normal photosynthesis cells (which takes up atmospheric CO_2 and breaks it down to give out oxygen), they contain some larger cells (heterocysts), where hydrogen can be formed in presence of certain enzymes that are naturally present in blue green algae. In the presence of atmospheric nitrogen, however, the hydrogen released can combine with nitrogen to produce ammonia. By preventing access of nitrogen (i.e., in an inert argon gas atmosphere), blue green algae decompose water in sunlight to yield hydrogen and oxygen. Instead of live algae to produce hydrogen from water, it would be more convenient to utilize biological material, that is, chloroplasts from plants for this purpose as the conditions can then be controlled and varied to optimize the production of hydrogen. Chloroplasts are small bodies containing chlorophyll in green plants and can retain the property of photosynthesis even when extracted from the plant. The water can be decomposed to oxygen and hydrogen when chloroplasts are exposed to sun light in the presence of the enzyme hydrogenase and the electron carrier material 'Ferredoxin' (a biological origin material).

a. Main Obstacles to Photobiological Production of Hydrogen

The main obstacle to the production of H_2 from the photobiological process is the inhibition of hydrogenase enzymes that evolves H_2 from water by the presence of 'oxygen'. Overcoming the 'Oxygen sensitivity' of hydrogenase enzymes, and out-competing other metabolic pathways for photosynthetic reductants, dissipating the proton gradients across the photosynthetic membrane, and ensuring adequate efficiency when capturing and converting solar energy are the major challenges. The present efficiency of hydrogen production by solar energy is only 5 to 6%, which needs to be increased. By genetic engineering, the activity of oxidation of hydrogen may be brought down and the production of hydrogen may be enhanced.

b. Developments in the Photobiological Production of Hydrogen

Enzymes that produce 'hydrogen' and 'oxygen' have been identified. Some of the bacteria that have been identified have 'Enzymes more tolerant of oxygen'. Such 'enzymes' when extracted from the suitable strains of bacteria and genetically introduced in the green algae of the 'Chlamydomonas strain' (which has oxygen-evolving enzymes), can create a new genetically engineered strain of green algae that may produce both oxygen and hydrogen simultaneously. Suitable strains of cyanobacteria are under development/have been developed by genetic engineering process. One such strain of bacteria created is named 'Nostoc puntiforme ATCC strain 29133', which has been found to produce H_2 with increased efficiency.

As per the discovery in the late 1990s, algae start producing hydrogen when exposed to light instead of producing oxygen (by normal photosynthesis), if the algae are deprived of sulphur. It may be possible to produce 'Biological hydrogen' in an 'Algae Bioreactor'.

Another method to produce 'biological hydrogen' in a bioreactor can use feedstocks other than algae; the most common feedstock being waste streams involving bacteria feeding on hydrocarbons and excreting 'hydrogen and CO_2'. The CO_2 can be sequestered, leaving behind hydrogen gas. A prototype hydrogen bioreactor using waste as a feedstock is in operation at Welch's grape juice factory in North East, Pennsylvania.

More biological research is needed to optimise the process in the organisms. A good amount of engineering and development work will be required to use these processes for making large hydrogen generation systems.

c. Development of Cell-Free Systems for the Production of Hydrogen

Such a system may have the following characteristics:

o Cell-free systems just use 'Enzymes' taken from microorganisms.

o Both the 'oxygen evolving' and 'hydrogen evolving' enzymes are immobilised on to the opposite sides of a solid conducting surface.

o Light is used by 'oxygen evolving' enzymes to oxidize water creating flow of 'electrons' to the 'hydrogen evolving enzyme' where hydrogen is produced.

7.5.3 *Photoelectrocatalytic Production of Hydrogen* [4]

A new method of hydrogen production has been found by Thomas Nann and his team at the University of East Anglia. In this method, a 'Gold electrode' covered in layers of 'Indium phosphide (InP) Nano particles' is used with the introduction of an iron-sulphur complex into the layered arrangement. When submerged in water and irradiated with light under a small electric current, produces hydrogen with an efficiency of 60%. The discovery is promising, but further development work is needed before the process can be utilized commercially.

7.6 Thermolysis of Water—Thermochemical Production of Hydrogen [56]

Many thermochemical cycles which can be used for water splitting are in the research, development, and testing phase to produce hydrogen and oxygen from water and heat without using electricity. These processes can be more efficient than high-temperature electrolysis. Several thermochemical hydrogen production processes have been demonstrated.

The thermochemical splitting of water into hydrogen and oxygen can be done by a series of chemical reactions driven by thermal energy. A good example of such a thermochemical splitting is the sulphur–iodine (S–I) cycle. The cycle is shown Figure 7.2.

The heat needed for the thermochemical splitting of water for the production of hydrogen can be provided from any high-temperature source such as a gas-cooled nuclear reactor or a concentrated solar heat. The thermochemical production of hydrogen using chemical energy from coal or natural gas is not environment friendly and so is not being considered here. It may be mentioned here that sulphuric acid, iodine, and hydrogen iodide are all corrosive substances. Therefore, special corrosion-resistant materials are required to be used in the cycle.

The term 'cycle' is used because aside from water, hydrogen, and oxygen, the chemical compounds used in these processes are continuously recycled. If electricity is partially used as an input, the resulting thermochemical cycle is defined as a 'hybrid' one.

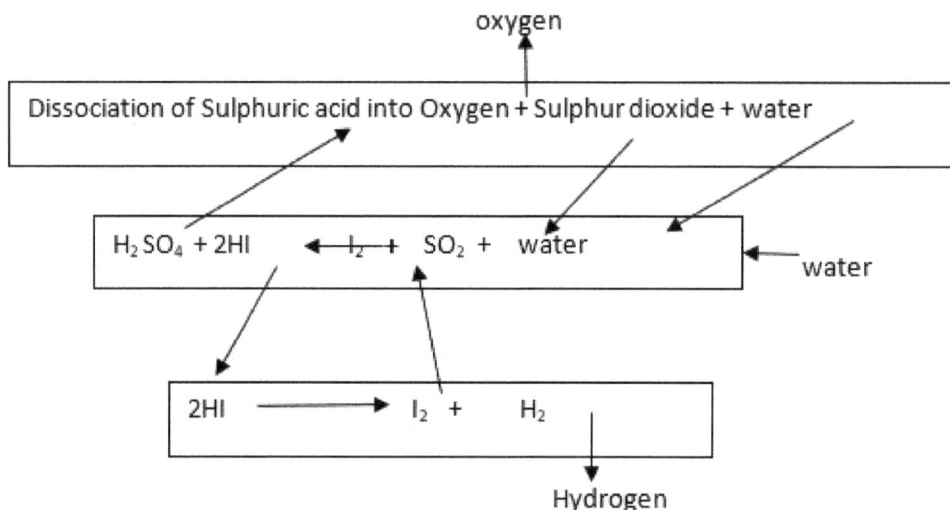

Fig.7.2. Thermochemical Splitting of Water.

The S–I cycle is a thermochemical cycle that generates hydrogen from water with an efficiency of approximately 50%. The sulfur and iodine used in the process are recovered and reused, and not consumed by the process. The cycle can be performed with any source of very high temperatures, approximately 950°C, such as by concentrating solar power systems (CSP), and is regarded as being well suited for the production of hydrogen by high-temperature nuclear reactors. As such, it is being studied in the high-temperature engineering test reactor in Japan.

The sulfur–iodine cycle (S–I cycle) is a three-step thermochemical cycle used to produce hydrogen. The S–I cycle consists of three chemical reactions whose net reactant is water and whose net products are hydrogen and oxygen. All other chemicals are recycled. The S–I process requires an efficient source of heat. The three reactions that produce hydrogen are as follows:

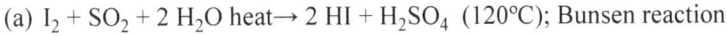

(a) $I_2 + SO_2 + 2\ H_2O$ heat$\rightarrow 2\ HI + H_2SO_4$ (120°C); Bunsen reaction

   HI is then separated by distillation or liquid/liquid gravitic separation.

(b) $2\ H_2SO_4$ heat$\rightarrow 2\ SO_2 + 2\ H_2O + O_2$ (830°C or 1,530°F)
   o The water, SO_2, and residual H_2SO_4 are separated from the oxygen by-product by condensation.

(c) $2\ HI \rightarrow I_2 + H_2$ (450°C)
   o Iodine and any accompanying water or SO_2 are separated by condensation, and the hydrogen product remains as a gas.

Net reaction: $\mathbf{2\ H_2O \rightarrow 2\ H_2 + O_2}$

The sulfur and iodine compounds are recovered and reused. The difference between the heat entering and leaving the cycle exits the cycle in the form of the heat of combustion of the hydrogen produced.

7.6.1 *Hydrogen from Gas-cooled Nuclear Reactors* [184]

While 'Nuclear plant generated electric power' could be used for electrolysis, the 'Nuclear heat' can also be directly applied to split hydrogen from water. High temperature (950–1000°C) gas-cooled nuclear reactors have the potential to split hydrogen from water by a 'Thermo-chemical' reaction using 'Nuclear heat'. Developments into high-temperature nuclear reactors may eventually lead to a hydrogen supply that is cost-competitive with hydrogen produced by the 'Natural gas-steam reforming method' of hydrogen production. General Atomics had predicted that hydrogen produced in a 'High temperature gas cooled reactor (HTGR)' would cost less than the hydrogen from the reforming of natural gas.

Modular Helium Reactor (MHR)

The high helium outlet temperature also makes possible the use of the MHR for production of hydrogen using the S–I cycle. The hydrogen-production efficiency can exceed 50% at temperatures of about 900°C.

Gas-cooled Fast Reactor (GFR)

The GFR system is a high-temperature helium-cooled fast-spectrum reactor with a closed fuel cycle. It has the advantage of:

a. Long-term sustainability of the uranium resource.

b. Waste minimization through fuel multiple reprocessing and fission of long-lived actinides.

c. High-thermal efficiency due to a high-temperature system.

Industrial use of heat generated, for the *production of hydrogen*.

   Utilization of high-temperature outlet helium after cooling the reactor core <u>can</u> impart its heat to a secondary helium circuit in a heat exchanger. The secondary heated fluid can be linked to the production of hydrogen by the splitting of water through the sulphur–iodine cycle as shown Figure 7.3 and Figure 7.4.

Note: *In Figure 7.4, both the fluids on the primary and secondary sides of the intermediate heat exchanger are helium, but the fluid on the secondary side of this intermediate heat exchanger can, alternatively, be air.*

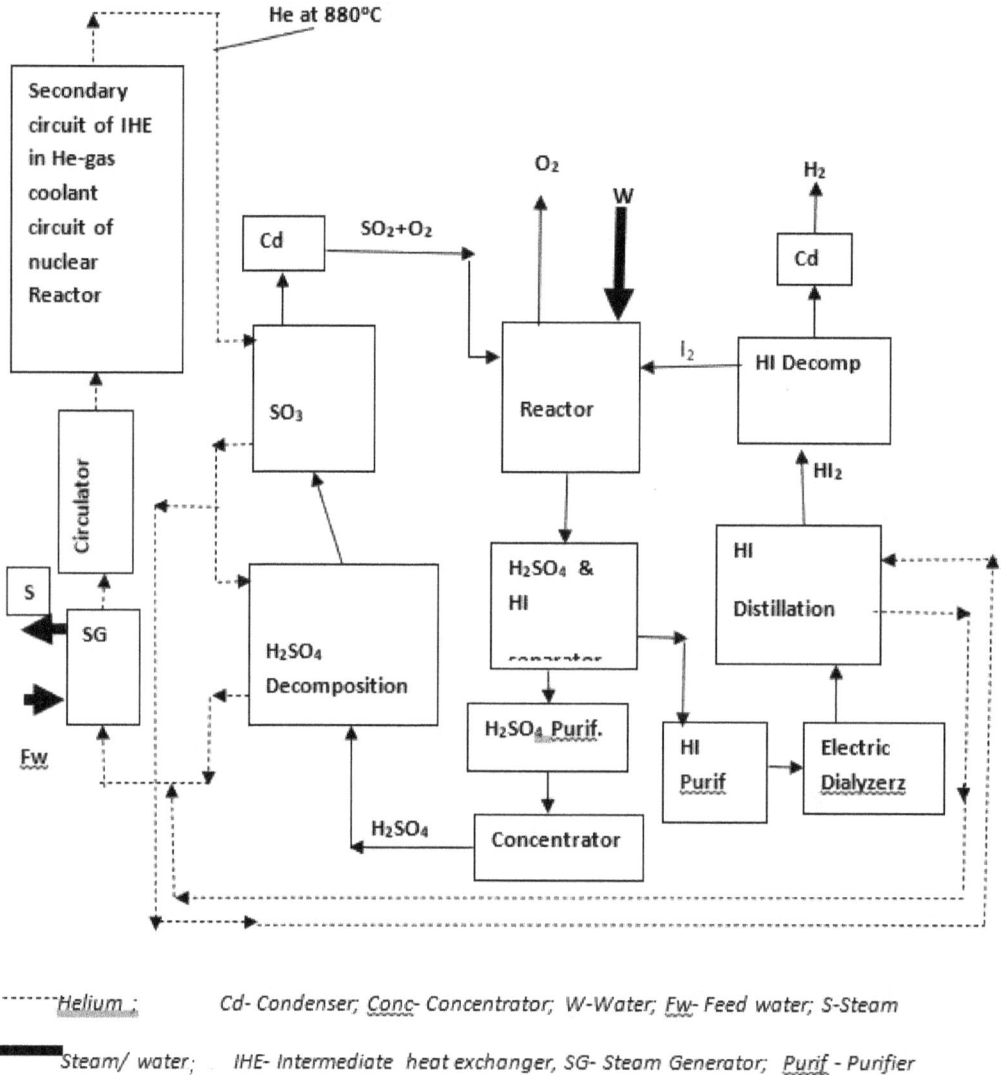

------Helium ;　　　*Cd- Condenser; Conc- Concentrator; W-Water; Fw- Feed water; S-Steam*

■■■■■Steam/ water;　*IHE- Intermediate heat exchanger, SG- Steam Generator; Purif - Purifier*

Fig. 7.3. Hydrogen Production Through Sulphur– Iodine Cycle Utilizing Heat from Helium Gas Cooled Nuclear Reactor.

7.6.2 Hydrogen by Solar Thermolysis [55,56]

The high-temperature heat can easily be provided from a concentrated solar heat source, where it will come in the category of *'Solar Thermolysis'*. Concentrated solar power (CSP) produces heat that can be used to drive thermochemical reactions for the production of hydrogen, or drive electrolysis at a very high temperature for the more efficient decomposition of water into oxygen and hydrogen. As very high temperatures are required to dissociate water into hydrogen and oxygen, a catalyst is required to make the process operate at lower temperatures.

The concentrated solar heat source can also be used for the production of hydrogen from water through the abovementioned *'Sulphur-Iodine cycle'*.

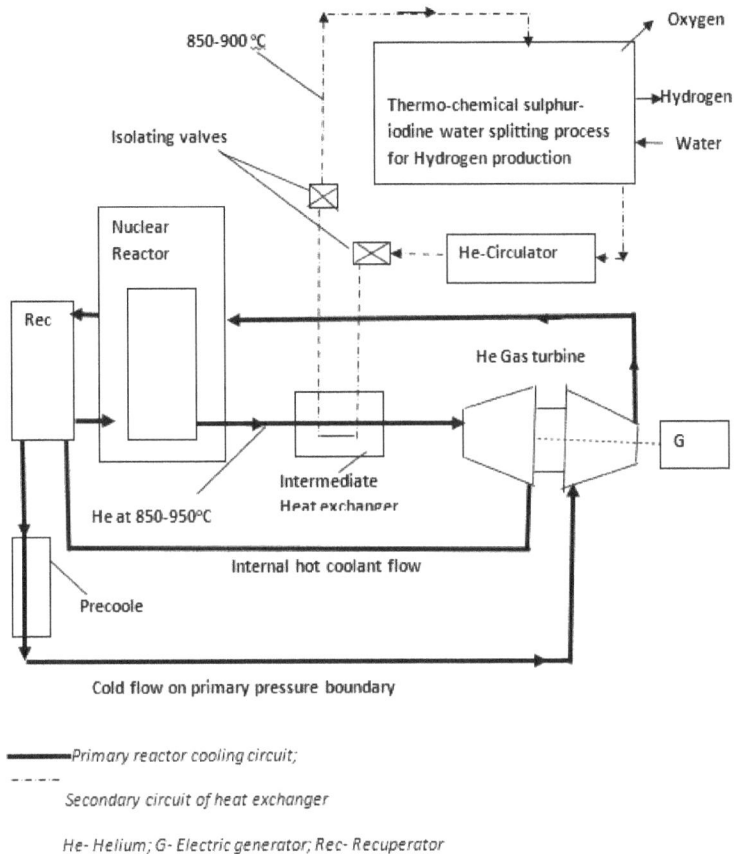

Fig. 7.4. Thermochemical Hydrogen Production by Heat from Gas Cooled High Temperature Nuclear Reactor.

7.7 Biomass Conversion to Hydrogen

Biomass conversion to hydrogen is quite complex. There can be several options for producing hydrogen from biomass, such as:

 i. Gasification

 ii. Fast pyrolysis

iii. Advanced gasification by breaking down the wastes by plasma torches

iv. Initial conversion of lignocellulosic materials to ethanol and then to hydrogen

 v. Hydrogen production through gasification of wet biomass by 'Super critical water'

vi. Biological production of hydrogen

7.8 Hydrogen Production from Biomass through the Gasification Route [109,154,218]

7.8.1 Thermal Gasification of Biomass

Biomass is subjected to elevated temperatures in the environment of pressurised 'Steam or air / oxygen' to break down organic matter into 'hydrogen and CO or CO_2'. These gases are accompanied by undesirable solids and gaseous by-products, which are to be removed. The 'Use of catalysts' improves the efficiency of the process. The carbon monoxide then reacts with water to form carbon dioxide and more hydrogen via a water-gas shift reaction. The 'hydrogen' can be separated after cleaning the gases by the membrane process or the chemical process. The gasification refineries should be large enough to achieve an economy of scale. *[Also see Para. 5.6, chapter 5, and Para. 6.6, chapter 6, for the biomass gasification technologies]. The chemical reaction will be like:*

Cellulose $+ O_2 + H_2O \rightarrow CO + CO_2 + H_2 +$ other species

Water-gas shift reaction

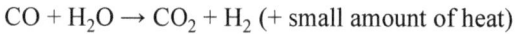

$CO + H_2O \rightarrow CO_2 + H_2$ (+ small amount of heat)

As biomass gasification is a mature technology, the following factors can make it a viable pathway for cost-competitive hydrogen production.

a. Biomass is an abundant domestic resource.

b. Biomass "recycles" carbon dioxide as plants consume carbon dioxide from the atmosphere as part of their natural growth process as they make biomass.

c. Improved agricultural practices and breeding efforts should result in low and stable feedstock costs.

d. Research focuses on overcoming the following challenges to lower capital costs:
   • Replacing the cryogenic process currently used to separate oxygen from air when oxygen is used in the gasifier with new membrane technology.
   • Developing new membrane technologies to better separate and purify hydrogen from the gas stream.
   • Combining steps into fewer operations.

7.8.2 *Gasification of Biomass by 'Supercritical Water'* [50,131,189]

Supercritical water (at a temperature of 647.29 K and pressure of 22.09 MPa) shows intermediate properties of a liquid and gaseous state. The gasification of wet biomass is *very inefficient*, but its gasification by high-pressure hot water is attractive, especially for raw sewage and aquatic biomass. The biomass decomposes rapidly by the treatment by supercritical water into syngas. A very low amount of char and tar are formed in the process. The supercritical water gasification can be carried out in the pressure and temperature range of 25–40 MPa/400–750°C. The product gas will contain hydrogen, CO_2, CO, methane, ethane, propane, and other hydrocarbons. All these hydrocarbons can be reformed to 'Hydrogen'. Supercritical reforming to hydrogen yields better results, but the energy costs involved are very high.

7.8.3 *Thermochemical Method—Fast Pyrolysis* [153]

Biomass is converted via thermal decomposition (fast pyrolysis) to 'bio-oil,' and small amounts of solid products in the absence of O_2 at temperature of 400–600°C and a vapor residence time of 0.5–3 s at a central place. The bio-oil can then be transported to several 'Bio refineries'. This bio-oil (liquid or its vapor) is subjected to 'Catalytic steam- reforming' to obtain 'Hydrogen'. This approach is better for smaller distributed 'Bio-refineries'.

7.8.4 *Advanced Gasification by Breaking down the Wastes by Plasma Torches*

This has already been described in *Para. 6.6.4, Chapter 6*. The flow diagram for the production of hydrogen from biomass through the gasification route is shown in Figure 7.5.

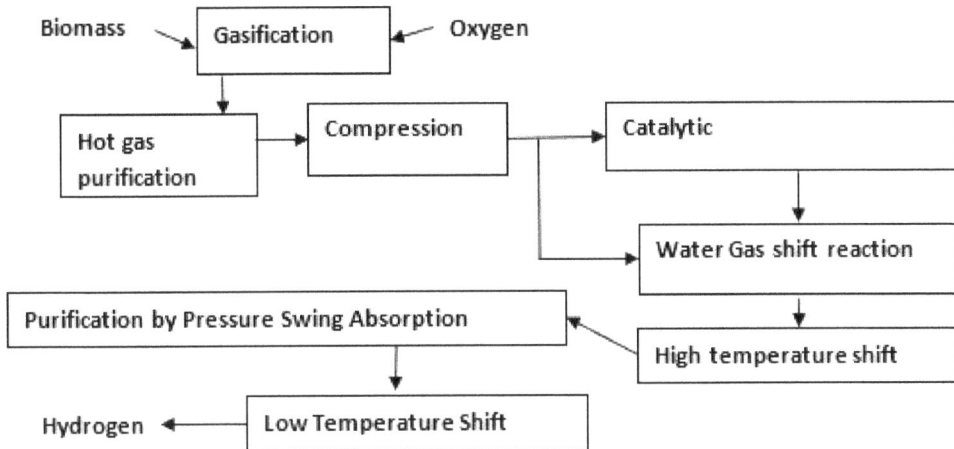

Fig. 7.5. Hydrogen Production from Biomass Through Gasification Route.

7.9 Hydrogen from Glycerol [19,183]

Biodiesel—motor fuel derived from vegetable oil—is a renewable alternative to rapidly depleting fossil fuels. It is biodegradable and nontoxic, and its production is on the rise. However, for each molecule of biodiesel produced, another of low-value crude glycerol is generated, and its disposal presents a growing economic and environmental problem. Besides hydrogen being a green fuel, it is in great demand for use in the production of fertilizers, chemicals and food products, etc.

It has been shown by researchers that glycerol can be converted to produce a hydrogen-rich gas. The novel process developed by Dr. Valerie Dupont and her co-investigators by the University's Faculty of Engineering, mixes glycerol with steam at a controlled temperature and pressure, separating the waste product into hydrogen, water, and CO_2, with no residues. A special absorbent material filters out the CO_2, which leaves a much purer product.

7.10 Hydrogen Production from Biomass Gasification by Disposal of Emitted CO_2 [35,37,54]

The production of hydrogen or biogas through the gasification of biomass or any other carbon-based fuel entails the emission of CO_2. This CO_2 can be separated during the production of hydrogen from biomass and can be compressed, used, or safely disposed. Figure 7.6 shows a typical scheme for hydrogen production and the recovery of CO_2 for use.

The recovered compressed CO_2 can be used for injection in oil fields to enhance oil recovery (EOR). The injected CO_2 gets sealed in the underground oil layer. This process would also help fight against the increasing CO_2 level in the atmosphere.

Note: on the use of CO_2 for enhanced oil recovery and disposal of CO_2 in voids created by the 'Recovery of oil and gas'

It is well known that CO_2 extinguishes fire and therefore, the recovered CO_2 is being used in its compressed form for safe injection in oil fields to enhance the oil recovery, a process dubbed as 'EOR'. The use of CO_2 for enhanced oil recovery (EOR) helps the efforts of reducing 'CO_2 in atmosphere' by sealing the injected CO_2 into the 'underground oil layer'. The CO_2 is also useful for injection in oil fields for the maintenance of pressure. The enhanced oil recovery (EOR) would need to be practiced widely because of the very high cost of oil/petroleum.

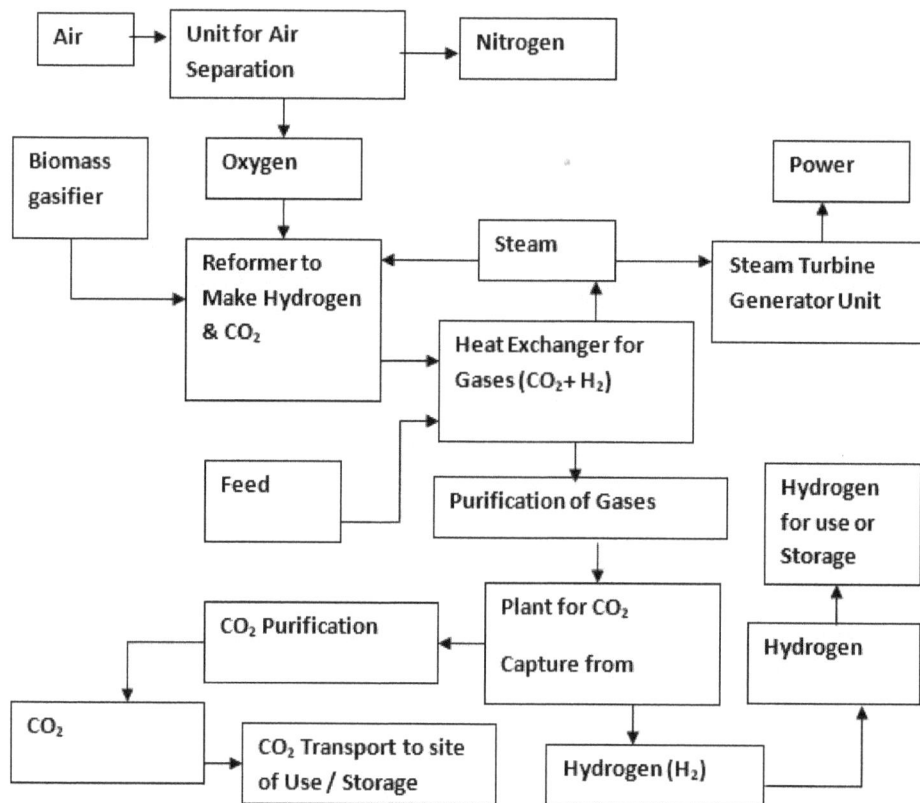

Fig. 7.6. Reforming Gases Produced by Biomass Gasifier into Hydrogen Using O_2 with CO_2 Capture and Power Generation from Generated Steam.

USA has been using the EOR technique since 1972. In USA, millions of tonnes of CO_2 have been used for injection in declining oil fields to enhance oil recovery. In Texas alone, there were more than 10,000 wells (2006) where CO_2 was injected for the enhanced recovery of oil.

Oil/gas fields have 'alternate laminations of porous and dense layers' where the oil and gas have collected. After taking out gas or oil from the 'field', the voids can be filled with compressed CO_2 if the field has a 'firm cap rock' over the porous layer. Such a disposal is possible in a porous layer without oil/gas where water has filled as the compressed CO_2 would push the water layer and occupy the space in such an aquifer, but a 'firm cap rock' must exist. It has become possible to store CO_2 in voids created in depleted oil and gas fields such as in 'Norway's Sleipner West Gas Field' and 'Indonesia's Natuna Gas Field'.

In 2000, a synthetic gas manufacturing plant (from coal) at Beulah, North Dakota (USA), became the first coal-fired plant to capture and store CO_2.

As per a study in 1996, the UK found that they had space for 5,300 million tons of CO_2 in depleted oil and gas fields, which was about ten years' capacity for storage of the CO_2 emissions of the UK.

In a study in the UK, they found that they can store CO_2 in deep saline aquifers to the extent of 716,000 million tons of CO_2. This capacity is sufficient for them for storage for 500 years. A geological formation 1 km below the seabed, sited above the Norwegian Sleipner field was found.

The geochemical trapping mechanism will prevent the escape of CO_2 to the surface if there is an impervious rock cap on the storage formation.

7.11　Initial Conversion of Lignocellulosic Materials to Alcohols and then to Hydrogen [119]

The 'Ethanol' or butanol produced from lignocellulosic material can be reformed to hydrogen. *[See Para 3.4, 3.5, 3.6 and 3.9, Chapter 3.]* Various types of lignocellulosic feedstocks are available in different regions and this route of hydrogen production may become viable.

Production of hydrogen from butanol is a promising alternative when it is obtained from biobutanol or bio-oil due to the higher hydrogen content compared to other oxygenates such as methanol, ethanol, or propanol. Catalysts and operating conditions play an important role in hydrogen production. Nickel and rhodium are mainly used for butanol steam reforming, oxidative steam reforming, and partial oxidation. Catalytic activity can be increased manyfold by additives such as Copper. The steam reforming technique is an option that is more frequently used due to the higher hydrogen production capability in comparison to other thermochemical techniques.

7.12 Hydrogen Production from Biomass through Biological Fermentation [118,166,226]

In this method, hydrogen is produced by the 'dark fermentation' of organic substrate by a diverse group of bacteria using multienzyme systems involving steps similar to the anaerobic process.*

In this process, 'anaerobic micro-organisms' are used to produce hydrogen directly, much in the same way that bacteria or yeast can produce ethanol via fermentation. There are a number of microorganisms that can produce hydrogen directly. The performance of hydrogen production through dark fermentation can be improved by the use of genetic modification of such microorganisms by 'genetic engineering'.

Such organisms typically start to work on 'monomeric sugars'. Therefore, the initial task is to convert lignocellulosic material to sugars. The pretreatment and hydrolysis techniques have been developed to break down 'Cellulose/hemi cellulose' into sugars, as already discussed in *Para 3.4*, and *Para 3.5, Chapter 3* in connection with the production of alcohol from biomass/cellulose. Once the biomass is converted to sugars, its fermentation to hydrogen by the dark fermentation pathway may be done by the aid of special microorganisms. It may be mentioned that the *dark fermentation* reactions do not require light energy, so they are capable of constantly producing hydrogen from organic compounds throughout the day and night. [*Photofermentation differs from dark fermentation because it only proceeds in the presence of light.*]

The fermentative production of hydrogen is fundamentally dependent on hydrogen-evolving enzymes. Hydrogenases constitute a family of enzymes found throughout the bitota. These enzymes catalyze the reversible oxidation of hydrogen gas: $H2 \leftrightarrow 2H^+ + 2e^-$. There are hosts of organisms of a hydrogen producing hydrogenase, such as the saccharolytic species of clostridia, *Clostridium pasteurianum, Clostridium saccharobutylicum, Clostridium thermolacticum,* * *etc.*

It has been found that addition of some substances to the growth medium can change the conversion of cellulose to desired products as indicated below:

(i) Addition of ethanol to the growth medium at the initiation of the fermentation process results in significant increases in hydrogen and acetate.

(ii) Formate addition increases hydrogen and ethanol and decreases acetate production.

(iii) The addition of CO in the bioreactor headspace leads to lower synthesis of hydrogen, CO_2, and acetate, but significantly increases ethanol synthesis.

Production of hydrogen using the fermentative biological process seems to potentially be one of the most attractive strategies as it is not as energy intensive as other means and could potentially utilize refuse or agricultural waste streams as the raw material.

Scientist at the Agrotechnological Research Institute, Netherlands, found that the agricultural waste fermented in the anaerobic reactor by the extremophilic bacteria, '*Caldicellulosiruptor saccharolyticus*' to release hydrogen and also form acetates. The brew of acetates, when transferred to a glass- encased reactor and after the action of the photosynthetic bacteria *Rhodopseudomonas species*, this brew of acetates breaks down into hydrogen and carbon dioxide.

Attempts are also being made to create a new organism capable of producing hydrogen from organic waste through bioengineering.

Scientist at the Murugappa Chetiyar Research Centre (MCRC), Chennai, are engaged in working in the area of biological production of hydrogen from organic distillery waste by using a cocktail of bacteria.

Microwave-assisted Production of Hydrogen from Rice straw

Rice straw is pretreated with microwaves and dilute caustic soda/sulphuric acid, and then sugar is released by enzymatic hydrolysis. The sugar is converted to hydrogen by bacterial dark fermentation. The microorganisms mentioned in the previous paragraphs can be used. It has been found that bacterial strain '*Bacillus coagulans* 2323' can also be used.

Notes

(i)* Clostridium thermocellum: Clostridium thermolacticum (C. thermocellum) is a thermophilic bacterium (optimum growth at 60°C) that utilizes cellulose as the sole carbon source and carries out mixed product fermentation, synthesizing various amounts of lactate, formate, acetate, ethanol, H_2, and CO_2 under different growth conditions. C. thermocellum expresses a cellulosome on its surface and is known to ferment pretreated hardwood or avicel (crystalline cellulose) in batch and continuous cultures.

(ii) Hydrogen Production from the Microwave-assisted Pretreatment of Rice Straw
Microwave-assisted chemically pretreated rice straw is treated with enzymes to release sugars. These sugars are then fermented in the absence of light by a suitable bacterial culture for conversion to hydrogen. Hydrolyzate can be used in the production of hydrogen via dark fermentation. (Caustic soda/sulphuric acid/hydrogen peroxide may be used as chemicals.)

7.13 Hydrogen Production from Fossil Fuels [119]

Fossil fuels are the dominant source of industrial hydrogen. As of 2020, the majority of hydrogen (~ 95%) was produced from fossil fuels by the steam reforming of natural gas, partial oxidation of methane, and coal gasification.

7.13.1 Steam Methane Reforming

Steam reforming is a method of hydrogen production from natural gas. Presently, this is the cheapest source of industrial hydrogen. The process is 65–75 % efficient and has established infrastructure. The process consists of heating the gas to between 700–1100°C in the presence of steam and a Ni catalyst. The resulting endothermic reaction breaks up the methane molecules and forms CO and H_2. The CO is then passed with steam over the iron oxide or other oxides and undergoes a water-gas shift reaction to yield more quantity of H_2. The negative aspects of this process are its by-products, namely the atmospheric release of CO_2, CO, and other greenhouse gases. Based on the quality of the feedstock (natural gas, rich gases, naphtha, etc.), one tonne of hydrogen produced will also produce 9 to 12 tonnes of CO_2, which may be captured.

In this process, high temperature (700–1100°C) steam (H_2O) reacts with methane (CH_4) in an endothermic reaction to yield syngas.

$$CH_4 + H_2O \rightarrow CO + 3\ H_2$$

In the 2nd stage, additional hydrogen is generated through the lower-temperature, exothermic, water-gas shift reaction, at about 360°C:

$$CO + H_2O \rightarrow CO_2 + H_2$$

Essentially, the oxygen (O) atom is stripped from the additional water (steam) to oxidize CO to CO_2. This oxidation also provides the energy to maintain the reaction. The additional heat required to drive the process is generally supplied by burning some portion of the methane.

7.13.2 Methane Pyrolysis

The pyrolysis of methane is another process of producing hydrogen from natural gas. The separation of hydrogen occurs in one step via flow through a molten metal catalyst in a 'bubble column'. This is a 'no greenhouse gas' approach for potentially low-cost hydrogen production. It has the capability of scaling up for operations. The process is conducted at high temperatures (1065°C or 1950°F).

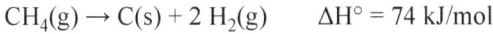

$$CH_4(g) \rightarrow C(s) + 2\ H_2(g) \qquad \Delta H° = 74\ kJ/mol$$

The industrial quality solid carbon can be used as manufacturing feedstock or for landfill. It is not released into the atmosphere and no groundwater pollution takes place.

7.13.3 Partial Oxidation of Natural Gas or Hydrocarbons

Hydrogen production from natural gas or other hydrocarbons can be achieved by partial oxidation. A fuel-air or fuel-oxygen mixture is partially combusted resulting in a hydrogen-rich syngas. Hydrogen and CO are obtained via the water-gas shift reaction. Carbon dioxide can be co-fed to lower the hydrogen to CO ratio.

Partial oxidation occurs when a sub-stoichiometric fuel-air or fuel-oxygen mixture is partially combusted in a reformer or partial oxidation reactor. A distinction is made between *thermal partial oxidation* (TPOX) and *catalytic partial oxidation* (CPOX). The chemical reaction takes the general form as:

$$C_nH_m + {}^n/_2\ O_2 \rightarrow n\ CO + {}^m/_2\ H_2$$

Idealized examples for heating oil and coal, assuming compositions $C_{12}H_{24}$ and

$C_{24}H_{12,}$ respectively, are as follows:

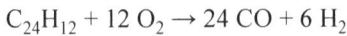

$$C_{12}H_{24} + 6\ O_2 \rightarrow 12\ CO + 12\ H_2$$
$$C_{24}H_{12} + 12\ O_2 \rightarrow 24\ CO + 6\ H_2$$

7.13.4 Plasma Reforming of Liquid Hydrocarbons

The Kværner-process is a plasma reforming method, developed in the 1980s by a Norwegian company of the same name, for the production of hydrogen and carbon black from liquid hydrocarbons (C_nH_m). Approximately 48% of the available energy of the feed is contained in hydrogen, 40% in activated carbon, and 10% in superheated steam. CO_2 is not produced in the process.

A variation of this process was presented in 2009 using plasma arc waste disposal technology for the production of hydrogen, heat, and carbon from methane and natural gas in a plasma converter.

7.13.5 Coal Gasification

In order to produce hydrogen from coal, coal gasification is used. The process of coal gasification uses steam and oxygen at high temperature and pressure to break molecular bonds in coal and form a gaseous mixture of hydrogen and CO. The process is 45% efficient and produces CO_2 emissions. Carbon dioxide and other pollutants are easily removed from the gas obtained from coal gasification. Another method for conversion is low-temperature and high-temperature coal carbonization. Carbon sequestration, however, would increase costs.

The coke oven gas made from pyrolysis (oxygen-free heating) of coal has about 60% hydrogen, the rest being CH_4, CO, CO_2, NH_3, N_2 and H_2S. Hydrogen can be separated from other impurities by the pressure swing adsorption process. Japanese steel companies have carried out production of hydrogen by this method.

7.13.6 Petroleum Coke

Petroleum coke can also be converted to hydrogen-rich syngas via coal gasification. The produced syngas consists mainly of hydrogen, carbon monoxide, and H_2S from the sulfur in the coke feed. Gasification is an option for producing hydrogen from almost any carbon source.

7.14 Brief Summary of Hydrogen Production Methods

Table 7.3 gives a brief summary of the various methods used for producing hydrogen. There are four main sources for the commercial production of hydrogen—natural gas, oil, coal, and electrolysis—that account for 48%, 30%, 18%, and 4% of the world's hydrogen production, respectively. Other methods of hydrogen production included biomass gasification, no CO_2 emission, methane pyrolysis, and electrolysis of water. Details on these methods are given in the relevant subsections.

Table 7.3. Summary of the Various Methods of Hydrogen Production. [2,3,4,7,14,16,19,20,23,25,29,30,31,34,35,37,38,41, 42,45,50,51,54,55,56,68,73,74,77,78,79,80,82,84,85,86,87,102,104,109,114,115,116,117,118,119,130,131,133,134,136, 137,138,139,142,143,144,145,147,148,150,153,154,157,158,160,166,168,174,175,179,180,181,183,184,186,187,188,189, 197,199,201,202,203,206,207,208,209,216,218,225,226,228,229,230,231,232,234,238, 239]

Se. No.	Method	Brief Details of the Method
Hydrogen from Water by Water splitting (Electrolysis)- Not in widespread use due to the high cost of electricity		
1.	Standard electrolysis of water	Water containing a small quantity of H_2SO_4 or KOH of 10 to 25% strength is electrolysed in a cell using an iron sheet as the cathode and nickel-plated iron sheet as the anode. The anode and cathode are separated from each other by a porous diaphragm. On passing a direct current through the cell, oxygen is liberated at the anode and the hydrogen is liberated at the 'Cathode'. Actual average consumption of power in practice with the best efficiency achievable is of the order of 48—50 kWh/kg of hydrogen. The energy consumption in electrolysis falls with the increase in surface area of electrodes and pair of electrodes.
	Hydrogen from brine electrolysis	Large quantities of H_2 and Cl_2 are obtained during the production of NaOH by the electrolysis of brine.
2.	High pressure electrolysis	Hydrogen can be generated by 'High pressure electrolysis'. Compressed hydrogen output is at around 120–200 bar. The porous diaphragm separating the anode and the cathode presents a problem (the mixing of O_2 and H_2 and the safety issues need to be tackled). The hydrogen's purity also decreases on increasing the pressure.
3.	Proton exchange membrane water electrolysis	In this method, both the electrodes are pressed against a proton conducting polymer electrolyte, forming a membrane electrode assembly that is immersed in de-ionized water. When the current is passed, the splitting of water takes place, the hydrogen ions move across the membrane and appear at the cathode, while the O_2 evolves at the anode. The most popular membrane material is 'Nafion' (co-polymer of tetrafluoroethylene). Despite the protons remaining inside the membrane, the acidity of the material is high and the noble metal catalysts are required at both the anode and cathode. The catalyst is deposited on the surface of the membrane with the ion-exchange electrolyte. The catalyst at anode and cathode may be Ir and Pt, respectively.
4.	High temperature electrolysis	This method uses a high-temperature electrolyzer, which is a 'Solid-oxide water electrolysis cell', where the electrolyte is zirconia (ZnO_2) stabilized with yttrium and scandium oxides. The solid oxide electrolysis cell normally works in the temperature range of 800–1000°C. The bi-polar plates are usually of stainless steel and the solid electrolyte is manganite-coated stabilized zirconia. The oxide ions diffuse across the zirconia electrolyte. Thin ceramic membranes are required for the electrolysis process to reduce resistance.

Table 7.3 contd. ...

...Table 7.3 contd.

Se. No.	Method	Brief Details of the Method
5.	Hydrogen production by biocatalyzed electrolysis	Biocatalyzed electrolysis is a novel biological hydrogen production process with the potential to efficiently convert a wide range of dissolved organic materials in wastewaters. Even substrates formerly regarded to be unsuitable for hydrogen production due to the endothermic nature of the involved conversion reactions can be converted with this technology. Biocatalyzed electrolysis achieves this by utilizing electrochemically active microorganisms that are capable of generating an electrical current from the oxidation of organic matter. When this biological anode is coupled to a proton reducing cathode by means of a power supply, hydrogen is generated.
	Solar Photolysis for Hydrogen Production Solar photolysis involves the use of 'Solar photons' to produce hydrogen directly via 'photoelectrochemical systems' or 'photobiological systems'.	
6.	Photoelectrochemical (PEC) Production of Hydrogen	Solar PEC photolysis involves the use of 'Solar photons' to produce H_2 directly via: 'PEC Systems'. It involves the dissociation of water into hydrogen and oxygen directly at the surface of the semiconductor through irradiation by 'solar photons'. The semiconductor material acts as a 'catalyst' to produce hydrogen directly at the 'semiconductor and water surface interface'.
7.	Photobiological Method of Hydrogen Production	The photobiological methods use photosynthetic organisms like some 'cyanobacteria' species (blue green algae) and 'green algae' to photo-produce H_2. Hydrogen production in green algae occurs within the chloroplast and it is a light-dependent process. It may be noted that 'Hydrogenase enzymes' catalyze production in this prototroph (green algae). The algal hydrogenase has an iron-iron cluster (Fe-hydrogenase) at its catalytic site, while the cyanobacteria enzyme contains a nickel-iron cluster. The algae for its growth does not need any carbon-based material as it can take CO_2 and nitrogen from air. Other needs for growth are water, minerals (available in water itself), and light energy.
8.	Photoelectrocatalytic Production of Hydrogen	In this method, a 'gold electrode' covered in layers of 'Indium phosphide (InP) Nano particles' is used with the introduction of an iron-sulphur complex into the layered arrangement. This when submerged in water and irradiated with light under small electric current, produces hydrogen with an efficiency of 60%.
	Thermolysis of Water: Thermochemical Production of Hydrogen – Thermochemical splitting of water into hydrogen and oxygen can be done by a series of chemical reactions driven by thermal energy. This is a long-term technology pathway, with potentially low or no greenhouse gas emissions.	
9.	Hydrogen from Gas Cooled Nuclear Reactors	'Nuclear heat' from a Nuclear Power Plant can be directly used to split hydrogen from water. High temperature (950–1000°C) gas cooled nuclear reactors have the potential to split hydrogen from water by 'Thermochemical' reaction. This method eventually may be economical. High temperature outlet helium after cooling the reactor core can impart its heat to a secondary Helium circuit in a heat exchanger. The secondary heated fluid can be linked to the production of hydrogen by splitting of water through sulphur-iodine cycle.

Table 7.3 contd. ...

...Table 7.3 contd.

Se. No.	Method	Brief Details of the Method
10.	Hydrogen by Solar Thermolysis or Thermochemical Production of Hydrogen	A concentrated solar heat source can provide for 'Solar Thermolysis' which drives thermochemical reactions for production of hydrogen, or drive electrolysis at a very high temperature for more efficient decomposition of water into oxygen and hydrogen. As very high temperatures are required to dissociate water into hydrogen and oxygen, a catalyst is required to make the process operate at lower temperatures. The sulfur–iodine cycle (S–I cycle) is a three-step thermochemical cycle used to produce hydrogen. The S–I cycle consists of three chemical reactions whose net reactant is water and whose net products are hydrogen and oxygen. All other chemicals are recycled. The S–I process requires an efficient source of heat.
	Hydrogen Production from Biomass through Gasification Route	
11.	Thermal Gasification of Biomass—Hydrogen Production	Biomass is subjected to elevated temperatures in the environment of pressurised 'steam or air /oxygen' to break down organic matter into 'hydrogen and CO or CO_2'. These gases are accompanied by undesirable solids and gaseous by-products ,which have to be removed. The carbon monoxide then reacts with water to form carbon dioxide and more hydrogen via a water-gas shift reaction. The 'hydrogen' can be separated after cleaning the gases by a membrane process or by a chemical process. The chemical reaction will be like: Cellulose $+ O_2 + H_2O \rightarrow CO + CO_2 + H_2$ + other species Water-gas shift reaction $CO + H_2O \rightarrow CO_2 + H_2$ (+ small amount of heat) Biomass gasification is a mature technology. It can be a viable alternative for cost-competitive hydrogen production.
12.	Hydrogen Production by Gasification of Biomass by 'Supercritical Water'	The gasification of wet biomass is very inefficient, but its gasification by supercritical water (at 647.29 K and 22.09 MPa) is attractive, especially for raw sewage and aquatic biomass. The biomass decomposes rapidly into syngas. A very low amount of char and tar are formed in the process. This gasification can be carried out in the pressure and temperature range of 25–40 MPa; 400–750°C. The product gas will contain hydrogen, CO_2, CO, CH_4, C_2H_6, C_3H_8, and other hydrocarbons (HCs). All these HCs can be reformed to Hydrogen. Supercritical reforming to hydrogen yields better results, but the energy costs involved are very high.
13.	Thermo Chemical Method - Fast Pyrolysis	Biomass is converted to 'bio-oil' via thermal decomposition (fast pyrolysis) at a central place. The bio-oil is then transported to several 'biorefineries' and subjected to 'catalytic steam reforming' to obtain H_2. This approach is better for smaller distributed 'biorefineries'.
14.	Advanced gasification by breaking down the wastes by plasma torches	The waste is shredded and is reacted with oxygen or air with the help of plasma torches in a reactor. It is not burnt but actually gets broken down to its elemental components in their gaseous form. Any type of waste material can be gasified which, otherwise, is difficult to be gasified. The resulting syngas is cooled and cleaned of dust. The syngas can be used as fuel directly or may be further converted to other forms of fuels. The CO in the syngas when reacted with steam can produce hydrogen (See Para 7.13.1).
15.	H_2 production from biomass gasification by disposal of emitted CO_2	Production of hydrogen or biogas through the gasification of biomass or any other carbon-based fuel entails the emission of CO_2. This CO_2 can be separated, compressed, and sent for suitable use during the production of hydrogen from biomass.

Table 7.3 contd. ...

...Table 7.3 contd.

Se. No.	Method	Brief Details of the Method
16.	Hydrogen from glycerol—a by-product in production of biodiesel	For each molecule of biodiesel (derived from vegetable oil), another molecule of crude glycerol is generated. This glycerol can be mixed with steam at a controlled temperature and pressure and converted to produce a hydrogen rich gas separating the waste product into water, and CO_2 with no residues. A special absorbent material filters out the CO_2, which leaves a much purer product.
17.	Initial conversion of lignocellulosic materials to alcohols and then to H_2	Alcohol can be made from sugar-containing materials, starch or starch-containing materials, and even lignocellulosic biomass. Various types of lignocellulosic feedstocks are available in different regions. The process for producing ethanol from biomass through biological fermentation involves pretreatment, hydrolysis, fermentation, and product purification. Ethanol or butanol produced can be reformed to hydrogen. This method of hydrogen production may become viable. The production of hydrogen from butanol is a promising alternative when it is obtained from biobutanol or bio-oil due to the higher hydrogen content compared to other oxygenates such as methanol, ethanol, or propanol.
18.	Production of H_2 from biomass through biological fermentation	The lignocellulosic material is converted to sugars, which are fermented to by a dark fermentation pathway by the aid of special microorganisms. These reactions do not require light energy, so they are capable of constantly producing hydrogen from organic compounds throughout the day and night. The fermentative production of hydrogen is fundamentally dependent on hydrogen-evolving enzymes.
	Microwave- assisted production of Hydrogen from rice straw	Rice straw is pretreated with microwaves and dilute $NaOH/H_2SO_4$ to release sugar by enzymatic hydrolysis. The sugar is converted to hydrogen by bacterial dark fermentation.
	Hydrogen Production from fossil fuels	
19.	Steam methane reforming	Steam reforming is currently a major source of industrial hydrogen, apart from being cheapest. The process is 65–75 % efficient and has established infrastructure. It consists of heating the gas to between 700–1100°C in the presence of steam and a Ni catalyst. The resulting endothermic reaction breaks up the methane molecules and forms CO and H_2. The CO is then passed with steam over iron oxide or other oxides and undergoes an exothermic water-gas shift reaction at 360°C to yield more H_2.
20.	Methane pyrolysis	The pyrolysis of methane is another process of producing hydrogen from natural gas. The separation of hydrogen occurs in one step via flow through a molten metal catalyst in a 'bubble column'. This is a 'no greenhouse gas' approach for a potentially low-cost of hydrogen production. It has the capability of scaling up for operations. The process is conducted at high temperatures (1065°C or 1950°F). $CH_4(g) \rightarrow C(s) + 2\,H_2(g)$ $\Delta H° = 74$ kJ/mol The industrial-quality solid carbon can be used as manufacturing feedstock or for landfill. It is not released into the atmosphere and no groundwater pollution takes place.
21.	Partial oxidation of natural gas or hydrocarbons	Hydrogen production can be achieved by partial oxidation of natural gas or other hydrocarbons. A fuel-air or fuel-oxygen mixture is partially combusted, resulting in hydrogen-rich syngas. Hydrogen and CO are obtained via the water-gas shift reaction. Carbon dioxide can be co-fed to lower the hydrogen to CO ratio. Partial oxidation occurs when a sub-stoichiometric fuel-air or fuel-oxygen mixture is partially combusted in a reformer or partial oxidation reactor. $C_nH_m + \frac{n}{2}\,O_2 \rightarrow n\,CO + \frac{m}{2}\,H_2$

Table 7.3 contd. ...

...Table 7.3 contd.

22.	Plasma reforming of liquid hydrocarbons	The Kværner process is a plasma reforming method for the production of hydrogen and carbon black from liquid hydrocarbons (C_nH_m). Approximately 48% of the available energy of the feed is contained in hydrogen, 40% in activated carbon, and 10% in superheated steam. CO_2 is not produced in the process. A variation of this process was presented in 2009 using plasma arc waste disposal technology for the production of hydrogen, heat, and carbon from methane and natural gas in a plasma converter.
23.	Coal gasification	Coal gasification is used to produce hydrogen from coal. The process uses steam and oxygen at high temperature and pressure to break molecular bonds in coal and form syngas—a gaseous mixture of hydrogen and CO. Carbon dioxide and other pollutants are easily removed from the gaseous mixture. Another method for conversion is low-temperature and high-temperature coal carbonization. Coke oven gas made from the pyrolysis (oxygen-free heating) of coal has about 60% hydrogen, the rest being CH_4, CO, CO_2, NH_3, N_2, and H_2S. Hydrogen can be separated from other impurities by the pressure swing adsorption process.
24.	Petroleum coke	Petroleum coke can also be converted to hydrogen-rich syngas via coal gasification. The produced syngas consists mainly of hydrogen, carbon monoxide, and H_2S from the sulfur in the coke feed. Gasification is an option for producing hydrogen from almost any carbon source.

*Note: Hydrogen producers categorize hydrogen according to the energy sources used for its production. For example, hydrogen produced using renewable energy might be referred to as **renewable hydrogen** or **green hydrogen**. Hydrogen produced from coal may be called **brown hydrogen**, and hydrogen produced from natural gas or petroleum might be referred to as **grey hydrogen**. **Brown** or **grey hydrogen** production combined with carbon capture and storage/sequestration might be referred to as **blue hydrogen**.*

7.15 Hydrogen Separation [8,24,26,57,125,126,156,164,169,176,195]

Before the generated hydrogen gas can be put to some use, it is necessary that it is separated from the other gases present in it. The impurities present in the gas/gases generated depends on the method of production and resource used for production. There are four technologies that are commonly used for the separation of hydrogen and purification of hydrogen.

a. Separation by 'pressure swing adsorption,' that is, the adsorption of gases other than hydrogen, leaving pure hydrogen in the end.

b. Physical or chemical recovery of CO_2 from the mixture, leaving hydrogen with a small amount of impurities and producing pure CO_2 as a by-product.

c. Membrane separation.

d. Cryogenic separation, which can produce multiple pure by-products along with pure hydrogen.

7.15.1 Separation by Adsorption (PSA) [120,169,190,195,210,235]

The pressure swing adsorption (PSA) technology is based on the physical binding of gas molecules to an adsorbent material. The force of attraction acting between the gas molecules and the adsorbent material depends on the property of the gas component, type of adsorbent material, partial pressure of the gas component, and operating temperature. The force that causes the adsorption varies for different gases. The increasing order of adsorption for different gases is as follows:

Hydrogen; oxygen; argon; nitrogen; carbon monoxide; methane; carbon dioxide; ethane; ethylene; propane; butane; propylene; ammonia; hydrogen sulphide; water vapor.

The separation of the gases is based on the different behavior of the binding forces vis-à-vis the adsorbent material. Highly volatile components with low polarity, such as hydrogen, are practically non-adsorbable as opposed to molecules of nitrogen, CO, CO_2, hydrocarbons, and water vapor. These impurities can be adsorbed from a stream of gases containing hydrogen, and thus, high-purity hydrogen can be obtained.

a. Adsorption and Regeneration

The PSA process works at a fairly constant temperature and uses the effects of alternating pressure and partial pressure to perform adsorption and desorption. No heating or cooling is required, resulting in short cycles (within a range of minutes), and impurities can be removed economically and quite efficiently.

The adsorption decreases as the temperature rises, while it increases with increase in partial pressure of the component gas in the mixture and the pressure of the gas mixture.

Adsorption is carried out at a high pressure (and hence, high respective partial pressure) typically in the range of 10 to 40 bar until the equilibrium loading of the adsorption material (saturation level at that pressure) is reached. No further adsorption capacity is available and the adsorbent material requires regeneration. The regeneration of this material is done by decreasing the pressure to slightly above atmospheric pressure, which results in a respective decrease in equilibrium loading and desorption of the impurities from the adsorbent material, thus restoring the adsorption capacity of the material. The amount of impurities removed from a gas stream within one cycle corresponds to the difference of adsorption to desorption loading. After termination of regeneration, the pressure is increased back to the adsorption pressure level and the process starts again from the beginning.

A plant for the PSA system would basically consist of the adsorber vessels containing the adsorbent material, tail gas drum(s), valve skid(s) with interconnecting piping, control valves and instrumentation, and the control system.

b. Sequence of operation in PS

The pressure swing adsorption process has four basic process steps:

- Adsorption
- Depressurization
- Regeneration
- Repressurization

At least four adsorber vessels would be required for providing a continuous supply of hydrogen. The sequence of operation in four vessels is shown in the following Figure 7.7.

A Vessel: Adsorption — E_1 —   PP — D — Regeneration   — R_1/R_o

B Vessel: R_1/R_o — Adsorption — E_1 —   PP   —   D   — Regeneration

C Vessel: D — Regeneration — R_1/R_o —   Adsorption — E_1 —   PP

D Vessel: E_1 —   PP — D — Regeneration   — R_1/R_o — Adsorption

Fig. 7.7. Sequence of Operations in a Four Vessel PSA System.

c. Adsorption

The adsorption of impurities is carried out at high pressure, which is determined by the pressure of the feed gas. The feed gas flows through the adsorber vessels from the bottom to the top, and the impurities like water vapor, heavy hydrocarbons, light hydrocarbons, CO_2, CO, and nitrogen are selectively adsorbed on the surface of the adsorbent material. The pure hydrogen exits from the top of the adsorber vessel. After a defined time, the adsorption phase of this vessel stops and regeneration starts. Another adsorber vessel takes over the task of adsorption to ensure a continuous hydrogen supply.

d. Regeneration

The regeneration phase consists of five consecutive steps namely:

Pressure equalization (E_1), Providing of the purge (PP), Dumping (D), Purging and Repressurization R_1/R.

1. *Pressure equalization (step E1)*—Depressurization starts in the co-current direction from the bottom to the top. The hydrogen still stored in the void space of the adsorbent material in this vessel is used to pressurize another adsorber vessel having just finished its regeneration. The pressure equalizing is performed in each vessel. Each additional pressure equalization step minimizes hydrogen losses and increases the hydrogen recovery rate.

2. *Providing the purge (step PP)*—Purging is the final depressurization step in the co-current direction for the pure hydrogen to be purged or the regeneration of another adsorber vessel.

3. *Dump (step D)*—After purging the hydrogen, the remaining pressure has to be released (in the countercurrent direction) so that the impurities absorbed in the adsorbent material get desorbed and leave the adsorber vessel at the bottom and flow to the tail gas system of the PSA plant. It is, in fact, the first step of the regeneration phase.

4. *Purging (final)*—It is the final purging of the vessel at the lowest pressure by pure hydrogen from step 2 (providing the purge) to drive away impurities to the tail gas system of the PSA plant. The residual loading on the adsorbent material is reduced to a minimum to achieve high efficiency of the PSA cycle. The adsorbent material is now fully regenerated.

5. *Repressurization (steps R1/R0)*—For restarting the adsorption function, the vessel containing regenerated adsorbing material has to be pressurized. This is achieved in two steps. First, the pure hydrogen (from the adsorber vessel under depressurization) is used for pressure equalization. Then, the final pressurization up to the required adsorption pressure is carried out by a split pipeline from the pressurized hydrogen (final) product line. After this repressurization of the vessel, it takes over the task of adsorption again.

7.15.2 *Recovery of CO_2 from Syngas Leaving H_2 with a Small Amount of Impurities* [17,53,190]

By reforming hydrocarbons or the gasification of biomass/coal, a mixture of CO and hydrogen with some amount of other gases as impurities is produced. In such a case, the CO can be converted to CO_2 with the production of more hydrogen by the water-gas shift reaction.

$$CO + H_2O = H_2 + CO_2$$

The CO_2 can be removed by the methods like PSA, chemical absorption, physical absorption, physical adsorption, and membrane separation.

Physical or chemical absorption/scrubbing of CO_2 is used in plants for ammonia synthesis from hydrogen and nitrogen. Various amines that may be used for the absorption of CO_2, also absorb sulphur compounds (including hydrogen sulphide).

(1) Chemical Absorption of CO_2

The absorption chemicals that are used for removal of CO_2 from a mixture of gases are:

* Amino alcohols like monoethanolamine (MEA)

- Aqueous ammonia
- Alkali carbonates
- Hot potassium carbonate solution (110°C)
- 2.5 molar diethanolamine (DEA) solution (50°C)
- 3 molar amisol deta solution
- Sulfinol* solution (50°C)

[Sulfinol is a Shell trademark for a gas-treating solvent, consisting of sulfolane, water, and one or more amines. The solvent is formulated based on such factors as feed gas composition and desired treated gas specification. The formulation combining the chemical absorption properties of amines and physical absorption properties of sulfolane, allows for the cost-effective contaminants removal from gases. Three different solvents are available for the Sulfinol process, namely: (i) Sulfinol-X consists of sulfolane, methyldiethanolamine (MDEA), piperazine, and water; (ii) Sulfinol-M consists of sulfolane, MDEA, and water; (iii) Sulfinol-D consists of sulfolane, dipropanolamine (DIPA), and water.]

The binding between the above sorbent molecules and CO_2 is quite strong and therefore, the capture of CO_2 is quite effective. The relation between the CO_2 partial pressure and the amount of CO_2 absorbed chemically is not linear and therefore, the increase in the partial pressure of CO_2 will not increase the CO_2 gas loading of the sorbent after a certain limit. Sorbents are degraded by impurities like SO_2 and O_2 and therefore, their removal is essential before the chemical absorption process of CO_2 is initiated. Generally, the sorbents are corrosive and therefore they are used in their diluted form.

The various types of liquid amines (amino alcohols) that are being used for the capture of CO_2 from gases include monoethanolamine (MEA), methyldiethanolamine (MDEA), Diglycolamine (DGA), etc.

Some proprietary activators are also being used to enhance/accelerate the CO_2 removal. Employing heated-flash-regeneration improves absorption and the removal of CO_2 by achieving a high loading ratio of moles of acid gas to moles of amine. This process named as the 'formulated methyldiethanolamine (MDEA) process,' requires less energy compared to the conventional amine process.

(2) *Physical Absorption of Carbon Dioxide*
Carbon dioxide is captured by the physical absorption by a solvent without any chemical bonding. The bond between CO_2 and solvent is weaker than in case of chemical absorption. The amount of CO_2 absorbed is linearly proportional to the partial pressure of CO_2, therefore the more the partial pressure, the better is the absorption. The physical absorption depends on the temperature of the sorbent (the absorption rises with the fall in temperature because of the higher solubility of gas at lower temperatures). The physical absorption requires a high partial pressure of CO_2 and a lower temperature.

The examples of physical solvents are: menthol (used in Rectisol process); N-Methyl-2-pyrrolidone/NMP (used in the Purisol process); dimethylether of polyethylene glycol/DMPEG (used in the Selexol process); n-oligo ethylene glycol methyl isopropyl ethers (used in the Sepasolv process); N-Methylcaprolactarn/NMC (used in the Gaselan process).

The desorption of CO_2 can be done by the lowering of pressure (the pressure swing process) or alternatively, by raising the temperature (the temperature swing process).

(3) *Physical Adsorption of Carbon Dioxide*
In physical adsorption, a gas is adsorbed on the solid's surface. The separation is based on the difference in the size of the molecules of the various gases (the steric effect) and difference in binding forces between a particular gas and adsorbent (equilibrium effect or kinetic effect).

The examples of some adsorbents materials are: activated carbon; zeolite; silica gel; aluminium oxide.

Special adsorption media like 'Zeolite–13X' can be used for the adsorption and recovery of CO_2 on a small scale. In this process, the 'cleaned' gases are pressurised to about 3–5 psig and are led to a reactor containing Zeolite–13X, which selectively adsorbs CO_2. The adsorbed CO_2 is recovered by desorption from Zeolite–13X by the pressure swing operation (reduction of pressure). In the physical adsorption method of separation, both the pressure swing (decrease of pressure) and temperature swing methods (increase of temperature) can be used for the recovery of CO_2.

This method may, however, be useful when used in combination with other capture methods (like membrane technology) for obtaining a pure product. If future research finds new adsorbents having a higher capacity of adsorption, low-pressure gradient, and increased selectivity/improved reactivity for CO_2, this technology for the capture of CO_2 may become attractive even on its own.

(4) *Membrane Separation of CO₂*

Membranes are microscopic sieves. Under applied pressure, some molecules will pass through the micropores in the membrane and some molecules will be stopped. The driving forces for gas separation by the membrane are pressure and concentration gradient. There are two mechanisms at work in membrane separation, namely:

(a) Molecular sieving, that is, the separation of smaller molecules from larger ones via a very fine mesh.

(b) Solution diffusion, that is, the gas dissolves and diffuses through the membrane by mass transfer.

This gives rise to two types of membranes: (i) gas separation membranes, and (ii) gas absorption membranes.

Gas Separation Membranes—In this case, the pressure is applied and the difference in permeability of the gas species leads to the separation of gas from the mixture of gases. This technology may be useful for the separation of CO_2 from light hydrocarbons such as methane/natural gas but is not useful for separating CO_2 from flue gases.

Gas Absorption Membranes—This type of membrane is used in conjunction with a liquid sorbent to carry away the molecules of CO_2 that diffuse through the membranes by a mechanism called 'solution diffusion and mass transfer' without the use of hydrostatic pressure. The membrane is an interface between the feed gas (for example, flue gas) and liquid sorbent. This membrane system can be used in the form of parallel hollow fibers to provide better liquid–gas contact. In this configuration, the independent control of gas and liquid flows can be exercised, and the minimization of flooding, channelling, and foaming can be affected. Further research on the problems of choking/clogging of hollow fibers is being carried out in the world.

(5) *Cryogenic Separation of Carbon Dioxide*

Cryogenic separation works on the principle of difference in the boiling points of various liquefied gases. Nearly total separation can be achieved by this method. The flue gas/mixture of gases have to be compressed and cooled. The carbon dioxide can be liquefied between 31.6°C and (–) 56.8°C and then separated. Since there is only about 15% carbon dioxide in the flue gas, the total flue gas has to be compressed. The process is energy intensive and consumes lot of energy and is, therefore, very costly.

7.15.3 *Purification of Hydrogen by Membrane Separation*
[8,24,26,57,125,126,150,156,169,176,202,206,207,210]

Typically, in a membrane system, the incoming feed stream is separated into two components: permeant and retentate. The permeant is the gas that travels across the membrane and the retentate is what is left of the feed. On both sides of the membrane, a gradient of the chemical potential is maintained by a pressure difference, which is the driving force for the gas molecules to pass through. The general principle of membrane purification is shown in Figure 7.8.

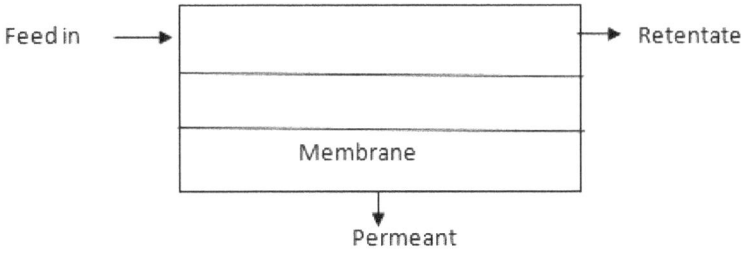

Fig. 7.8. Schematic of Membrane Separation.

The mechanism of molecular sieving refers to the case where pores of the membrane are too small to let one component pass. The separation of gases by the mechanism of diffusion through a membrane at very low pressures will take place because lighter molecules move faster across the stable pores of the membrane, like that in a pressure-driven convective flow through capillaries.

However, the more general model for the membrane gas separation is the solution-diffusion mechanism. In this mechanism, the gas molecules are first dissolved onto the membrane and then diffused through it at different rates.

The permeability of each gas component is different and the ease of movement of the gas component through the membrane would depend on the component's permeability. The greater the permeability of a gas component, faster is its flow through the membrane. The permeability of a gas component is proportional to its diffusivity (in an ideal gas) and is given by Henry's law.

$$P_1 = D_1 K_1$$

Where P_1 is the permeability, D_1 is diffusivity, and the K_1 is the Henry coefficient of gas component No.1. [The product of $D_1 K_1$ is often expressed as the permeability of the species 1, on the specific membrane being used.]

The flux of a component gas can be related to the difference in partial pressures of the components in the mixture of ideal gases by Fick's Law.

$$J_1 = [D_1 K_1 (p_{f1} - p_{p1})]/L \ or \ = [P_1(p_{f1} - p_{p1})]/L$$

Where: J_1 is the molar flux, P_1 is permeability, D_1 is the diffusivity, K_1 is the Henry coefficient of gas component No.1. The p_{f1} and p_{p1} are the partial pressures of component No.1 on the feed side and permeate side, respectively of the membrane, whose thickness is L.

The flow of a second species, J_2 is given by: $J_2 = [P_2 (p_{f2} - p_{p2})]/L$

The ratio of feed pressure to permeate pressure is called as the membrane pressure ratio. The total flow across the membrane is strongly dependent on the relation between the feed and the permeate pressures. The membrane will experience flow across it when there exists a concentration gradient between the feed and the permeate. If the gradient is positive, the flow will go from the feed to the permeate and gas component No.1 will be separated from the feed.

If n_{f1} and n_{p1} denote the concentration of component on the feed side and permeate side of gas component No.1 and p_{f1} and p_{p1} are the partial pressures of component No.1 on feed side and permeate side, respectively,

$$n_{f1} \cdot p_{f1} > n_{p1} \cdot p_{p1} \ or \ (n_{f1}/n_{p1}) > (p_{p1}/p_{f1})$$

The product of the concentration and partial pressure of the component on the feed side should always be greater than the product of the concentration and partial pressure of the component on the permeate side. Another factor is the selectivity of the membrane for the gas component. The maximum separation of gas component No.1, that is, the maximum concentration on the permeate side results under the condition as follows:

Max. concentration on the permeate side $n_{p1} = (p_{f1}/p_{p1}) \cdot n_{f1}$

Another important factor is the selectivity coefficient for the membrane, which is the ratio of permeability of the membrane for gas component No. 1 (P_1) to the permeability of the membrane for gas component No. 2 (P_2). *[Membrane Coefficient = (P_1/P_2)]*

For the optimum selection of membrane for separating gas component No.1 from gas component No. 2, the ratio (P_1/P_2), that is,the membrane coefficient should be high. If the ratio is 1, there will not be any separation as both the component would permeate equally through the membrane.

7.15.4 *Types of Membranes* [8,17,24,26,57,125,169,176,195]

The properties of some hydrogen selective membranes are given in Table 7.4. Each type of membrane has advantages and disadvantages. Further research on the suitability of membranes for hydrogen separation is being carried out. There are many types of membrane, namely:

- Dense polymer membranes
- Microporous ceramic membranes
- Dense ceramic membranes
- Porous carbon membranes
- Dense metallic membranes

However, only some of them are suitable for the purpose of hydrogen separation.

Palladium–Copper (Pd–Cu) Alloy Composite Membranes

It has been found that palladium (Pd) and its alloys, nickel, platinum, and the metals in Groups 3 to 5 of the periodic table are all permeable to hydrogen. Pd composite membranes are stable in operation at 450°C for over 70 days. Biomass- or coal-derived synthesis gas will contain H_2S as well as CO, CO_2, N_2, and other gases. High selectivity membranes are necessary to reduce the H_2S concentration to acceptable levels for solid oxide and other fuel cell systems. Pure Pd membranes are poisoned by sulphur, and suffer from mechanical problems caused by thermal cycling and hydrogen embrittlement. However, recent advances have shown that Pd–Cu composite membranes are not susceptible to the mechanical, embrittlement, and poisoning problems. These membranes consist of a thin film of metal deposited on the inner surface of a porous metal or ceramic tube.

Fabrication of thin, high flux Pd–Cu alloy composite membranes using a sequential electroless plating approach is feasible. Thin Pd60–Cu40 films exhibit a hydrogen flux more than ten times

Table 7.4. Properties of Some Hydrogen Selective Membranes.

Property	Dense polymer	Microporous ceramic	Dense ceramic	Porous carbon	Dense metallic
Temperature	Less than 100°C	200–600°C	600–900°C	500–900°C	300–600°C
Hydrogen selectivity	Low	Moderate	Very high	Low	Very high
Hydrogen flux	Low	High	Moderate	Moderate	High
Known poisoning by	HCl, SOx, CO_2	-	H_2S	Strong vapours & organics	H_2S, HCl, CO
Examples of Materials	Polymers	Silica, Alumina, Zirconia, Titania, Zeolites			
Transport Mechanism	Solution diffusion	Molecular sieving	Solution diffusion	Surface diffusion, Molecular sieving	

Source: Hydrogen Separation Membranes, NCHT, EERC Energy & Environmental Research Centre, www.undeerc.org/NCHT/pdf/EERCMH36028.pdf

larger than commercial polymer membranes for hydrogen separation and resist poisoning by H_2S and other sulphur compounds typical of biomass/coal gasification. Such membranes can withstand temperatures even higher than 450°C. To sum up, these membranes have the following advantages:

   i. Give high hydrogen flux.
   ii. Tolerant to hydrogen sulphide and sulphur-containing gases, even at very high total sulphur levels (for example, 1000 ppm),
   iii. Operation at high temperatures is feasible
   iv. Resistant to embrittlement and degradation by the thermal cycling process [176]

Composite alloy membranes are fabricated by sequential electroless plating of palladium and copper onto symmetric 0.2 mm cut-off α-alumina tubes (for example, CoorsTek GTC-998) and asymmetric 0.05 and 0.02 mm cut-off zirconia coated α-alumina tubes *(for example,* US Filter T1-70).

A seeding procedure is required to be used prior to the palladium plating for ensuring adhesion between the metallic film and the ceramic surface. This involves impregnation of the ceramic support using an organic Pd salt solution, followed by calcination and reduction by flow of hydrogen. Pd and Cu electroless plating baths are used in combination with osmotic pressure gradients to deposit films ranging from 1 to 25 μm in thickness. The objectives are: reduction of porosity, promoting surface homogeneity and de-densification of plated Pd film. The osmotic pressure can be generated by any suitable technique (for example, circulating concentrated sucrose solutions on the outside of the tubes).

After sequential plating of Pd and Cu films, the membranes are annealed during the hydrogen permeation tests at temperatures greater than 350°C to produce homogeneous alloy membranes.

Note: [121]

Thin (less than 2 μm thickness) and pinhole-free palladium–copper (Pd–Cu) alloy composite membranes with a diffusion barrier can also be fabricated on mesoporous stainless steel supports by vacuum electrodeposition. The deposition film can be fabricated by multilayer coating and diffusion treatment and the formation of Pd–Cu alloys can be achieved by annealing the 'as-deposited' membranes at 723 K in a nitrogen atmosphere.

The final composition and phase structures of the alloy film can be studied by energy-dispersive electronic analysis (EDS) and X-ray diffractometry (XRD).

The Pd–Cu alloy composite membrane can achieve excellent separation performance for hydrogen (for example, hydrogen permeance of 2.5×10^{-2} cm³/cm² cm. Hg s and hydrogen/nitrogen (H_2/N_2) selectivity was above 70,000 at 723 K.

Examples of Commercially Available Membranes

A few commercially available membranes for different applications are indicated below:

(1) Air Liquide has technology called MEDAL™ that is typically used in refinery applications for hydrotreating. The membrane is selective to components other than hydrogen, including water/vapor, ammonia, and carbon dioxide. It is not suitable for application in gas separation for gasification of biomass and coal.

(2) Air Products offer a line of hydrogen recovery membranes referred to as PRISM® membrane systems. The PRISM membrane is suitable for separations in hydrocracker and hydrotreater systems or for CO purification in reformer gases. The systems maintain a low temperature and are not suitable for biomass and coal gasification.

(3) Wah-Chang offers small-scale Pd–Cu membranes for commercial sale that are capable of producing an ultrapure stream of hydrogen from syngas. One drawback of the membrane is that it has a very low tolerance to hydrogen sulphide and hydrochloric acid; both of these contaminants are found in biomass/coal-derived syngas.

7.16 Hydrogen Storage [8,12,15,18,21,40,42,45,46,48,61,64,76,79,83,97, 102,114,115,127,128,132,135,148,149,151,159,161,167,185,186,202,215, 217,237]

7.16.1 General Problems in Storage of Hydrogen [8,18,61,128,135,148,159,161, 167,185,237]

Hydrogen can be stored in a gaseous, liquid, or solid state. Although molecular hydrogen has very high energy density on a mass basis, partly because of its low molecular weight, but 'as a gas at ambient conditions,' it has very low energy density by volume. If hydrogen is to be used as a fuel stored onboard the vehicle, it has to be pressurized or liquefied to provide sufficient driving range. *[Onboard vehicles, the hydrogen gas can be stored at a pressure of about 700 bar (70 MPa).]*

Increasing the gas pressure reduces the volume and improves the energy density by volume, but thicker and heavier container tanks (pressure vessels) have to be used. High-pressure compression consumes a large quantity of external energy.

Alternatively, higher volumetric energy density liquid hydrogen or slush hydrogen may be used. Liquid hydrogen is cryogenic and boils at 20.268 K (–252.882°C or –423.188°F). Cryogenic storage cuts weight, but it requires large liquefaction energies. (*Liquefaction of hydrogen would involve pressurizing and cooling steps for large amount of energies would be required*). Liquid hydrogen storage tanks are required to be thoroughly well insulated to minimize boil off. Ice may form around the tank and promotes its corrosion further if the liquid hydrogen tank insulation fails. Liquid hydrogen storage is very difficult because of low temperatures and the above problems involved.

[The liquefied hydrogen has lower energy density by volume than gasoline by approximately a factor of four, because of the low density of liquid hydrogen–there is actually more hydrogen in a liter of gasoline (116 g) than there is in a liter of pure liquid hydrogen (71 g).]

The heavy thick tanks needed for compressed hydrogen reduce the fuel economy of the vehicle. Also, heavy tanks are costlier. As hydrogen is a small molecule, it tends to diffuse through any liner material intended to contain it, leading to the phenomena of 'hydrogen embrittlement' of the material, and weakening of the container vessel.

7.16.2 Chemical Storage of Hydrogen [40,61,64,128,237]

a. Storage as Hydride

Distinct from storing molecular hydrogen, hydrogen can be stored as a hydrogen-containing chemical compound (for example, as metal hydride), which can be transported with relative ease. Metal hydrides are used for storing hydrogen because they have good hydrogen binding properties, which is released only at high temperatures (close to 120°C). Another important feature which favours metal hydrides for hydrogen storage is the lesser volume occupied by them. At the point of use, the hydrogen-containing material can be made to decompose by heating (an endothermic reaction), yielding hydrogen gas.

Metal hydrides are compounds of one or more metal cations (M+) and one or more hydride anions (H−). When pressurised, most metals bind strongly with hydrogen, resulting in stable metal hydrides that can be used to store hydrogen conveniently onboard vehicles. Some examples of metal hydrides are LaNi5H6, MgH2, and NaAlH4.

An important property of metal hydrides is that the pressure of the gas released by heating a particular hydride depends mainly on the temperature and not the composition. At a fixed temperature, the gas pressure remains essentially constant until the hydrogen content is almost exhausted.

The mass and volume density problems are associated with molecular hydrogen storage. The use of hydrides involves the requirement of high pressure and temperature conditions for hydride formation and hydrogen release. For hydriding *(making metal hydrides)* and dehydriding *(releasing hydrogen from hydrides for use)*, kinetics and heat management problems have also to be overcome.

For hydride formation, the hydrogen has to dissociate into atomic hydrogen to be absorbed in metal. For physical reasons, the hydride formation always takes place at constant pressure. A hydride is formed with the evolution of heat, that is, the reaction is exothermic, while the release of hydrogen is endothermic. However, the metal hydride is formed at high temperatures and pressures only.

At present, however, solid state storage in the form of metallic hydrides is most attractive. To achieve this, the hydrogen gas is reacted with powdered metal to form the hydride in a closed evacuated pressure vessel at room temperature. Generally, the metal used for making hydride is *powdered elemental magnesium (Mg)*. The storage capacity of magnesium is about 8% by weight. The drawback in the process is the high heat of formation *(exothermic reaction)*. The heat has to be removed by the heat exchanger to keep the temperature constant. The drawback of the high heat needed for the hydride formation can be minimised by adding *nickel;* for example, the powdered magnesium and nickel alloy *(with formula Mg_2Ni)*. However, hydrogen absorption is primarily associated with the formation of MgH_2. To get back hydrogen from the *hydride* the pressure vessel or cylinder can be heated according to the required rate of flow.

In connection with the storage of hydrogen, research is going on in India vis-à-vis the development of new and complex compounds such as *sodium aluminium hydride, magnesium aluminium hydride, lithium and magnesium amides, graphite nano fibers*, etc., at Banaras Hindu University, Varanasi. The development of liquid organic hydrides has also been taken up at the National Environmental Engineering Research Institute (NEERI), Nagpur.

Sodium alanates or sodium aluminium hydrides ($NaAlH_4$) are also promising materials for hydrogen as their theoretical reversible capacity is about 5.5%. Their reversibility kinetics improves with the addition of titanium or zirconium dopements. Lithium Alanates $LiAlH_4$ and Li_3AlH_6 have a hydrogen storage theoretical capacity of 10.5% and 11.2%, respectively.

Boron hydrates are very promising for hydrogen storage. Lithium tetrahydroborate $Li(BH_4)$ has a hydrogen storage capacity of 18.4%, which is the highest among hydrides.

Some more hydrides mentioned below are found to be promising, but their hydrogen storage capacity is smaller:

1. Hydride of lanthanum-nickel (*La Ni$_5$*) H_6, containing theoretically 1.35% of hydrogen by weight.
2. Hydride of iron-titanium *(Fe +Ti) H$_2$*, containing theoretically 1.9% hydrogen by weight.
3. Zirconium-manganese hydride ($Zr Mn_2 H_4$) containing theoretically 1.7% hydrogen by weight.
4. Lithium hydride (LiH) has a hydrogen storage capacity of 7.7%
5. Sodium hydride (NaH) has a hydrogen storage capacity of 4.9%

An ideal metal/metal hydride for use for the chemical storage of hydrogen should have the following desirable properties:

• High hydrogen storage capacity per unit mass and per unit volume
• Moderate dissociation pressure and low dissociation temperature
• Low heat of formation to minimize the energy necessary for the release of hydrogen and low heat dissipation during exothermic hydride formation
• Reversibility with small energy loss during charge and discharge of hydrogen
• Fast kinetics
• High stability against oxygen and moisture for a long life cycle
• Cycle ability
• Low cost of recycling and charging infrastructures
• High safety

b. Hydrogen Absorbed/Adsorbed on Solid Sorbent

If a suitable absorbing or adsorbing solid material is found that can hold very high quantities of molecular hydrogen, it would be advantageous. Unlike in the hydrides mentioned above, in this system of storage, no dissociation, recombination, etc., would be involved and hence, would not suffer from the kinetic limitations of many hydride storage systems.

As reported in the press,* British scientists have developed a new low-cost, hydrogen-based fuel, which is actually hydrogen densely packed in tiny beads that can be poured or pumped like a liquid. The scientist also noted that a tankfull of these hydrogen-packed beads (called 'artificial petrol' as it has not been given any brand name) is expected to last 300 to 400 miles (480 to 640 kilometers) in line with conventional fuel.

7.16.3 Underground Storage of Hydrogen [32,33,66,128,178,220,227,233,237]

Hydrogen gas can be stored in 'abandoned natural gas underground storage sites' as is being planned in Denmark, where large-scale wind energy is available to generate hydrogen. Hydrogen storage can be done in underground caverns, salt domes, and depleted oil and gas fields. The storage of large quantities of hydrogen underground can function as grid energy storage, which is essential for the hydrogen economy.

Large quantities of gaseous hydrogen were stored in underground caverns by the ICI for many years without any difficulties.

7.17 Greenhouse-Gas-Neutral Fuel from Hydrogen

7.17.1 Greenhouse-gas-neutral Alcohol [71,233]

A theoretical alternative to the direct use of elemental hydrogen in vehicles would be to use a liquid fuel like alcohol made from a CO_2 source and centrally produced hydrogen. Thus, hydrogen would be used captively to make fuel, and would not require expensive hydrogen transportation or storage.

To be greenhouse gas neutral, the CO_2 source can be from: (i) air, (ii) flue gases of biomass/waste burning, (iii) flue gases of fossil-fuel-fired power plants or industries, or (iv) any source where the CO_2 is released into the atmosphere.

Captive hydrogen production can be utilized by converting it to more easily transportable and storable transportation fuels like alcohols or even methane, using CO_2 as input. This can thus be seen as the artificial, or "non-biological green" analogue of biomass, biodiesel, and vegetable oil technologies. *[Green plants, in a sense, already use solar power to produce organic material containing hydrogen, which is then used to make 'easier-to-store-and-use fuels'.]*

Moreover, models of methanol fuel cells are beginning to be demonstrated, so methanol may eventually compete directly with hydrogen in the fuel cell and hybrid market.

Recycling of CO_2 to Make Methanol/Methyl Alcohol

A company in Iceland is turning captured CO_2 into methyl alcohol for use in automobiles. The generated hydrogen from renewable electricity from the geothermal power station in Iceland is reacted with captured CO_2 to make methyl alcohol. The existing engines running on gasoline require modification for making them suitable for running on methyl alcohol, which has a high octane rating and can be a good fuel. In fact, some race cars in USA already run on methyl alcohol. This technology can remove the difficulty in the storage and transport of hydrogen, while also recycling CO_2.

7.17.2 Synthetic Methane Production from Hydrogen

In a similar way to synthetic alcohol production, hydrogen produced with the help of solar or wind energy sources can be used to directly (non-biologically) produce greenhouse-gas-neutral gaseous fuels.

Liquid methane has a higher boiling point, higher energy density (3.2 times that of hydrogen), and it is easier to store.

The infrastructural facilities for natural gas (which is mainly methane) already exist. Natural-gas-powered vehicles already exist, and are known to be easier to adapt from existing internal engine technology, than internal combustion autos running directly on hydrogen. Experience with CNG (compressed natural gas)-powered vehicles shows that methane storage is easier and not so expensive.

Captive hydrogen and CO_2/CO obtained from any source may be used to *synthesize* methane as mentioned below:

(a) Methane can be formed on passing gasifier gases CO and H_2 (largely free from nitrogen) over nickel at about 300°C. The gasifier gases are a mixture of CO and hydrogen:

$$CO + 3H_2 = CH_4 + H_2O \text{ ------- (Sabatier and Senderens reaction)}$$

(b) The methane can also be formed on passing a mixture of CO_2 and hydrogen (H_2) over nickel (Ni) deposited on magnesia (MgO) at about 328°C.

$$CO_2 + 4H_2 = CH_4 + 2H_2O$$

However, the cost of alcohol storage is even lower, so this technology would need to produce methane at a considerably cheaper price to compete with the cost of alcohol production and storage. The ultimate mature prices of fuels in the competing technologies cannot be presently estimated, but both are expected to offer substantial infrastructural savings over attempts to transport and use hydrogen directly.

7.18 Hydrogen Infrastructure [8,9,46,58,60,96,108,146,170,174,218]

The infrastructure needed for hydrogen mainly consists of:

- Hydrogen piping system
- Hydrogen tanks
- Compressed hydrogen tankers-trailers or liquid hydrogen tanker-trucks or tanker-trailers
- Hydrogen filling stations
- Dedicated onsite production facility

There are certain disadvantages with hydrogen transportation and storage such as:

 i. Hydrogen pipelines are more expensive than long-distance electric lines

 ii. Hydrogen is about three times bulkier in volume than natural gas for the same heat value

iii. As pointed out in Para 7.5.1, hydrogen causes *hydrogen embrittlement* in steel tanks and pipelines, accelerating the cracking of steel, which increases maintenance costs, leakage rates, and material costs. Thus, pipelines or tanks have to be coated by protective layer. (For the same reason, even natural gas pipelines are required to be coated).

Pipeline is the cheapest way to transport hydrogen. Theoretically, hydrogen piping can be drastically cut in distributed systems of hydrogen production, where hydrogen is routinely made onsite using medium- or small-sized generators that would produce enough hydrogen for personal use or perhaps a neighborhood. The long-distance transportation of hydrogen by high-pressure pipelines is a very costly affair and therefore, in practice, other methods such as a combination of transport of hydrogen, onsite generation of hydrogen from basics, or the regeneration of hydrogen from hydrides or other chemicals (in which the hydrogen is converted for easy storage and transport) can be implemented.

Hydrogen production can be central or distributed or more practically, a mixture of both. *Hydrogen's conversion to alcohol or other chemicals as mentioned in previous paragraphs can be resorted to for overcoming transport problems.*

7.19 Hydrogen Leak Detection and Pipeline Integrity Monitoring [6,43,75,89,92,95,97,102,111,155,162,165,171,172,173,200,205]

7.19.1 Hydrogen Detectors/Sensors for Leak Detection [6,27,89,92,95,100,123,124,152,155,162,171,182,196,200,205,219]

Hydrogen, being odorless and tasteless, is undetectable by human senses. The addition of odorant to hydrogen is an option for making it detectable by smell. Some odorants contain sulphur, which may poison membranes. However, the odorants may contaminate fuel cells and therefore, are not suited for use in hydrogen detection.

Sensors based on optical fiber, using chromogenic material as indicators of the presence of hydrogen can instead be used for leak detection. These chromogenic materials undergo optical changes like change in color or change in transmittance through a film when exposed to atomic hydrogen. An example of chromogenic materials is a complex oxide of tungsten, nickel, and vanadium (WO_3 NiO_x V_2O_5).

In such a sensor, the dissociation of hydrogen is accomplished on the surface of a thin catalyst top layer like Pd. Some atomic hydrogen diffuses rapidly through the catalyst and into the chromogenic layer, resulting in change of its optical properties. The optical changes are read by a light beam by 'measuring transmission through the stack of thin films', or 'measuring reflectance from the catalyst layer'. Such an optical layer can be deposited on the end of a fiber-optic cable. When a beam of light is propagated through the cable, the intensity of the reflected beam or transmitted beam helps detect hydrogen.

At present, the application of the thin film optic fiber sensor leak detection technology to a long-distance hydrogen pipeline is still a great challenge.

7.19.2 Development of Suitable Composite Materials for Hydrogen Pipes [5,18,44,70,97,102,211,212,221,242]

Because of hydrogen embrittlement in steel and high capital costs of the pipeline system, alternative materials need to be developed. Research is going on globally in this direction.
Several composites are potential materials for hydrogen service such as:

• *Fiber-reinforced polymer (FRP) composite*, which may consist of:
(i) an inner nonpermeable barrier tube for high pressure, (ii) a protective layer on the barrier tube, (iii) an interface layer over the protective layer, (iv) multiple glass fiber or carbon fiber composite layers, (v) an outer pressure barrier layer, (vi) an outer protective layer. Suitable materials for the inner nonpermeable barrier are being developed such as polyethylene terephthalate (PET) mixed with organo-modifier clay.

• *Composite reinforced line pipe (CRLP)* mainly consists of composite material and a thin-wall, high-strength low alloy (HSLA) steel pipe of X42-X80. The steel and composites together create a hybrid, providing an economical alternative to a high-strength all steel pipe. The hydrogen embrittlement problem would exist though somewhat reduced due to the lower thickness of the pipe.

However, the adoption of any composite material technology for hydrogen service would only be possible only when:

(i) Developing the inner nonpermeable layer of the pipe with satisfactory performance.

(ii) Hydrogen compatibility is finally established after evaluating it on a long-term model.

(iii) Developing an efficient defect-free method of manufacture of high-pressure large diameter pipe and their suitable jointing method.

7.19.3 *Pipeline Integrity Monitoring* [5,102,173,211,212]

For a long-distance pipeline, there are a number of threats such as:

i. Manufacture and construction defects

ii. Internal and external corrosion

iii. Hydrogen embrittlement and structural changes in the metal of the pipeline

iv. Soil movements including ground settling and soil erosion

v. Earthquake

vi. Interference by outsiders and outside force

vii. Incorrect operation

viii. Extremes of weather

ix. Third-party damage, mainly due to damage by contractors inadvertently damaging the pipeline during construction and excavation

x. Security threat

The mechanical integrity of the hydrogen pipeline and pressure vessels has to be monitored. For the natural gas system, there are fiber-optic sensors as well as other types of sensors available that can monitor time-dependent defects, including material corrosion and structural defects, pipe movement, pipe stress, buckling strains, ground settling, and other defects. These sensing systems may be useful for hydrogen pipeline integrity as well with some modifications.

7.20 Application of Hydrogen as a Fuel [1,2,28,39,45,47,51,79,91,103, 193,204,213]

7.20.1 *Use of Hydrogen as a Fuel for 'Fuel Cell' Applications* [1,40,45,51,60,62,64,69,91,103,105,113,140,159,213]

It may be mentioned here that the generation of electricity by fuel cells is very efficient; the conversion efficiency of heat to power is about 54%. Although the fuel cells can use methane or purified natural gas as fuel, hydrogen is very suitable as fuel for fuel cells as on chemical reaction with oxygen, it creates only steam/water, which do not create problems for fuel cells. The principle of the fuel cell is illustrated in Figures 7.9a, 7.9b, 7.10, and 7.11.

Figure 7.10 shows the complete electrochemistry of the fuel cell. Hydrogen fed to a fuel cell is converted to negatively charged particles or electrons. The electrons (e^-) flow through an external load to the 'cathode' while the 'hydrogen ion' H^+ from anode migrate through 'electrolyte' to 'cathode' and get 'oxidised' to produce water or steam. The individual cells produce very low voltage, for example, 0.55–0.75 Volts. Hence, a number of cells are arranged in stacks to provide the required voltage and direct current. The area of the electrode determines the total amperage, the current density being 100–500 mA/sq. cm.

Hydrogen Purity

Hydrogen gas must be distinguished as "technical-grade" (five nines pure), which is suitable for applications such as fuel cells. Commercial-grade hydrogen has carbon- and sulfur-containing impurities and can be produced by the much cheaper steam-reformation process. Fuel cells require high purity hydrogen because the impurities would quickly degrade the life of the fuel cell stack.

Hydrogen as Fuel for IC Engines or for Power Generation through Fuel Cells

Although hydrogen can be used in conventional internal combustion engines, the fuel cells, being electrochemical, have an advantage of efficiency over heat engines. Fuel cells are much costlier than common internal combustion engines, but their cost is expected to come down eventually as

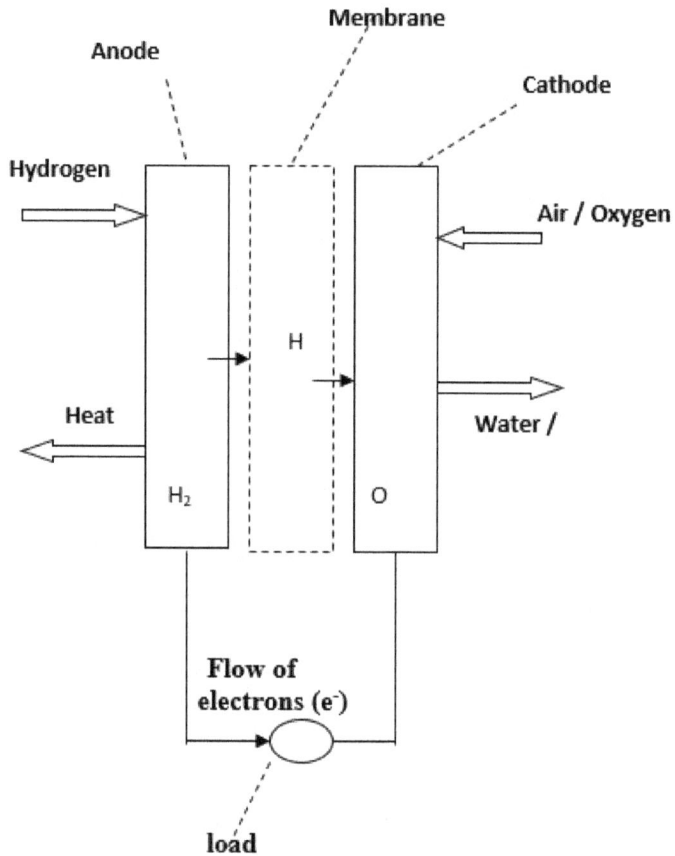

Fig. 7.9a. Typical Fuel Cell – Principle of Operation.

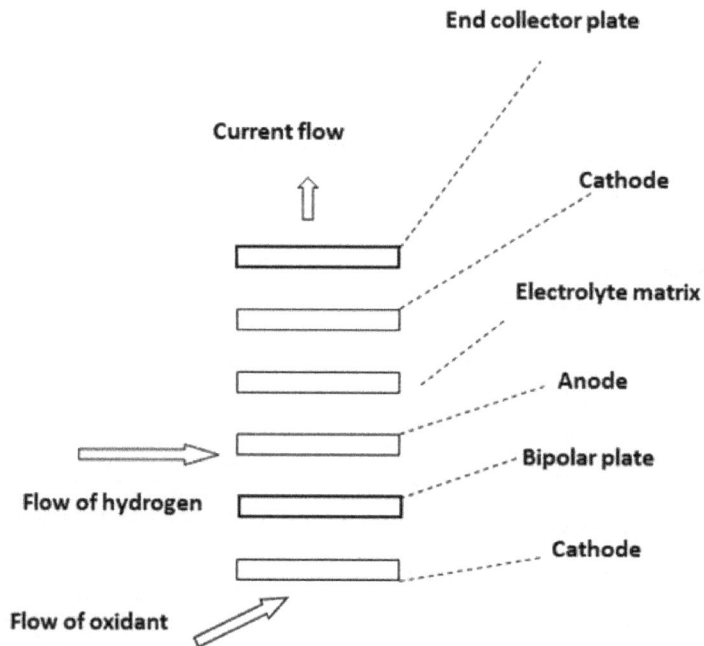

Fig. 7.9b. Typical Repetitive Elements of Fuel Cell Stack.

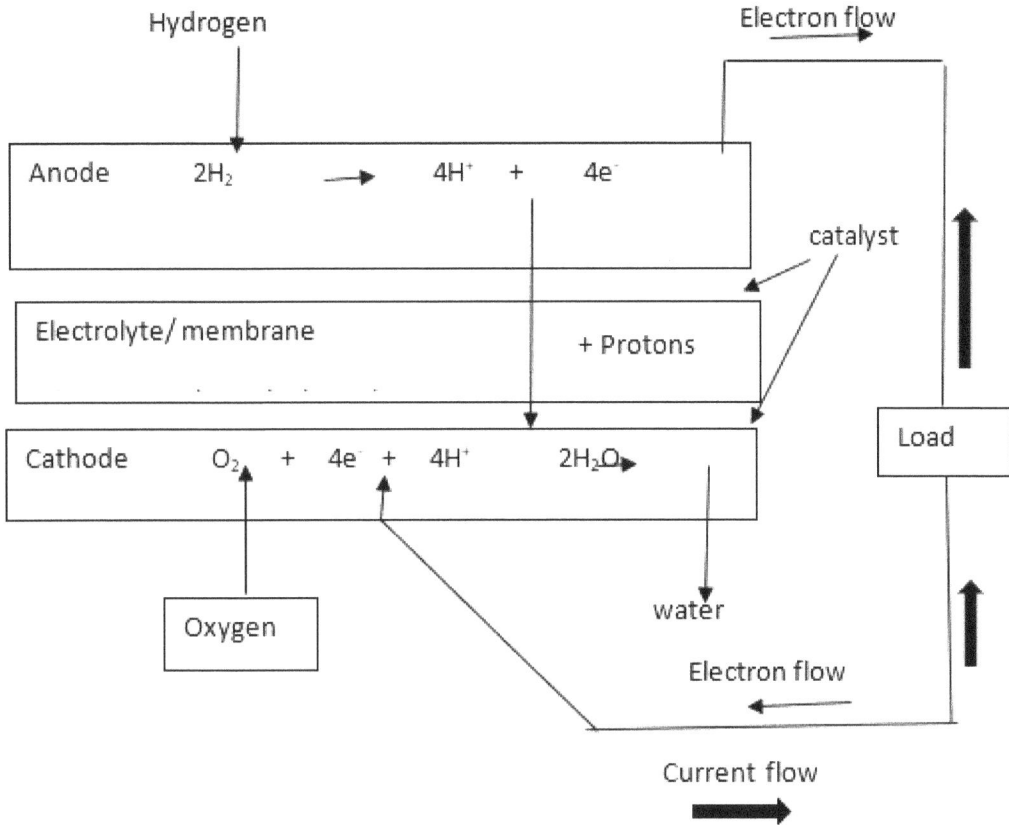

Fig. 7.10. Electro-Chemistry of Fuel Cell.

new technologies and production systems develop. Some types of fuel cells work with hydrocarbon fuels, while all can be operated on pure hydrogen.

The application of hydrogen as fuel for vehicles via fuel cells has already been discussed in *Para 7.2*, while a few examples of hydrogen-based power generation are given below.

7.20.2 *Various Types of Fuel Cells* [12,22,47,49,62,69,72,81,88,90,94,101,107, 177,213]

Many different types of fuel cells are in use such as:

Low-temperature Fuel Cells

1. Proton exchange membrane cell (PEM) fuel cell or polymer electrolyte membrane fuel cell
2. Direct menthol fuel cells (DMFC)
3. Alkaline fuel cells (AFC)

Medium-temperature Fuel Cell

4. Phosphoric acid fuel cells (PAFC)

High-temperature Fuel Cell

5. Molten carbonate fuel cell (MCFC)/direct fuel cell (DFC) technology
6. Solid oxide fuel cell (SOFC)

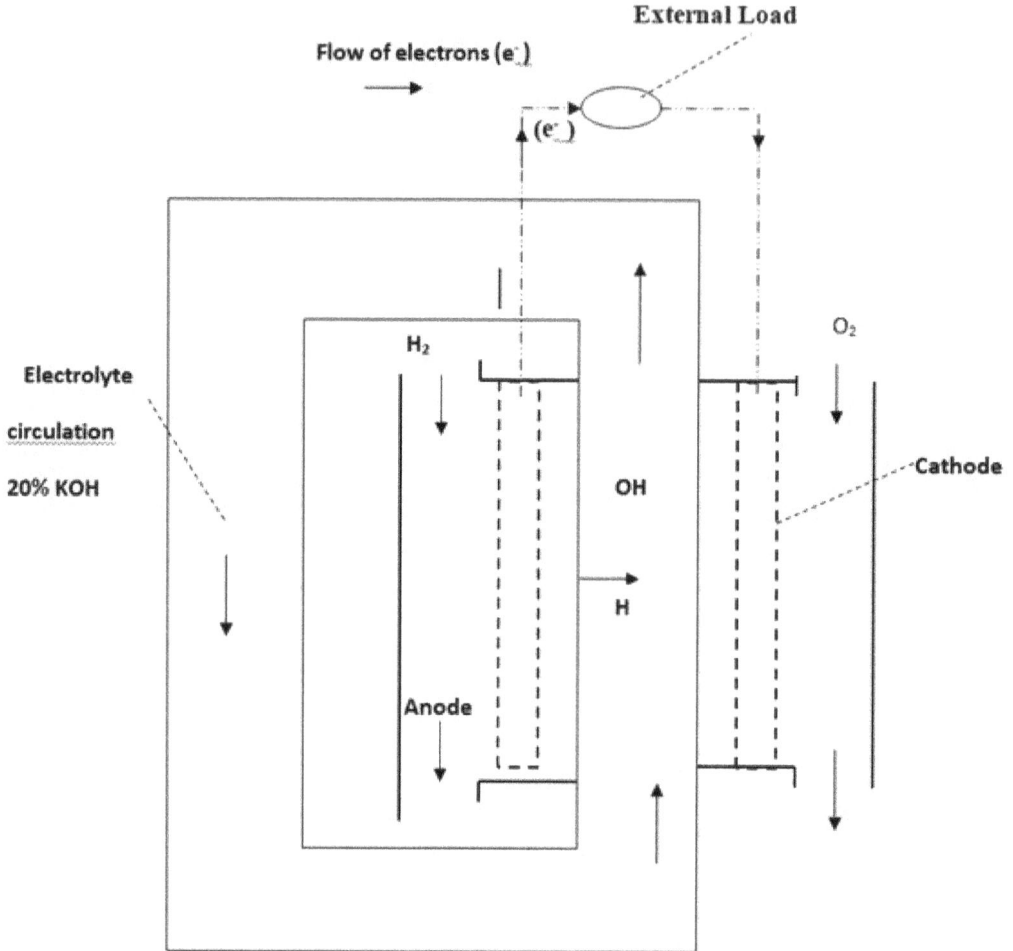

Fig. 7.11. Hydrogen – Oxygen Fuel Cell.

(1) Proton Exchange Membrane (PEM) Fuel Cell

The proton exchange membrane fuel cell 'or briefly PEM fuel cell', uses hydrogen as the fuel and oxygen from the air as the oxidant. This fuel cell is also called the polymer electrolyte membrane fuel cell. The places where the chemical reactions take place are shown in Figure 7.10. Electrodes of the fuel cell are made of porous carbon, which allows the active gases to pass through. A thin sheet of acidic solid organic polymer is used as an electrolyte. The thickness of polymer sheet is only about 50 microns, which is impermeable to electrons, but permits protons to pass through it.

Hydrogen is supplied as fuel at the anode while oxygen is supplied at the cathode as an oxidant. The positive hydrogen ions pass through the electrolyte membrane, while electrons travel through the external circuit through the electrical load to cathode. The hydrogen gets oxidized forming water, which goes out as shown in Figure 7.10.

The electrode surfaces support platinum catalysts. The function of the catalyst is to increase the rate of oxidation at the anode and the rate of reduction at the cathode. The temperature at which the reactions take place can be kept low by the use of catalysts, otherwise, without the use of catalyst, the cell has to be designed to work at high temperatures. Platinum is used as a catalyst in PEM fuel cells, but it is very costly and sensitive to CO. *(It is poisoned by even a small amount of CO).* The hydrogen supplied as fuel therefore must be very pure. Additional filtering/purification has to be included in the system for eliminating potential contaminants.

The operating temperature of the PEM fuel cells is low (for example, about 80°C (range 50–120°C)). A higher operating temperature may damage the fragile polymer membrane.

As the PEM fuel cells have an efficiency of about 32 to 40% (60% is the maximum achievable). *As they have higher output and work at lower temperatures, they are considered very suitable for automobile applications.* These fuel cells also have the advantages of having high power density and a quick response, and being lightweight, while they have the disadvantages such as requiring humidification, intolerance to CO, and a high cost due to the use of platinum.

Research and development for indigenous membranes and catalysts for PEM fuel cells has been taken up by IIT Kanpur, Kanpur and IIT Madras, Chennai.

(2) Direct Methanol Fuel Cells (DMFC)

Previously, hydrogen was generated in the reformer from methanol for use in the fuel cell of the PEM type, but now direct use of methanol in fuel cells has been made possible by the development of the direct methanol fuel cell (DMFC). In this case also, like PEM fuel cells, the electrolyte is a polymer. The hydrogen ions carry the positive charge. Methanol liquid is the primary fuel fed at the anode of the fuel cell. It is oxidized there in the presence of water to generate hydrogen ions, electrons, and CO_2. The hydrogen ions pass on to the cathode through the solid polymer electrolyte, where they are oxidized by oxygen from air to form water. The reactions which take place at the anode and cathode are shown below:

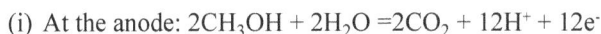

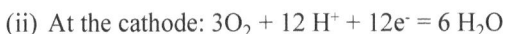

(i) At the anode: $2CH_3OH + 2H_2O = 2CO_2 + 12H^+ + 12e^-$

(ii) At the cathode: $3O_2 + 12 H^+ + 12e^- = 6 H_2O$

The overall chemical reaction in the fuel cell is as follows:

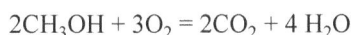

$2CH_3OH + 3O_2 = 2CO_2 + 4 H_2O$

Direct methanol fuel cells work at low temperatures, like in PEM fuel cells. The operating temperature range is 50 to 120°C. *The efficiency of these fuel cells is low and presently, the output of the fuel cell is limited to 1.5 kW, which may suffice for most of the uses except for automobiles.*

The feature of these fuel cells of avoiding the requirement of an external reformer for producing hydrogen and thereby, problems of an external hydrogen supply is very attractive. These types of fuel cells are *very suitable for use in portable electronic equipment because of the above features and their operation at a low temperature.*

(3) Alkaline Fuel Cells (AFC)

In alkaline fuel cells (AFC), a potassium hydroxide solution (in water) is used as an electrolyte and their operating temperature is the same as that for PEM fuel cells. They can use a cheaper catalyst than PEM type fuel cells. They have high efficiency (50–60%). CO can be tolerated by them, but the catalyst of these cells are prone to poisoning by *atmospheric CO_2.* They require the inlet air scrubber removal of CO_2 and suspended particulate matter and topping of circulating alkali.

These fuel cells are not expensive like PEM fuel cells but their output is low although their efficiency is higher than that of PEM fuel cells. *These type of fuel cells were used in the Apollo space programme for providing electrical power and drinking water.*

(4) Phosphoric Acid Fuel Cells (PAFC)

The phosphoric acid electrolyte fuel cells (PAFC) *operate at a temperature of about 220°C. Their efficiency is about 35% (maximum 42%), but they can be constructed to provide a high output of about one MW or more capacity.* The cause for the lower conversion efficiency is the generation of more heat in fuel stacks and a higher working temperature. If the heat emitted is used for heating purposes such as for heating water and air, and raising low-pressure steam, the overall efficiency can be raised. These fuel cells are tolerant to CO up to 1.5% and have a multifuel capability. Platinum is used as a catalyst so they are expensive; their size is large and they are heavy.

(5) Molten Carbonate Fuel Cells (MCFC)

The molten carbonate fuel cells (MCFC) work at high temperatures (for example, between 650–1000°C). These fuel cells can be constructed for high output, for example, 1 MW size or more and their efficiency is higher (about 45% to 60% (maximum)).

Catalysts are used here, but they are much less expensive than those used in PEM fuel cells. Nickel can be used as a catalyst. As the electrolyte is in molten carbonate form, the catalysts have special requirements of anti-corrosion measures and containment. They can be operated directly with *hydrocarbon gases and do not need an exclusive hydrogen supply*. The hydrocarbon gases when fed to the fuel cell get reformed internally.

Molten carbonate fuel cells can be integrated with gasifiers and electric power generation. They are typically used in grid supply applications.

(6) Direct Fuel Cell (DFC) Technology with Molten Carbonate Electrolyte

Molten carbonate salts serve as an electrolyte in the Direct Fuel Cell technology of Fuel Cell Energy Inc., USA. This technology is very efficient. The fuel cell operates at a high temperature of about 649°C (1200°F). At this temperature, the catalyst in the fuel cell is able to extract hydrogen from a hydrocarbon fuel internally within the fuel cell module itself. The DFC does not necessarily require pure hydrogen. The variety of fuels like natural gas, propane, gaseous fuel from coal gasification, anaerobic digester gas (biogas), etc., can be used. *The high temperature fuel cells produce heat along with electricity. The heat produced can be usefully used for water heating, the steam generation for power production or for industrial processes, air conditioning system, absorption chiller, etc.*

At a *waste water treatment plant in King County, Washington State*, the heat generated by the fuel cell is used for the anaerobic treatment of sludge in its conversion to biogas, which is then used as a fuel for a 1 MW DFC module (Fuel cell) power plant.

(7) Solid Oxide Fuel cells (SOFC)

The operating temperature of solid oxide fuel cells is very high (800–1000°C).

The electrolyte is a thin ceramic material that conducts *oxygen ions*. The charge carrier in this type of fuel cell is *oxygen*. The oxygen molecules from air are split into *oxygen ions* with the addition of *electrons* at the cathode. The oxygen ions go through the solid electrolyte and combine with hydrogen at the anode releasing electrons, which travel through the external load circuit providing power, and the by-product of heat.

The efficiency of the SOFCs is 60%, which is the highest of all the fuel cells. With the utilization of the high pressure steam generated in the process, the system becomes more efficient. The combined efficiency of more than 70% can be obtained when the fuel cell generation is integrated with steam turbine electric generators. These fuel cells can accept any hydrocarbon fuel and no reformer is necessary, like in molten carbonate fuel cells. They can also be integrated with gasifiers and an electric power generation system.

The advantages of SOFC's are:

(i) High efficiency, (ii) use of impure gases (for example, those obtained from gasifiers or the industrial processes), and (iii) the capability of being integrated in power generation or heat utilization systems (for space heating, industrial processes, etc.).

The necessity of the use of costly materials that can withstand high temperatures is a *disadvantage* of SOFC.

Some materials suited to the high temperatures of these fuel cells are:

For electrolyte: Yattria-stabilized zirconia (referred as 'YSZ')

For cathode: $(La, Sr)MnO_2$

For the anode: Ni-YS

For the IC: $(La,Sr)CrO_3$ (referred to as 'LSC')

For support: Partially stabilized Zirconia (referred as 'PSZ')

7.20.3 *Power Generation by H₂ through Fuel Cells—Some Examples* [45,49,60,62,72,101,107]

A pilot project demonstrating a hydrogen economy is operational on the Norwegian island of Ustrina, with the installation of a wind power plant and hydrogen production and storage. In periods when there is surplus wind energy, the excess power is used for generating hydrogen by electrolysis. The hydrogen is stored, and is available for power generation in periods when there is little wind. Similarly, a joint venture between NREL and Xcel Energy has combined wind power and hydrogen power is ongoing in Colorado.

A project on Stuart Island uses solar power to generate electricity. When excess electricity is available after the batteries are full, hydrogen is generated by electrolysis and stored for production of electricity by the fuel cell later.

In Italy, at Fusin, near Venice (in the Veneto region), a 12 MW power plant fueled by 100% hydrogen is operating since October 2009. Hydrogen is produced as a by-product in the adjacent petrochemical complex of Porto Marghera.

Many other plants (*like the 250 MW Hydrogen energy California power project in Western Kern County*) have been planned.

The Hydrogen Expedition is developing a hydrogen fuel cell-powered ship and using it to circumnavigate the globe, as a way to demonstrate the capability of hydrogen fuel cells. Some hospitals have installed combined electrolyzer-storage fuel cell units for local emergency power.

7.20.4 *Challenges to be Addressed for Hydrogen Use* [2,8,11,22,39,51,67,69,103,119,122,139,140,145,159,218]

Apart from the environmental concerns (see Para 7.21 ahead), following important challenges have to be addressed satisfactorily before we can go full steam ahead with hydrogen use.

- Where will the energy to extract the hydrogen from the water come from? (Hydrogen is an energy carrier, not an energy source, so the energy it delivers would ultimately have to be provided by a power-generating plant.)
- Increasing energy efficiency in production, storage, and transport of hydrogen.
- Development of suitable composite materials for pipelines because of the problem of *hydrogen embrittlement in steel* and the evaluation of these materials for hydrogen compatibility and constructing large diameter pipes suitable for pressure.
- Development of a suitable leak detection system and pipeline integrity monitoring system for long-distance hydrogen pipelines.
- Solving safety issues satisfactorily for end-use consumers.
- Reducing cost of production, storage, and transportation of hydrogen.
- Reducing cost of fuel cells and other systems needed for hydrogen utilization.
- Development of high-pressure cylinders (~700 bars) or other types of cheap, convenient, and safe storage systems for onsite or onboard utilization.
- Establishment of hydrogen gas pipeline network in high-demand areas.
- Development of hydrogen-fuelled IC engine having an operating life > 30,000 hours and costs comparable to existing petroleum-based IC engines.
- Efficiency improvement for different types of fuel cell systems in the range of 40–80 %.
- Development of fuel cell stacks having an operating life > 5,000 hours and a cost comparable to existing vehicles for transport applications.

7.21 Environmental Concerns in Hydrogen Production, Storage, and Use [2,11,20,23,39,51,59,77,98,119,139,174,194,204,216,222]

1. Hydrogen is produced generally from the electrolysis of water or by reforming fossil fuels. If it is made from electrolysis using power from a fuel-burning power station, more CO_2 is emitted than when the power from the fossil fuel station is directly used.

2. If the hydrogen is generated with renewable power for use in driving gas engines or using the same for production of power by other methods, it entails the more inefficient use of energy than the direct use of renewable power because of an extra conversion stage losses and distribution losses.

3. When the prime purpose of the electrolyser is to store surplus electricity generated by solar or wind power for subsequent use in a fuel cell, the "round trip efficiency" of the storage process (electricity to hydrogen and back to electricity) is between 30% and 50%. This compares unfavorably with battery storage where the round trip efficiency, known as the 'coulombic efficiency' in battery parlance, is over 90% for a lead acid battery and even more for a lithium battery.

4. If hydrogen is produced by reforming a fossil fuel and then using the hydrogen in internal combustion engines, it would produce more CO_2 than the direct use of the fossil fuel in internal combustion engines.

5. It may also be noted that the manufacture of hydrogen via steam reforming has a thermal efficiency of 75 to 80%. Additional energy will be required to liquefy or compress the hydrogen, and to transport it to the filling station via a truck or pipeline. The compressor has an efficiency of 90% or so.

6. Although the engines running on hydrogen do not produce CO_2 and particulate matter, but would produce oxides of nitrogen including nitrous oxide, and other air pollutants like any other fuel. Pollutants may include nitric acid and hydrogen cyanide.

7. Hydrogen leakage is dangerous and must be avoided. Apart from the danger of fire, significant leakage of molecular hydrogen may produce atomic hydrogen, which, on reaching the stratosphere, would be catalyst for ozone depletion. This will cause more ultraviolet radiations to reach earth, causing problems for life on earth. However, as reported, the leakage of hydrogen in the modern installations is lower than that for natural gas. *(As reported, the leakage rate for hydrogen is 0.1%, while the leakage for natural gas is 0.7% in Germany.)*

Therefore, the energy efficiency for hydrogen production, storage, and transport is low, although its utilization efficiency through fuel cells is quite high as mentioned earlier.

7.22 Hypthane—A Hybrid Fuel [39,51,65,93,106,204,236,240]

Compressed natural gas (CNG) is often thought of as the most promising alternative fuel for vehicles. Natural gas is a much more abundant fuel than petroleum and is often described as the cleanest of the fossil fuels, producing significantly less CO and non-methane hydrocarbon emissions than gasoline. When compared to diesel fuel for vehicles, the use of CNG almost eliminates particulate matter and SO_2 and significantly reduces other emissions like oxides of nitrogen, CO, and non-methane hydrocarbons. The generation of CO_2 is less in the case of combustion of natural gas than in the cases of gasoline and diesel fuels. Natural gas has 90% methane and has small quantities of other combustible gases like ethane, propane, and higher hydrocarbons and some incombustible gases like CO_2 and nitrogen. Natural gas fuel has a high H/C ratio (ratio of hydrogen to carbon) and a high research octane number (RON), causing the exhaust to be clean and allowing for high anti-knocking properties. Hydrogen is the most abundant element on earth, and is often thought of as the ideal alternative fuel. However, the current infrastructure does not support the widespread use of hydrogen as a fuel. In order to expand the role of hydrogen in the near term, one option is to use hydrogen for transportation by mixing it with natural gas and use it in IC engines. This new blended fuel is known as HCNG, or 'Hyphane' (i.e., hydrogen + methane).

Enrichment of natural gas with hydrogen for use in an internal combustion engine is a good method to improve the burn velocity because the laminar burning velocity of hydrogen is 2.9 m/s compared to the laminar burning velocity for methane of 0.38 m/s. Therefore, the use of hypthane as fuel would improve the cycle-by-cycle variations caused by the relatively poor lean-burn capabilities of the methane/natural gas engine. The hydrogen has rapid combustion speed characteristics with a wider combustion limit and low ignition energy. These characteristics of hydrogen would be imparted to enriched methane (Hypthane) and also the hydrogen to carbon ratio (H/C) would improve compared to CNG. The exhaust emissions (*especially the methane (CH_4), CO, and CO_2 from an engine using Hypthane (HCNG)*) would be lower as compared to the engine using CNG. The HCNG fuel will help to avoid problems associated with evaporative emissions and cold start enrichment as seen in gasoline engines. The high antiknock properties of the CNG portion in hydrogen-enriched fuel (due to the high activation energy) helps resist the self-ignition of fuel.

The fuel economy and thermal efficiency of a Hypthane-fueled engine would also be higher than that of CNG-fueled engine. Methane and hydrogen are complimentary fuels for vehicles. Methane has a relatively narrow flammability range that limits the fuel efficiency improvements in NOx emissions at lean air fuel ratios. The addition of a small amount of hydrogen extends the flammability range significantly and accelerates methane combustion. This is good for hybrid engines.

The existing infrastructure for CNG can be used for Hypthane (HCNG) by small amendments and precautions. In the meantime, hydrogen infrastructure can be readied using the experience gained. This would allow time for lowering the production cost of hydrogen and establishing the efficiency demands for the hydrogen economy. The experience and research on Hypthane-fueled engines may be applied to a hydrogen-fueled engine.

Problems Associated with the Hydrogen Enrichment of CNG

- One of the biggest challenges in the use of enriched compressed natural gas is to determine the most suitable hydrogen/natural gas ratio. The increase of the hydrogen fraction above a certain extent may give rise to abnormal combustion, pre-ignition, knock, and back-fire. This may be due to the low quench distance and higher burning velocity of hydrogen, which causes the combustion chamber walls to become hotter. The heat loss of the cooling water also increases.

- By increasing the hydrogen addition, the lean operation limit extends and the maximum brake torque (MBT) decreases. This points towards the possible interactions among the hydrogen fraction, ignition timing, and excess air ratio. Therefore, finding the optimal combination of the hydrogen fraction, ignition timing, and excess air ratio along with other parameters is a challenge.

- The NOx emissions from CNG are extremely low compared to traditional fuels (gasoline/diesel), but the addition of hydrogen to CNG causes an increase in NOx emissions. The addition of hydrogen has the opposite effect on the hydrocarbon emissions. It is therefore necessary to compromise at a hydrogen ratio for which the NOx and hydrocarbon emissions are equally low (*as the level of emissions from the use of fuels are probably the most important factor in determining whether the fuel is suitable as an alternative*).

- The current cost of hydrogen is higher than that of CNG and therefore, Hypthane (HCNG) is more expensive than CNG. The cost of hydrogen production would need to be lowered to improve the economics of producing HCNG.

Chemical reactions

Combustion reactions are:

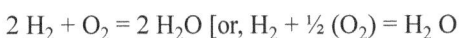

$CH_4 + O_2 = CO_2 + 2H_2O$

$2 H_2 + O_2 = 2 H_2O$ [or, $H_2 + \frac{1}{2} (O_2) = H_2 O$

If the fraction of hydrogen in HCNG is x by volume, taking the molecular weight of carbon as 12, hydrogen as 2, and oxygen as 32, content of oxygen in the air as 23.2% by weight, molar mass of methane as 16, and molar mass of water/steam as 18, the *mass fraction q of hydrogen to methane* will be given by the following equation:

$$q = (2x)/[2x + 16(1–x)] = (x)/(8 – 7x)$$

The combustion equation of such a gas mixture is as follows:

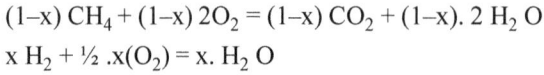

$$(1–x) CH_4 + (1–x) 2O_2 = (1–x) CO_2 + (1–x). 2 H_2 O$$
$$x H_2 + ½ .x(O_2) = x. H_2 O$$

Summing up:

$$(1–x) CH_4 + x H_2 + (1–x).2O_2 + ½. x(O_2) = (1–x) CO_2 + (1–x) 2H_2 O + xH_2O$$
$$\text{Or, } (1–x) CH_4 + xH_2 + (2–3/_2 .x) O_2 = (1–x) CO_2 + (2–x) H_2O$$

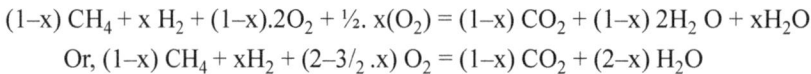

(For the [(1-x) CH$_4$ +x H$_2$] gas mixture, oxygen required by weight = 32.(2– 3/2 .x)
Stoichiometric ratio of oxygen to fuel $= 32 (2 – 3/_2 .x)/[2x + 16(1–x)]$
$$= (64 – 48x)/(16 – 14x)$$
$$= (32 – 24x)/(8 – 7x)$$
Stoichiometric air/fuel ratio $= (100/23.2)[(32 – 24x)/(8 – 7x)] =$
$$= 34.48 [(4 – 3x)/(8 – 7x)]$$

Assuming the lower calorific value (LCV) of hydrogen as 120 MJ/kg and that of methane as 50 MJ/kg fuel, the lower calorific value of Hypthane fuel would be worked out as:
LCV of Hypthane $= 120 q + 50. (1-q)$
$$= 120.\{(x)/(8 – 7x)\} + 50.[1– .\{(x)/(8 – 7x)\}]$$
$$= \{120 x + 50.(8-8x)\}/(8 – 7x) \text{ MJ/kg}$$

The NOx in the engine exhaust is higher in the case of Hypthane than in the case of methane, as already mentioned. The use of excess air reduces the generation of NOx because of a reduction in temperature. There may be a rise of the hydrocarbons in the engine exhaust at higher excess air, but the rise is smaller in the case of Hypthane than in the case of methane. It may be mentioned that, at present, natural gas is being enriched by 5 to 7% hydrogen.

7.23 The Future of Hydrogen and Recommendations [119c]

A report was prepared by the IEA at the request of the Govt. of Japan under its G20 Presidency in June 2019. This study provides an extensive and independent assessment of hydrogen where things stand now, the ways in which hydrogen can help achieve a clean, secure, and affordable energy future, and how we can go about realizing its potential. The IEA was given the responsibility of shaping the global policy on hydrogen. A rigorous analysis was conducted in close collaboration with governments, industries, and academia.

Hydrogen is enjoying unprecedented momentum around the world and could finally be set on a path to fulfil its longstanding potential as a clean energy source. Governments and companies need to take advantage of this opportunity.

The report on hydrogen will hopefully inform discussions and decisions among G20 countries, as well as those among other governments and companies across the world. Beyond this report, the IEA will remain focused on hydrogen, further expanding our expertise, monitoring the progress, and provide guidance on technologies, policies, and market design.

The number of policies and projects around the world is expanding rapidly. The time is now to scale up technologies and bring down costs to allow hydrogen to become widely used.

The IEA examines the full spectrum of energy issues including oil, gas, and coal supply and demand, renewable energy technologies, electricity markets, energy efficiency, access to energy, demand side management, and much more. Through its work, the IEA advocates policies that will enhance the reliability, affordability, and sustainability of energy in its 30 member countries, eight association countries, and beyond. **Please note that this publication is subject to specific restrictions that limit its use and distribution. The terms and conditions are available online at www.iea.org/t&c/Source: IEA.**

7.23.1 *Hydrogen Can Help Tackle Various Critical Energy Challenges*

1. It offers ways to decarbonize a range of sectors—including long-haul transport, chemicals, and iron and steel—where it is proving difficult to meaningfully reduce emissions.

2. Hydrogen is versatile. Technologies are already available today to produce, store, and move hydrogen and use energy in different ways. A wide variety of fuels can produce hydrogen, including renewables, nuclear energy, natural gas, coal and oil. It can be transported as a gas by pipelines or in liquid form by ships, much like LNG.

3. It can also help improve air quality and strengthen energy security. Outdoor air pollution also remains a pressing problem, with around 3 million people dying prematurely each year.

4. It can be transformed into electricity and methane to power homes and feed industry, and into fuels for cars, trucks, ships, and planes.

5. The recent successes of solar PV, wind, batteries, and electric vehicles have shown that policy and technology innovation have the power to build global clean energy industries.

6. Today, hydrogen is used mostly in oil refining and for the production of fertilizers. For it to make a significant contribution to clean energy transitions, it also needs to be adopted in sectors where it is almost completely absent at the moment, such as transport, buildings, and power generation.

7. However, the clean, widespread use of hydrogen in global energy transitions faces several challenges:
   i. Producing hydrogen from low-carbon energy is costly at the moment.
   ii. The development of hydrogen infrastructure is slow and holding back widespread adoption.
   iii. Hydrogen prices for consumers are highly dependent on how many refueling stations there are, how often they are used, and how much hydrogen is delivered per day.
   iv. Hydrogen is almost entirely supplied from natural gas and coal today. Hydrogen is already being used at an industrial scale all around the world, but its production is responsible for annual CO_2 emissions. We need to capture the CO_2 and provide more hydrogen.
   v. Regulations currently limit the development of a clean hydrogen industry.

7.23.2 *Key Recommendations of IEA to Scale up Hydrogen*

1. **Establish a role for hydrogen in long-term energy strategies.** Key sectors include refining, chemicals, iron and steel, freight and long-distance transport, buildings, and power generation and storage.

2. **Stimulate commercial demand for clean hydrogen.** Clean hydrogen technologies are available but costs remain challenging.

3. **Address investment risks of first movers.** New applications for hydrogen, as well as clean hydrogen supply and infrastructure projects should be encouraged.

4. **Support R&D to bring down costs.**

5. **Eliminate unnecessary regulatory barriers and harmonize standards.**

6. **Engage internationally and track progress.** Hydrogen production and use need to be monitored and reported on a regular basis to keep track of the progress towards long-term goals.

7. **Focus on four key opportunities to further increase momentum over the next decade.** By building on current policies, infrastructure, and skills, these mutually supportive opportunities can help to scale up infrastructure development, enhance investor confidence, and lower costs:
   • Make the most of existing industrial ports to turn them into hubs for lower-cost, lower-carbon hydrogen.
   • Use existing gas infrastructure to spur new clean hydrogen supplies.
   • Support transport fleets, freight, and corridors to make fuel-cell vehicles more competitive.
   • Establish the first shipping routes to kickstart the international hydrogen trade.

7.23.3 *Increasing Global Spending on Hydrogen Energy*

Supplying hydrogen to industrial users is now a major business around the world. The demand for hydrogen, which has grown more than threefold since 1975, continues to rise almost entirely supplied from fossil fuels, with 6% of global natural gas producing around three-fourths of the annual global 70 million tons of hydrogen and the remaining from 2% of global coal, including a small fraction from the use of oil and electricity. As a consequence, the production of hydrogen is responsible for CO_2 emissions of around 830 million tons of CO_2 per year.

The number of countries with polices that directly support investment in hydrogen technologies is increasing, along with the number of sectors they target. There are around 50 targets, mandates, and policy incentives in place today that directly support hydrogen, with the majority focused on transport. Over the past few years, global spending on hydrogen energy research, development, and demonstration by national governments has risen.

The production cost of hydrogen from natural gas is influenced by a range of technical and economic factors, with gas prices and capital expenditures being the two most important. Fuel costs are the largest cost component, accounting for between 45% and 75% of production costs. While less than 0.1% of the global dedicated hydrogen production today comes from water electrolysis, with declining costs for renewable electricity, in particular, from solar PV and wind, there is a growing interest in electrolytic hydrogen.

Dedicated electricity generation from renewables or nuclear power offers an alternative to the use of grid electricity for hydrogen production. In **power generation**, hydrogen is one of the leading options for storing renewable energy, and hydrogen and ammonia can be used in gas turbines to increase power system flexibility. Ammonia can also be used in coal-fired power plants to reduce emissions.

7.23.4 *Hydrogen's Status as a Future Fuel may be in Doubt* [119d,e]

For many, hydrogen is seen as the clean energy of the future. However, new research raises doubts. Industry has been promoting hydrogen as a reliable, next-generation fuel to power cars, heat homes, and generate electricity. It may, in fact, be worse for the climate than previously thought. However, a new peer-reviewed study on the climate effects of hydrogen casts doubt on its role in tackling the greenhouse gas emissions that are driving the catastrophic global warming.

Most hydrogen used today is extracted from natural gas in a process that requires a lot of energy and emits vast amounts of CO_2. Producing natural gas also releases methane, a potent greenhouse gas. While the natural gas industry has proposed capturing CO_2—creating emissions-free, "blue" hydrogen—even that fuel still emits more across its entire supply chain than simply burning natural gas, according to the paper published in the *Energy Science & Engineering* journal by researchers from the universities Cornell and Stanford.

"To call it a zero-emissions fuel is totally wrong", said Robert W. Howarth, an ecosystem scientist at Cornell. "What we found is that it's not even a low-emissions fuel, either."

To arrive at their conclusion, Dr. Howarth and Professor Mark Z. Jacobson of Stanford examined the life cycle greenhouse gas emissions of blue hydrogen. They accounted for both CO_2

emissions and the methane that leaks from wells and other equipment during natural gas production. The researchers assumed that 3.5% of the gas drilled from the ground leaks into the atmosphere, an assumption that draws on mounting research that has found that drilling for natural gas emits far more methane than previously known.

They also accounted for the natural gas required to power the carbon capture technology. In all, they found that the greenhouse gas footprint of blue hydrogen was more than 20% greater than that created by burning natural gas or coal for heat. Considering even a lower gas leak rate of 1.54% only reduced emissions slightly, but emissions from blue hydrogen remained higher than from simply burning the natural gas.

Over the past few years, the natural gas industry has begun heavily promoting hydrogen as a reliable, next-generation fuel to be used to power cars, heat homes, and burn in power plants. The industry has also pointed to hydrogen as a justification for continuing to build gas infrastructure like pipelines, saying that pipes that carry natural gas could in the future carry a cleaner blend of natural gas and hydrogen.

Many experts agree that hydrogen could eventually play an important role in energy storage or powering of certain types of transportation but, there is an emerging consensus that a wider hydrogen economy that relies on natural gas could be damaging to the climate. At current costs, it would also be very expensive.

The latest United Nations report found that slashing emissions of methane is much more important in tackling global warming than previously thought. In a new report, the U.N. warned that essentially all the rise in global average temperatures since the nineteenth century has occurred by burning of fossil fuels. An industry group founded in 2017 including BP, Shell, and other big oil and gas companies, did not provide immediate comment.

The hydrogen would ultimately need to be made using renewable energy to produce what the industry calls 'green hydrogen,' which uses renewable energy to split water into its constituent parts, hydrogen, and oxygen. That, they said, would eliminate the fossil and the methane leaks. Hydrogen made from fossil fuels could still act as a transition fuel but would ultimately be "a small contributor" to the overall sustainable hydrogen economy. First, we use blue, then we make it all green. Today, very little hydrogen is green because the process involved—electrolyzing water is very energy intensive.

For the foreseeable future, most hydrogen fuel will very likely be made from natural gas through an energy-intensive and polluting method called the 'steam reforming process', which uses steam, high heat, and pressure to break down the methane into hydrogen and carbon monoxide. Blue hydrogen uses the same process but applies carbon capture and storage technology, which involves capturing carbon dioxide before it is released into the atmosphere and then pumping it underground in an effort to lock it away. But that still doesn't account for the natural gas that generates the hydrogen, powers the steam reforming process, and runs the CO_2 capture.

Hydrogen is an important technology that will allow utilities to adopt much greater levels of renewables. National Grid referred to its Net Zero Plan, which says hydrogen will play a major role in the next few decades and that producing hydrogen from renewable energy was the plan's backbone.

7.24 Summary

This chapter is devoted to hydrogen as the renewable fuel. It is a secondary renewable fuel. Hydrogen, on combustion, yields only water or steam and no greenhouse gas is evolved so it is an absolutely clean and pollution-free fuel. [$2H_2 + O_2 = 2H_2O$]

However, hydrogen is not available in nature as a gas and it is to be produced by releasing it from water or other compounds of hydrogen. It requires energy, which should preferably be provided from green sources, preferably solar or wind power source so as not to add to the greenhouse gases to the atmosphere.

The characteristics and properties of hydrogen are quite favorable for its use as a green fuel, but in practice, there are many problems in adopting it.

Hydrogen has the widest explosive/ignition mix range with air of all the gases except acetylene; if a spark occurs, it will lead to an explosion, not merely a flame. Hydrogen-oxygen flames burn in the ultraviolet (pale blue) color range and are nearly invisible to the naked eye. Thus, it requires a flame detector to detect if a hydrogen leak is burning. Hydrogen is colorless and odorless and leaks cannot be detected by color and smell. Hydrogen gas explosions are more destructive and carry further. Thus, hydrogen fires burn quickly and will cause a severe explosion, but the fires tend to be relatively shortlived. The quenching gap of hydrogen is approximately one third that of other fuels and therefore, the hydrogen flames travel closer to the cylinder wall before they get extinguished, making them more difficult to quench than others and increase the tendency for backfire. Hydrogen flow or agitation may generate electrostatic charges resulting in sparks, which is sufficient to ignite flammable mixtures of hydrogen and air. Enclosed-area leaks of hydrogen pose the danger of asphyxiation since the hydrogen diffuses quickly to fill the volume.

Liquid hydrogen is required to be maintained at very low temperatures, its critical temperature being (–) 241°C and requires very effective thermal insulation. Filling or emptying a hydrogen tank or system requires purging by an inert gas. A number of precautions and preventive measures have been mentioned to avoid fire risks in the processes of generation, compressing, filling, transport, emptying/unloading, change of physical state, or use of hydrogen in any form/state.

The reasons for the promotion of hydrogen as an environmentally friendly fuel include: (i) it being an abundantly available element, which can be released from water (seawater/freshwater) to which it returns after burning; (ii) its availability as a by-product from some industries; (iii) its use as fuel for fuel cells/internal combustion engines; (iv) its ensured availability even after the fossil fuels get exhausted; (v) it providing more energy per kg weight compared to all other fuel gases and (vi) the nontoxicity of the products of its combustion; etc.

Hydrogen can be used in vehicles in several ways, like blending hydrogen with CNG (compressed natural gas), for example, up to 30% as gas engine fuel, onboard generation of hydrogen from metal hydrides and its use as fuel for fuel-cell-powered vehicles, or through establishing a compressed hydrogen distribution network for fuel-cell-powered vehicles. The application of hydrogen as fuel for aircraft and rocket engines is also actively being developed. Developments in all these directions have been mentioned.

The technologies and pathways for the production of hydrogen that are discussed in detail include: (i) standard electrolysis of water, preferably by using electricity from a renewable source only so that no carbon dioxide is added to the atmosphere; (ii) high-pressure electrolysis; (iii) proton exchange membrane water electrolysis; (iv) high-temperature electrolysis; (v) biocatalyzed electrolysis; (vi) Photo-Electro-Chemical Splitting of Water; (vii) Photo-Biological Production of Hydrogen from Algae; (viii) photoelectrocatalytic splitting of water; (ix) thermolysis of water or thermochemical production of hydrogen; (x) use of nuclear heat from gas-cooled nuclear reactors for splitting water; (xi) solar thermolysis; (xii) biomass conversion to hydrogen through gasification by fast pyrolysis/plasma torches/supercritical water; (xiii) H_2 production from biomass gasification by disposal of emitted CO_2; (xiv) hydrogen from glycerol—a by-product in the production of biodiesel; (xv) initial conversion of lignocellulosic materials to alcohols and then to H_2; (xvi) production of H_2 from biomass through biological fermentation; (xvii) microwave-assisted production of H_2 from rice straw; (xviii) hydrogen production from fossil fuels, including steam methane reforming, methane pyrolysis, partial oxidation of natural gas or hydrocarbons, plasma reforming of liquid hydrocarbons, coal gasification, and petroleum coke.

Several technologies for the separation of hydrogen from gasification products are discussed such as: (i) separation by pressure swing adsorption (PSA) and regeneration; (ii) chemical recovery of CO_2 from syngas (a mixture of CO and H_2) leaving H_2 with a small amount of impurities and producing pure CO_2 by chemical absorption by alkalo-amine (like MEA, DEA), aqueous ammonia, alkali carbonates, AMISOL DETA solution, or Sulfinol solution; (iii) physically absorption or adsorption of CO_2, membrane separation of CO_2 from syngas, or cryogenic separation of CO_2. Details of membrane separation/purification of hydrogen and suitable membranes for it are also given.

Various methods of storage of hydrogen including (i) absorption/adsorption, (ii) hydride formation, (iii) storage in underground formations, (iv) the alternative way of transporting it as alcohol and onboard generation of H_2 or synthetic methane production from hydrogen and using the same have been detailed.

Use of hydrogen as fuel for fuel cells, construction, and working of various types of fuel cells and their applications are discussed at length. Fuel cells in use for various applications are:

(1) proton exchange membrane (PEM) fuel cell/polymer electrolyte membrane fuel cell—(low temperature FC), (2) direct methanol fuel cells (DMFC)—(low temperature FC), (3) alkaline fuel cells (AFC)—(low temperature FC), (4) phosphoric acid fuel cells (PAFC)—(medium temperature FC), (5) molten carbonate fuel cell (MCFC)/direct fuel cell (DFC) Technology—(high temperature FC), and (6) solid oxide fuel cell (SOFC)—(high-temperature fuel cell).

The high-temperature fuel cells, which can use impure hydrogen, like molten carbonate fuel cells or solid oxide fuel cells can be conveniently used for power plants and also for some industrial applications. The phosphoric electrolyte fuel cells (PAFC) can be constructed for high output, for example, for one MW or more capacity and their efficiency can be raised by using the heat usefully. Direct Menthol Fuel Cells (DMFC) are very good for application in portable electronic equipment. PEM fuel cells, which have high efficiency and can be constructed for higher outputs, are considered very suitable for powering automobiles using hydrogen as fuel.

Problems of material failures by hydrogen embrittlement in pipelines and vessels, hydrogen leak detection and pipeline integrity monitoring, development of suitable composite materials for hydrogen pipes, and environmental concerns and challenges in the production and use of H_2 have been dealt with.

Enrichment of compressed natural gas along with associated problems and use of Hypthane (hybrid fuel CNG + hydrogen) have also been discussed.

Hydrogen is enjoying unprecedented momentum around the world and could finally be set on a path to fulfil its longstanding potential as a clean energy source. IEA has given key recommendations to scale up hydrogen. New research raises doubts. It may, in fact, be worse for the climate than previously thought. Climate scientists conclude this so-called clean alternative has a 20% larger carbon footprint than natural gas and coal when used for heat and approximately 60% greater than burning diesel oil for heat.

A Cornell Scientist says, "It is totally wrong to call it a zero-emissions fuel, it's not even a low-emissions fuel".

QUESTIONS

1. *Enumerate the properties of hydrogen.*
2. *What is meant by (a) burning speed, (b) quenching gap, (c) asphyxiation,*
   (d) flammability range with reference to hydrogen?
3. *Does the electrostatic charge generation affect the safety aspects of hydrogen?*
4. *Why is hydrogen being promoted as a future fuel?*
5. *Why is hydrogen leak detection a problem?*
6. *Enumerate the basic safety considerations for hydrogen systems as per ISO/TR 15916:2004.*
7. *Enumerate preventive measures for hydrogen accidents. How can you reduce the potential of deflagration leading to detonation?*
8. *What are the advantages and disadvantages of using hydrogen as a fuel for vehicles?*
9. *How can hydrogen be used as an onboard fuel for vehicles?*

10. *Give some instances of progress in fuel-cell-powered transport applications?*

11. *Compare the overall efficiencies of a battery-driven electric motor with a hydrogen fuel-cell-powered electric motor.*

12. *What are the different methods of producing hydrogen by splitting water?*

13. *Enumerate different pathways of producing hydrogen from renewables.*

14. *Write a short note on the standard electrolysis of water. How much energy is approximately consumed in producing hydrogen by electrolysis?*

15. *Compare the merits and demerits of standard electrolysis, high-pressure electrolysis, and high-temperature electrolysis.*

16. *What are the advantages of using mixed fuels in vehicles? Give an example of a hybrid fuel.*

17. *What are the merits of proton exchange membrane water electrolysis?*

18. *Describe the principle of hydrogen production by biocatalyzed electrolysis.*

19. *What is the role of the photocatalyst material of a semiconductor in a photoelectrochemical cell for the production of hydrogen?*

20. *Give examples of photosemiconductor materials which can split water (photoelectrochemical photolysis) under (i) UV radiation, (ii) a wide range of solar spectrum.*

21. *What is the role of hydrogenase enzymes in the photobiological production of hydrogen in green algae and cyanobacteria (called 'blue green algae')?*

22. *What are the differences between the hydrogenase enzymes of green algae and cyanobacteria?*

23. *What are the main problems in the production of hydrogen by blue green algae (cyanobacteria) and green algae?*

    What are the approaches to solve these problems?

24. *What is the thermolysis of water? What is the principle of thermochemical hydrogen production using the sulphur–iodine cycle?*

25. *Write short notes on (a) the photoelectrochemical production of hydrogen, and (b) the photobiological production of hydrogen.*

26. *Illustrate by diagram how you would use heat from a nuclear reactor to generate hydrogen.*

27. *What is meant by the solar thermolysis of water?*

28. *What are the options available for the production of hydrogen from biomass?*

29. *How would you convert biomass to hydrogen by (a) the gasification route, (b) biological fermentation?*

30. *What are the basic differences among the normal gasification, plasma gasification, and fast pyrolysis of biomass in connection with the production of hydrogen from biomass?*

31. *Explain the principle of hydrogen production by gasification of biomass by supercritical water.*

32. *What is a water-gas shift reaction? Explain the difference between a high-temperature water-gas shift reaction and a low-temperature water-gas shift reaction.*

33. *Can you use glycerol (which is produced as a by-product in biodiesel production) for the production of hydrogen?*

34. *How can you use the CO_2 from biomass gasification and reforming to hydrogen to enhance oil recovery from oil wells? Is it being done somewhere?*

35. *How would you produce hydrogen directly from biomass by using a biological method like dark fermentation?*

36. *How does photofermentation differ from dark fermentation in relation to hydrogen generation?*

37. How can you increase the production of hydrogen using biological methods and also extract the maximum energy from biomass?

38. Enumerate the different ways to separate hydrogen from the gases evolved during the production of hydrogen from biomass?

39. Describe the principle of pressure swing adsorption (PSA) method of separation of CO_2 from a mixture of H_2 and CO_2.

40. Enumerate the different technologies to produce pure CO_2 from the gases produced from the gasification of biomass with a view to produce H_2.

41. What are the differences among the chemical absorption, physical absorption, and physical adsorption technologies?

42. Enumerate the different chemicals used for the removal of CO_2 by chemical absorption, physical absorption, and physical adsorption technologies, respectively.

43. State the temperature conditions necessary for the cryogenic separation of CO_2 from a mixture of H_2 and CO_2.

44. What are molecular sieving and solution diffusion in relation to gas separation by membrane?

45. What do you understand by 'permeability', and 'diffusivity'? State Henry's law for ideal gas. What is the relation between the partial pressure and molar flux of a component gas in a mixture of gases according to Fick's law in the case of a membrane separation of gases?

46. What are the types of membranes in relation to hydrogen separation?

47. Name a palladium–copper alloy composite membrane used for hydrogen separation. What is electroless plating?

48. What are the different technologies used for the storage of hydrogen?

49. What are the various mechanisms by which hydrogen deteriorates the container/pipeline materials?

50. Describe the phenomena of hydrogen embrittlement of steel in relation to hydrogen storage and transport.

51. What are the differences between the phenomena of cold embrittlement and general hydrogen embrittlement of metals?

52. What is meant by the chemical storage of hydrogen? Give some examples of solid hydrides in which hydrogen is stored for use as a vehicle fuel.

53. Enumerate the advantages and disadvantages of the storage of hydrogen as compressed gas, as liquid hydrogen, or as a chemical hydride.

54. Describe a stand-alone system of the solar hydrogen power system.

55. What is meant by 'greenhouse neutral alcohol' or 'synthetic methane' made from 'elemental hydrogen'? How do they solve the problem of transport and distribution of hydrogen?

56. What would be the main constituents of the hydrogen infrastructure needed for replacing petroleum fuels by hydrogen?

57. Enumerate the types of hydrogen leak detectors/hydrogen sensors.

58. Briefly describe (a) a Chemo-Chromic hydrogen leak indicator and (b) an optical-fiber-based hydrogen sensor using chromogenic material.

59. What are the methods of leak testing for hydrogen pressure vessels and pipelines?

60. Write a brief note on pipeline materials for hydrogen, pointing out the metal deterioration mechanisms and problems in welded joints and parent material of pipelines.

61. What are the limitations of composite materials (like polymer/ fiberglass based materials) developed for hydrogen pipes?

62. *What threats exist for a long-distance hydrogen pipeline?*

63. *Enumerate the safety measures needed for a hydrogen system.*

64. *Write a brief note on hydrogen compressors.*

65. *Describe the principle of a fuel cell. How does it differ from other modes of power generation?*

66. *Enumerate the various types of fuel cells. Which of these fuel cells can use fuels other than hydrogen?*

67. *Briefly describe: (a) PEM fuel cell and (b) solid oxide fuel cell.*

68. *Describe a fuel cell hybrid power generation system with CO_2 recovery, the hydrogen as fuel for fuel cell being obtained from reforming the natural gas.*

69. *What are the challenges to be addressed before hydrogen can become a dominant fuel?*

70. *State the effects of enriching the natural gas with small amounts of hydrogen.*

71. *What are environmental concerns in hydrogen production, storage, and use?*

72. *Write a short note on hydrogen enriched compressed natural gas as fuel, pointing out the associated problems with CNG enrichment.*

References

1. A Basic over view of fuel cell technology- Smithsonian Institution American History.si.edu/fuelcells/basics. htm.

2. Abbasi, Tasneem and S.A. Abbasi. August 2011. Renewable hydrogen: Prospects and challenges. Renewable and Sustainable Energy Reviews 15(6): 3034-3040. https://doi.org/10.1016/j.rser.2011.02.026.

3. Acar, C. and I. Dincer. 2014. Comparative assessment of hydrogen production methods from renewable and non-renewable sources. Int. J. of Hydrogen Energy 39: 1–12.

4. Adamopoulos, Panagiotis Marios, Ioannis Papagiannis, Dimitrios Raptis and Panagiotis Lianos. 2019. Photoelectrocatalytic hydrogen production using a TiO2/WO3 bilayer photocatalyst in the presence of ethanol as a fuel. Catalysts 9(12): 976. https://doi.org/10.3390/catal9120976.

5. Adams, T. and G. Rawals. 2009. Hydrogen piping of fiber reinforced composite pipelines. Savannah river National Lab. [https://www.hydrogen.energy.gov/pdfs/review09/pd_42_adams.pdf].

6. Adegboye, Mutiu Adesina, Wai-Keung Fung and Aditya Karnik. 2019. Review recent advances in pipeline monitoring and oil leakage detection technologies: principles and approaches. Sensors 19: 2548. doi:10.3390/s19112548 www.mdpi.com/journal/.

7. Adhikari, Sushil, Sandun Fernando, Steven R.Gwaltney, S.D. Filip To, R .Mark Bricka, Philip H. Steele et al. September 2007. A thermodynamic analysis of hydrogen production by steam reforming of glycerol. Int. J. of Hydrogen Energy 32(14): 2875–2880. https://doi.org/10.1016/j.ijhydene.2007.03.023.

8. Advances in Hydrogen Separation and Purification with Membrane Technology. pp. 245–268. In book: Gandía, Luis M., Gurutze, Gurutze Arzamendi and Pedro M. Diéguez (eds.). Renewable Hydrogen Technologies Production, Purification, Storage, Applications and Safety. Chapter: 11. December 2013. Elsevier. DOI: 10.1016/B978-0-444-56352-1.00011-8 b.

9. Agnolucci, Paolo. October 2007. Hydrogen infrastructure for the transport sector. Int. J. of Hydrogen Energy, 32(15): 3526–3544. https://doi.org/10.1016/j.ijhydene.2007.02.016.

10. Akansu, S.O., Z. Dulger, N. Kahranman and N.T. Veziroglu. 2004. Internal combustion engines fueled by natural gas--hydrogen mixtures. Int. J. of Hydrogen Energy 29: 1527–1539.

11. Ali, Rosnazri, Tunku Muhammad Nizar Tunku Mansur, Nor Hanisah Baharudin and Syed Idris Syed Hassan. 2016. Environmental impacts of renewable energy. In Electric Renewable Energy Systems, pp. 519–546, Academic Press. https://doi.org/10.1016/B978-0-12-804448-3.00021-9.

12. Ali, Suhaib Muhammad and John Andrews. Jul 2006. Report Number: INIS-FR-6789. Low-cost storage options for solar hydrogen systems for remote area power supply. Conference: WHEC16: 16. World Hydrogen Energy Conference, Lyon (France), 13–16 Jun 2006. https://www.osti.gov/etdeweb/biblio/20946751.

13. Allenby, S., W.C. Chang, A. Megaritis and M.L. Wyszynski. 2001. Hydrogen enrichment: a way to maintain combustion stability in a natural gas fuelled engine with exhaust gas recirculation- the potential of fuel reforming. Proc. Institution of Mechanical Engineers: 215 Part D: 405–418.

14. Anastasios, Melis, Liping Zhang, Marc Forestier, Maria L. Ghirardi and Michael Seibert. Jan. 2000. Sustained photobiological hydrogen gas production upon reversible inactivation of oxygen-evolution in the green

alga *Chlamydomonas reinhardtii*. DOI: https://doi.org/10.1104/pp.122.1.127. [http://www.plantphysiol.org/content/122/1/127].

15. Andersson, Joakim and Stefan Grönkvist. May 2019. Large-scale storage of hydrogen. Int. J. of Hydrogen Energy 44(23): 11901–11919. https://doi.org/10.1016/j.ijhydene.2019.03.063.

16. Arregi, A., G. Lopez, M. Amutio, I. Barbarias, J. Bilbao and M. Olazar. 2016. Hydrogen production from biomass by continuous fast pyrolysis and in-line steam reforming. J. RSC Advances. 31.

17. Atsonios, K., A.K. Koumanakos, K.D. Panopoulos, A. Doukelis and E. Kakaras. 2015. Using palladium membranes for carbon capture in natural gas combined cycle (NGCC) power plants: process integration and techno-economics. pp. 247–285. Chapter 12 – In Palladium Membrane Technology for Hydrogen Production, Carbon Capture and Other Applications, Principles, Energy Production and Other Applications. Woodhead Publishing Series in Energy. https://doi.org/10.1533/9781782422419.2.247.

18. Attia, Nour F., Mini M. Menemparabath, Sivaram Arepalli and Kurt E. Geckeler. July 2013. Inorganic nanotube composites based on polyaniline: Potential room-temperature hydrogen storage materials. Int. J. of Hydrogen Energy 38(22): 9251–9262. https://doi.org/10.1016/j.ijhydene.2013.05.049.

19. Avasthi, Kartik S, Ravaru Narasimha Reddy and Sanjay Patel. 2013. Challenges in the production of hydrogen from glycerol-a biodiesel byproduct via steam reforming process. Procedia Engineering 51: 423–429.

20. Balat, Mustafa and Mehmet Balat. May 2009. Political, economic and environmental impacts of biomass-based hydrogen. Int. J. of Hydrogen Energy 34(9): 3589–3603. https://doi.org/10.1016/j.ijhydene.2009.02.067.

21. Baldwin, D. 2009. Design and Development of High Pressure Hydrogen Storage Tank for Storage and Gaseous Truck Delivery. US Department of Energy Program Review – Hydrogen Delivery, Contract DE-FG36-08GO18062.

22. Barbir, Frano. 2013. Chapter Ten - Fuel Cell Applications. PEM Fuel Cells (Second Edition). Pages 373–434, Academic Press. https://doi.org/10.1016/B978-0-12-387710-9.00010-2.

23. Barbir, Frano. May 2005. PEM electrolysis for production of hydrogen from renewable energy sources. Solar Energy 78(5): 661–669. https://doi.org/10.1016/j.solener.2004.09.003.

24. Basile, A. and P. Millet. 2013. Inorganic membrane reactors for hydrogen production: an overview with particular emphasis on dense metallic membrane materials. In Handbook of Membrane Reactors: Fundamental Materials Science, Design and Optimization.

25. Beck, Fiona J. 2019. Rational Integration of Photovoltaics for Solar Hydrogen Generation. ACS Appl. Energy Mater. 2(9): 6395–6403. https://doi.org/10.1021/acsaem.9b01030.

26. Bernardo, P. and J.C. Jansen. 2015. Polymeric membranes for the purification of hydrogen. In Compendium of Hydrogen Energy.

27. Bévenot, X., A. Trouillet, C. Veillas, H. Gagnaire and M. Clément. December 2001. Surface plasmon resonance hydrogen sensor using an optical fibre. Measurement Science and Technology 13(1). [http://iopscience.iop.org/article/10.1088/09570233/13/1/315/meta].

28. Bhattacharya, R.K. Oct 2004. Emerging Technologies. First Int. Conf. on Renewable energy, CBIP, New Delhi. Corp. R&D Division, BHEL, Hyderabad, India.

29. Biological production of hydrogen. [https://en.wikipedia.org/wiki/Biological_hydrogen_production_(algae)].

30. Bose, Debajyoti, Meenal Arora and Amarnath Bose. 2017. Renewable Electrolysis using Graphene electrodes for Solar water splitting. Int. J. of Chem Tech Research 10(4): 103–114. IJCRGG. ISSN: 0974-4290.

31. Broda, Marcin, Vasilije Manovic, Qasim Imtiaz, Agnieszka M. Kierzkowska, Edward J. Anthony and Christoph R. Müller. April 2013. High-Purity Hydrogen via the Sorption-Enhanced Steam Methane Reforming Reaction over a Synthetic CaO-Based Sorbent and a Ni Catalyst. Environ. Sci. Technol. 47(11): 6007–6014. https://doi.org/10.1021/es305113p.

32. Bünger, U., J. Michalski, F. Crotogino and O. Kruck. 2016. Large-scale underground storage of hydrogen for the grid integration of renewable energy and other applications. Pages 133-163. In compendium of Hydrogen Energy. Volume 4: Hydrogen Use, Safety and the Hydrogen Economy. Woodhead Publishing Series in Energy. https://doi.org/10.1016/B978-1-78242-364-5.00007-5Ge.

33. Carbon-neutral fuel, From Wikipedia, the free encyclopedia. https://en.wikipedia.org/wiki/Carbon-neutral_fuel.

34. Carmo, Marcelo et al. A review on PEM water electrolysis. [http://www.sciencedirect.com/science/article/pii/S0360319913002607?via%3Dihub].

35. Carpentieri, Matteo, Andrea Corti and Lidia Lombardi. July 2005. Life cycle assessment (LCA) of an integrated biomass gasification combined cycle (IBGCC) with CO2 removal. Energy Conversion and Management. 46(11–12): 1790–1808. https://doi.org/10.1016/j.enconman.2004.08.010.

36. CBIP (Central board of irrigation & power). 2008. Water & Energy Digest Energy section. April–June.

37. Chaiwatanodom, Paphonwit, Supawat Vivanpatarakij and Suttichai Assabumrungrat. February 2014. Thermodynamic analysis of biomass gasification with CO2 recycle for synthesis gas production. Applied Energy 114: 10–17. https://doi.org/10.1016/j.apenergy.2013.09.052.

38. Chang, Pao-Long, Chiung-WenHsu and Po-ChienChang. October 2011. Fuzzy Delphi method for evaluating hydrogen production technologies. Int. J. of Hydrogen Energy 36(21): 14172–14179. https://doi.org/10.1016/j.ijhydene.2011.05.045.

39. Chaparro, Antonio M. and Alba M. Fernández-Sotillo. 2018. Research trends in Portable Hydrogen Energy Systems.

40. Chemical Hydrogen Storage Materials. https://www.energy.gov/eere/fuelcells/chemical-hydrogen-storage-materials.

41. Chen, Qi, Guozheng Fan, Hongwei Fu, Zhaosheng Li and Zhigang Zou. 2018. Tandem photoelectrochemical cells for solar water splitting. J. Advances in Physics: X. 3(1). Published online: 15 Sep 2018. https://doi.org/10.1080/23746149.2018.1487267.

42. Chen, W., L.Z. Ouyang, J.W. Liu, X.D. Yao, H. Wang, Z.W. Liu et al. August 2017. Hydrolysis and regeneration of sodium borohydride (NaBH4) – A combination of hydrogen production and storage. J. of Power Sources 359(15): 400–407. https://doi.org/10.1016/j.jpowsour.2017.05.075.

43. Clarke, Dan. 2009. Pipeline Integrity Monitoring: Developments in Non-Intrusive Flow-Through Devices. Proc. Pipeline Technology Conference.https://www.pipelineconference.com/sites/default/files/papers/PTC%202009%202.1%20Clarke.pdf.

44. Clyne, T.W. and D. Hull. 2019. An Introduction to Composite Materials, Cambridge University Press.

45. Cohen, M. and G.C. Snow. 2008. Hydrogen Delivery and Storage Options for Backup Power and Off-Grid Primary Power Fuel Cell Systems. Proc. IEEE Intelec.

46. Compendium of Hydrogen Energy: Hydrogen Storage, Distribution and Infrastructure. Volume 2: 2015, Ram Gupta, Angelo Basile, T. Nejat Veziroglu (eds.). A Volume in Woodhead Publishing Series in Energy.

47. Comprehensible Renewable Energy: Volume 4: Fuel Cells and Hydrogen Technology.

48. Cooper, A.C., D.E. Fowler, A.R. Scott et al. 2005. Hydrogen storage and delivery by reversible hydrogenation of liquid-phase hydrogen carriers. Abstracts of Papers of the American Chemical Society 229. U868–U868. Part:1 Meeting Abstract: 113-FUEL.

49. Corbo, Pasquale, Fortunato Migliardini and Ottorino Veneri. January 2011. Hydrogen Fuel Cells for Road Vehicles. pp. 71–102. Springer.

50. Correa, Catalina Rodriguez and Andrea Kruse. March 2018. Supercritical water gasification of biomass for hydrogen production – Review. The Journal of Supercritical Fluids 133(Part 2): 573–590. https://doi.org/10.1016/j.supflu.2017.09.019.

51. Crabtree, George W. Mildred S. Dresselhaus and Michelle V. Buchanan. Dec. 2004. The Hydrogen Economy. Physics Today: 39. http://scitation.aip.org/journals/doc/PHTOAD-ft/vol_57/iss_12/39_1.shtml. Retrieved 2008-05-09.

52. Das, L.M. 2016. Seven Hydrogen-fueled internal combustion engines. Compendium of Hydrogen Energy. Volume 3: Hydrogen Energy Conversion. pp. 177–217. Woodhead Publishing Series in Energy. https://doi.org/10.1016/B978-1-78242-363-8.00007-4.

53. Dave, Ashok, Medha Dave, Ye Huang, Sina Rezvani and Neil Hewitt. June 2016. Process design for CO2 absorption from syngas using physical solvent DMEPEG. Int. J. of Greenhouse Gas Control. 49: 436–448. https://doi.org/10.1016/j.ijggc.2016.03.015.

54. Detchusananard, Thanaphorn, Karittha Im-orb, Pimporn Ponpesh and Amornchai Arpornwichanop. September 2018. Biomass gasification integrated with CO2 capture processes for high-purity hydrogen production: Process performance and energy analysis. Energy Conversion and Management 171: 1560–1572. https://doi.org/10.1016/j.enconman.2018.06.072.

55. Dincer, I. and A.S. Joshi. 2013. Solar based hydrogen production systems. Springer, New York.

56. Direct water splitting by new oxide semiconductor. https://www.electrochem.org/dl/ma/203/pdfs/2827.pdf.

57. Douglas Way, J. Palladium/Copper Alloy Composite Membranes for High Temperature Hydrogen, Separation from Coal-Derived Gas Streams. Final Technical Progress Report. Chemical Engineering and Petroleum Refining Department, Colorado School of Mines, Golden, CO 80401-1887. www.fischertropsch.org/DOE/DOE_reports/40585/40585.

58. Driving the hydrogen infrastructure. Jan /Feb.2003. Refocus. Project up to date.

59. Dutta, Dr. Sukanya, Parvinder Chawla et al. June 2003. Hydrogen –the ultimate fuel. Science Reporter, CSIR, India.

60. Eberle, Ulrich, Bernd Mueller and Rittmar von Helmolt. 2012. Fuel cell electric vehicles and hydrogen infrastructure: status 2012. Energy & Environmental Science 5(10): 8780. doi:10.1039/C2EE22596D.

61. Eberle, Ulrich, Michael Felderhoff and Ferdi Schüth. December 2009. Chemical and Physical Solutions for Hydrogen Storage. Angewandte Chemie International Edition 48(36): 6608–30. DOI: 10.1002/anie.200806293. https://www.researchgate.net/publication/26666827_Chemical_and_Physical_Solutions_for_Hydrogen_Storage.

62. Edwards, P.P., V.L. Kuznetsov, W.I.F. David and N.P. Brandon. December 2008. Hydrogen and fuel cells: Towards a sustainable energy future. Energy Policy 36(12): 4356–4362. https://doi.org/10.1016/j.enpol.2008.09.036

63. EERE Information Centre, 1-877-EERE-INFO (1-877-337-3463). www.eere.energy.gov/informationcenter.

64. Eftekhari, Ali and Fang Baizeng. 2017. Electrochemical hydrogen storage: Opportunities for fuel storage, batteries, fuel cells, and supercapacitors. Int. J. of Hydrogen Energy 42(40): 25143–25165. doi:10.1016/j.ijhydene.2017.08.103.

65. El-Ghafour, S.A.A., A.H.E. El-dein and A.A.R. Aref. 2010. Combustion characteristics of natural gas-hydrogen hybrid fuel turbulent diffusion flame. Int. J. of Hydrogen Energy 35: 2556e65.

66. Evangelopoulou, Stavroula, Alessia De Vita, Georgios Zazias and Pantelis Capros. July 2019. Energy System Modelling of Carbon-Neutral Hydrogen as an Enabler of Sectoral Integration within a Decarbonization Pathway. Energies 12(13): 2551. https://doi.org/10.3390/en12132551.

67. Faraji, Ghader, Hyoung Seop Kim and Hessam Torabzadeh Kashi. 2018. Severe Plastic Deformation, Methods, Processing and Properties. Chapter 8 - Physical, Chemical, and Functional Properties of UFG and NS Metals. pp. 259–274. Elsevier. https://doi.org/10.1016/C2016-0-05256-7.

68. Fedorov, A., A. Tsygankov, K. Rao and D. Hall. November 1998. Hydrogen photoproduction by Rhodobacter sphaeroides immobilised on polyurethane foam. Biotechnology Letters 20(11): 1007–1009.

69. Ferreira-Aparicio, Paloma and Antonio M. Chaparro (eds.). 2018. Portable Hydrogen Energy Systems: Fuel Cells and Storage Fundamentals and Applications 1st Edition, Academic Press.

70. Fiber reinforced composite pipelines—DOE Hydrogen Program. [https://www.hydrogen.energy.gov/pdfs/review13/pd022_adams_2013_o.pdf] and [https://www.hydrogen.energy.gov/pdfs/review15/pd022_rawls_2015_o.pdf.

71. Forsberg, Charles W. May 2009. Meeting U.S. liquid transport fuel needs with a nuclear hydrogen biomass system. Int. J. of Hydrogen Energy 34(9): 4227–4236. https://doi.org/10.1016/j.ijhydene.2008.07.110.

72. Fuel cell from Wikipedia, the free encyclopedia.https://en.wikipedia.org/wiki/Fuel_cell.

73. Galvita, V.V., G.L. Semin, V.D. Belyaev, T.M. Yurieva and V.A. Sobyanin. August 2001. Production of hydrogen from dimethyl ether. Applied Catalysis A: General. 216(1-2): 85–90; https://doi.org/10.1016/S0926-860X(01)00540-3.

74. Gardner, Dale. Jan /Feb 2009. Hydrogen production from renewables. Renewable energy focus.

75. Gąsior, Paweł and Jerzy Kaleta. July 2014. Waste Hydrogen Pipelines Monitoring in Modern Power Plant. Seventh European Workshop on Structural Health Monitoring, La Cité, Nantes, France.

76. Graetz, J. and J.J. Reilly. 2007. Kinetically stabilized hydrogen storage materials. Scripta Materialia 56(10): 835–839.

77. Granovskii, M., I. Dincer and M.A. Rosen. 2006. Environmental and economic aspects of hydrogen production and utilization in fuel cell vehicles. Journal of Power Sources 157: 411–421.

78. Grätzel, M. and A.J. McEvoy. March 2004. Hydrogen Production by Solar Photolysis of Water. presented at the American Physical Society Symposium on Basic Research for the Hydrogen Economy, Montreal, Canada.

79. Gupta, Ram B. (Ed.). 2008. Hydrogen Fuel, Production, transport and Storage. CRC Press.

80. Hajjaji, N., M.N. Pons, V. Renaudin and A. Houas. 2013. Comparative life cycle assessment of eight alternatives for hydrogen production from renewable and fossil feedstock. Journal of Cleaner Production 44: 177–189.

81. Hauch, A., S.H. Jensen et al. 2005. Electrodes in solid oxide electrolyzer cells. Proc. of the 26th Riseo int. symposium on materials science: solid state electrochemistry, Roside, Denmark. pp. 203–208.

82. Herring, J. Stephen, James E.O'Brien, Carl M. Stoots, G.L. Hawkes, Joseph J. Hartvigsen and Mehrdad Shahnam. March 2007. Progress in high-temperature electrolysis for hydrogen production using planar SOFC technology. Int. J. of Hydrogen Energy 32(4): 440–450. https://doi.org/10.1016/j.ijhydene.2006.06.061.

83. Heung, L. Kit. Using Metal Hydride to Store Hydrogen. Savannah River Technology Center, Aiken, SC 29808 USA. [https://en.wikipedia.org/wiki/Hydrogen_storage].

84. High Pressure Electrolysis for Hydrogen generation. [https://en.wikipedia.org/wiki/High-pressure_electrolysis].

85. Holladay, J.D., J. Hu, D.L. King and Y. Wang. 2009. An overview of hydrogen production technologies. Catalysis Today 139(4): 244–260.

86. Holladay, J.D., Y. Wang and E. Jones. 2004. Review of developments in portable hydrogen production using micro reactor technology. Chemical Reviews 104(10): 4767–4790.

87. Honnery, Damon and Patrick Moriarty. March 2011. Energy availability problems with rapid deployment of wind-hydrogen systems. Int. J. of Hydrogen Energy 36(5): 3283–3289. https://doi.org/10.1016/j.ijhydene.2010.12.023.

88. How Fuel Cells Work- Fuel Economy. [www.fueleconomy.gov/feg/fcv_PEM.shtml].

89. http://www.pipeline-leak-detection-congress-usa.com/8/venue/53/venue-information/.

90. https://en.wikipedia.org/wiki/Fuel_cell.

91. https://hydrogeneurope.eu/hydrogen-applications.

92. https://www.emerson.com/en-ca/industries/automation/oil-gas/pipeline-integrity-leak-detection.

93. https://www.google.co.in/search?q=HYPTHANE+%E2%80%94+A+HYBRID+FUEL&tbm=isch&source=univ&sa=X&ved=2ahUKEwiR-ZeghL_mAhXayDgGHeJ3ArkQsAR6BAgHEAE&biw=1366&bih=657.

94. https://www.google.com/search?q=Various+Types+of+Fuel+Cells&tbm=isch&source=univ&sa=X&ved=2ahUKEwjZ1pqXk9bnAhVn8XMBHVR4BpEQsAR6BAgBEAE&biw=819&bih=542.

95. https://www.pipeline-journal.net/articles/leak-detection.

96. Hugo, André, Paul Rutter, Stratos Pistikopoulos, Angelo Amorelli and Giorgio Zoia. December 2005. Hydrogen infrastructure strategic planning using multi-objective optimization. Int. J. of Hydrogen Energy. 30(15): 1523–1534.

97. Huitt, William M. Piping material for hydrogen service. http://www.wmhuittco.com/images/Hydrogen Piping.pdf.

98. Hunt, John. 4th May 2017. Hydrogen-powered transport: a zero-emission future. http://environmentjournal.online/articles/hydrogen-powered-transport-zero-emission-future/.

99. Hussain, Fayaz, Saidur Rahman, N. Razali and Fadhilah Shikh Anuar. October 2012. An overview of hydrogen as a vehicle fuel. Renewable and Sustainable Energy Reviews 16(8): 5511–5528. DOI: 10.1016/j.rser.2012.06.012.

100. HY-ALERTA™ 500 Handheld Hydrogen Leak Detector. http://h2scan.com/products/hy-alerta/500.

101. Hydrogen based power plants
     (a) https://en.wikipedia.org/wiki/Fusina_hydrogen_power_station.
     (b) https://hydrogenenergycalifornia.com/projectnews/page/4.
     (c) www.businesskorea.co.kr/.../19106-world's-first-hydrogen-power-plant-hanwha-ener.

102. Hydrogen economy From Wikipedia, the free encyclopedia. https://en.wikipedia.org/wiki/Hydrogen_economy.

103. Hydrogen economy. [https://en.wikipedia.org/wiki/Hydrogen_economy]

104. Hydrogen from solar energy. July /August 2003. Refocus. General News.

105. Hydrogen Fuel Cells. [https://www.hydrogen.energy.gov/pdfs/doe_fuelcell_factsheet.pdf].

106. Hydrogen fuel enhancement from Wikipedia, the free encyclopedia. https://en.wikipedia.org/wiki/Hydrogen_fuel_enhancement.

107. Hydrogen Fuelled Electricity Generation (Fuel cells). [www.mpoweruk.com/hydrogen_fuel.htm].

108. Hydrogen Infrastructure. https://www.nproxx.com/hydrogen-storage/hydrogen-infrastructure/

109. Hydrogen internal combustion engine vehicle from Wikipedia. the free encyclopedia. https://en.wikipedia.org/wiki/Hydrogen_internal_combustion_engine_vehicle.

110. Hydrogen internal combustion engine vehicle. [https://en.wikipedia.org/wiki/Hydrogen_vehicle].

111. Hydrogen leak testing. https://en.wikipedia.org/wiki/Hydrogen_leak_testing.

112. Hydrogen Powered Buses. [http://www.airproducts.com/industries/Transportation/Mass-Transit/Hydrogen-Powered-Buses.aspx].

113. hydrogen powered rockets. https://en.wikipedia.org/wiki/Hydrogen_technologies#Hydrogen_powered_rockets.

114. Hydrogen production and storage R&D, IEA. https://www.iea.org/publications/freepublications/publication/hydrogen.pdf.

115. Hydrogen production and storage. https://scholar.google.co.in/scholar?q=hydrogen+production+and+storage+ppt&hl=en&assdt=0&as_vis=1&oi=scholart.

116. Hydrogen production from steam reforming of bioethanol using Cu=Ni=K= -Al2O3 catalysts. Effect of Ni. http://www.di.fcen.uba.ar/diq/lpc/produccion.pdf.

117. Hydrogen production from steam reforming of ethanol and glycerol over ceria-supported metal catalysts. https://www.sciencedirect.com/science/article/pii/S0360319906005453.

118. Hydrogen production html. Chapter 5—Renewable biological systems for alternative sustainable energy production-, FAO Corporate repository, Agriculture and Consumer Protection. [http://www.fao.org/docrep/w7241e/w7241e0g.htm].

119. Hydrogen Production Technologies. Chapter 8. The Hydrogen Economy: Opportunities, Costs, Barriers and R&D needs. 2004. Also see
    (a) https://academic.oup.com/ce/article/1/1/90/4743500,
    (b) Hydrogen production – Wikipedia, https://en.wikipedia.org › wiki › Hydrogen_production,
    (c) https://www.iea.org/reports/the-future-of-hydrogen),
    (d) [https://indianexpress.com/article/technology/science/hydrogen-fuel-future-new-research-7452650/],
    (e) https://www.smithsonianmag.com/smart-news/blue-hydrogen-20-worse-burning-coal-study-states-180978451/

120. Hydrogen Recovery by Pressure Swing Adsorption. [https://www.lindeengineering.com/.../HA_H_1_1_e_09_150dpi_NB19_6130.pdf?v...].

121. Hydrogen Safety Fact Sheet (PDF). National Hydrogen Association. http://www.hydrogenassociation.org/general/factSheet_safety.pdf.

122. Hydrogen safety. [https://en.wikipedia.org/wiki/Hydrogen_safety].

123. Hydrogen sensor from Wikipedia, the free encyclopedia. https://en.wikipedia.org/wiki/Hydrogen_sensor.

124. Hydrogen Sensor-Wikipedia. [https://en.wikipedia.org/wiki/Hydrogen_sensor].

125. Hydrogen Separation Membrane. https://www.sciencedirect.com/topics/engineering/hydrogen-separation-membrane

126. Hydrogen Separation Membranes. NCHT- EERC Energy & Environmental Research Center at www.undeerc.org/NCHT/pdf/EERCMH36028.pdf.

127. Hydrogen storage From Wikipedia, the free encyclopedia. https://en.wikipedia.org/wiki/Hydrogen_storage.

128. Hydrogen storage From Wikipedia, the free encyclopedia. https://en.wikipedia.org/wiki/Hydrogen_storage

129. Hydrogen vehicles. (https://en.wikipedia.org/wiki/Hydrogen_technologies#Hydrogen_vehicles.

130. Ji, F., C. Li, and J. Zhang. 2010, ACS Appl. Mater. Interface 2,1674.

131. Jin, Hui, Youjun Lu, Leijin Guo, Ximin Zhang and Aixia Pei. 2014. Hydrogen Production by Supercritical Water Gasification of Biomass with Homogeneous and Heterogeneous Catalyst. State Key Laboratory of Multiphase Flow in Power Engineering, Xi'an Jiaotong University, Xi'an 710049, China. [https://www.hindawi.com/journals/acmp/2014/160565/].

132. Jones, Lawrence W. 1970. Toward a liquid hydrogen fuel economy. Engineering Technical Report UMR2320. University of Michigan.

133. José Alves, Helton, Cícero Bley Junior, Rafael RickNiklevicz, Elisandro PiresFrigo, Michelle SatoFrigo, Carlos HenriqueCoimbra-Araújo. May 2013. Overview of hydrogen production technologies from biogas and the applications in fuel cells. Int. J. of Hydrogen Energy 38(13): 5215–5225. https://doi.org/10.1016/j.ijhydene.2013.02.057.

134. Kalamaras, M. Christos and Angelos M. Efstathiou. 2013. Hydrogen Production Technologies: Current State and Future Developments. Conference Papers in Energy Volume: Article ID 690627, 9 pages. http://dx.doi.org/10.1155/2013/690627.

135. Kan, H.M., M. Sun and N. Zhang. 2014. Nanomaterials for Hydrogen Storage. Applied Mechanics and Materials 587: 216–219.

136. Kojima, Y., Y. Kawai, M. Kimbara, H. Nakanishi and S. Matsumoto. 2004. Hydrogen Generation by Hydrolysis Reaction of Lithium Borohydride. Int. J. of Hydrogen Energy 29: 1213–1217.

137. Korobtsev, S.V. Feb 8, 2006. Development of fundamental technologies of production and use of hydrogen based on solid oxide electrochemical reversible high temperature systems. report on round table Meeting-Russian Research and Development in the field of Hydrogen Technologies, Moscow.

138. Kosar, Sonya, Yuriy Pihosh, Ivan Turkevych and K. Mawatari. April 2016. Tandem photovoltaic-photoelectrochemical GaAs/InGaAsP-WO3/BiVO4 device for solar hydrogen generation. Japanese J. of Applied Physics 55(4). DOI: 10.7567/JJAP.55.04ES01.

139. Kothari, Richa, D. Buddhi and R.L. Sawhney. February 2008. Comparison of environmental and economic aspects of various hydrogen production methods. Renewable and Sustainable Energy Reviews 12(2): 553–563. https://doi.org/10.1016/j.rser.2006.07.012.

140. Kundu, P.P. and K. Dutta. 2016. Hydrogen fuel cells for portable applications. In Compendium of Hydrogen Energy.

141. Kurup, Vinod, K.D. Chudasama, G.S. Grewal, V. Shrinet and A.K. Singh. Oct 2004. An Effort Towards Hydrogen Energy. 1st Int. Conf. on Renewable Energy. CBIP, New Delhi.

142. Laguna-Bercero, M.A. April 2012. Recent advances in high temperature electrolysis using solid oxide fuel cells: A review. J. of Power Sources 203: 4–16. https://doi.org/10.1016/j.jpowsour.2011.12.019.

143. Lakaniemi, Aino-Maija, Perttu E.P. Koskinen, Laura M. Nevatalo, Anna H.Kaksonen and Jaakko A. Puhakka. February 2011. Biogenic hydrogen and methane production from reed canary grass. Biomass and Bioenergy 35(2): 773–780. https://doi.org/10.1016/j.biombioe.2010.10.032.

144. Le Valley, L. Trevor, Anthony R. Richard, Maohong Fan. October 2014. The progress in water gas shift and steam reforming hydrogen production technologies – A review. Int. J. of Hydrogen Energy 39(30): 16983–17000. https://doi.org/10.1016/j.ijhydene.2014.08.041.

145. Levin, B. David and Richard Chahine. May 2010. Challenges for renewable hydrogen production from biomass. Int. J. of Hydrogen Energy 35(10): 4962–4969; https://doi.org/10.1016/j.ijhydene.2009.08.067.

146. Li, Zheng, Dan Gao, Le Chang, Pei Liu and Efstratios N. Pistikopoulos. October 2008. Hydrogen infrastructure design and optimization: A case study of China. Int. J. of Hydrogen Energy 33(20): 5275–5286. https://doi.org/10.1016/j.ijhydene.2008.06.076.

147. Lipman, T.E. July 2004. What Will Power the Hydrogen Economy? Present and Future Sources of Hydrogen Energy. Institute of Transportation Studies. University of California – Davis. UCD-ITS-RR-04-10.

148. Lipman, Timothy. May 2011. An Overview of Hydrogen Production and Storage Systems with Renewable Hydrogen Case Studies. Prepared for: Clean Energy States Alliance telipman@berkeley.edu. [https://www.cesa.org/assets/2011-Files/Hydrogen-and-Fuel-Cells/CESA-Lipman-H2-prod-storage-050311.pdf].

149. Liquid Hydrogen. [https://en.wikipedia.org/wiki/Liquid_hydrogen].

150. Liu, Ke, Chunshan Song and Velu Subramani. 2010. Hydrogen and Syngas Production and Purification Technologies. (Ebook). John wiley.

151. Luo, W., P.G. Campbell, L.N. Zakharov and S.Y. Liu. 2013. A single-component liquid-phase hydrogen storage material. J. of the American Chemical Society 135(23): 8760–8760.

152. Lupan, Oleg, Vasilii Cretu, Vasile Postica, Mahdi Ahmadi, Beatriz Roldan Cuenya, Lee Chow et al. February 2016. Silver-doped zinc oxide single nanowire multifunctional nanosensor with a significant enhancement in response. Sensors and Actuators B: Chemical 223: 893–903. https://doi.org/10.1016/j.snb.2015.10.002.

153. Magrini-Bair, Kimberly, Stefan Czernik, Richard French, Yves Parent, Marc Ritland and Esteban Chornet. 2002. Fluidizable catalysts for producing hydrogen by steam reforming biomass pyrolysis liquids. Proc. of the U.S. DOE Hydrogen Program Review NREL/CP-610-32405.

154. Mandotra, S.K., M.R. Suseela and P.W. Ramteke. July 2013. Algae: the source of hydrogen energy. Environews.

155. Manekiya, H. Mohammedhusen and Pachiyappan Arulmozhivarman. 2016. Leakage detection and estimation using IR thermography. Computer Science, Int. Conf. on Communication and Signal Processing (ICCSP).

156. Membrane gas separation. Separation of hydrogen from gases like nitrogen and methane; Wikipedia, the free encyclopedia. en.wikipedia.org/wiki/Membrane_gas_separation.

157. Milbrandt, A. and M. Mann. 2007. Potential for Hydrogen Production from Key Renewable Resources in the United States. NREL/TP-640-41134, Golden CO: National Renewable Energy Laboratory. www.nrel.gov/docs/fy07osti/41134.pdf.

158. Milbrandt, A. and M. Mann. February 2009. Hydrogen Potential from Coal, Natural Gas, Nuclear, and Hydro Resources. Technical Report NREL/TP-560-42773.

159. Milliken, JoAnn, John Petrovic and Walter Podolski. o November 2002. Hydrogen Storage Issues for Automotive Fuel Cells. Int. Workshop on Hydrogen in Materials and Vacuum Systems, Virginia. https://www.jlab.org/hydrogen/talks/Milliken.pdfStorage.

160. Molburg, J.C. and D. Richard. 2003. Hydrogen from steam methane reforming with CO2 capture. 20th Annual Int. Coal Conference, Pittsburgh. 15–19.

161. Moradi, Ramin and Katrina M. Groth. May 2019. Hydrogen storage and delivery: Review of the state of the art technologies and risk and reliability analysis. Int. J. of Hydrogen Energy 44(23): 12254–12269. https://doi.org/10.1016/j.ijhydene.2019.03.041.

162. Murvay, Pal-Stefan and Ioan Silea. November 2012. A survey on gas leak detection and localization techniques. J. of Loss Prevention in the Process Industries 25(6): 966–973. https://doi.org/10.1016/j.jlp.2012.05.010.

163. Nagalim, B., F. Duebel and K. Schmillen. 1983. Performance study using natural gas, hydrogen-supplemented natural gas and hydrogen in AVL research engine. Int. J. of Hydrogen Energy 8(9): 715–720.

164. Nam, Seung-Eun and Kew-Ho Lee. 'Hydrogen separation by Pd alloy composite membranes: introduction of diffusion barrier at Membranes and Separation Center, Korea Research Institute of Chemical Technology, P.O. Box 107, Yusung, Taejon 305-606, South Korea Received 3 November 2000, Revised 20 April 2001, Accepted 8 May 2001, Available online 17 August 2001, www.sciencedirect.com/science/article/pii/S0376738801004999.

165. NASA-Chemochromic hydrogen leak detectors. [https://www.nasa.gov/topics/technology/hydrogen/chem_hydro_leak.html].

166. Nath, Kaushik, Manoj Muthukumar, Anish Kumar and Debabrata Das. February 2008. Kinetics of two-stage fermentation process for the production of hydrogen. Int. J. of Hydrogen Energy 33(4): 1195–1203. https://doi.org/10.1016/j.ijhydene.2007.12.011.

167. Niemann, U. Michael, Sesha S. Srinivasan, Ayala R. Phani, Ashok Kumar, D. Yogi Goswami and Elias K. Stefanakos. 2008. Nanomaterials for hydrogen storage applications: a review. J. of Nanomaterials. Article ID 950967. 9 pages. https://doi.org/10.1155/2008/950967.

168. Nirmal, V., B.V. Reddy and M.A. Rosen. 2010. Hydrogen production from coal gasification for effective downstream CO2 capture. Int. J. of Hydrogen Energy 35: 4933–4943.

169. Ockwig, W. Nathan, M. Tina and Nenoff. 2007. Membranes for Hydrogen Separation. Chem. Rev. 107(10): 4078-4110. https://doi.org/10.1021/cr0501792.

170. Ogden, Joan M. 1999. Prospects for Building a Hydrogen Energy Infrastructure. Annual Review of Energy and the Environment 24: 227–279. https://doi.org/10.1146/annurev.energy.24.1.227t.

171. Okazaki, S., H. Nakagawa, S. Asakura, Y. Tomiuchi, N. Tsuji, H. Murayama and M. Washiya. August 2003. Sensing characteristics of an optical fiber sensor for hydrogen leak. Sensors and Actuators B: Chemical. 93(1-3): 142–147. https://doi.org/10.1016/S0925-4005(03)00211-9.

172. Ossai, Chinedu I., Brian Boswell and Ian J.Davies. July 2015. Pipeline failures in corrosive environments– A conceptual analysis of trends and effects—Review. Engineering Failure Analysis 53: 36–58. https://doi.org/10.1016/j.engfailanal.2015.03.004.

173. Overview of Interstate Hydrogen Piping System. Argonne National Lab. (US). Environment Science Division. [http://corridoreis.anl.gov/documents/docs/technical/APT_61012_EVS_TM_08_2.pdf].

174. Padró, Grégoire, E. Catherine and Francis Lau (eds.). Advances in Hydrogen Energy. Springer. https://link.springer.com/book/10.1007%2Fb118796.

175. Pagliaro, Mario and Athanasios G. Konstandopoulos. 2012. Solar Hydrogen-Fuel of the Future. Royal Society of Chemistry (RSC), Cambridge.

176. Palladium/Copper Alloy Composite Membranes for High Temperature Hydrogen Separation from Coal-Derived Gas Streams. [https://www.osti.gov/scitech/biblio/811445].

177. Pandey, Amit. 2019 . Progress in Solid Oxide Fuel Cell (SOFC) Research. Advancement in Solid Oxide Fuel Cell Research, JOM 71: 88–89.

178. Panfilov, Mikhail. December 2016. Underground and pipeline hydrogen storage. In book: Compendium of Hydrogen Energy, DOI: 10.1016/B978-1-78242-362-1.00004-3.

179. Photo biological production of hydrogen; https://www.nrel.gov/docs/fy08osti/42285.pdf.

180. Photoelectrochemical cell. (https://en.wikipedia.org/wiki/Photoelectrochemical_cell).

181. Pin-Ching, Maness, Jianping Yu, Carrie Eckert and Maria L. Ghirardi. Photobiological Hydrogen Production – Prospects and Challenges. [https://www.asm.org/ccLibraryFiles/FILENAME/000000004925/znw00609000275.pdf].

182. Pitts, R.R., D. Smith, S. Lee and E. Tracy. 2006. Interfacial stability of thin film hydrogen sensors. Annual Progress Report, VI.3. National renewable energy laboratory.

183. Plácido, Jersson and Sergio Capareda. 2016. Conversion of residues and by-products from the biodiesel industry into value-added products. Bioresources and Bioprocessing. 3: Article number 23.

184. Portal-Gas-Cooled Fast Reactor (GFR) gen-4.org. [www.gen-4.org/.../gas-cooled-fast-reactor-gfr].

185. Preuster, Patrick, Alexander Alekseev and Peter Wasserscheid. 2017. Hydrogen storage technologies for future energy systems. Annual Review of Chemical and Biomolecular Engineering. 8:445-471.

186. Putten, Robbert van, Tim Wissink, Tijn Swinkels and Evgeny A. Pidko. November 2019. Fuelling the hydrogen economy: Scale-up of an integrated formic acid-to-power system. Int. J. of Hydrogen Energy 44(53): 28533–28541; https://doi.org/10.1016/j.ijhydene.2019.01.153.

187. Rajeshwar, Krishnan, Robert McConnell and Stuart Licht. 2008. Solar Hydrogen Generation-Toward a Renewable Energy Future. Springer.

188. Reber, J.F. and K.J. Meier. 1986. Photochemical production of hydrogen. Phys Chem. 90: 824.

189. Reddy, Sivamohan N., Sonil Nanda, Ajay K. Dalai and Janusz A. Kozinski. April 2014. Supercritical water gasification of biomass for hydrogen production. Int. J. of Hydrogen Energy 39(13): 6912–6926. https://doi.org/10.1016/j.ijhydene.2014.02.125.

190. Reddya, Satish and Sunil Vyasa. 2009. Recovery of Carbon Dioxide and Hydrogen from PSA Tail Gas. GHGT-9. Energy Procedia 1: 149–154.

191. Refocus Nov/Dec 2005.

192. Report from UK PTI. 2011. Times of India Lucknow dated 1st Feb.

193. Report on Transportation Through Hydrogen Fuelled Vehicles in India. Oct 2002. Prepared by Sub-Committee on Transportation through Hydrogen Fuelled Vehicles of the Steering Committee on Hydrogen Energy and Fuel Cells, Ministry of New and Renewable Energy, Government of India, New Delhi.

194. Richa, K., D. Biddhi and R.L. Shawney. 2008. Comparison of environmental and economic aspects of various hydrogen production methods. Renewable and Sustainable Energy Reviews 12: 553–563.

195. Ritter, James A. and Armin D. Ebner. 2007. State of the art adsorption and membrane separation processes for hydrogen production in the chemical and petrochemical industries. J. Separation Science and Technology. 42(6): 1123–1193. https://doi.org/10.1080/01496390701242194.

196. Ritterbusch, Joern, Emily Hu, Jos Lenders and Hakim Meskine (eds.). February, 2020. Advanced Functional Materials. Volume 30(7). © WILEY-VCH Verlag GmbH & Co. KGaA, Weinheim.

197. Roberts, Z., R.R. Wang, G.F. Naterer and K.S. Gabriel. November 2012. Comparison of thermochemical, electrolytic, photoelectrolytic and photochemical solar-to-hydrogen production technologies. Int. J. of Hydrogen Energy 37(21): 16287–16301. https://doi.org/10.1016/j.ijhydene.2012.03.057.

198. Rodjaroen, S. et al. High biomass production and starch accumulation in native green algal strains and cyanobacterial strains of Thailand. [https://www.researchgate.net/publication/279621740_High_biomass_production_and_starch_accumulation_in_native_green_algal_strains_and_cyanobacterial_strains_of_Thailan].

199. Rozendal, Rene A., H.V.M. Hamelers, Gert-Jan Euverink and Sybrand J. Metz. September 2006. Principle and perspectives of hydrogen production through bio catalyzed electrolysis. Int. J. of Hydrogen Energy 31(12): 1632–1640. DOI: 10.1016/j.ijhydene.2005.12.006.

200. Rui, Zhenhua, Guoqing Han, He Zhang, Sai Wang, Hui Pu and Kegang Ling. October 2017. A new model to evaluate two leak points in a gas pipeline. J. of Natural Gas Science and Engineering 46: 491–497. https://doi.org/10.1016/j.jngse.2017.08.025.

201. Safari, Farid and Ibrahim Dincer. December 2018. Assessment and optimization of an integrated wind power system for hydrogen and methane production. Energy Conversion and Management 177: 693–703. https://doi.org/10.1016/j.enconman.2018.09.071.

202. Sazali, Norazlianie, 2020. A comprehensive review of carbon molecular sieve membranes for hydrogen production and purification. The Int. J. of Advanced Manufacturing Technology 107: 2465–2483.

203. Schultz, K.R. September 2003. Use of the Modular Helium Reactor for Hydrogen Production. General Atomics Project 04962, Hydrogen Production by Thermochemical Water-Splitting, General Atomics, San Diego, California 92186-5608, USA. fusion.gat.com/pubs ext/MISCONF03/A24428.pdf.

204. Sharma, Sunita and Sib Krishna Ghoshal. March 2015. Hydrogen the future transportation fuel: From production to applications. Renewable and Sustainable Energy Reviews 43: 1151–1158. https://doi.org/10.1016/j.rser.2014.11.093.

205. Sheltami, Tarek R., Abubakar Bala and Elhadi Shakshuki. March 2016. Wireless sensor networks for leak detection in pipelines: a survey. J. of Ambient Intelligence and Humanized Computing 7(3). DOI: 10.1007/s12652-016-0362-7.

206. Shirasaki, Y. and I. Yasuda. 2013. Membrane reactor for hydrogen production from natural gas at the Tokyo Gas Company: a case study. In Handbook of Membrane Reactors: Reactor Types and Industrial Applications. https://www.sciencedirect.com/topics/engineering/hydrogen-separation-membrane.

207. Shirasaki, Y., T. Tsuneki, Y. Ota et al. 2009. Development of membrane reformer system for highly efficient hydrogen production from natural gas. Int. J. of Hydrogen Energy 34(10): 4482–4487.

208. Silveira, José Luz, Lúcia bollini Braga, Antonio Souza and Julio Santana Antunes. December 2009. The benefits of ethanol use for hydrogen production in urban transportation. Renewable and Sustainable Energy Reviews 13(9): 2525–2534. DOI: 10.1016/j.rser.2009.06.032.

209. Singh, Renu et al. Feb 2015. Utilization of microwave assisted chemical pretreated rice straw hydrolysate for hydrogen production via dark fermentation. 5th Int. conf. on plants and environmental pollution. ISEB & NBRI, Lucknow.

210. Sircar, S. and T.C. Golden. 2000. Purification of Hydrogen by Pressure Swing Adsorption. J. Separation Science and Technology 35(5): 667–687. https://doi.org/10.1081/SS-100100183.

211. Smith, B. and C. Eberle et al. 2006 . FRP hydrogen Pipelines. Annual progress report II.A.2.

212. Smith, D. Barton et al. 2012. Feasobility of using glass-fiber-reinforced polymer pipeline for hydrogen delivery. J. of nondestructive and prognostics of engineering systems. ASME. PVP2016-63683 pp. V06BT06A036.   doi:10.1115/PVP2016-63683.   [http://proceedings.asmedigitalcollection.asme.org/proceeding.aspx?articleid=2590540].

213. Solid oxide Fuel Cell. [https://en.fcc.gov.ir/Fuel-Cell-Solid-Oxid.aspx; and https://en.wikipedia.org/wiki/Solid_oxide_fuel_cell].

214. Sorensen, Bent. 2009. '2024: energy storage tale'. Renewable energy focus May/June.

215. Srivastava, O.N., T.P. Yadav, Rohit R. Shahi, Sunita K. Pandey, M.A. Shaz and Ashish Bhatnagar. September 2015. Hydrogen Energy in India: Storage to Application. Proc. Indian Natnl. Sci. Acad. 81(4): 915–937. Hydrogen Energy Centre, BHU, India and MNNIT, Allahabad, India. [insa.nic.in/writereaddata/ UpLoadedFiles/PINSA/Vol81_2015_4_Art17.pdf].

216. Stagner, Jacqueline A. and David S.-K. Ting (eds.). Sustaining Resources for Tomorrow, Part of the Green Energy and Technology book series (Green), Springer.

217. Sunandana, C.S. May 2007. Nanomaterials for Hydrogen Storage - The van 't Hoff Connection. RESONANCE. / https://www.ias.ac.in/article/fulltext/reso/012/05/0031-0036 .

218. Suppes, Galen J. and Truman S. Storvick. 2016. Energy Conversion and Storage in Sustainable Power Technologies and Infrastructure, Energy Sustainability and Prosperity in a time of Climate Change.

219. Surface Plasmon Resonance. [https://en.wikipedia.org/wiki/Surface_plasmon_resonance].

220. Tarkowski, Radoslaw. May 2019. Underground hydrogen storage: Characteristics and prospects. Renewable and Sustainable Energy Reviews 105: 86–94. https://doi.org/10.1016/j.rser.2019.01.051.

221. Tenney, Darrel R., John G. Jr. Davis, R. Byron Pipes and Norman Johnston. October 2009. NASA Composite Materials Development: Lessons Learned and Future Challenges. NATO RTO AVT-164 Workshop on Support of Composite Systems; Bonn, Germany. https://ntrs.nasa.gov/search.jsp?R=20090037429.

222. The Hydrogen Economy: Opportunities, Costs, Barriers, and R&D Needs. 2004. National Academy of Science and National Academy of Engineering. National Academies Press.

223. The Hydrogen Expedition (PDF). January 2005. http://www.atti-info.org/HydrogenVeh/prospectus.pdf. Retrieved 2008-05-09.

224. T-Raissi, Ali and D.L. Block. December 2004. Hydrogen: Automotive fuel of the future. IEEE Power and Energy Magazine 2(6): 40–45. DOI: 10.1109/MPAE.2004.1359020.

225. UC Berkeley and Colorado scientists find valuable new source of hydrogen fuel, produced by common algae. News Feb 2000. [http://www.berkeley.edu/news/media/releases/2000/02/02-21-2000.html].

226. Ueno, Yoshiyuki, Tatsushi Kawai, Susumu Sato, Seiji Otsuka and Masayoshi Morimoto. 1995. Biological production of hydrogen from cellulose by natural anaerobic microflora. J. of Fermentation and Bioengineering. 79(4): 395–397. https://doi.org/10.1016/0922-338X(95)94005-C.

227. Underground hydrogen storage, From Wikipedia, the free encyclopedia. https://en.wikipedia.org/wiki/ Underground_hydrogen_storage.

228. Vijayaraghavan, Krishnan, Karthik Rajendran and S.P. Kamala Nalinini. June 2010. Hydrogen Generation from Algae: A Review. J. of Plant Sciences 5(1). DOI: 10.3923/jps.2010.1.19.

229. Volgusheva, A., S. Styring and F. Mamedov. 2013. Increased photosystem II stability promotes H2 production in sulfur-deprived Chlamydomonas reinhardtii. Proc. of the National Academy of Sciences 110(18): 7223–7228. ISSN 0027-8424. PMC 3645517 Freely accessible. PMID 23589846. doi:10.1073/pnas.1220645110. [https://www.ncbi.nlm.nih.gov/pmc/articles/PMC3645517/].

230. Wang, Jianlong and Yanan Yin. 2018. Fermentative hydrogen production using pretreated microalgal biomass as feedstock. Microbial Cell Factories. Volume 17, Article number 22.

231. Wang, X., K. Maeda, A. Thomas, K. Takanabe et al. Jan. 2009. A metal-free polymeric photocatalyst for hydrogen production from water under visible light. Nat Mate. 8(1): 76–80. DOI: 10.1038/nmat2317.

232. Water Splitting by Visible Light: A Nanophotocathode for Hydrogen Production. Feb. 9, 2010. Nanowerk News Posted J. Angewandte Chemie. [http://www.nanowerk.com/news/newsid=14748.php?utm_source=feedburner&utm_medium=em.

233. Why is biomass greenhouse gas neutral? https://www.quora.com/Why-is-biomass-greenhouse-gas-neutral.

234. Yanagida, Tsakata Y. et al. 2004. Chem. Lett. 33: 7(6)726-727.

235. Yang, Jaeyoung, Chang-Ha Lee and Jay-Woo Chang. 1997. Separation of Hydrogen mixtures by a two-bed pressure swing adsorption process using Zeolite 5A. Ind. Eng. Chem. Res. 36(7): 2789–2798. https://doi. org/10.1021/ie960728h.

236. Yon, S. and J.C. Sautet. 2012. Flame lift-off height, velocity flow and mixing of hypthane in oxycombustion in a burner with two separated jets. Applied Thermal Engineering 32(1): 83–92.

237. Zacharia, Renju and Sami ullah Rather. October 2015. Review of Solid State Hydrogen Storage Methods Adopting Different Kinds of Novel Materials. J. of Nanomaterials (9): 1–18. DOI: 10.1155/2015/914845.

238. Zhang, Percival, Y.H., Barbara R. Evans, Jonathan R. Mielenz, Robert C. Hopkins and Michael W.W. Adams. May 2007. High-Yield Hydrogen Production from Starch and Water by a Synthetic Enzymatic Pathway. PLOS ONE. 2(5): e456. Anastasios Melis (Academic Editor), Published online. doi: 10.1371/journal.pone.0000456.

239. Zhao, Ming, Nicholas H. Florin and Andrew T. Harris. June 2010. Mesoporous supported cobalt catalysts for enhanced hydrogen production during cellulose decomposition. Applied Catalysis B: Environmental 97(1-2, 9): 142–150. https://doi.org/10.1016/j.apcatb.2010.03.034.

240. Zhen, H.S., C.S. Cheung, C.W. Leung and Yatsze Choy. April 2012. Effects of hydrogen concentration on the emission and heat transfer of a premixed LPG-hydrogen flame. Int. J. of Hydrogen Energy 37(7): 6097–6105. DOI: 10.1016/j.ijhydene.2011.12.130.

241. Zhou, Wei, Xiaoxiao Meng and HyungKuk Ju. Oct. 2021. Chemicals-assisted Water Electrolysis for Hydrogen Production Frontiers in Energy Research-Hydrogen Storage and Production. https://www.frontiersin.org/ research-topics/19597/chemicals-assisted-water-electrolysis-for-hydrogen-production.

242. Zimmerman, T., G. Stephen and A. Glover. Oct 2002. Composite reinforced line pipe (CRLP) for on shore gas pipeline. IPC. paper No. 2725. Proc. of IPC2002, Calgary.

Subject Index

About the Authors

Er. Rajni Kant

Er. Rajni Kant retired from UP State Electricity Board as Chief Engineer, Thermal Design and Engineering following which he remained Consultant/Advisor to a number of Design and Engineering Organizations namely Premier Engineering Technologies-Engineering and Management Consultants, New Delhi, M/S Premier Mott Macdonald (Consulting Engineers) New Delhi and other organizations.

Throughout his career both in UPSEB as well as a Consultant/Adviser (after superannuation), he has actively participated in Planning, Design and Engineering, Development, Construction & Erection, Testing & Commissioning and Operation & Maintenance of Thermal Power Plants of various capacities. He was also deeply involved in Environmental studies, Planning and Environmental design of various power plants.

Er. Rajni Kant, an Electrical Engineering graduate from BHU in 1954 (now IIT BHU Varanasi), received an extensive training in the works of M/S Hitachi Ltd. (Japan) and also at large capacity Power Stations and Projects of Georgia Power Company (USA). He has more than 50 years of experience to his credit in the aforesaid fields and has authored about 40 papers in various fields including Power Plants and Environmental Engineering, which were published in National and International Journals and Proceedings of National and International Conferences. He is an author of 2 books entitled, '*Water Pollution, Management, Control and Treatment* authored by Er. Rajni Kant and Dr. Keshav Kant, New Age International Publishers' and '*Air Pollution and Control*' authored by Er. Rajni Kant and Dr. Keshav Kant, Khanna Book Publishing Co. He is a member of several professional Organizations namely; (1) Member of the Association of Technical Scholarship (Japan), (2) Member of the Association on Electrical Generation, Transmission & Distribution (Afro-Asian Region), (3) Member Central Board of Irrigation & Power (India) and (4) Member of the International Society of Environmental Botanists.

Dr. Keshav Kant

Dr. Keshav Kant is a former Professor of Mechanical Engineering from Indian Institute of Technology, Kanpur, who after superannuation in June 2005, served as a Professor and Head of Mechanical Engineering Department and later as the Director in a number of private Engineering Colleges/Institutes. He has also been Professor emeritus of Mechanical Engineering at PSIT Kanpur. He has almost 50 years of total teaching and research experience to his credit. He has 3 Monographs and 95 publications to his credit in National and International Journals and Proceedings of National and International Conferences. He has also been a Consultant to Rural Technology Action Group of IIT Kanpur for almost 3 years.

He is a graduate in Mechanical Engineering from Birla Institute of Technology Mesra, Ranchi in 1964, Master of Engineering with Hons. from the University of Roorkee (now IIT, Roorkee) in 1966, and Ph.D. Degree in 1983 from McGill University, Montreal, Canada. He served as a faculty of Mechanical Engineering mostly at IIT Kanpur since 1967. He has been the recipient of several awards from his Alma-maters. He is Included in 'Who's Who in the World', 2010 edition. He has significantly contributed in developing Undergraduate and Graduate Courses in the Department and Thermal Science Laboratories at I.I.T. Kanpur. He has supervised 52 graduate students including 7 Ph.D.s and 45 M.Tech.s doing cutting edge research in the area of Thermal Sciences. He has also supervised 35 senior UG Design Projects and 45 AMIE Students. His research interests include Heat Transfer, Heat Exchangers, Cooling Towers, Air Pollution & Control, Renewable Fuels and Thermal Environmental Control.

Professor Keshav Kant is the co-author of 2 books namely *'Water Pollution Management, Control and Treatment'*, New Age International Publishers, and *'Air Pollution and Control'*, Khanna Book Publishing Co. and co-editor of *'Strategies for Rural Development'*, Arnold Publishing Co. He has been involved with Technology Development related to Surfactants, Simulated Moving bed Heat Regenerators and Retrofitting Small Capacity Refrigeration Systems using Eco-friendly Refrigerants. He has also been involved with several sponsored research projects, industrial consultancy and Design projects. He has been a Visiting Professor to several U.S. Universities, an invited speaker to many universities in India and abroad and keynote speaker in many National and International conferences in India. He was accredited as one of the most distinguished teachers at IIT Kanpur and U.S. Universities. As a Chairman of the Kanpur Centre of the Institution of Engineers, India for 2 years and member of its executive for 23 years, he has organized several technical events related to 'Water Resources' and 'environment' for the awareness of the people in the society. He is member of several Professional Organizations namely: (1) Fellow of the Institution of Engineers (India), (2) Member of the Indian Society for Technical Education, (3) Member of the Indian Society for Heat and Mass Transfer and (4) Member of Indian Science Congress.

For Product Safety Concerns and Information please contact our EU
representative GPSR@taylorandfrancis.com
Taylor & Francis Verlag GmbH, Kaufingerstraße 24, 80331 München, Germany

* 9 7 8 1 0 3 2 0 5 9 8 4 6 *